Excited States of Proteins and Nucleic Acids

CONTRIBUTORS

W. B. Dandliker
Department of Biochemistry
Scripps Clinic and Research Foundation
La Jolla, California

J. Eisinger
Bell Telephone Laboratories, Inc.
Murray Hill, New Jersey

Robin M. Hochstrasser
Department of Chemistry
University of Pennsylvania
Philadelphia, Pennsylvania

Edward P. Kirby
Laboratory of Physical Biochemistry
Naval Medical Research Institute
Bethesda, Maryland

Eloise Kuntz
Biophysics Department
Michigan State University
East Lansing, Michigan

A. A. Lamola
Bell Telephone Laboratories, Inc.
Murray Hill, New Jersey

J. W. Longworth
Biology Division
Oak Ridge National Laboratory
Oak Ridge, Tennessee

Everett T. Meserve
TRW Instruments
TRW, Inc.
El Segundo, California

A. J. Portmann
Department of Chemistry
Scripps Clinic and Research Foundation
La Jolla, California

R. F. Steiner
Department of Chemistry
University of Maryland
Baltimore County
Baltimore, Maryland

I. Weinryb
Department of Biochemical Pharmacology
Squibb Institute for Medical Research
New Brunswick, New Jersey

STEINER, ROBERT FRANK
EXCITED STATES OF PROTEINS AN
000221474

HCL QP551.S82

Excited States of Proteins and Nucleic Acids

Edited by

Robert F. Steiner
University of Maryland
Baltimore County
Baltimore, Maryland

and

Ira Weinryb
Department of Biochemical Pharmacology
Squibb Institute for Medical Research
New Brunswick, New Jersey

MACMILLAN

© 1971 Plenum Press, New York
A Division of Plenum Publishing Corporation
227 West 17th Street, New York, N.Y. 10011

United Kingdom edition published by
THE MACMILLAN PRESS LTD
London and Basingstoke

SBN 333 13671 3

All rights reserved

No part of this publication may be reproduced in any form
without written permission from the publisher

Printed in the United States of America

PREFACE

The choice of title for this collective volume reflects the desire of the editors and authors to make clear that, while the bulk of the material is concerned with luminescence, other aspects of the excited state have not been excluded. In the five years which have elapsed since the publication of the classical monograph of Konev, a wealth of new information has appeared on the emission properties of proteins and nucleic acids. Indeed, since new publications in this area appear to be proliferating in a geometric ratio, this may be the last opportunity to provide a comprehensive summary of the field in a book which is not of prohibitive length. This is what we have attempted to do here.

While the orientation of each chapter naturally reflects the interests and point of view of the author, there has been a general effort to present a critical assessment of existing results and interpretations, rather than a compendium of data with minimal comment.

Finally, it should be stressed that the rapid evolution of the subject at the time of writing makes it inevitable that the book will age to some degree over the next few years, although this will occur at differing rates for the various chapters. We can only hope that most of the material in this interim summing-up will prove resistant to the erosion of time and provide a solid foundation for further progress.

Robert F. Steiner
Ira Weinryb

October 1970

CONTENTS

Chapter 1
Some Principles Governing the Luminescence of Organic Molecules
by Robin M. Hochstrasser

1. Introduction 1
2. Spontaneous Emission 2
   2.1. Relationship Between Lifetime and Absorption
        Coefficient 2
   2.2. Influence of Multiplicity on Observed Lifetime 4
   2.3. Luminescence from Nearby States 5
   2.4. Multiple-State Decay: An Example 6
3. Molecular Luminescence Characteristics 7
   3.1. The Transition Dipole Moment 8
   3.2. Spontaneous Luminescence in Aggregates 10
4. The Adiabatic Approximation 11
   4.1. Dependence of Transition Moment on Nuclear
        Displacements 12
   4.2. Effect of Nuclear Displacements on the Emission
        Spectrum and Lifetime 13
   4.3. Numerical Estimates of Vibronic Effects 14
5. Triplet–Singlet Transitions and Selection Rules 15
   5.1. The Mixing of $\pi\pi^*$ and $n\pi^*$ States 18
6. Relaxation Processes in Molecules 20
   6.1. Vibrational Relaxation 20
   6.2. Electronic Relaxation 21
   6.3. Intersystem Crossing 24
   6.4. Spin Polarization 27
References 27

Chapter 2
Experimental Techniques

Part A
Fluorescence Instrumentation and Methodology 31
by Edward P. Kirby

1. Basic Considerations 31
    1.1. General Description of a Spectrofluorimeter 31
    1.2. Representation of Spectra 33
    1.3. Calculation of Quantum Yields 34
    1.4. Polarization Spectra 35
2. Methodology 36
    2.1. Instrument Calibration 36
    2.2. Correction for Sample Variations 39
    2.3. Cuvettes 43
3. Criteria for a Spectrofluorimeter 44
    3.1. Sensitivity 44
    3.2. Resolution 45
    3.3. Sample Compartment 49
    3.4. Photomultipliers 50
    3.5. Amplifiers 52
    3.6. Summary 53
References 53

Part B
Direct Measurement of Fluorescence Lifetimes 57
by Everett T. Meserve

1. Introduction 57
2. Instrumentation 58
    2.1. Instrument Considerations 58
    2.2. Oscilloscope Techniques 64
    2.3. Curve Normalization Techniques 66
    2.4. Gated Photomultiplier Detection 78
    2.5. Single-Photon Counting 82
References 84

Part C
Phosphorescence Instrumentation and Techniques 87
by Eloise Kuntz

1. General Instrumentation 87
    1.1. Light Choppers 89
    1.2. Photomultipliers 90

	1.3.	Sample-Cooling Devices	90
2.	Matrices	92
3.	Population of the Triplet State	94
	3.1.	Spectra	95
	3.2.	Quantum Yields	98
	3.3.	Lifetimes	99
	3.4.	Polarization of Phosphorescence	102
References	104

Chapter 3
The Excited States of Nucleic Acids
by J. Eisinger and A. A. Lamola

1. History and Introduction 107
2. Structures, Nomenclature, and Abbreviations 109
3. Excited States of Monomers 112
   - 3.1. Relevance of Low-Temperature Experiments 112
   - 3.2. Emission Spectra and Other Experimental Parameters .. 114
   - 3.3. Sensitized Phosphorescence Spectra 120
   - 3.4. Wavefunctions of the Excited States 121
4. Excited States of Oligonucleotides and Polynucleotides at Low Temperature 125
   - 4.1. Types of Interactions 125
   - 4.2. Excited States of Dinucleotides 125
   - 4.3. Excited States of Polynucleotides 134
5. Excited States at Room Temperature 139
   - 5.1. Energy Levels 139
   - 5.2. Nonradiative Rates in Aqueous Solution 140
   - 5.3. Triplet-State Molecules in Aqueous Solution 143
   - 5.4. Temperature Dependence of Fluorescence 148
   - 5.5. Speculations about Fluorescence Quenching and Temperature Effects 152
6. Excited-State Precursors of Photoproducts 154
   - 6.1. Photohydrates 154
   - 6.2. The Cytosine–Thymine Adduct 155
   - 6.3. Photodimers of Pyrimidines 157
   - 6.4. Sensitized Pyrimidine Dimers in Polynucleotides 163
7. Energy Transfer in Polynucleotides 165
   - 7.1. General Considerations 165
   - 7.2. Theory of Energy Transfer 167
   - 7.3. Förster Energy Transfer 169
   - 7.4. Experiments and Calculations 171

8. Transfer RNA 185
   8.1. The Role of Odd Bases 185
   8.2. $tRNA^{phe}$ Studies 189
References 194

Chapter 4
Fluorescent Protein Conjugates
by W. D. Dandliker and A. J. Portmann

1. Introduction 199
2. Chemistry of Conjugation 200
   2.1. Functional Groups in Proteins and in the Label 200
   2.2. Dye Structures 213
3. Experimental Procedures for Labeling 216
   3.1. Conditions of Labeling 216
   3.2. Isolation of the Labeled Conjugate 217
   3.3. Determination of the Degree of Labeling 218
   3.4. Fractionation According to the Degree of Labeling 222
4. Effect of the Label on the Properties of the Protein 224
5. Excitation and Emission Spectra 226
   5.1. Spectral Data 226
   5.2. Changes Due to Alterations in Environment of the Dye Molecule 226
   5.3. Electronic Mechanisms Responsible for Changes 233
   5.4. Changes Due to Photochemical Reactions 235
6. Lifetime, Decay Time, and Quantum Yield 235
7. Energy Transfer 237
8. Polarization of Fluorescence 239
9. Visible Tracing 253
   9.1. Coons Fluorescent Antibody Technique 253
   9.2. Quantitative Precipitation Test 257
   9.3. N-Terminal Analysis 258
10. Noncovalently Bound Labels 258
References 262

Chapter 5
The Luminescence of the Aromatic Amino Acids
by I. Weinryb and R. F. Steiner

1. Introduction 277
2. Excitation of the Aromatic Amino Acids 278
   2.1. Excitation by Near-Ultraviolet Radiation: Ultraviolet Absorption Spectra 278

      2.2. Excitation by Higher-Energy Radiation 281
3. Environmental Effects upon the Fluorescence of the
      Aromatic Amino Acids 282
      3.1. Temperature... 282
      3.2. Physical State 283
      3.3. Solvent 284
4. Fluorescence Quantum Yields and Lifetimes for the
      Aromatic Amino Acids 288
5. Fluorescence of Derivatives of the Aromatic Amino Acids .. 293
      5.1. Tryptophan Derivatives...................................... 293
      5.2. Tyrosine Derivatives 294
      5.3. Phenylalanine Derivatives 295
      5.4. Oligopeptides Containing Tryptophan and/or Tyrosine 295
6. Radiationless Deactivation of the Excited State 296
7. Phosphorescence of the Aromatic Amino Acids 299
      7.1. Temperature Dependence and Solvent Dependence 299
      7.2. Tryptophan 300
      7.3. Tyrosine 303
      7.4. Phenylalanine 304
8. Polarization of Luminescence 304
      8.1. Theory 304
      8.2. Phenol and Tyrosine 305
      8.3. Indole and Tryptophan 306
9. Energy Transfer in Oligopeptides 308
      9.1. Radiationless Exchange 308
      9.2. Intermolecular Transfer 309
      9.3. Intramolecular Transfer in Tyrosine Oligopeptides 309
      9.4. Intramolecular Transfer in Oligopeptides Containing
           Tryptophan and Tyrosine 311
10. Thermoluminescence of the Aromatic Amino Acids 312
References 312

Chapter 6
Luminescence of Polypeptides and Proteins
by J. W. Longworth

1. Historical Survey.. 319
      1.1. Existence of Excited States 320
      1.2. Protein Fluorescence.. 323
2. Luminescence of Synthetic Polypeptides 334
      2.1. Chemistry and Stereochemistry of Polypeptides 334
      2.2. Homopolypeptide Luminescence 336

- 2.3. Heteropolymer Luminescence: Aromatic Amino Acid Systems ... 345
- 2.4. Quenching Studies ... 354
- 2.5. Photochemistry of Polytyrosine ... 356
3. Luminescence of Natural Polypeptides: Hormones and Antibiotics ... 357
    - 3.1. Phenylalanine Systems ... 357
    - 3.2. Tyrosine-Containing Polypeptides ... 361
    - 3.3. Tryptophan-Containing Polypeptides ... 364
    - 3.4. Summary ... 378
4. Luminescence of Proteins—Class A Proteins ... 378
    - 4.1. Fluorescence Spectra ... 379
    - 4.2. Fluorescence Quenching ... 380
    - 4.3. Phosphorescence ... 386
    - 4.4. Fluorescence Lifetime ... 389
    - 4.5. Phosphorescence Lifetime ... 390
    - 4.6. Temperature-Induced Quenching ... 390
    - 4.7. Acid Denaturation ... 392
    - 4.8. Muscle Proteins ... 393
    - 4.9. Summary ... 395
5. Luminescence of Proteins—Class B Proteins ... 396
    - 5.1. Introduction ... 396
    - 5.2. Tyrosine Fluorescence ... 403
    - 5.3. Tyrosine Fluorescence and Phosphorescence Spectra ... 408
    - 5.4. Tyrosine Quantum Yield ... 411
    - 5.5. Excitation Spectra of Tyrosine ... 414
    - 5.6. Tyrosine Phosphorescence Yield and Decay Time ... 418
    - 5.7. Electronic Energy Transfer ... 420
    - 5.8. Fluorescence Polarization Spectra ... 422
    - 5.9. Phosphorescence Polarization Studies ... 427
    - 5.10. Tryptophan Excitation Spectra ... 428
    - 5.11. Quantum Yields of Tryptophan Residues ... 436
    - 5.12. Solvent Perturbation ... 443
    - 5.13. Solvent Isotopic Effect ... 443
    - 5.14. Temperature Dependence of Quantum Yields ... 445
    - 5.15. Energy Loss at 77°K ... 450
    - 5.16. Luminescence Lifetimes ... 453
    - 5.17. Fluorescence Spectra of Protein Tryptophan Residues .. 456
    - 5.18. Phosphorescence Spectra of Tryptophan ... 460
    - 5.19. Stokes' Shift of Fluorescence ... 462
    - 5.20. Heterogeneity of Environment ... 467

5.21. Heterogeneity of Phosphorescence 472
5.22. Transfer and Heterogeneity 473
References 474

Index 485

Chapter 1

SOME PRINCIPLES GOVERNING THE LUMINESCENCE OF ORGANIC MOLECULES

Robin M. Hochstrasser

*Department of Chemistry
The University of Pennsylvania
Philadelphia, Pennsyhania*

1. INTRODUCTION

The reader will discover that this article contains rather more theory than is usually found in a "biological" exposition of physical phenomena. My intention is to get down to some of the formal details of relatively well-established chemical physics, since I believe that the increasingly wide pertinence of quantum chemistry to biology requires that the researcher more and more make his own rationale. Some of the basic concepts of large-molecule spectroscopic phenomena, especially luminescence, have been outlined in a series of articles by Kasha,[1-4] and many of the working procedures of the large-molecule spectroscopist are now available in books.[5,6]

Following electronic excitation, it may or may not be possible to observe the luminescence of molecules in a particular medium, at a given temperature, and under special conditions of excitation, detection sensitivity, time scale, and other experimental constraints, but in principle the molecules have an emission spectrum corresponding to any given set of conditions. It turns out, however, that the currently interesting aspects of emission spectra are concerned with "nonluminescence," namely, the understanding of those features that cause certain systems to luminesce more or less effectively than others.

The actual emission spectra of molecules, corresponding to electronic transitions, are usually a result of spontaneous transitions from higher- to lower-energy electronic states. The spontaneous emission probability is a fundamental property of a system which, in the simplest cases, is calculable from a knowledge of the absorption spectrum. In general, excited molecules will undergo dissipative processes in addition to spontaneous emission; thus if the source of excitation of the system is suddenly shut off at $t=0$, the manner of decay of the luminescence, which is the decay of the excited molecules in the system, is normally described by a relation of the type

$$I(t) = I(0) \exp\left[-(k_s + k_d)t\right] \qquad [1]$$

where k_s and k_d are, respectively, the *rate constant* for spontaneous emission and the *rate* of dissipation by other processes. The mean lifetime $\tau[I(\tau)=(1/e)I(0)]$ for decay of luminescence is therefore given by

$$\frac{1}{\tau} = \frac{1}{\tau_s} + \frac{1}{\tau_d} \qquad [2]$$

These considerations are expected to apply to systems that contain effectively only one electronically excited state that can radiate. Obviously, Eq. [2] is not to apply in general to systems in thermal equilibrium having other radiative excited states nearby to the emitting state, nor is it to apply to spontaneous radiative decay from or into more than one electronic state. Nevertheless, the exponential decay is so common in practice that we will first discuss systems that display this behavior and reserve multiple-state systems until later.

The phenomenon of luminescence involves the first term in Eq. [2], and monluminescence is concerned with the second term. We first deal with the spontaneous process characterized by τ_s.

2. SPONTANEOUS EMISSION

The relationship between the natural luminescence lifetime and the absorption spectrum is not so obvious for molecules as for atoms. In the first place, molecular electronic absorption and emission spectra do not occur in the same spectral position. Secondly, emission may occur from an excited state that is not adiabatically connected with the absorbing state, and the ground and excited states may not have the same equilibrium nuclear coordinates.

2.1. Relationship between Lifetime and Absorption Coefficient

We now show the relationship between the theoretical parameters of quantum mechanics and the customary parameters of spectrophotometry.[7-9] In an absorption experiment, the Beer–Lambert law is assumed:

$$\varrho(\nu, x) = \varrho(\nu, 0)\, e^{-k(\nu)x} \qquad [3]$$

The absorption coefficient $k(\nu)$, defined at each frequency ν, is $2.303\,\varepsilon(\nu)C$, where ε is the molar extinction coefficient (liter mole^{-1} cm^{-1}) and C is the molarity. The density of radiation of frequency ν that is found at a penetration depth x is $\varrho(\nu,x)$ and the density of incident radiation is $\varrho(\nu,0)$; ϱ is measured in energy per unit volume per unit frequency range, that is, erg cm^{-3} per unit frequency range. The number of quanta per unit volume is $\varrho/h\nu$. The incremental change in $\varrho(\nu,x)$ corresponding to a slice of sample between x and $x+dx$ is $d\varrho(\nu,x)$

$$d\varrho(\nu, x) = -k(\nu)\varrho(\nu, x)\,dx \qquad [4]$$

The total number of quanta absorbed in unit time is $d\varrho(\nu,x)\,dxds/h\nu t$, where t is the time spent by the radiation in the increment dx; i.e., $t = dx/V(\nu)$ where $V(\nu)$ is the velocity of light in the medium for frequency ν; ds is the surface area. The total number of excited molecules in the volume increment is thus given by N_e^ν, where

$$N_e^\nu = \frac{V(\nu)k(\nu)\varrho(\nu, x)\,dxds}{h\nu} \qquad [5]$$

Since the total number of molecules in the volume increment, N_0, is $C\mathbf{N}dxds/10^3$, we have for the fraction that are excited

$$\left(\frac{N_e^\nu}{N_0}\right) = \left[\left(\frac{2303C}{\mathbf{N}h}\right)\left(\frac{\varepsilon(\nu)}{n(\nu)\nu}\right)\right]\varrho(\nu, x) \qquad [6]$$

where $n(\nu)$ is the frequency-dependent refractive index of the sample and $\mathbf{N}$ is Avogadro's number. The fraction (N_e^ν/N_0) is actually the probability that a molecule in the volume element at depth x becomes excited, but it refers to a fixed frequency. The total probability of transitions to the excited state is obtained by integrating [6] over the whole frequency range chosen such that the radiation density at depth x is constant over the range of integration. We write the total probability of transitions as $W(i \leftarrow 0)$, where

$$W(i \leftarrow 0) = \left[\left(\frac{2303C}{\mathbf{N}h}\right)\int \frac{\varepsilon(\nu)\,d\nu}{n(\nu)\nu}\right]\varrho(\nu, x) \qquad [7]$$

The range of integration can be chosen to encompass the complete vibrational–electronic spectrum of a given state, but in so doing one must use with caution the connection between spontaneous and induced processes.

The Einstein coefficients for induced emission B_{io} and for spontaneous emission A_{io} are related for a two-state system plus radiation if the whole system is considered to be at thermal equilibrium. The two key relations that we will require are first the definition of B [Eq. 8] and then the connection between A and B [Eq. 9]

$$W(i \leftarrow 0) = B_{oi}\varrho(\nu_A) \qquad [8]$$

$$\frac{1}{\tau_s} = A_{io} = \left[\frac{8\pi h \nu_E^3 n^3(\nu_E)}{c^3}\right] B_{io} \qquad [9]$$

The B coefficient is obtained from the comparison of Eq. [7] and [8], and the value of τ_s for a two-state system is then obtainable from [9] in terms of experimentally measurable quantities. For the two-state system, ν_A and ν_E (absorption and emission frequencies) are equal. Strickler and Berg[9] have given a method of obtaining τ_s when the mean absorption and emission frequencies are not equal, as will generally be the case in polyatomic molecule spectra. In that case, the induced absorption coefficient is $\sum_{v'} B_{v'}(i \leftarrow 0)$, where v' represents excited-state vibrational quantum numbers, whereas the spontaneous emission rate is $\sum_{v''} A_{v''}(i \rightarrow 0)$, where v'' are ground-state vibrational quantum numbers. The problem is readily solved for the case when all the emission originates on a single excited level, i, and all the absorption originates from a single ground-state level, 0. This is the usual state of affairs for molecules in condensed media (but *not* in gases) where rapid internal conversion to the lowest excited state normally occurs prior to any significant emission. The lifetime of the emission is then given by

$$\frac{1}{\tau_s} = \frac{2.88 \times 10^{-9} n^2}{\langle \bar{\nu}_E^{-3} \rangle} \left(\frac{g_0}{g_i}\right) \int \varepsilon(\bar{\nu}) d \ln \bar{\nu} \qquad [10]$$

To obtain [10] we have assumed no dispersion in the refractive index, n. The integral is to be taken over the whole absorption band corresponding to the electronic state that will ultimately emit; $\langle \bar{\nu}_E^{-3} \rangle$ is the mean value of $\bar{\nu}_E^{-3}$, and $\bar{\nu}$ is to be measured in cm^{-1} ($\bar{\nu} = \nu/c$).

2.2. Influence of Multiplicity on Observed Lifetime

The factors g_0 and g_i are the spatial degeneracies of the ground and excited states; (g_0/g_i) is necessary in Eq. [10] to take into account that emission can occur to or from more than one electronic state. For molecules having low symmetry, the g's never differ from unity; if the excited state has a net spin, the effect of spin degeneracy may be lumped into the factor g_i. In that case, the degeneracy factor, g_i, can be nonintegral or even unity under certain circumstances, since for a spin degeneracy there is no reason to suppose that transitions will occur equivalently to or from all the components of a spin multiplet. We shall describe below the most common situation for a triplet state of an organic molecule in which absorption and emission each involve only *one* component state of the triplet level but the factor (g_0/g_i) is still approximately equal to $\frac{1}{3}$!

The case of a triplet–singlet emission (phosphorescence) holds some interesting features in regard to the relationship between the value of $1/\tau_s$ obtained from the decay curve and that obtained from the integrated absorption spectrum. At normal temperatures, say 77°K or above, the rate of spin–lattice relaxation is much larger than the rates of phosphorescence from most types of organic molecules. This means that in condensed systems we can expect that all three spin states of the triplet will be equally populated and they will be interconverting rapidly compared with the mean time that a molecule remains in its excited state. In a luminescence decay experiment, at all times after the exciting light is shut off, the same situation prevails—namely, all three states remain equally populated at all subsequent times. If only one of these three states can radiate into the ground state, then the system will decay at one third of the rate that would prevail if only that one state were populated. Thus for normal phosphorescence experiments involving a large class of organic molecules in condensed inert media, the factor (g_0/g_i) in Eq. [10] should be approximately equal to $\frac{1}{3}$. In general, when all three spin substates are active, and when the rate of interconversion between them is very fast compared with the phosphorescence lifetime, the excited system will decay with a $1/\tau$ of $\frac{1}{3}(k_1+k_2+k_3)$, where k_1, k_2, and k_3 are the spontaneous rates for each transition taken separately. The singlet–triplet absorption $\int \varepsilon d(\ln \bar{\nu})$ measures the sum of the intensities to the unresolved components 1, 2, and 3. Thus once again the factor $g_0/g_i = \frac{1}{3}$ is necessary, but the A and B refer to sums over what are actually three different electronic states. At much lower temperatures the spin–lattice relaxation may be relatively slow, and it follows that the decay of phosphorescence may be quite complex in general.[10] Quite frequently for triplets we find $k_2 \approx k_3 \approx 0$, so that τ_s is given quite accurately by $3/k_1$. Nonradiative processes complicate this analysis.

2.3. Luminescence from Nearby States

The decay of emission from a group of closely spaced states is a fairly common occurrence in condensed-phase spectra. Whenever the medium can be considered to provide a quasi-continuum of levels such as those derived from lattice vibrations of any type—in ordered or in random media—we may assume that an equilibrium distribution over excited states can be maintained under conditions of steady excitation and at temperatures far from 0°K. This depends on the spontaneous emission rates (and also on $1/\tau_A$) being much slower than the rate of exchange of energy between excited molecules and the lattice. In these circumstances, the decay of emission will be always exponential, with $1/\tau$ dependent on temperature if more than one electronic state of the system is thermally accessible. This is a common situation in aggregates since the weak interactions between chromophores usually result in a number of optically accessible electronic states being present in a relatively narrow energy region. Under conditions of low spectral resolution ($\sim kT$), the emission spectra at different temperatures would be indistinguishable. A similar situation will arise in the luminescence spectra of molecules having thermally accessible nonradiative pathways: The decay would be exponential with a half-life corresponding to a Boltzmann averaged mean of the rate constants. It follows that molecules having nearby states (a few kT) should show an exponential decay of emissions at all temperatures but with a lifetime that depends on the temperature.

2.4. Multiple-State Decay: An Example

It is quite common, among heterocyclic molecules, to find very close-lying states that may correspond to different orbital classifications, such as $n\pi^*$ and $\pi\pi^*$. The two kinds of states often have very different natural lifetimes with respect to transitions into the ground state. A similar situation arises with two relatively weakly coupled chromophores that are different, for example, two nearby bases in the DNA molecule. With reference to a situation in which the two states are A (lower) and B (upper), a Boltzmann distribution is maintained when the interchange $B \rightleftharpoons A$ is at least much faster than the individual spontaneous rates. The relative amounts of A and B at all times are controlled by the temperature only:

$$A(t) = \left(\frac{1}{Z}\right) X(t) \qquad [11]$$

$$B(t) = \left(\frac{1}{Z}\right) e^{-E/kT} X(t) \qquad [12]$$

where $X(t) = A(t) + B(t)$ is the total excited state concentration, and Z is the partition function. The excited system must now decay at a rate given by

$$\frac{1}{\tau} = \left(\frac{1}{Z}\right)(k_A + k_B e^{-E/kT}) \approx \frac{1}{2}[(k_A + k_B) + \frac{E}{2kT}(k_A - k_B)] \qquad [13]$$

The spontaneous rate constants for separated A and B are k_A and k_B. Note that the *system* decays at $1/\tau$, and this is true whether one makes observations of A, or of B, or of both A and B simultaneously. When $kT \gg E$, we find

$$\frac{1}{\tau} = \frac{1}{2}(k_A + k_B) \qquad [14]$$

Even if k_B is close to zero the lifetime is different for the two-state system than for isolated A. If the state having smallest k is at lower energy (k_A), then a long lifetime of emission will persist even at high temperatures—and both A and B transitions will show this long decay time. For example, if $k_B = 10^3 k_A$, we have

$$\frac{1}{\tau} = k_A \left[\frac{1 + 10^3 e^{-E/kT}}{1 + e^{-E/kT}} \right] \qquad [15]$$

which does not start to exceed significantly the value of k_A until kT greatly exceeds $E/6$.

These kinetic considerations beg the mechanistic questions that ask how and to what extent states may interact with one another. So a more detailed look at the molecular nature of the emitting states and the topic of radiationless transitions are the next items for discussion.

3. MOLECULAR LUMINESCENCE CHARACTERISTICS

A spectroscopist who deals frequently with the luminescence of organic molecules would seldom have difficulty estimating the expected natural lifetime of emission from almost any molecule provided he could guess the orbital characteristics of the emitting state. These order-of-magnitude estimates turn out to be very reliable, indicating that the qualitative theoretical ideas and the experimental experience on which they depend are reliably connected. The theoretical basis is concerned with two main points:

(a) The vast majority of organic molecule (and other) luminescences involve the spontaneous emission of dipole radiation. As such, the probability of emission is proportional to the electric dipole matrix element connecting the initial and final states.
(b) Certain properties of the low-energy states of organic (and other) molecules are apparently quite well described by rather simple molecular orbital approximations. Thus it appears that the emissive states of most molecules can be considered, to a first approximation, to have specific atomic orbital characteristics.[11] Indeed, it turns out that two-

electron configurations are often quite suitable to find reliable estimates of certain properties. Hence the wide usage of terms such as $\pi\pi^*$, $n\pi^*$, $\sigma\pi^*$, $\pi\sigma^*$, $\sigma\sigma^*$, and so on.

From these approximations one may proceed to estimate absorption intensities and lifetimes with the help, ultimately, of perturbation theory. The perturbations of interest are, first, the electron interactions that cause the simple configurations in (a) above to become scrambled: These interactions must be taken into account whenever the model molecule is subjected to any new electric fields such as are caused by substituents, heteroatoms, deviations from planarity, or solvent interactions. Second, the effect of nuclear motions (molecular vibrations) on the simple configurations, and hence on the transition moment, must be taken into account. Third, the spin–orbit interaction must be considered; otherwise, we could not so readily account for transitions between states having different total electron spin. It is the essence of the perturbation approach that each of the aforementioned interactions, all of which lead to superpositions of configurations for the states of interest, can be estimated qualitatively by employing relatively simple features of the orbitals in the two-electron configurations that were given as the model first approximations.

3.1. The Transition Dipole Moment

The transition dipole moment, $\mathbf{m}_{oi}$, is given by

$$\mathbf{m}_{oi} = e\int \psi_i \psi_o^* \mathbf{r} d\mathbf{r} \qquad [16]$$

$\mathbf{m}_{oi}$ is the first (or dipole) moment of the charge distribution $\psi_i \psi_o^*$—known as the transition density—where ψ_i and ψ_o are the wavefunctions for the excited state and ground state, respectively. The position vectors of all the electrons are lumped together in the factor $\mathbf{r}$, so that $\mathbf{r} = \sum_n \mathbf{r}_n$ for an n-electron system, and $\mathbf{r}_n = x_n \hat{i} + y_n \hat{j} + z_n \hat{k}$. The substitution of two-electron configurations for ψ_i and ψ_o can give an approximate idea of the magnitude of $\mathbf{m}_{oi}$. A zero or nonzero judgment can be made by inspecting the multipole structure of the transition density, for unless the transition density has the structure of a dipole the first moment of the charge distribution will vanish. Sometimes it is only necessary to examine the multipole character of $\psi_i \psi_o^*$ at a single atomic center in the molecules: Such is the case when the initial or the final state involves a mostly localized molecular orbital, for example, a nonbonding orbital. Some common transition density multipoles are shown diagrammatically in Figure 1. These local transition

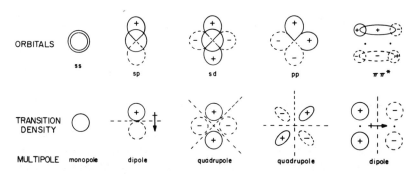

Figure 1. Multipole character of transition densities. The top line shows angular distributions for pairs (θ_1, θ_2) of atomic or molecular orbitals. A diagrammatic representation of the transition density $\theta_1 \theta_2^*$ is shown below each pair. In L.C.A.O. theory each product $\psi_0 \psi_i^*$ can be reduced to products like $\theta_1 \theta_2^*$.

density concepts may be used successfully with molecules that have no symmetry if the electronic transition is a localized one.

The transition dipole moment, $\mathbf{m}_{oi}$, has a direction in the molecule fixed-axis system. Emitted radiation from nonspherical, or nonaveraged, systems is therefore polarized. The transition density multipole notion also predicts properly the polarization of the emitted radiation for a usefully large variety of molecules.

The relationship between the transition dipole moment, $\mathbf{m}_{oi}$, and experimental spectroscopic parameters is as follows:

$$|\mathbf{m}_{oi}|^2 = \left(\frac{3\hbar^2}{2\pi}\right) B_{oi} = \frac{9.18 \times 10^{-39}}{n} \int \varepsilon d(\ln \lambda) \qquad [17]$$

$$\mathbf{m}_{oi} \equiv e\mathbf{d}_{oi} \times 10^8 \qquad [18a]$$

$$|\mathbf{d}_{oi}|^2 = \frac{3.98 \times 10^{-4}}{n} \int \varepsilon d(\ln \lambda) \qquad [18b]$$

The units of $\mathbf{m}_{oi}$ are esu cm in Eq. [17]. Equation [18b] gives the dipole length $\mathbf{d}_{oi}$ in Ångstrom units. The relationship between τ and $\mathbf{d}_{oi}$ involves the frequency $\bar{\nu}_E$

$$\frac{1}{\tau} = [7.14 \times 10^{-6} n^3 \langle \bar{\nu}_E^{-3} \rangle^{-1}] d_{oi}^2 \qquad [19]$$

where again d_{oi} is expressed in Å and $\bar{\nu}_E$ is expressed in cm^{-1}. For fairly symmetric, narrow emission spectra, $\langle \bar{\nu}_E^{-3} \rangle^{-1}$ is roughly $\bar{\nu}_E^3$—the cube of the mean emission frequency (expressed in cm^{-1}). Thus for usual organic solvents and near ultraviolet emission, the lifetime is given approximately by

$$\frac{1}{\tau} = 3.82 \times 10^8 d_{oi}^2 \qquad [20]$$

Thus we see that d_{oi} is approximately 1 Å for a τ of approximately 3 nsec.

3.2. Spontaneous Luminescence in Aggregates

The spontaneous luminescence from aggregates of potential emitters is dominated by the effects of interference among the transition moments for the individual chromophores. In general, if there are n chromophores in an aggregate, the moment $\boldsymbol{m}_k$, for transtitions to the kth state of the aggregate, is given by

$$\boldsymbol{m}_k = \sum_{l=1}^{n} C_{kl} \boldsymbol{m}_{oi}(l) \qquad [21]$$

where $\boldsymbol{m}_{oi}(l)$ is the transition moment for the lth molecule. In ordered systems, the coefficients C_{kl} are such that $|C_{kl}|=1/n$ for all k, l. Thus all moments have the same weight in the superposition, but there may be either constructive or destructive interference occurring in the product $|\boldsymbol{m}_k|^2$. For example, in a dimer, for each monomer state there are two aggregate states: The two transition moments are $2^{-\frac{1}{2}}[\boldsymbol{m}_{oi}(1) \pm \boldsymbol{m}_{oi}(2)]$. The averaged values of $|\boldsymbol{m}|^2$ for real transition moments are given by

$$|\boldsymbol{m}|^2 = m_{oi}^2 \pm \boldsymbol{m}_{oi}(1) \cdot \boldsymbol{m}_{oi}(2) \qquad [22]$$

If the molecular transition moments happen to be parallel, then the spontaneous emission rate for the destructively interfering state is zero, while that for the other state is just twice the value of m_{oi}. If the interactions between molecules in the aggregate are not sufficient to spectrally separate any of the n-states of the system, then the spontaneous emission rate is not influenced, since what is observed is

$$\frac{1}{n} \sum_{k} |\boldsymbol{m}_k|^2 = m_{oi}^2 \qquad [23]$$

If the aggregate states are energetically separated, then at low temperatures the lowest-energy aggregate states will be the principal emitters because a Boltzman population of the various levels should be maintainable within the excited-state lifetime. Under these circumstances, the spontaneous emission rate is a function of temperature. In principle, the spontaneous lifetime may be used to probe structural features of aggregates, and this can be quite useful since the absorption spectrum may not clearly display the lowest state.

All of the foregoing features neglect to take into account the fact that molecules are not rigid. The vibrational–electronic interactions in complex molecules can be severe, and the next section deals with an approximate method of including these effects.

4. THE ADIABATIC APPROXIMATION

The transition moment is actually a function of the nuclear displacements or normal coordinates, R^μ, of which there are $3N-6$ (R^1, R^2, ..., R^μ, ..., R^{3N-6}) for every electronic state. In a simplified approach to vibronic structure, it is convenient to use a crude form of the adiabatic approximation.[12] In this approach, we imagine that we know the electronic states of a system for a fixed set of coordinates R, and it turns out to be convenient to choose R to be R_0—the set of equilibrium coordinates for the ground electronic state. The crude electronic states $\varphi_i(x)$ are then functions only of the electronic coordinates x, and the nuclei are assumed to have zero momentum and hence zero kinetic energy. The states $\varphi_i(x)$ are conceptually distinct orthonormal eigenstates of the Hamiltonian in which the nuclear kinetic energy is set equal to zero, and in which the electron–nuclear attractive potential is considered for the configuration R_0 only. The question now arises as to what description we should give to the state, say $i=1$, when the nuclei are not at R_0. Clearly the electronic-state numbering system 1...i...., etc., applies only to an R_0 configuration, but since that set of states is complete in a mathematical sense it must be that the states at $R \neq R_0$ are just superpositions of the R_0 states. In other words, if the nuclei are placed at $R \neq R_0$, the state $\varphi_1(x)$ becomes mixed with states having all other indices; the amount of mixing must depend on the magnitude of the displacements so that the coefficients in the superposition must be functions of $\Delta R = R_0 - R$. In this way, we have created a set of adiabatic (the nuclear velocities are still zero) electronic functions $\varphi_i(x,R)$

$$\varphi_i(x, R) = \sum_j C_{ij}(\Delta R)\varphi_j(x) \qquad [24]$$

The coefficients $C_{ij}(\Delta R)$ are the so-called Herzberg–Teller interaction coefficients and they are easily obtained from perturbation theory. If we made a good choice by defining the basis functions at R_0—and if our states are far apart so that the nonadiabatic corrections are very small—then $C_{ii} \approx 1$ will be by far the largest coefficient, and $C_{ij}(\Delta R)$ is given by

$$C_{ij}(\Delta R) = \sum_\mu \left(\frac{V_{ij}(\mu)}{E_i - E_j}\right) \Delta R_\mu \qquad [25]$$

where $V_{ij}(\mu)$ is the energy of interaction between the zero-order states i

and j due to the first-order R-dependent part of the electron–nucleus potential.

4.1. Dependence of Transition Moment on Nuclear Displacements

The electronic transition dipole moment can now be written as a function of ΔR using Eqs. [16] and [21]:

$$\boldsymbol{m}_{oi}(R) = \sum_j C_{ij}(\Delta R)\boldsymbol{m}_{oj} \qquad [26a]$$

or

$$\boldsymbol{m}_{oi}(R) \equiv \sum_{\mu,j} \boldsymbol{m}_{ij}^{\mu}\Delta R_{\mu} + \boldsymbol{m}_{oi} \qquad [26b]$$

where $\boldsymbol{m}_{oi}$ and $\boldsymbol{m}_{oj}$ are defined at R_0. The equivalent expression [26b] displays the transition moment in terms of $\boldsymbol{m}_{ij}$, a vector-induced moment for a unit displacement along the coordinate that mixes the states i and j. The vector $\boldsymbol{m}_{ij}$ is parallel to $\boldsymbol{m}_{oj}$. The states of nuclear motion are determined by the nuclear kinetic energy and by the extent to which the electronic energy changes as the nuclei move. For a description of many vibronic effects in large molecules, it appears to be reasonable to assume that the adiabatic potential-energy curves are harmonic along each of the normal coordinates. Under these circumstances, each electronic state of the system consists of a manifold of vibronic levels with harmonic spacing, and each vibronic state is characterized by exactly two quantum numbers at R_0: The two quantum numbers refer to the electronic and the harmonic oscillator indices. The vibrational index must contain a specification of the type or types of mode—specified by μ in ΔR_{μ}—and a specification of the harmonic.

It is important to realize that R_0^{μ} will not in general have the same value in every electronic state. However, in the most common large-molecule situation there is no change in *symmetry* following electronic excitation; there is nevertheless a change in the equilibrium nuclear coordinates. This means that R_0^{μ} is the same in all electronic states when μ is a unique non–totally symmetric vibration: *The shifts in equilibrium configuration occur along the totally symmetric coordinates.* Strictly speaking, it is only necessary that the potential energy curves along non–totally symmetric coordinates be symmetric about R_0, so R_0^{μ} must represent a turning point of the curve.

We are now in a position to write a more precise expression for the transition dipole moment by incorporating the modulation of $\boldsymbol{m}_{oi}$ due to nuclear motions and also the effects of mixing in other electronic states. The emission spectrum originates from the zero-point level $F_{io}(R')$ of the ith electronic state. It is easy to show that the total transition moment squared for transitions into all possible ground vibrational levels is just the average value of $\mathbf{m}_{oi}(R) \cdot \mathbf{m}_{oi}^*(R)$ in the vibrational level $F_{io}(R')$. The function $F_{io}(R')$ is a product of $3N-6$ (μ) harmonic oscillator zero-point wavefunctions:

$$F_{1v}(R') = f_{1n_1}(R_1')f_{1n_2}(R_2')f_{1n_3}(R_3')\ldots, \text{ etc.} \qquad [27]$$

where n_1, n_2, etc., are the number of quanta of harmonic mode 1, 2, etc., involved in the composite quantum number v. In the cases considered, $F_{10}(R')$ means $n_1=n_2=n_3\ldots=0$; $F_{1\mu}(R')$ means $n_1=n_2=n_3\ldots=0$, $n_\mu=1$, and so on. The argument R' is $[\Delta R_\mu - \delta_\mu]$, where δ_μ is the shift (increase assumed) of the equilibrium R_0^μ along the coordinate μ following electronic excitation. If μ is a non–totally symmetric vibration, then $\delta_\mu=0$; otherwise, the emitting state has different symmetry from the ground state—in that case, the emission spectrum will show a progression of non–totally symmetric vibrations.

4.2. Effect of Nuclear Displacements on the Emission Spectrum and Lifetime

Equation [26b] may be written in a simplified form for one vibrational mode:

$$\mathbf{m}_{oi}(R) = \mathbf{m}_{oi} + \mathbf{m}_{ij}\Delta R \qquad [28]$$

The displacement $(R-R_0)$ is taken from the ground-state equilibrium configuration, and m_{ij} is the moment induced by mixing with states j for a unit nuclear displacement. The emission intensity is given by $\mathbf{m}_{oi}(R) \cdot \mathbf{m}_{oi}(R)$ averaged over the zero-point motions *in the upper state* (i); the absorption intensity is given by $\mathbf{m}_{oi}(R) \cdot \mathbf{m}_{oi}(R)$ averaged over the zero-point motions *in the ground state*. We note that these two quantities are not equal in general, since from Eq. [28], the properties of harmonic oscillators, and the definitions, we find by averaging that

$$I_{\text{ABS}} \propto m_{oi}^2 + \frac{m_{ij}^2 \hbar}{4\pi mc\bar{\nu}_0} \qquad [29]$$

$$I_{\text{EMIS}} \propto m_{oi}^2 + \frac{m_{ij}^2 \hbar}{4\pi mc\bar{\nu}_i} + m_{ij}^2 \delta^2 \left[\frac{1-m_{oi}}{m_{ij}\delta}\right] \qquad [30]$$

$\bar{\nu}_0$ and $\bar{\nu}_i$ refer to the perturbing vibrational frequency (cm^{-1}) in ground and excited states. These results show that a change in geometry, δ, along a perturbing normal mode will result in a discrepancy between the lifetime of emission measured and that calculated from absorption data. This situation should be especially prevalent in molecules having low or no symmetry, since now δ can be nonzero for every vibrational motion.[13]

For simpler molecules, most commonly $\delta=0$ unless the vibration being considered is totally symmetric. Whether the observed lifetime (from emission) is greater or less than that calculated from the absorption is determined by whether m_{oi} is, respectively, less than or greater than $m_{ij}\delta$. If the transition $0 \rightarrow i$ is strong, then changes of geometry on excitation will most probably lengthen the lifetime of emission, but the effect will be small. For unsymmetrical molecules and medium-strength transitions, it is readily shown that the mirror symmetry between absorption and emission may be lost—not only is the total intensity changed if δ is not zero, but what remains may be distributed differently in the absorption and emission spectra.[13]

If the vibration for which m_{ij} is nonzero is non–totally symmetric, then usually the terms in δ vanish, and the observed and calculated lifetimes are not expected to be different from this cause. The luminescence polarization is now more seriously affected since $\mathbf{m}_{oi}$, is necessarily perpendicular to $\mathbf{m}_{oj}$. For the case of unsymmetrical molecules, the changes in luminescence polarization are not so obvious since $\mathbf{m}_{oi}$ and $\mathbf{m}_{oj}$ may make any angle with one another. The only criterion for the nonvanishing of the extra term in [28] is that $\mathbf{m}_{oi}\cdot\mathbf{m}_{oj}\neq 0$. When there are a number of non–totally symmetrical modes having the same symmetry, then these motions become mixed because of the existence of the Herzberg–Teller interaction.

4.3. Numerical Estimates of Vibronic Effects

The values of $m_{ij}^2(\hbar/4\pi mc\bar{\nu})$ range from about $10^{-3}\, m_{oj}^2$ to about $10^{-5}\, m_{oj}^2$ over a wide class of aromatic, heteroaromatic and carbonyl compounds. The value of $\hbar/4\pi mc\bar{\nu}$ is about 10^{-3} (Å)^2 for the mass of a carbon atom (m), and $\bar{\nu}=1000$ cm^{-1}. This factor corresponds to a root mean square displacement by a carbon atom of approximately 0.03 Å; thus a median value of $|m_{ij}|$ is approximately $3\times 10^{-2}\, m_{oj}$ for each 0.1 Å displacement from equilibrium. The transition moment dipole lengths for medium-strength transitions are in the range 0.05–0.5Å [see Eq. 20]. These numerical factors work very roughly for extinction coefficients as well; thus we usually find approximately $10^{-3}\, \varepsilon_j$ of vibronic coupling if the mixing states are separated by about 5000–10,000 cm^{-1}. The ratio $m_{oi}/m_{ij}\,\delta$ will often be in the range $(3.3/\delta)\,(m_{oi}/m_{oj})$, where δ is the coordinate displacement in angstroms. This ratio can be greater or less than unity: If $m_{oi}^2 > 10^{-2}\, m_{jo}^2$, then, for reasonable values of δ, the ratio will be greater than unity, implying a lengthening of the luminescence lifetime.

These results indicate that the molecular emission spectrum is a much more complex result than one would project from the Einstein two-state model. They also show that the emission spectrum can display transitions that terminate on non–totally symmetric vibrational levels of the ground state, and that there are at least two mechanisms available for the appearance of totally symmetric vibrations in the emission.

For strongly allowed transitions the emission is expected to be simple, consiting of a pattern of ground-state vibrational intervals, with each vibronic transition having an intensity determined by the square of the Franck–Condon overlap integral $\langle F_{io}(R-\delta_\mu)/F_{0\mu}(R)\rangle$. Those vibrations that appear in the luminescence spectrum therefore are specifically those associated with the normal modes that are not the same in the ground and excited states. The emission spectrum therefore displays—albeit in a rather obscure manner—the relative equilibrium geometries of the initial and final electronic states. The length of a vibrational progression in the emission spectrum, in particular the number of ground-state vibrational quanta associated with the intensity maximum of a progression, is directly concerned with the extent of the change in equilibrium geometry along that normal coordinate.

5. TRIPLET–SINGLET TRANSITIONS AND SELECTION RULES

Practically everything said previously about the influence of nuclear motion on electronic spectra applies also to transitions for which the combining states do not have the same spin. The only difference is that for these cases the pure spin angular momentum has to be somewhat quenched before electric dipole transitions can occur at all. In principle, this quenching can be effected by taking into account the fact that the spinning electrons in our simple configurations are actually moving in very strong electric fields due to the nuclear charges and the other electrons in the system. The result is that states having zero spin are mixed with states having net spin: The real states of the system are superpositions of states having different spins. For most organic molecules with closed-shell ground states, the zero-order one-electron excitations give rise to singlet ($s=0$) and triplet ($s=1$) zero-order states. The spin–orbit interaction mixes $s=1$ and $s=0$ states, such that the mixed state $\varphi'_{i\sigma}$ is a superposition of the zero-order states $^3\varphi_{j\sigma'}$ and $^1\varphi_j$:

$$^3\varphi'_{i\sigma} = {}^3\varphi_{i\sigma} + \sum_k \sum_{j\sigma'\neq i\sigma} \alpha_k(i\sigma|j\sigma')^k\varphi_{j\sigma'} \qquad [31]$$

In this expression the (presuperscript) multiplicity k is either 1 or 3 ($s=0$ or $s=1$) and j runs over every electronic state index. The indices σ and σ' are used to identify which substate of the triplet is under consideration: For singlets ($k=1$) the subscript σ' can be dropped. The coefficient

$a_k(i\sigma|j\sigma')$ represents the first-order perturbation correction due to spin–orbit interaction between the σth substate of the triplet state i with the σ'th substate of the triplet $j(k=3)$ or with the singlet $j(k=1)$. There are some remarkable generalities found from experiments that suggest a much simpler form for Eq. [30] when it is to be applied to relatively symmetric planar aromatics, heteroaromatics, ketones, etc. The experiments indicate the following:

(a) Singlet–triplet transitions have very low intensity, so all the $k=3$ terms can be eliminated from Eq. [30] if one is only considering the intensity sources. The σ' designations need not, therefore, be considered in the transition moment expression. It should be noted that the $k=3$ and σ' terms cannot be dropped if it is the spin–orbit energy shift $\Delta E_{i\sigma}^{(2)}$ that is being sought, since the second-order energy shift is given by

$$\Delta E_{i\sigma}^{(2)} = \sum_k \sum_{j\sigma'} a_k(i\sigma|j\sigma')a_k^*(i\sigma|j\sigma')(E_i - E_{j\sigma'})^{-1} \quad [32]$$

and the triplet states $j\sigma'$ contribute in an important way.

(b) Very often among a wide class of organic molecules it is found that it is mainly only one of the σ states of $^3\varphi_{i\sigma}$ that will couple with higher states: Transitions to or from the ground state to the other two σ states frequently constitute only a small fraction of the total transition probability. This means that the three transition moment amplitudes for triplet to ground-state electric dipole transitions are very often given quite accurately by

$$|\mathbf{m}_{oi}(\sigma_1 \to 0)| = |\sum_j a_1(i\sigma_1|j)\mathbf{m}_{oj}|$$

$$|\mathbf{m}_{oi}(\sigma_2 \to 0)| \approx |\mathbf{m}_{oi}(\sigma_3 \to 0)| \approx 0 \quad [33]$$

It has to be realized that these simplifications *as written* are pertinent because σ_1, σ_2, and σ_3 are distinguishable substates for a free molecule in the absence of external fields. The substate indices σ_1, σ_2, and σ_3 refer to the spin properties of the molecule after taking into account the anisotropy of the internal magnetic interactions. The three principal axes of the magnetic anisotropy constitute a molecule fixed principal axis system x, y, z and the spin symbols are usually written σ_x, σ_y, and σ_z rather than σ_1, σ_2, and σ_3. The function $^3\varphi_{i\sigma}$ can be any of the product functions $\varphi_i\sigma_x$, $\varphi_i\sigma_y$, or $\varphi_i\sigma_z$, where φ_i (as before) is a purely spatial function and involves no spin property.

The spin functions σ_x, σ_y, and σ_z have very simple properties, and for conceptual purposes perhaps their most useful property is that they are known to have the symmetry characteristics of *axial vectors* along the noted axes; that is, they transform under symmetry operations just like the axial rotations do. The singlet spin function σ_0 has the full symmetry of the molecule. Since the spin–orbit interaction operator is a scalar quantity, it also has the full symmetry of the molecule. The question of the vanishing or nonvanishing of $a_1(i\sigma_x|j)$ is therefore simply answered by discovering whether a is antisymmetric to some or symmetric to all the symmetry operations of the molecular frame. This procedure requires knowledge only of the symmetry properties of φ_i, σ_x, and φ_j.

A further point it that because of experimental observation (b) above, all the singlet states j in Eq. [33] must have the same symmetry. Equation [33] can therefore be written in the simplified form:

$$|\boldsymbol{m}_{oi}(\sigma_1 \leftarrow 0)| = |a\hat{m}_1| = a \qquad [34]$$

where $\hat{m}_1$ is a unit vector corresponding to the direction of polarization of the singlet–triplet transition and a is the magnitude of the moment induced by spin–orbit interaction. The direction of $\hat{m}_1$ is determined by experiment and referred to theory using the symmetry of σ_1, i, and j as above: The polarization of the transition is parallel to $\boldsymbol{m}_{oj}$.

The statement (b) above has rather far-reaching consequences since such a generality as this should clearly focus on a molecular property that is common to a very wide class of molecules. Indeed, it will not surprise this author to find that Eq. [33] is approximately obeyed for an even wider class of molecules than has been studied to date. For example, there is a good possibility than these equations will hold even for unsymmetrical and larger molecules. The reason, we believe, is the very large difference between single-center and two-center spin–orbit interaction matrix elements. In the single-center case, the excited electron is associated with an atomic orbital centered on the same atom as the unexcited electron, and these two atomic orbitals must have nonzero L value, such as different *p*-orbitals. The spin–orbit interaction increases as a high power of the nuclear charge, so, if a molecule contains a variety of atoms, the heaviest one will usually dominate the spin–orbit coupling. Even if the state under observation does not satisfy the one-center criteria, the nuclear motions will provide enough mixing of the electronic states to enable any one of them to "feel out" a one-center spin–orbit interaction. The magnitude of a will depend on how far away in energy are these appropriately spin–orbit coupled states. Of course, in symmetrical or in planar molecules there are additional symmetry restrictions. Precisely the same kind of thing may happen in the external heavy-atom effect where a singlet–triplet transition can be significantly intensified simply by a heavy-atom solvent.[14] In that case, the mixing of singlet and triplet states is so great in the solvent molecules that the solute molecules "feel out" this source of dipole strength by weak electrostatic coupling with the state of nearby solvent molecules. In this case, the relative electric dipole transition probabilities for transitions to the three substates of the solute triplet are determined by the relative orientations of solvent and solute. Since in a solution these orientations are perhaps random, we expect all spin substates to be about equally active. This has never been verified experimentally.

5.1. The Mixing of $\pi\pi^*$ and $n\pi^*$ States

The strict separability of π- and σ- type states is lost whenever a molecule is nonplanar. Even in a molecule that is ostensibly planar, the π, σ distinction will be lost in states of either type that are nearby to the other type.[15] Because π and σ have different symmetry with respect to reflection in a molecular plane, the two types of states (e.g., $\pi\pi^*$ and $n\pi^*$) become mixed not only when the molecule is nonplanar in its equilibrium geometry, but also if they are mixed by out-of-plane oscillations of the nuclei. It follows that if two such states are quite close in energy, there is liable to be a static distortion from planarity in the excited state even if the molecular ground state is planar.[15]

The effect of $n\pi^*$–$\pi\pi^*$ mixing on the luminescence is to introduce progressions of out-of-plane vibrations into the spectrum. The luminescence spectra, being dependent upon the separation of $n\pi^*$ and $\pi\pi^*$ states, are sensitively dependent upon solvent since the $n\pi^*$ and $\pi\pi^*$ states are usually subject to different solvent shifts. In addition, the luminescence lifetime is expected to be solvent sensitive, as are radiationless processes and therefore quantum yields of all photoprocesses.

The spin–orbit interaction between $n\pi^*$ and $\pi\pi^*$ states is very often much stronger than between $n\pi^*$ and $n\pi^*$, or $\pi\pi^*$ and $\pi\pi^*$. This long-recognized fact[16-18] is the primary cause of a number of the important photoproperties of N-heterocyclics, ketones, and aromatic hydrocarbons. It is supposed that the large one-center spin–orbit interaction between $n\pi^*$ (or equivalently $\sigma\pi^*$ or $\pi\sigma^*$) and $\pi\pi^*$ states is dominant in all these classes of molecules. In aromatics the C–Hσ orbitals are the analogue of the nitrogen n-orbitals in heterocyclics. However, in aromatics the $\pi\pi^*$–$\sigma\pi^*$ energy gaps are large and the resultant spin–orbit effect at the (lighter) carbon atom is rather small. Similarly, in aliphatic ketones the $n\pi^*$–$\pi\pi^*$ gap is relatively large so that the effectiveness of oxygen spin–orbit induced processes is reduced over those in aromatic ketones. For N-heterocyclics the $n\pi^*$ and $\pi\pi^*$ states are both at low energy and are quite strongly spin–orbit mixed via the fields of the nitrogen nucleus; accordingly, nonradiative pathways are relatively well defined, and spectroscopic characteristics such as lifetime and polarization are easily rationalized. The energy diagrams and coupling schemes shown in Figure 2 are intended to illustrate this discussion (see also Section 6.3): In these diagrams the spin–orbit interaction is always of the type σ mixing with π. The resultant predictions of luminescence polarization, spin-state dipole activity, and intersystem crossing pathways, and rough ideas of the relative rates of various radiationless processes involving spin–orbit interaction, are mostly in agreement with experiment. The intervention of states other than those shown in the diagram can modify this oversimplified view but perhaps not so much as to destroy the usefulness of these concepts in many situations. To use these concepts effectively it must be remembered that the spin–orbit interaction varies directly as a high power of the effective atomic number of the atom whose nuclear field is involved. For transition probabilities, and all other things being equal, the spin–orbit contributions from $2p$-electrons in oxygen, nitrogen, and carbon are expected to be in the approximate ratio 15 : 4 : 1. These points are taken up in more detail in Section 6.3, following a discussion of the mechanisms of electronic and vibrational relaxation.

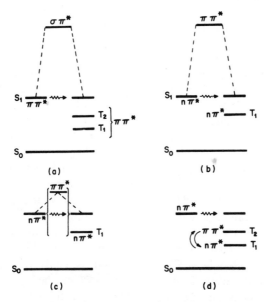

Figure 2. Various state-coupling schemes that can account for the radiative and nonradiative properties of $n\pi^*$ and $\pi\pi^*$ states. Note that the diagrams vary only in that the $n\pi^*$ or the equivalent $\sigma\pi$ states have different energies relative to $\pi\pi^*$ states.

6. RELAXATION PROCESSES IN MOLECULES

The depth of appreciation of the mechanistic detail of relaxation processes in polyatomic molecules has widened considerably over the past few years[19-23] following the earlier important and stimulating work of Ross and coworkers[24,25] and Robinson and coworkers.[26-28] Recent reviews of radiationless transitions in molecules[29-31] are available, so in this article I will present instead what appears to me to be a simplified practical account of relaxation phenomena in condensed media.

The general appearance of conventional molecular spectra in the condensed phase indicates that the Born–Oppenheimer adiabatic approximation is basically justified even in large molecules having relatively closely spaced electronic states. This approximation seems to be, therefore, a quite appropriate starting point to discuss relaxation phenomena. In addition, the adiabatic approximation is an appropriate starting point to discuss phenomena that are initiated by optical transitions in which all optically prepared states have essentially the same nuclear configuration because of the relatively slow motion of nuclei compared with electrons.

6.1. Vibrational Relaxation

Following excitation into some vibronic level $\varphi_1(x,R)F_{1v}(R')$ of the first electronic state, it is generally true that a rapid relaxation occurs, followed by emission from thermally equilibrated system levels into the ground state. As far as the observation of luminescence spectra is concerned—and their relative constancy regardless of the excitation energy in the region of $\varphi_1(x,R)$— this process is one that is most important. If $F_{1v}(R')$ is a high vibrational level of $\varphi_1(x,R)$, then there will be a large number of other levels $F_{1w}(R')$ in the immediate energy vicinity of $F_{1v}(R')$. Anharmonic perturbations that exist in the real molecule will have caused these nearby levels to mix. Thus the two-quantum-number description is not really adequate for higher vibrational levels, and the exclusive excitation of $\varphi_1(x,R)F_{1v}(R')$ cannot always be effected under normal conditions, without concomitantly exciting the nearby levels. This effect gives rise to the broadening of absorption bands, and of emission bands $\varphi_1 F_{10} \rightarrow \varphi_0 F_{0v}$, and is equivalent to the decay (at the bandwidth over $\hbar$ times per second) of the optically prepared state into the dense manifold of optically inactive nearby states. We note that all of this is occurring in one molecule, and that the effect is not a statistical average over many molecules. Already we have conditions for a radiationless transition, but the dissipation cannot occur without the intervention of the lattice, or the states of the condensed medium, with which the molecule is weakly coupled. However, an exactly analogous situation exists between each of the actual vibrational levels and the finely spaced lattice states as we have just described as existing between the state $\varphi_1(x,R)F_{1v}(R')$ and its nearby vibrational levels. Thus in reality the system decays into lattice–vibronic type states. The vibrational energy can now be transported into the bulk of the lattice, and a dissipation process occurs in the molecule. Ultimately, the rate of relaxation among high vibrational levels depends on the degree of coupling with the lattice.

Among the lower-energy vibrational bands the vibrational relaxation is completely controlled by interactions between the molecule and its surroundings and by the manner in which this interaction is modified by the details of the normal displacements. For example, if $\varphi_1(x, R)F_{1v}(R')$ refers to the state having one quantum of the lowest vibrational frequency in the molecule, then there is no way of getting to the state $\varphi_1(x,R)F_{10}(R')$ without incorporating the states of the medium. In glassy media, in solutions, and even in ordered solids consisting of molecules having reasonable size, it appears that the vibrational relaxation occurs rapidly compared with luminescence rates. Much work remains to be done on vibrational relaxation processes in the condensed phase, and this will be greatly assisted by the ultrafast methods being developed by Rentzepis.[32]

6.2. Electronic Relaxation

In the condensed phase the rate at which the lattice can dispose of excess energy is clearly an important feature of any relaxation process. Having admitted that vibrational dissipation is rapid compared with luminescence rates, we will expect that the dissipation of electronic excitation energy should certainly not occur any faster than this. However, the electronic relaxation processes—involving no change in the total energy of the system—may proceed much faster than vibrational relaxation.

In most polyatomic molecules, the zero-point, or vibrationless, level of the second electronic state of a given multiplicity $\varphi_2(x,R)F_{20}(R'')$ will be at essentially the same energy as many other vibronic states $\varphi_1(x,R)F_{1v}(R')$ corresponding to high vibrational levels of the first electronic excited state (refer to Figure 3). The adiabaticity and the inadequacy of the Born–Oppenheimer approximation for nearby states cause the actual states in the overlapping region to be superpositions of the states corresponding to both of the electronic, and all the appropriate vibrational, quantum numbers. There are many actual cases for which direct transitions from the ground state to $\varphi_1(x,R)F_{1v}(R')$ cannot occur with a probability even approaching transitions to $\varphi_2(x,R)F_{20}(R'')$. [If the energy gap (E_2-E_1) is large enough, the Franck–Condon factor $|\langle F_{00}(R)|F_{1v}(R')\rangle|^2$ will be extremely small for many large molecules, or the lower energy state may be a triplet.]

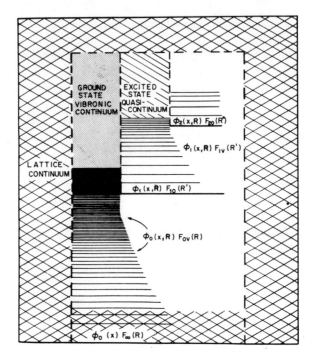

Figure 3. Schematic diagram indicating the complexity of of the vibronic spectrum in the neighborhood of excited electronic states.

Transitions to the region of E_2 will normally excite characteristics of the state $\varphi_2(x)$ such as polarization, electronic distribution, and polarizability. The electronic relaxation can usually be thought of as occurring at times subsequent to optical excitation, although there are very special conditions of excitation—involving highly monochromatic coherent sources—for which this may not be a proper description.[33] The observed spectral width of the 0,0-band of state 2 will be determined by the extent and number of the interactions between the two electronic states in the energy region E_2. In condensed media these spectral widths range from the barely measurable 10^{-4}–10^{-1} cm^{-1} broadening of singlets due to mixing with the lower triplets, to the very evident diffuseness ($\sim 10^3$ cm^{-1}) of the higher excited states of large molecules.[34]

Associated with the spread of electronic energy due to vibronic mixing of two electronic states is a time uncertainty $(2\pi c\bar{\nu})^{-1}$ where $\bar{\nu}$ is the half-width in cm^{-1} of the uniform line broadening due to state mixing. Values of $\bar{\nu} = 100$ cm^{-1} are not uncommon, and this corresponds to a time uncertainty of 5×10^{-13} sec. We call this the time of *electronic relaxation:* It is not the time for *dissipation* of energy since that is controlled by vibrational relaxation, which may be much slower. This latter comment is most crucial since it points out the difficulty of measuring, with present technologies, any decay process that is faster than vibrational relaxation.

The discussion presented here refers to large molecules in condensed media: A large molecule in this context is one for which a quasi-continuum of effective vibronic levels is isoenergetic with all the so-called vibrationless electronic states. The effectiveness of a vibronic level $\varphi_i F_{iv}$ in assisting the nonradiative transitions from j to i is concerned with whether in the first approximation $\langle \varphi_i | (\partial V/\partial R)_0 | \varphi_j \rangle$ is nonzero. The degree of electronic relaxation is determined by the number of effective states and the strength of the interaction for each of them. In the language of the vibronic approximations being adopted here, the extent of electronic relaxation is determined by four factors:

(a) The electronic interaction $C'_{ij}(R)$:

$$C'_{ij}(R) = \langle \varphi_i | (\partial V | \partial R)_0 | \varphi_j \rangle \Delta E_{ij}^{-1} \quad [35]$$

This matrix element [compare with Eq. 25] actually expresses the extent to which the electronic states i and j are mixed by nuclear displacements about equilibrium. In the most usual case, only a few non–totally symmetric modes R will contribute to the electronic relaxation. ΔE_{ij} is the energy gap between the zero-order electronic states.

(b) The number of effective states of the type $\varphi_i F_{iv}$ in the neighborhood of $\varphi_j F_{jo}$: This is determined by ΔE_{ij}, the energy gap to be made up by combinations of vibrations; by the number of non–totally symmetric displacements capable of providing the electronic interaction; and by the number of totally symmetric modes available to make up the combinations. One expects the actual number of effective states to be dependent upon the anharmonicity.

(c) The Franck–Condon factor between $F_{jo}(R)$ and each of the interacting vibrational levels $F_{iv}(R)$: The factor $|\langle F_{jo}/F_{iv}\rangle|^2$ decreases as v becomes very large but it increases as the difference in equilibrium geometries for the two electronic states increases. Thus high-frequency vibrations are preferred unless the two states are quite distorted along lower-frequency vibrational coordinates.

(d) The magnitude of the nuclear matrix element $|\langle f_{jo} | d/dR | f_{i1} \rangle|^2$, where f_{jo} and f_{i1} correspond to the harmonic oscillator wavefunctions for zero and one quantum of the effective non–totally symmetric vibration, respectively: If the molecule has the same symmetry in both states, then this term has the value ($2\pi c\bar{\nu}/\hbar$), with $\bar{\nu}$ being the wavenumber of the vibration. Thus low-frequency vibrations provide a larger effect than high-frequency vibrations.

Clearly, all of these features do not provide the same tendencies, and the electronic relaxation turns out to be, on this basis, a rather complex combination of offsetting effects. However, the Franck–Condon factor and the density of effective states seem to be the two quantities that control the main features of most nonradiative transition experiments. Furthermore, large differences in geometry along certain totally symmetric coordinates should provide a favorable situation for electronic relaxation, although only if the vibrational frequency is large enough such that not too many quanta are required to make up the energy gap. It is important to recognize that the importance of C–H vibrations in determining the extent of electronic relaxation in planar aromatic hydrocarbons is not necessarily expected to be carried over into other classes of molecules. Those vibrations that will be effective for relaxation into the ground state will often be those that have large Franck–Condon factors in the emission spectrum, and this will become more exact as the electronic energy gap gets smaller. At very large energy gaps, a mode with a small Franck–Condon factor in emission may dominate the electronic relaxation because its vibrational frequency is higher than modes along which the molecule is distorted on excitation. In principle, it should be possible to test which modes are active in electronic relaxation by investigating the effects of isotopic substitution on the luminescence lifetime.

6.3. Intersystem Crossing

The factors influencing the electronic relaxation from singlet to triplet states or vice versa may be slightly different in principle from those influencing relaxation between states having the same multiplicity. The difference lies in the fact that non–totally symmetric vibrations are not necessarily involved in the intersystem mixing between states having different symmetry. Of course, non–totally symmetric vibrations are not necessary in internal conversion if the combining electronic states have the same symmetry, but this situation is uncommon for the first two singlet states of most organic molecules. In any event, vibronic interactions even via totally symmetric modes are necessary to couple two singlet states. The lowest singlet excited state, S_1, always has at least one triplet state, T_1, at lower energy, and if these two electronic states S_1 and T_1 can undergo significant spin–orbit interaction, there is no need for non–totally symmetric vibrational participation in the radiationless process. This means that the vibronic interaction need not be involved in the intersystem crossing: The vibronic levels (a quasi-continuum perhaps) of the state T_i in the vicinity of S_1 can be coupled to S_1 directly using spin–orbit interaction. The excitation of S_1 will initiate the growth in population of T_1 (with conservation of energy) at a rate $k[S_1]$. The transition from the levels of the quasi-continuum back into the state S_1 will be slow. In terms of the adiabatic basis model given here, the radiationless transition probability will be proportional to (a) the square of the spin–orbit interaction between S_1 and a lower-energy triplet state T, (b) the number of vibrational levels of T in the region of S_1, and (c) the differences in geometry between the S_1 and T states that determine the Franck–Condon factor.

Quite by chance the mechanism described in the previous paragraph is not common among symmetric organic molecules. The reason is that very few classes of molecules have the capability for spin–orbit interactions between the lowest singlet and lower-energy triplet state. This is a consequence of the *dominance of single-center spin–orbit interactions*—the same reason why only one spin state is predominantly radiatively active in many organic molecules as described in section 5, Eq. [33]. Put more simply, if only one spin substate of a triplet is radiatively active, then the triplet level is presumably only interacting strongly with one type of singlet, and it turns out in practice that it is unusual to find S_1 interacting with lower triplets. For the cases of aromatic hydrocarbons, for molecules in which the $\sigma\pi^*$ states are at higher energy than the $\pi\pi^*$ states, and for aromatic and often alphatic ketones, the reasons why the simple mechanism does or does not work are readily documented:

(a) Aromatic hydrocarbons: The spin–orbit interaction between singlet and triplet $\pi\pi^*$ states is extremely weak[16] and is at most about 10^{-2} of the spin–orbit interactions between $\pi\pi^*$ and $\sigma\pi^*$ states. If the intersystem crossing in aromatic hydrocarbons was controlled by direct spin–orbit interaction between $^1\pi\pi^*$ and $^3\pi\pi^*$ states, then probably phosphorescence would never be observed in these molecules. The $\pi\pi^*$ and $\sigma\pi^*$ states are mixed during nuclear motions, so the intersystem crossing turns out to be assisted in the sense that the $\sigma\pi^*$–$\pi\pi^*$ spin–orbit interaction can then be used to couple two $\pi\pi^*$ states. This process of intersystem crossing (see Figure 2a) requires the intervention of a non–totally symmetric vibration. A similar mechanism should persist in molecules where the $\sigma\pi^*$ (and $\pi\sigma^*$) states are always at higher energy than the $\pi\pi^*$ states.

(b) Organic ketones: From a fixed-nucleus viewpoint and for an ideal ketone having precise C_{2v} symmetry, the carbonyl singlet and triplet $n\pi^*$ states cannot couple by spin–orbit interaction. Small deviations from C_{2v} should hardly change the radiationless transition mechanisms. The dominant contribution[35] to the spin–orbit interaction in carbonyls involves the single-center (oxygen) matrix element $\langle p_x(0)|L_z|p(0)_y\rangle$, where $p_x(0)$ and $p_y(0)$ are $2p_x$ and $2p_y$ orbitals on oxygen, with $2p_x$ being the nonbonding orbital and z the carbonyl direction: This term represents a $\pi\pi^*-n\pi^*$ interaction. The spin–orbit mixing of singlet and triplet $n\pi^*$, that cannot occur in C_{2v} symmetry, can occur if the symmetry is reduced to C_2—which happens during the motion of the nuclei along an a_2 coordinate. However, the lowering of the symmetry usually has a much larger effect on the mixing of the electronic states due to adiabaticity: The modified electronic states may then interact directly by means of the one-center spin–orbit interaction. This mechanism works even if the equilibrium geometry is C_{2v} in both electronic states except that then a vibration is required. However, it is likely that ketones will always have lower symmetry than C_{2v} in at least one electronic state. A C_s distortion of the formaldehyde type is insufficient to cause a mixing of $n\pi^*$ and $\pi\pi^*$ states appropriate for enhancement of the radiationless transition. The intersystem crossing in ketones is also assisted by the fact that the $^1n\pi^*$ and $^3n\pi^*$ states generally have different geometries so the Franck–Condon factors are large,[36] and this is probably a significant factor in determining the absence of fluorescence from many aromatic ketones. If $^3\pi\pi^*$ states are located between the $^1n\pi^*$ and $^3n\pi^*$ states, the intercombinations may be enhanced a little, but it is unlikely that this intervention will make much difference to the intersystem crossing which can already be so fast due to the factors mentioned.

(c) Nitrogen heterocyclics: In nitrogen heterocyclics the situation is rather similar to that in carbonyl compounds, with one important difference: The $n\pi^*$ and $\pi\pi^*$ states are usually intermeshed such that intersystem crossing *out of* $^1n\pi^*$ or *out of* $^1\pi\pi^*$ states can usually be *into* $^3\pi\pi^*$ or $^3n\pi^*$ states, respectively. Thus the one-center (p_x or p_y interacting with p_z, out of plane, at the nitrogen center) interactions can be utilized without necessarily invoking nuclear motions (see Figure 2d). In special cases, these one-center interactions may nearly cancel (say on two nitrogens) or they may vanish by local symmetry (the π orbital may have a node through the nitrogen), in which case vibronic interactions are needed and the intersystem crossing probability is lessened. The one-center interactions on nitrogen are roughly two times less than those on oxygen, so usually (but not exclusively) we expect slower intersystem crossing rates in N-heteroaromatics compared with aromatic ketones if $n\pi^*-\pi\pi^*$ mixing is involved. The hybridization of the n-orbital in N-heterocyclics increases this tendency, and the net calculated ratio of probabilities for ketones versus N-heterocyclics is approximately 10:1 for the spin–orbit interaction contribution.

6.4. Spin Polarization

Because of spin–orbit coupling selection rules, the final states of the intersystem crossing process have well-defined spin. In many instances, as discussed above, one spin–orbit interaction predominates over all the others, and the effect of this on the intersystem crossing is to ensure the excitation of only one type of spin state (i.e., σ_x, σ_y, or σ_z) in each excited molecule: The system thereby becomes spin-polarized. The depolarization of the spins in a system is achieved by nonradiative transitions between the spin substates—the energy being conserved by means of the very low-frequency lattice vibrations of the medium. This process is termed spin–lattice relaxation, and it is naturally slowest at the lowest temperatures. In molecular condensed-phase systems, the spin–lattice relaxation times range on the order of 1 sec in the neighborhood of 2°K. Spin–lattice relaxation times at 77°K are maybe as slow as 10^{-6}–10^{-7} sec, and should perhaps not be disregarded in kinetic schemes involving triplet-state intermediates. Times in the range 1–10^{-7} sec are very long compared with vibrational relaxation times; thus we may expect, following intersystem crossing, that the excited molecules will come into thermal equilibrium much faster than the spins can be interchanged. So the system remains spin-polarized for times long compared with intersystem crossing and, at very low temperatures, for times long compared with the phosphorescence lifetime. Accordingly, as shown by de Groot et al.,[10] the phosphorescence decay is temperature dependent at very low temperatures.

REFERENCES

1. M. Kasha, in "Comparative Effects of Radiation" (M. Burton, J. S. Kirby-Smith, Smith, and J. L. Magee, eds.), pp. 72–96, John Wiley and Sons, Inc., New York (1960).
2. M. Kasha, Paths of molecular excitation, *Radiation Res.* 2, 243 (1960).
3. M. Kasha, in "Light and Life" (W. M. McElroy and B. Glass, eds.), pp. 31–64, The Johns Hopkins Press, Baltimore (1961).
4. M. Kasha, Theory of molecular luminescence, Proceedings of the International Conference on Luminescence, pp. 166–182, Budapest (1966).
5. S. P. McGlynn, T. Azumi, and M. Konoshita, "Molecular Spectroscopy of the Triplet State," Prentice-Hall, Inc., Englewood Cliffs, N.J. (1969).
6. R. S. Becker, "Fluorescence," John Wiley and Sons, Inc., New York (1969).

7. W. T. Simpson, "Theories of Electrons in Molecules," pp. 171–178, Prentice-Hall, Inc., Englewood Cliffs, N.J. (1962).
8. W. B. Fowler and D. L. Dexter, Relation between absorption and emission probabilities in luminescent centers in ionic solids, *Phys. Rev. 128*, 2154–2165 (1962).
9. S. J. Strickler and R. A. Berg, Relationship between absorption intensity and fluorescence lifetime of molecules, *J. Chem. Phys. 37*, 814–822 (1962).
10. M. S. de Groot, I. A. M. Hesselmann, and J. H. van der Waals, Phosphorescence and spin polarization: A preliminary report, *Mol. Phys. 12*, 259–264 (1967).
11. M. Kasha, Characterization of electronic transitions in complex molecules, *Disc. Farad. Soc. 9*, 14–19 (1950).
12. H. C. Longuet-Higgins, Theory of molecular energy levels, *Adv. Spectroscopy 2*, 429–472 (1961).
13. D. P. Craig and G. J. Small, Totally symmetric vibronic perturbations, *J. Chem. Phys. 50*, 3827–3834 (1969).
14. M. Kasha, Collisional perturbation of spin–orbital coupling and the mechanism of fluorescence quenching, *J. Chem. Phys. 20*, 71–76 (1952).
15. R. M. Hochstrasser and C. J. Marzzacco, in "Luminescence" (E. Lim, ed.), pp. 631–656, W. A. Benjamin, Inc., New York (1969).
16. D. S. McClure, Spin–orbit interaction in aromatic molecules, *J. Chem. Phys. 20*, 682–686 (1952).
17. J. W. Sidman, Spin–orbit coupling in pyrazine, *J. Mol. Spectroscopy 2*, 333–343 (1958).
18. J. W. Sidman, Spin–orbit coupling in the $^3A_2$–$^1A_1$ transition of formaldehyde, *J. Chem. Phys. 29*, 644–652 (1958).
19. M. Bixon and J. Jortner, Intramolecular radiationless transitions, *J. Chem. Phys. 48*, 715–726 (1968).
20. J. Jortner and R. S. Berry, Radiationless transitions and molecular quantum beats, *J. Chem. Phys. 48*, 2757–2766 (1968).
21. W. Siebrand, Radiationless transitions in polyatomic molecules; I: Calculation of Franck–Condon factors, *J. Chem. Phys. 46*, 440–447 (1967).
22. S. H. Lin, Rate of interconversion of electronic and vibrational energy, *J. Chem. Phys. 44*, 3759–3767 (1966).
23. W. Siebrand and D. F. Williams, Radiationless transitions in polyatomic molecules III, Anharmonicity, isotope effects, and single-to-ground state transitions, *J. Chem. Phys. 49*, 1860–1871 (1968).
24. G. R. Hunt, E. F. McCoy, and I. G. Ross, Excited states of aromatic hydrocarbons: Pathways of internal conversion, *Austral. J. Chem. 15*, 591–604 (1962).
25. J. P. Byrne, E. F. McCoy, and I. G. Ross, Internal conversion in aromatic and N-heteroaromatic molecules, *Austral. J. Chem. 18*, 1589–1603 (1965).
26. G. W. Robinson and R. P. Frosch, Theory of electronic energy relaxation in the solid phase, *J. Chem. Phys. 37*, 1962–1973 (1962).
27. G. W. Robinson and R. P. Frosch, Electronic excitation transfer and relaxation, *J. Chem. Phys. 38*, 1187–1203 (1963).
28. G. W. Robinson, in "The Triplet State" (A. B. Zahlan, ed.), pp. 213–227, The Cambridge University Press, Cambridge (1967).
29. R. M. Hochstrasser, Analytic and structural aspects of vibronic interactions in U. V. spectra of organic molecules, *Acc. Chem. Res. 1*, 266–274 (1968).
30. B. R. Henry and M. Kasha, Radiationless molecular electronic transitions, *Ann. Rev. Phys. Chem. 19*, 161–192 (1968).
31. J. Jortner, S. A. Rice, and R. M. Hochstrasser, *in* "Advances in Photochemistry"

(W. A. Noyes, Jr., G. S. Hammond, and J. N. Pitts, Jr., eds.), Interscience, New York (1969).
32. P. M. Rentzepis, Direct measurements of radiationless transitions in liquids, *Chem. Phys. Letters* **2**, 117–120 (1968).
33. W. Rhodes, B. R. Henry, and M. Kasha, A stationary state approach to radiationless transitions: Radiation bandwidth effect on excitation processes in polyatomic molecules, *Proc. Nat. Acad. Sci.* **63**, 31–37 (1969).
34. R. M. Hochstrasser and C. J. Marzzacco, Perturbations between electronic states in aromatic and heteroaromatic molecules, *J. Chem. Phys.* **49**, 971–984 (1968).
35. S. Dym, R. M. Hochstrasser, and M. Schafer, Assignment of the lowest triplet state of the carbonyl group, *J. Chem. Phys.* **48**, 646–652 (1968).
36. S. Dym and R. M. Hochstrasser, Spin-orbit coupling and radiationless transitions in aromatic ketones, *J. Chem. Phys.* **51**, 2458–2468 (1969).

Chapter 2: Experimental Techniques, Part A

FLUORESCENCE INSTRUMENTATION AND METHODOLOGY*

Edward P. Kirby

Laboratory of Physical Biochemistry
Naval Medical Research Institute
Bethesda, Maryland

1. BASIC CONSIDERATIONS

The general methodology and instrumentation utilized in fluorescence spectroscopy have been reviewed by several authors,[1-8] and perhaps most definitively covered by Parker.[9] Therefore, the basic aspects of the instrumentation will be covered here only briefly. Emphasis will be placed on the considerations most relevant to the fluorescence of proteins and nucleic acids.

1.1. General Description of a Spectrofluorimeter

A basic spectrofluorimeter† is diagrammed in Figure 1. Light from a high-intensity source (L) is dispersed by the first monochromator (M_1) and the desired wavelength selected to excite the sample (S). Fluorescence is emitted by the sample in all directions, but is usually collected at 90° from the incident beam so as to minimize the effects of scatter and reflection of the exciting light. The fluorescent emission, which represents a broad

---

*From Bureau of Medicine and Surgery, Navy Department, Research task MR005.06-0005A. The opinions in this paper are those of the author and do not necessarily reflect the views of the Navy Department or the naval service at large.

†In line with the suggestion of Parker,[9] the term "spectrofluorimeter" is used to describe an instrument designed to measure the *intensity* of fluorescence and equipped with monochromators for selection of the excitation and observation wavelengths.

spectrum of wavelengths, is dispersed by the second monochromator (M_2), and the desired wavelength is focused on the photomultiplier tube (PM). The output of the photomultiplier is amplified and displayed on a meter, oscilloscope, or some type of $X-Y$ recorder.

The light source is generally a high-pressure xenon arc. This provides a continuous distribution of light over the ultraviolet and visible spectrum so that light of any desired wavelength may be selected by the first monochromator for excitation of the sample. Alternatively, a mercury arc may be used as a light source to provide high intensity at certain wavelengths. If the sample absorbs significantly at one of the mercury lines, it is often possible to obtain much higher light levels at these specific wavelengths with a mercury arc than with a xenon arc of comparable wattage.

The monochromators may use either diffraction gratings (as shown) or quartz prisms to disperse the light. Most commercial spectrofluorimeters utilize grating monochromators, which generally are more efficient and have higher dispersion than comparably priced prism monochromators. An added advantage of grating monochromators is that the dispersion is approximately constant throughout the ultraviolet and visible range. Spectra obtained with grating instruments are linear in wavelength. Variable slits at the entrance and exit of each monochromator and near the sample

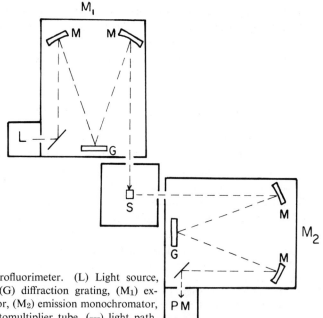

Figure 1. Basic spectrofluorimeter. (L) Light source, (M) concave mirror, (G) diffraction grating, (M_1) excitation monochromator, (M_2) emission monochromator, (S) sample, (PM) photomultiplier tube, (---) light path.

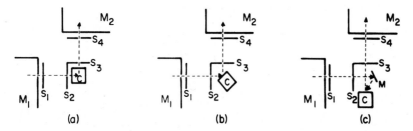

Figure 2. Various arrangements of sample compartment. (a) Right-angle viewing of fluorescence. (b) Front-surface viewing. (c) Front-surface viewing with adjustable mirror (M) to allow variation in the angle of viewing. S_1, S_2, S_3, and S_4 are adjustable slits; M_1 and M_2 are the excitation and emission monochromators, respectively. The sample is contained in cuvette C.

(Figure 2a) control the amount of light transmitted and also its spectral bandwidth.

For most work, the fluorescence of the sample is viewed at right angles to the exciting beam, as shown in Figure 2a. For good spectral resolution, the slit width must be rather small (1–2 mm or less), which means that only the center of the cell is viewed by the emission monochromator. In this case, the solutions must be very dilute or else some correction must be applied for attenuation of the exciting beam due to absorption by the portion of the sample not viewed by the emission monochromator. For concentrated solutions in which all of the exciting light is absorbed in the first few microns of solution, the emitted light may be viewed from the front surface of the cuvette. The angle of viewing may be fixed, as in Figure 2b, or a movable mirror may be introduced, as in Figure 2c, so that the angle can be adjusted to achieve minimum interference from the reflected exciting beam.

1.2. Representation of Spectra

Two types of spectra are obtained which are characteristic of the fluorescence of a sample—its excitation and emission spectra. The excitation spectrum is obtained by keeping the observation wavelength constant and measuring the intensity of the fluorescence as a function of the wavelength of the exciting light. For solutions of simple compounds this should correspond to the absorption spectrum of the molecule. For proteins and nucleic acids the excitation and absorption spectra are usually not identical because not all of the species which absorb the incident light are fluorescent.

The emission spectrum of a compound is obtained by keeping the excitation wavelength constant and measuring the wavelength dependence

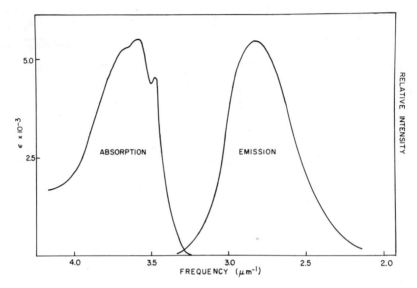

Figure 3. Absorption and emission spectra of tryptophan in water.

of the fluorescence intensity. The emission spectrum is generally broad, unstructured, and red-shifted with respect to the absorption spectrum (Figure 3), especially in aqueous solutions.

Since fluorescence emission is a quantized process, the best way to plot emission spectra is in terms of intensity (quanta/sec) versus wavenumber (the reciprocal of the wavelength). It is not uncommon, however, to plot intensity versus wavelength. While this not as useful from a photochemical standpoint, it has the advantage of familiarity and is the form in which spectra are presented by most commercial instruments.

1.3. Calculation of Quantum Yields

Determination of the absolute quantum yield (Q) of a substance can be accomplished in several ways,[10-12] but generally is too involved for routine work. It is usually much easier to obtain a relative quantum yield in comparison with a known standard. The standard most commonly used is quinine in 1 N H_2SO_4 ($Q=0.55$).[13] A very convenient standard for work with proteins is trytophan in water. The UV absorption (between 260 and 300 nm) and the fluorescence emission properties of most proteins are primarily due to the tryptophan chromophores. The quantum yield of tryptophan in water has recently been redetermined[14-16] to be 0.14. The accepted standards are probably accurate to within 10%, and, since one is most often concerned with the magnitude of the *changes* in quantum yield under various conditions rather than the actual quantum yield itself, errors introduced by inaccuracies in the standard are minimized.

The relative quantum yield of a substance can be determined by exciting both the sample and the standard at the same wavelength and incident intensity, and plotting the emission spectrum of each. The quantum yield of a material is directly proportional to the area under its emission spectrum. Once the areas of the two spectra (A_1 and A_2) and the optical densities of the two solutions (OD_1 and OD_2) are known, the ratio of their quantum yields may be calculated:

$$\frac{Q_1}{Q_2} = \frac{A_1}{A_2} \times \frac{OD_2}{OD_1} \qquad [1]$$

The emission spectra used for this determination must be the true emission spectra of the sample and the standard, corrected for any instrumental or other variations.

1.4. Polarization Spectra

As described in Chapter 5, a great deal of additional information can be obtained from the polarization of the excitation and emission spectra. Polarizing units are commercially available, and several instruments have either been modified or specifically designed for obtaining polarization spectra. Most notable are those of Weber[17, 18] and of Ainsworth and Winter.[19]

In general, the only modification which needs to be made is the introduction of two rotatable polarizing filters into the paths of the exciting and emitted beams (Figure 4). These are usually Glan–Thompson prisms made with special UV-transparent resin, since the Canada balsam used for cementing most of this type of prism is not transparent in the UV. More recently, Polacoat filters[20,21] (Polacoat, Inc., Cincinnati, Ohio) have proven very successful.

If H is used to describe a beam of polarized light which has its electric vector parallel to the plane formed by the exciting and emitted beams, and

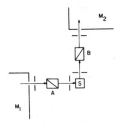

Figure 4. Arrangement for polarization spectra. (M_1) Excitation monochromator, (A and B) polarizers, (S) sample cuvette, (M_2) emission monochromator.

V is used to describe a beam whose electric vector is perpendicular to this plane, then the observed polarization of the emitted light is

$$P = \frac{I_V - I_H}{I_V + I_H} \qquad [2]$$

where I_V and I_H are the intensities observed when the polarizer in the emitted beam (polarizer B in Figure 4) is oriented either vertically or horizontally.

2. METHODOLOGY

In a simple spectrofluorimeter—similar to the one shown in Figure 1—the spectra obtained do not represent the true excitation or emission spectra. Superimposed upon the true spectra are errors introduced by variations in instrument sensitivity and variations due to the optical density, refractive index, etc., of the sample itself. Variations in the sensitivity of an instrument to light of different wavelengths, for example, can cause serious distortions in the apparent excitation and emission spectra.[22] Although these distortions may be acceptable for routine characterization work within a laboratory, correction is necessary not only for accurate determination of quantum yields, but also to allow comparison of spectra with results from other laboratories.

Considerable attention has been devoted to methods of correcting spectra.[22-29] Once the correction factors for a given instrument are known, each spectrum may be corrected by doing the necessary calculations on a point-by-point basis, but this is a rather laborious procedure. Computer programs have been written to perform these calculations,[30] but this still requires that the operator do point-by-point readings of the intensity, and only the calculations are done by the computer. In general, it is much more satisfactory to have the correction factors built into the instrument, so that all compensation is done automatically for each spectrum as it is being run. Variations arising from the method of sample preparation will still have to be corrected by hand, but these calculations are simpler and certain precautions in sample preparation can often eliminate them.

2.1. Instrument Calibration

2.1.1. Excitation Spectra

For excitation spectra the most important correction which must be made is for the wavelength dependence of the intensity of the exciting light.

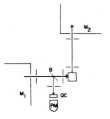

Figure 5. Arrangement for automatic correction of excitation spectra. (M_1) Excitation monochromator, (B) beam splitter, (QC) quantum counter solution, (PM) photomultiplier for monitoring output of quantum counter.

The most convenient method for obtaining corrected excitation spectra directly has been described by Parker.[22] The intensity of the incident beam is monitored by placing a beam splitter just prior to the sample, so that a fraction of the exciting light is diverted to a quantum counter* (see Figure 5). The output of the quantum counter is proportional to the intensity of the incident beam (in quanta/sec). The magnitude of the signal from the quantum counter then controls the amplification of the signal from the sample fluorescence, by means of either a ratio recorder[22] or an automatic gain-control amplifier.[31] In this way, as the excitation wavelength range is scanned, a corrected excitation spectrum is produced. This system has the added advantage that it automatically compensates for any changes in the total lamp output due to voltage fluctuations or arc shift.

2.1.2. Emission Spectra

The emission spectra must be corrected for the wavelength dependence of the efficiency of the emission monochromator and the photomultiplier. The most common method is to use a standard lamp whose output as a function of wavelength is known. Light from this lamp is directed onto a reflecting plate placed in the sample compartment, and an emission spectrum of the lamp is run. Comparison of the observed with the known emission spectrum yields the correction factor.

Once the correction function has been determined for a given monochromator–photomultiplier combination, this may be used to correct emission spectra after they have been taken. It is more convenient, however, to build this correction into the instrument, as described by Hamilton,[32] so that correction may be done automatically as the spectrum is being run. The function is incorporated into a tapped helical potentiometer, which is connected to the wavelength drive of the emission monochromator. As the spectrum is being scanned, a cam or potentiometer on the monochromator

---

*Melhuish[12] has suggested the use of a concentrated solution of Rhodamine B as a standard which is completely absorbing in the wavelength region 250–600 nm and whose quantum yield is constant throughout this range.

drives the tapped potentiometer, which puts out a voltage proportional to the correction function. The ratio recorder then divides the amplified output of the photomultiplier by this correction voltage to give the corrected intensity. Because of the large variation in phototube sensitivities, even between two tubes of the same type made by the same manufacturer, the correction function must be determined for each tube. It is probably also wise to redetermine the correction periodically to check for aging of the photomultiplier or deterioration of the mirrors of the monochromator.

2.1.3. Polarization Spectra

Calculation of polarization using Eq. [2] is not strictly correct, because preferential reflection from surfaces such as mirrors and gratings causes the efficiency of the monochromator and photomultiplier to be different for light beams polarized in different directions. If T is the efficiency of transmission of horizontally polarized light relative to that which is polarized vertically, then the correct form of Eq. [2] is

$$P = \frac{I_V - TI_H}{I_V + TI_H} \qquad [3]$$

Generally, T will be dependent upon wavelength, so that its value will have to be determined over the whole wavelength range of interest.

If a completely nonpolarized source is placed in the sample compartment, T can be determined from the ratio of the signals measured with the emission polarizer first oriented horizontally and then vertically. Two of the most convenient methods for obtaining an unpolarized source are the following:

2.1.3(a). Fluorescence from a Small Molecule in a Nonviscous Solution. If the small molecule has a long fluorescent lifetime compared to the time required for rotational relaxation, its fluorescent emission will be almost completely unpolarized, even if excited by polarized light. Azumi and McGlynn[33] found that the fluorescence of a solution of phenanthrene in methyl cyclohexane was completely unpolarized and used this to determine T in the region of 350–500 nm. For the 300–400 nm region used in protein and nucleic acid work, a convenient standard is skatole (or indole) in cyclohexane (for 300–340 nm) or in water (330–440 nm).

2.1.3(b). Excitation with Horizontally Polarized Light. If the exciting beam is polarized parallel to the plane of the exciting and emitted beams (i.e., H-polarized), then if fluorescence is observed at 90° to the exciting beam (as is almost always the case) the fluorescence will be unpolarized.

2.2. Correction for Sample Variations

After spectra have been corrected for the wavelength variations in instrument response, account must still be taken of inaccuracies introduced by the characteristics of the sample itself.

2.2.1. Inner-Filter Effects

The most common correction which must be made arises from what is termed the "inner-filter effect." As mentioned earlier (Figure 2a), when a standard 1 cm² cuvette is viewed at right angles to the exciting beam, only a small part of the solution is actually observed. If the sample solution has a sufficiently high optical density, an appreciable amount of the exciting light will be absorbed before it reaches the center of the cell. Generally, for a 1 cm² cuvette, if the optical density is greater than 0.04 (i.e., 5% of the exciting light is absorbed before reaching the center of the cell), correction

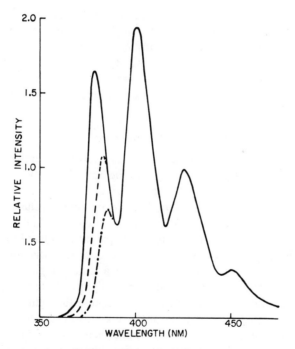

Figure 6. Self-absorption of anthracene fluorescence. Fluorescence emission spectra of anthracene in ethanol in 1 cm² cells. (———) 10^{-5} M, (— — —) 10^{-4} M, (—·—·—) 4×10^{-4} M.

must be made for this effective decrease in the light intensity in the region of observation.

Another difficulty may occur if the absorption and emission spectra overlap one another significantly. For solutions having appreciable optical density, a portion of the emitted fluorescence will be absorbed before it leaves the cuvette. The possible distortion of the emission spectrum which may result is demonstrated in Figure 6 for the case of anthracene in ethanol. At the higher concentrations, the fluorescence emitted near 380 nm is almost completely reabsorbed before leaving the cell because of the high absorption of the solution at these wavelengths. Above 400 nm there is little reabsorption by the solution, and these peaks are unaffected.

For a solution in which the absorbance does not exceed approximately 0.3, self-absorption and inner-filter effects can be corrected by

$$I_C = I_0 \times \text{antilog}_e \left(\frac{A_1 + A_2}{2} \right)$$

where I_0 and I_C are the observed and corrected fluorescence intensities, and A_1 and A_2 are the absorbances of the solution at the wavelengths of excitation and emission, respectively. In many cases, where overlap of the absorption and emission bands is low, A_2 will be essentially zero. As a general rule, the high sensitivity of most spectrofluorimeters makes it seldom necessary to work with solutions where the correction is greater than 10 or 20%.

2.2.2. Front-Surface Viewing of Fluorescence

The use of front-surface geometry for viewing fluorescence can also yield artifacts resulting from changes in the optical density of the solution. In the case where all of the light which strikes the cuvette is absorbed and re-emitted from only the first few millimeters of solution, there are no inner-filter effects. However, Melhuish[13] has pointed out that if the solution varies in absorbance, then in the more transparent regions of the absorption spectrum the exciting light will penetrate deeper into the solution and so the fluorescence emission will occur from a different position within the cell. This will affect not only the amount of light collected by the entrance slit of the emission monochromator, but, because the light will be entering the monochromator at a different angle, may also cause an apparent shift in the wavelength of the fluorescent light. Self-absorption can affect the apparent quantum yields when front-surface observation is used if a significant proportion of the fluorescence which is emitted *into* the bulk of the solution is absorbed and re-emitted. This can cause up to a 50% error in some cases.[13]

Any absorption of the fluorescent light and subsequent re-emission which could occur in more concentrated solutions would also cause a large decrease in the apparent polarization of fluorescence of the sample. For this reason, polarization measurements are almost always done on quite dilute solutions. Weber[18] has also demonstrated that in very concentrated solutions, where solute molecules are very close together, depolarization may occur by intermolecular energy transfer. If it is necessary to work with very concentrated solutions, Weber has described a sample holder in which the solution to be studied is held between two fused silica prisms as a very thin film.[18]

The high optical density required for front-surface viewing of fluorescence is a disadvantage, not only because of the larger amounts of often valuable samples required, but also because of the limited solubility of many proteins and nucleic acids, and their marked tendency to aggregate.

2.2.3. Refractive-Index Effects

When measuring the relative quantum yields of two samples in different solvents, correction must also be made for the difference in refractive index of the solvents. Hermans and Levinson[34] have pointed out that the refractive index of the solvent affects the angle at which the fluorescent rays emerge from the cuvette, and so affects the amount of light collected by the entrance slit of the emission monochromator. A suitable correction is

$$Q_C = Q_u \times \frac{\eta_1^2}{\eta_2^2}$$

where Q_C and Q_u are the corrected and uncorrected relative quantum yields, and η_1 and η_2 are the refractive indices of the sample and standard, respectively.

2.2.4. Effects of Contaminants

Certain precautions must be observed in handling fluorescence samples in order to obtain good quantitative results. Fortunately, the presence of small amounts of contaminating materials does not usually cause quenching of the fluorescence of most compounds, unless there is some specific binding between the fluorescent molecule and the contaminant. For a diffusion-controlled reaction between a quencher and a molecule with a fluorescent lifetime on the order of 10^{-9} sec, the concentration of quencher must be quite high before an appreciable decrease in quantum yield is seen. For the quenching of tryptophan, for example, the Stern–Volmer equation[35] predicts

$$\frac{Q_0}{Q} = 1 + k\tau_0[q]$$

$$= 1 + 6.5 \times 10^9 \times 2.7 \times 10^{-9}[q]$$

$$= 1 + 17.6[q]$$

where Q and Q_0 are the quantum yields of tryptophan in the presence and absence of the quencher, k is the rate constant for a diffusion-controlled reaction in water, τ_0 is the excited state lifetime of tryptophan in the absence of quencher, and $[q]$ is the concentration of quencher. For 1% quenching,

$$\frac{Q_0}{Q} = 1.01 = 1 + 17.6[q]$$

$$[q] = \frac{0.01}{17.6} = 5.7 \times 10^{-4} \text{ M}$$

From this it is seen that the quencher concentration must be on the order of 6×10^{-4} M to observe 1% quenching. Since samples are usually run at concentrations well below 10^{-4} M, quenching by contaminants in the sample would not be anticipated. The buffer used, however, could be a source of quenching molecules since it is usually present at quite high concentrations.

The one common contaminant of solutions which may cause quenching is atmospheric oxygen. Aqueous solutions saturated with air contain approximately 2.5×10^{-4} M oxygen, so that usually it does not affect fluorescence yields (unless by some interaction with the ground state). Oxygen quenching may play a more significant role in less polar solvents, where not only are fluorescence lifetimes longer, but also the solubility of oxygen is higher. For instance, oxygen is almost 12 times more soluble in hexane than in water.[36]

While oxygen may not affect fluorescence in aqueous solutions to a great extent, its high affinity for radicals and free electrons suggests that it may play a very significant role in other, longer-lived, photochemical processes. Certainly, studies of photodegradation and energy transfer ought to take oxygen effects into account.

A much more serious problem than quenching by contaminants is that the impurities may themselves be fluorescent. For this reason, it is generally necessary that both the solute and the solvent be extensively purified before use, and that solvent blanks be run routinely. At low concentrations of solute, Raman emission from the solvent may also make a significant contribution to the blank. It has also been noted[37] that even the cuvettes themselves may have appreciable fluorescence.

2.3. Cuvettes

There are certain advantages to be derived from using small (0.3 by 0.3 cm) cuvettes rather than the more conventional 1 cm² size. The obvious advantage is that only 10% as much sample solution is required. Because most high-resolution spectrofluorimeters view only the central few millimeters of the standard cuvette anyway (Figure 7), usually no sensitivity is lost in using a smaller cell. Moreover, since there is considerably less "non-utilized" solution for the beams to pass through in the small cells, higher concentrations of sample may be used before inner-filter or self-absorption effects become significant.

The low optical densities required for the solutions in standard 1 cm² cuvettes cannot be read accurately on a standard spectrophotometer. The solution must either be read in a spectrophotometer which can accommodate cells with 5 or 10 cm pathlengths, or else it must be a dilution of a more concentrated solution (with the assumption that Beer's law holds). Use of small cuvettes means that the optical density of the solution can be read in a standard 1 cm² cuvette and then the fluorescence spectrum determined on the same solution, using the cuvettes with the shorter pathlength.

One precaution, however, ought to be observed in the use of the smaller cells. When 1 cm² cuvettes are used with normal slits, only the center of the cell is observed, so that there is little opportunity for any possible fluorescence of the quartz of the cell to enter the emission monochromator. With the small cuvettes, however, almost all of the cell can be viewed by the entrance slit of the emission monochromator, so that any fluorescence of the walls of the cuvette can make a large contribution to the blank fluorescence. Exciting light scattered by the walls may also be much more of a problem.

An alternative to the use of small cuvettes has been proposed by Chen

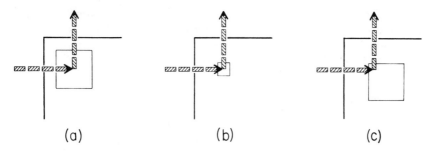

Figure 7. Use of various cuvette configurations. (a) Standard 1 cm² cuvette, showing that only a small portion of the solution is utilized, (b) 0.3 by 0.3 cm cuvette, (c) standard cuvette placed off-center.

and Hayes.[38] They suggested that an ordinary fluorescence cuvette could be used by placing it off-center within the sample compartment, as shown in Figure 7c. Since only a corner of the cell is illuminated and observed, the result is the same as if a smaller cuvette were used. Moreover, the position of the cuvette can be adjusted to minimize the effects of scatter and fluorescence from the walls, as long as correction is made for the resulting variations in pathlength.

3. CRITERIA FOR A SPECTROFLUORIMETER

Since the description of the first widely accepted spectrofluorimeter by Bowman et al. in 1955,[39] the availability and sophistication of the instrumentation has increased tremendously. Until very recently, however, anyone desiring to do truly high-resolution work in fluorescence generally either built his own instrument or specially adapted a commercial model. To some extent, this is still true, in that there is no commercially available instrument which will meet the most stringent specifications. Generally, one has to accept a trade-off between the desirable features of the various instruments and their prices. The decision must also be made as to whether one or more of these instruments is sufficiently sensitive, precise, and versatile to fit the needs of a given laboratory, or whether it would be better to adapt or completely redesign an instrument to satisfy more exacting specifications. The American Society for Testing and Materials has formulated a series of tests for evaluating the sensitivity, resolution, and accuracy of individual spectrofluorimeters.[8]

Parker[9] has discussed extensively the principles underlying the design of a spectrofluorimeter, and Udenfriend,[3, 8] Lott,[5] and Chen et al.[7] have described the features of some of the commercially available instruments. In addition, many workers[16,19,22,32,40–51] have published descriptions of their own instruments. The choice of which features are most important will, of course, depend upon the purposes, problems, and technical abilities of the individual laboratory; however, certain criteria may be established to guide in making a decision.

3.1. Sensitivity

High sensitivity in a spectrofluorimeter yields several advantages. For a given intensity of exciting light, a lower concentration of sample is required to give the same response. This is important not only because samples are often precious, but higher concentrations may cause aggregation, increased light scattering, or excessive absorption of the exciting beam.

Proteins and nucleic acids often have very low quantum yields, so that high sensitivity is required to give usable spectra at low concentrations.

Alternatively, high sensitivity means that for a given concentration of sample, less exciting light is required. For a given lamp wattage, narrower slits may be used, thus increasing the resolution of the instrument. Moreover, as Chen[52] has pointed out, protein samples may be subject to considerable photodecomposition if exposed to high light levels, so that the less incident intensity required, the better.

Since the corrected excitation spectrum of a compound is theoretically identical to its absorption spectrum, a high-sensitivity spectrofluorimeter may be used to determine the absorption spectrum at a concentration much less than that required by a spectrophotometer. By observing at a wavelength specific for the fluorescence of a compound, the absorption spectrum of this material may be determined in the presence of other absorbing species, provided that these other species (a) are either nonfluorescent or fluoresce in a different region of the spectrum, and (b) do not transfer their energy to the emitting species.

3.2. Resolution

The resolution of a monochromator is generally determined by three factors: the angular dispersion of the system, which is dependent upon the length of the optical path and the dispersing power of the prism or

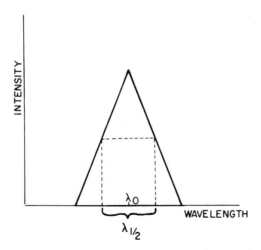

Figure 8. Triangular wavelength distribution o light intensity from monochromator. (λ_0) Central (nominal) wavelength, ($\lambda_{1/2}$) half-bandwidth.

grating; the width of the entrance and exit slits; and the quality of the optical components. For a given instrument, the only factor which is adjustable by the operator is the slit width. In general, as the width of the slits is decreased, the amount of light passed by the monochromator decreases but the spectral purity of this light increases. Reduction of the slit width can only increase the resolution up to a certain limiting value, because of inherent nonlinearities in the gratings and mirrors. Parker[9] demonstrated that optimal performance, in terms of maximal light flux with the highest resolution, is achieved when the entrance and exit slits of the monochromator are equal in width. Under these conditions, the wavelength distribution of the light coming out of the monochromator is triangular (see Figure 8). The spectral purity of the light can be described in terms of its half-bandwidth ($\lambda_{1/2}$), which is defined as the width of the band at one half of its maximum intensity. For a triangular intensity distribution, three fourths of all the light passed by the monochromator is within the region of the half-bandwidth.

The excitation and emission spectra of many compounds are highly structured, and it is, of course, important that the spectrofluorimeter be able to resolve this characteristic structure. Needless to say, the wavelength readings should be accurate as well as precise.*

While the spectra of proteins are often rather broad and unstructured, high resolution is still important. The low quantum yields of proteins mean that solutions concentrated enough to give useful spectra may scatter the exciting light to a considerable extent. The overlap of the scatter peak and the emission peak, especially when exciting at wavelengths greater than 290 nm, can cause considerable difficulty in the estimation of the emission spectrum. Higher resolution decreases the extent of this overlap and permits more accurate determination of the spectrum.

Since it is sometimes desirable to excite samples in a region where the absorption spectrum is changing rapidly and in a nonlinear manner, exact measurement of the quantum yield requires that the bandwidth of the exciting beam be the same as the bandwidth of the spectrophotometer used to measure the optical density of the sample. An example of this type of appli-

---

*It is often convenient to check the accuracy and resolution of the monochromators with standard compounds such as indole or anthracene in cyclohexane. The highly structured absorption spectra can be determined accurately on a recording spectrophotometer and compared to the observed excitation spectra. The emission spectra are also well structured and can be compared with published spectra. A mercury lamp is generally the best primary standard to check the calibration of the monochromators, but standard compounds are much easier to use routinely and have the additional advantage of simultaeously testing whether the devices used to correct the spectra are functioning properly.

cation could be in the investigation of protein structure where one might desire to excite one aromatic amino acid residue in preference to another. Usually, the desired wavelength will not be one where the absorption spectra of all the contributing chromophores are at maxima or minima, but rather where their absorptions are changing rapidly. As an example, consider a mixture of tyrosine and tryptophan in water at concentrations such that their optical densities are equal at 275 nm (Figure 9). Absorption spectra are usually run on spectrophotometers where $\lambda_{1/2}$ is considerably less than 1 nm. These spectra would predict that if the solution is irradiated at 275 nm, equal numbers of tyrosines and tryptophans will absorb quanta, whereas if the irradiation is done at 295 nm, almost 20 times as much of the absorption will be due to the tryptophan as to the tyrosine. However, for an excitation monochromator slit setting yielding a half-bandwidth much greater than 1 nm, the actual absorption ratios would differ significantly from the values predicted on the basis of optical densities.

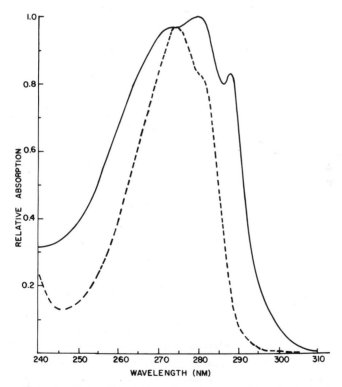

Figure 9. Absorption spectra of tryptophan and tyrosine in H_2O. Spectra of solutions normalized to have equal absorptions at 275 nm. (———) Tryptophan and (— — —) tyrosine.

The actual effect of using wider bandwidths can be determined. If a triangular intensity distribution is assumed, the apparent absorption spectrum of each component can be calculated for a series of different bandwidths. Measurement of the apparent absorptions at 275 and 295 nm from this series of spectra for each component gives markedly different ratios from those determined with the very narrow bandwidth used for absorbance measurements. The percent difference of the calculated effective absorbance ratio compared to the true value of the ratio (as determined from optical density measurements) is shown in Figure 10. The wider bandwidth of the excitation monochromator allows significant absorption of wavelengths considerably different from λ_0, and use of the ratios obtained by absorption spectroscopy would lead to erroneous quantum yields of fluorescence for the two species in solution. This demonstrates not only that one must excite at the optimal point, but that the half-bandwidth must also be narrow, and accurately known, so that the exact amount of each species which is excited can be determined.

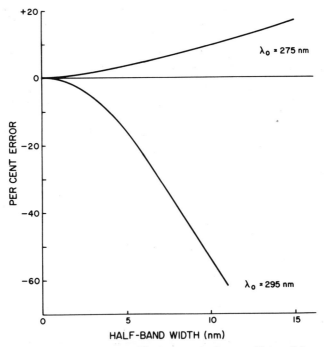

Figure 10. Effect of bandwidth on excitation specificity. Calculated percent error in apparent ratio of tryptophan to tyrosine absorption, relative to the ratio determined with a half-bandwidth of 0.25 nm.

An alternative would be to use a mercury arc light source and use the monochromator to isolate one of the sharp lines in the mercury emission spectrum. In this way one can achieve high intensity and high spectral purity of the exciting light, but this approach is of course limited to the wavelengths of the mercury lines. The use of the mercury lines is generally limited to lamps of less than 150 watts because the high pressure required for higher wattage lamps causes self-absorption and pressure broadening of the mercury lines.[9]

High resolution of the emission monochromator is similarly important not only for resolving fine structure in the emission spectra, but also for observing at a wavelength where more than one species emits.

3.3. Sample Compartment

The versatility of a spectrofluorimeter can be dependent in many ways upon the characteristics of the sample compartment. It is very often the case that one wishes to use a different type of sample holder, or introduce filters, polarizers, beam splitters, or other devices into the paths of the light beams. For this reason, the sample compartment should be large and easily accessible. The sample must, of course, be positioned accurately and reproducibly in the exciting beam and in line with the entrance slits of the emission monochromator. The compartment should be adaptable to a variety of cell styles and sample holders, and preferably should allow both right-angle and front-surface viewing. Since the focal lengths of the monochromators are generally not adjustable, provision must be made at the time the sample compartment is designed for all of the desired sample holders and accessories.

As most laboratories which have a high-quality spectrofluorimeter will want to use it for studying phosphorescence as well as fluorescence, the sample compartment ought to accommodate cell holders for work at liquid-nitrogen temperatures, as well as the necessary rotating shutters for isolating phosphorescence from fluorescence, and also a fast shutter for determining phosphorescent lifetimes.

In general, the sample holder ought to be thermostatted, even if most of the work is to be done at room temperature. Aside from the flexibility of working at other temperatures when desired, thermostatting is important because of the frequently great dependence of fluorescence quantum yield on temperature. For example, the quantum yield of tryptophan varies by about 2% per degree centigrade in the region around room temperature. Obviously, the estimation of quantum yields requires careful control and accurate measurement of the temperature of the sample.

3.4. Photomultipliers

The use of high-sensitivity photomultiplier tubes has been essential in the design of high-resolution spectrofluorimeters. As Udenfriend[3] has pointed out, it is often the characteristics of the photomultiplier which determine the effectiveness of an instrument for a certain type of analysis. The three most important characteristics in the selection of a photomultiplier for a specific instrument or application are its wavelength response, sensitivity, and noise level.

The blue-sensitive photomultipliers, such as those having an S-5 or S-13 response (see Figure 11) are most effective in the region of 250–500 nm. Most work in the fluorescence of proteins and nucleic acids is done in this region, so generally the blue-sensitive tubes will be quite adequate. However, for work with certain fluorescent conjugates or dyes used for quantum counters, the greater response of the red-sensitive tubes to light of longer wavelengths may be necessary. Red-sensitive photomultipliers are considerably more expensive, however, and generally have much higher noise levels.

The ultimate sensitivity of a photomultiplier is determined by two factors—its actual sensitivity, in terms of the number of microamperes of

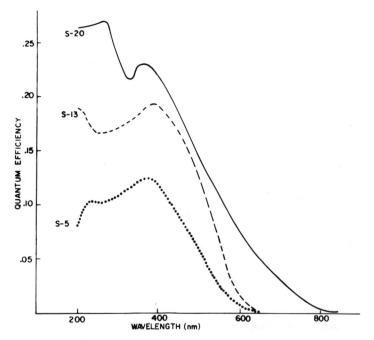

Figure 11. Wavelength response of three types of photomultiplier tube.

current output for a given amount of light incident on the cathode, and the ratio of this signal to the noise present in the absence of any incident light. The actual sensitivity of a photomultiplier may be greatly increased by increasing the voltage applied across it. Bartholomew[1] reports that the sensitivity is proportional to the eighth power of the applied voltage. This allows a great deal of flexibility in selecting the desired sensitivity of the photomultiplier, but also requires that for accurate work the high-voltage power supply of the photomultiplier tube must be stabilized to better than 0.01% to prevent gain fluctuations.

Unfortunately, the ultimate sensitivity of a photomultiplier cannot be increased indefinitely by raising the voltage, since the higher voltage also increases the dark current. The primary cause of the dark current is random release of electrons from the photocathode, either by thermal activation or by interaction with radioactive particles from the quartz of the tube or from the surroundings (such as cosmic rays). Increased voltage across the tube not only increases the amplification of the fluorescent signal, but also increases the amplification of the dark current.

An appreciable decrease in dark current may be achieved by cooling the photomultiplier, especially for the red-sensitive tubes. This is not without its disadvantages, however. The sensitivity of most photomultipliers is also very dependent on temperature, so for accurate measurements the temperature of the tube must be carefully controlled.[53] Moreover, the wavelength response of the tube is greatly affected by the temperature, so that calibrations done at room temperature are not applicable to a cooled tube, but must be repeated at the low temperature.

Actually, it is generally not the magnitude of the dark current itself, but the magnitude of the fluctuation in the dark current which is the limiting factor in determining the ultimate sensitivity of the tube. Most amplifiers have a provision for subtracting the d-c component of the dark current, so it is only the fluctuations in the dark current which represent noise. Since most of the noise in the red-sensitive photomultipliers is due to thermally activated emission, cooling of these tubes to very low temperatures generally leads to a significant improvement in the signal-to-noise ratio. For blue-sensitive tubes, however, Young[54] has demonstrated that the decreased sensitivity of the tube at low temperatures offsets the decreased noise level. Below 0°C, most of the noise in blue-sensitive tubes is due not to thermal activation of electrons in the cathode, but to cosmic rays. Probably for most routine work cooling even to 0°C is not required, especially in view of the difficulties involved in temperature control, calibration, and handling of cooled tubes.

Although most commercial instruments use side-window photomultipliers, the advantages of end-window tubes are sufficient so that for the more sophisticated instruments it is very worthwhile to use the end-window type. The flat end-window with the photocathode coated directly on the inside not only allows better mechanical positioning of the tube and focusing of the light beam on the cathode, but it is also much easier to cool the cathode when necessary. More importantly, however, end-window tubes usually not only have a higher quantum efficiency and greater gain, but also the current leakage which contributes to dark current is usually considerably less than that for side-window tubes.

3.5. Amplifiers

Most commercial spectrofluorimeters amplify the output from the photomultiplier tube with a d-c electrometer. This should be selected to be as stable as possible with high gain and with very little internal generation of noise. Moreover, its internal time constant should be variable over a wide range. Measurement of phosphorescence decay curves of components with short lifetimes often requires that the response of the amplifier be quite fast, on the order of 1 msec or less. On the other hand, it is often desirable to slow down the response of the amplifier—for instance, when measuring fluorescence signals of very low intensity—so that the noise resulting from the high gain required can be averaged over a longer period of time (0.1–1 sec). There should also be a control for subtracting the dark current of the photomultiplier tube.

While the d-c electrometers in most commercial instruments are rather crude, and significant improvement can often be gained by substituting a more sophisticated electrometer, for high-precision work it is probably better to use a phase-locked a-c amplifier. The advantages of an a-c versus a d-c amplifier are several: (a) Generally, higher gain can be achieved with less generation of internal noise; (b) there is a linear relationship between the signal-to-noise ratio and the signal amplitude; (c) there is no drift in the zero or the gain of the instrument; (d) generally, there is little change in the zero baseline with change in the gain setting; and (e) all constant, extraneous signals, such as those resulting from dark current of the photomultiplier tube or leakage of room light into the instrument, have no effect on the signal output.

Use of a phase-sensitive amplifier requires that the exciting and/or emitted light beam be chopped at a given constant frequency. This is usually done with a slotted disk rotated at a preset speed. For operation at a single frequency, it is often more convenient to use electromechanical tuning forks to interrupt the light beams. These are small, reliable, and easy to use, and do not present the hazard of a rapidly rotating disk.

With choppers in both the exciting and emitted light paths, three different types of spectra can be obtained from the same sample by varying the driving of the choppers: (a) Total luminescence—the exciting light is unmodulated, but the emitted light is chopped; thus all of the emitted light is detected by the phase-sensitive amplifier. (b) Fluorescence—no chopper is placed in the emitted beam, but the exciting light is chopped at a frequency which is high in comparison to the time constant of any phosphorescence. In this manner, the phosphorescence output of the sample is constant and so is not amplified by the phase-sensitive detector. Only the fluorescence, because of its very short lifetime, is modulated at the same frequency as the chopping and is detected. (c) Phosphorescence—the exciting light is chopped at a given frequency and the emitted beam chopped at the same frequency but 180° out of phase. The photomultiplier thus observes the sample only while it is not being illuminated.

3.6. Summary

It is apparent that selection of an instrument involves the consideration of several different criteria. In general, any given instrument will not be ideally suited for every purpose, and proper consideration must be given to the types of needs it will have to fill. It is also unfortunately true that often the more sensitive and versatile an instrument is, the more complex it is to maintain and operate, so that the training of the people who are to use it must be at a higher level. Of course, no matter how efficient an instrument is, the quality of the results obtained will be strictly dependent on the care observed in sample preparation and selecting the conditions of the experiment.

REFERENCES

1. B. J. Bartholomew, Spectrofluorimetry and its application to chemical analysis, *Rev. Pure Appl. Chem. 8*, 265–301 (1958).
2. C. E. White and A. Weissler, Fluorometric analysis, *Anal. Chem. 36*, 116R–129R (1964).
3. S. Udenfriend, "Fluorescence Assay in Biology and Medicine," Vol. 1, Academic Press, New York (1965).
4. L. Brand and B. Witholt, Fluorescence measurements, *Meth. Enzymol. 11*, 776–856 (1967).
5. P. F. Lott, Instrumentation for fluorometry, *J. Chem. Educ. 41*, A327, *et seq.*, A 421, *et seq.* (1967).
6. G. M. Barenboim, A. N. Damanskii, and K. K. Turoverov, "Luminescence of Biopolymers and Cells," Plenum Press, New York (1969).
7. R. F. Chen, H. Edelhoch, and R. F. Steiner, Fluorescence of proteins, *in* "Physical Principles and Techniques of Protein Chemistry" (S. J. Leach, ed.), Part A, pp. 171–244, Academic Press, New York (1969).

8. S. Udenfriend, "Fluorescence Assay in Biology and Medicine," Vol. 2, Academic Press, New York (1969).
9. C. A. Parker, "Photoluminescence of Solutions," Elsevier Publishing Co., Amsterdam (1968).
10. G. Weber and F. W. J. Teale, Determination of the absolute quantum yield of fluorescent solutions, *Trans. Faraday Soc. 53*, 646–655 (1957).
11. E. H. Gilmore, G. E. Gibson, and D. S. McClure, Efficiencies of luminescence of organic molecules in solid solutions, *J. Chem. Phys. 20*, 829–836 (1952).
12. W. H. Melhuish, The measurement of absolute quantum efficiencies of fluorescence, *New Zealand J. Sci. Technol. 37*, 142–149 (1955).
13. W. H. Melhuish, Quantum efficiencies of fluorescence of organic substances: Effect of solvent and concentration of the fluorescent solute, *J. Phys. Chem. 65*, 229–235 (1961).
14. R. F. Chen, Fluorescence quantum yields of tryptophan and tyrosine, *Anal. Letters 1*, 35–42 (1967).
15. H. C. Borresen, The fluorescence of guanine and guanosine. Effects of temperature and viscosity on fluorescence polarization and quenching, *Acta Chem. Scand. 21*, 920–936 (1967).
16. J. Eisinger, A variable temperature, u.v. luminescence spectrograph for small samples, *Photochem. Photobiol. 9*, 247–258 (1969).
17. G. Weber, Photoelectric method for the measurement of the polarization of the fluorescence of solutions, *J. Opt. Soc. Am. 46*, 962–970 (1956).
18. G. Weber, Fluorescence-polarization spectrum and electronic-energy transfer in tyrosine, tryptophan and related compounds, *Biochem. J. 75*, 335–345 (1960).
19. S. Ainsworth and E. Winter, An automatic recording polarization spectrofluorimeter, *Appl. Optics 3*, 371–383 (1964).
20. M. N. McDermott and R. Novick, Large aperture polarizers and retardation plates for use in the far ultraviolet, *J. Opt. Am. 51*, 1008–1010 (1961).
21. R. F. Chen and R. L. Bowman, Fluorescence polarization: Measurement with ultraviolet-polarizing filters in a spectrophotofluorometer, *Science 147*, 729–732 (1965).
22. C. A. Parker, Direct recording of fluorescence excitation spectra, *Nature 182*, 1002–1004 (1958).
23. C. A. Parker and W. T. Rees, Correction of fluorescence spectra and measurement of fluorescence quantum efficiency, *The Analyst 85*, 587–600 (1960).
24. W. H. Melhuish, A standard fluorescence spectrum for calibrating spectrofluorophotometers, *J. Phys. Chem. 64*, 762–764 (1960).
25. C. E. White, M. Ho, and E. Q. Weimer, Methods for obtaining correction factors for fluorescence spectra as determined with the Aminco-Bowman spectrophotofluorometer, *Anal. Chem. 32*, 438–440 (1960).
26. R. L. Christensen and I. Ames, Absolute calibration of a light detector, *J. Opt. Soc. Am. 51*, 224–236 (1961).
27. W. H. Melhuish, Calibration of spectrofluorimeters for measuring corrected emission spectra, *J. Opt. Soc. Am. 52*, 1256–1258 (1962).
28. R. J. Argauer and C. E. White, Fluorescent compounds for calibration of excitatilon and emission units of spectrofluorometer, *Anal. Chem. 36*, 368–371 (1964).
29. H. C. Borresen and C. A. Parker, Some precautions required in the calibration of fluorescence spectrometers in the ultraviolet region, *Anal. Chem. 38*, 1073–1074 (1966).
30. H. V. Drushel, A. L. Sommers, and R. G. Cox, Correction of luminescence spectra

and calculation of quantum efficiencies using computer techniques, *Anal. Chem. 40*, 2166–2172 (1968).
31. L. Eisenberg, P. Rosen, and G. M. Edelman, AGC amplifier to correct for effects of light source variations in spectrofluorometric measurements, *Rev. Sci. Instr. 33*, 1435–1440 (1962).
32. T. D. S. Hamilton, Absolute recording spectrofluorimeter, *J. Sci. Instr. 43*, 49–51 (1966).
33. T. Azumi and S. P. McGlynn, Polarization of the luminescence of phenanthrene, *J. Chem. Phys. 37*, 2413–2420 (1962).
34. J. J. Hermans and S. Levinson, Some geometrical factors in light scattering apparatus, *J. Opt. Soc. Am. 41*, 460–465 (1951).
35. D. Stern and M. Volmer, The extinction period of fluorescence, *Physik. Z. 20*, 183–188 (1919).
36. P. Pringsheim, "Fluorescence and Phosphorescence," Interscience, New York (1949).
37. C. A. Parker, Organic trace analysis by measurement of photoluminescence, Proceeding of the Society of Analytical Chemists Conference, Nottingham, pp. 208–223 (1965).
38. R. F. Chen and J. E. Hayes, Jr., Fluorescence assay of high concentrations of DPNH and TPNH in a spectrophotofluorometer, *Anal. Biochem. 13*, 523–529 (1965).
39. R. L. Bowman, P. A. Caulfield, and S. Udenfriend, Spectrophotofluorometric assay in the visible and ultraviolet, *Science 122*, 32–33 (1955).
40. F. R. Lipsett, Versatile apparatus for automatic recording of absolute fluorescent spectra, *J. Opt. Soc. Am. 49*, 673–679 (1959).
41. W. Slavin, R. W. Mooney, and D. T. Palumbo, Energy recording spectrofluorimeter, *J. Opt. Soc. Am. 51*, 93–97 (1961).
42. W. H. Melhuish and R. H. Murashige, Double beam spectrofluorimeter, *Rev. Sci. Instr. 33*, 1213–1215 (1962).
43. W. Kaye, A universal spectrophotometer, *Appl. Optics 2*, 1295–1302 (1963).
44. G. R. Haugen and R. J. Marcus, A spectrofluorophosphorimeter, *Appl. Optics 3*, 1049–1056 (1964).
45. G. K. Turner, An absolute spectrofluorometer, *Science 146*, 183–189 (1964).
46. G. Weber and L. B. Young, Fragmentation of bovine serum albumin by pepsin II. Isolation, amino acid composition, and physical properties of the fragments, *J. Biol. Chem. 239*, 1424–1431 (1964).
47. P. Rosen and G. M. Edelman, Versatile and stable recording spectrofluorometer for the measurement of corrected spectra, *Rev. Sci. Instr. 36*, 809–815 (1965).
48. R. B. Cundall and G. B. Evans, A fully compensated versatile spectrofluorimeter, *J. Sci. Instr. 1*, 305–310 (1968).
49. B. Witholt and L. Brand, Versatile spectrophotofluorometer–polarization fluorometer, *Rev. Sci. Instr. 39*, 1271–1278 (1968).
50. M. L. Franklin, G. Horlick, and H. V. Malmstadt, Basic and practical considerations in utilizing photon counting for quantitative spectrochemical methods, *Anal. Chem. 41*, 2–10 (1969).
51. J. Langelaar, G. A. de Vries, and D. Bebelaar, Sensitivity improvements in spectrophosphofluorimeter, *J. Sci. Instr. 2*, 149–152 (1969).
52. R. F. Chen, Photoinactivation of δ-glutamate dehydrogenase in a spectrophotofluorimeter, *Biochem. Biophys. Res. Commun. 17*, 141–145 (1964).
53. A. T. Young, Temperature effects in photomultipliers and astronomical photometry, *Appl. Optics 2*, 51–60 (1963).

54. A. T. Young, Undesirable effects of cooling photomultipliers, *Rev. Sci. Instr.* *38*, 1336 (1967).

Chapter 2: Experimental Techniques, Part B

DIRECT MEASUREMENT OF FLUORESCENCE LIFETIMES

Everett T. Meserve

TRW Instruments
TRW Inc.
El Segundo, California

---

1. INTRODUCTION

Observations of fluorescence have usually been concerned with intensity and spectral distribution, although fluorescence phenomena have long been known and investigated by naturalists, geologists, and other scientific researchers. In 1926, Gaviola[1,2] introduced the first method for measurement of fluorescence lifetimes, another basic fluorescence parameter. This new information has led to a better understanding of electronic structures, as well as energy transfer and loss, and has also made feasible the determination of molecular volume by combining lifetime data with fluorescence polarization information.

Two basic techniques are presently used for fluorescence-lifetime measurements—phase modulation and pulsed fluorometry. Gaviola's lifetime measurements employed the phase-fluorometry technique. This uses a continuous light source modulated at a high frequency to excite modulated fluorescence in the test sample. The phase of the modulated fluorescence is detected by a photomultiplier and compared to the phase or modulation of the exciting light. The degree of "phase shift" is then converted to a fluorescence-lifetime value.[3]

While phase or modulation fluorometry determines parameters proportional to fluorescence lifetime, pulse fluorometry determines lifetimes directly. This direct-measurement technique is the subject of this chapter. Direct measurement of phosphorescence lifetimes is not specifically discussed in this section, but most of the techniques are applicable to phosphorescence as well as fluorescence lifetimes.[38]

2. INSTRUMENTATION

Other investigators[4-15] since Gaviola have developed techniques which use pulsed fluorometry to determine fluorescence lifetimes directly. The first direct observations of fluorescence decay times in the nanosecond region were made in 1953 by Phillips and Swank,[7] and more accurate measurements were subsequently made in the nanosecond region by Brody.[5] The apparatus used by Brody became the forerunner of many similar instruments.

Measurement of fluorescence lifetimes did not become common practice, however, primarily because an instrument to perform this function had to be built by the investigator, or obtained from another experimenter. Not until 1966 did a commercial instrument become available.* Because the number of commercial instruments in use is limited, this section on lifetime-measurement techniques includes several well-known experimental methods of instrumentation as well as discussion of the one currently available commercial instrument.

2.1. Instrument Considerations

2.1.1. Major Components

As with other spectral instruments, a decay-time measuring system consists of major equipment blocks. These are (a) an exciting source, usually ultraviolet or visible light, (b) wavelength selection devices to limit exciting energy and luminescence energy to the regions of interest, (c) a sample holder or chamber, (d) a detector for luminescence emission, and (e) a data system to convert the luminescence into usable data. Such a block system is shown in Figure 1. Appropriate power supplies, triggering circuits, and electronic processing equipment are added as required. In the following sections, each of these major components will be discussed briefly in relation to its usage, operation, and selection.

2.1.2. Excitation Sources and Modulators

If asked to describe an ideal excitation source for decay-time fluoro-

---

*The TRW model 31A spectral source system for fluorescence-lifetime measurements was introduced commercially by TRW Instruments in 1966. In 1968, this original equipment was replaced by the TRW model 75A decay-time fluorometry system.

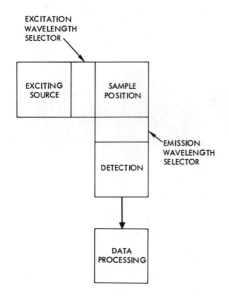

Figure 1. Major components block diagram for a decay-time measurement.

metry, most researchers would specify a small-diameter source with broad spectral coverage, especially rich in ultraviolet light, and having delta-function time characteristics. In practice, the first two characteristics may be feasible, but the delta pulse function* has not yet been achieved. It is primarily due to the lack of this delta-function behavior that various types of short-duration light sources, modulation methods, and detection and data-processing schemes have been developed.

Although a variety of exciting sources have been used for luminescence excitation, including high-energy radiation and X-rays, discussion here will be limited to the "light" region of the electromagnetic spectrum. Since the decay-time measurement techniques included in this section rely on pulsed excitation there will be no discussion of light sources used in d-c or quasi-d-c modes.

The most common pulsed light source used for decay-time fluorometry is the gaseous-discharge lamp. A typical example is shown in Figure 2. This light source consists of a glass or quartz envelope with two electrodes, between which a gaseous plasma is created when high voltage is applied. The time characteristics of the light emitted are governed by the high-voltage pulse characteristics and the gas in the envelope; the light spectrum is a

---

*A delta pulse function is defined as a pulse of zero width with zero rise and fall time. Therefore, the luminescence lifetime curve is much longer than the delta pulse function.

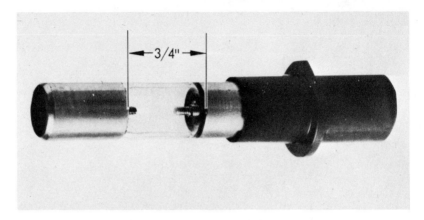

Figure 2. Gaseous-discharge lamp. Quartz envelope is approximately ¾ in. long with electrode spacing approximately ½ in. Gases used can be hydrogen, deuterium, nitrogen, carbon monoxide, krypton, oxygen, plus others. Depending upon gas used, a variety of special outputs from ultraviolet to near infrared can be obtained.

function of the type of gas. The first of such gaseous-discharge lamps employed in early experiments was hydrogen filled.[16] Other gases have subsequently been employed to obtain different pulse-time characteristics and spectral output.[8,15,17,18]

Xenon is one type of gaseous discharge usually associated with short light pulses. However, it is not commonly used for decay-time fluorometry because of its own long decay time. While xenon may have a decay time of 1 μsec or longer, the gaseous-discharge light sources commonly used in decay-time fluorometry have fall times between 1 and 10 nsec.

Despite the intrinsic shortcomings of xenon lamps, they have been adapted for use by employing a fast modulator (Kerr cells, rotating mirrors) at their output. Other modulators of various speeds for light sources include mechanical shutters and rotating slots.[3] Also, other types of excitation sources have been used for luminescence stimulation: electron beams, spark gaps, X-rays, mercury pulsers, and, more recently, lasers.[7,13,19,20]

2.1.3. Wavelength Selection

Selection of wavelength for decay-time fluorometry is required to limit the exciting light to the excitation or absorption curve and to allow only luminescence emission from the test sample to reach the detector. For most experiments, this can be achieved by using selected filters as shown in Figure 3. The obvious reason for using filters rather than a grating monochromator is that generally it is only necessary to separate luminescence emission and exciting light, rather than to resolve the fine structure of a particular spectrum.

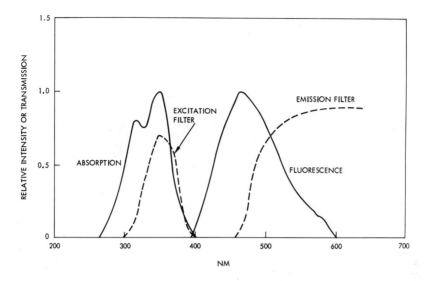

Figure 3. Wavelength selection for decay-time fluorometry of quinine sulfate using absorption filters. Solid curve is quinine sulfate absorption and fluorescence; dashed lines are filter transmission curves. Excitation filter is Corning Glass 7–60, and emission filter is Corning Glass 3–72.

Absorption and interference, as well as special metallic and chemical filters, are used for this purpose. However, when it is necessary to separate very close absorption and emission curves, special narrow-band filters, grating or prism monochromators, or variable bandpass interference filter monochromators can be employed successfully. Filters, when usable, are usually preferred over grating or prism monochromators because they provide higher intensities, an extremely important characteristic for achieving the highest luminescence output from short-duration, low-fluorescence light pulses.

The interference filter monochromator mentioned previously is a desirable method of wavelength selection for decay-time fluorometry because it offers all the advantages of filters, including high transmission and fast optical speed, as well as the ability to select or vary optical bandpass by tilting. The properties of variable interference filter wavelength selection are shown in Figure 4. The amount of wavelength shift varies with the type and the center wavelength employed, as discussed by Pollack.[21] Pollack also has described the background of tunable interference filters.

2.1.4. Detection

Photomultipliers are used almost exclusively in decay-time fluorometry instruments. While the choice of a photomultiplier depends primarily upon

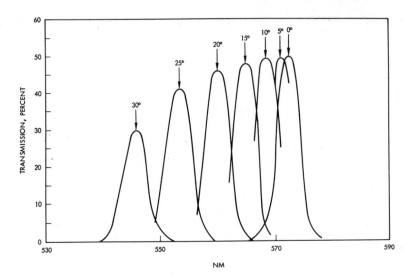

Figure 4. Transmission bands of a 575 nm interference filter at selected angles of tilt from the normal.

spectral sensitivity, other characteristics, which are not normally of paramount concern for spectral fluorometers, become very important for decay-time fluorometers. For example, since luminescence stimulation for lifetime measurements requires extremely fast pulsed operation, such properties as response time and current output assume importance. A fast response time would be desirable, of course, to preserve the integrity of the luminescence-time characteristics. For available photomultipliers with 2 nsec response times, this would imply a minimum decay-time measurement capability of 20 nsec. (This assumes the detector needs to be ten times faster than the parameter to be measured.) The potential limitations of response time are generally recognized, and a complete subsection below will be devoted to techniques used for resolving perturbations caused by limited response time and other sources of errors.

Most photomultiplier outputs for a decay-time fluorometer are designed for electrical impedance matching into 50 ohms to preserve the high-speed electrical pulses.* It is desirable for photomultipliers to have high current output so that maximum signal (voltage) with highest signal-to-noise ratio can be developed across the 50-ohm system impedance. A low-current photomultiplier output not only limits signal amplitude but, if excess current is

---

*The input may be an oscilloscope, current amplifier, pulse shaper, or other data-processing system.

drawn, phototube saturation can cause distortion of observed photomultiplier pulses.

It is also necessary to consider the more familiar photomultiplier characteristics such as current amplification, dark current, quantum efficiency, and spectral sensitivity in the region of luminescence emission. Some photomultipliers often used for decay-time fluorometry include the RCA 1P28, 931A, 6217, 7102, 1P21; ITT FW118; Hamamatsu R196; and Valvo 56UVP and 56AVP.

2.1.5. Potential Problems

The potential problems in decay-time fluorometry are of two major kinds—chemical and instrumental. Although chemical problems are of obvious importance, they will be mentioned only briefly in this chapter on instrumentation.

Errors associated with high concentration are not normally encountered in decay-time fluorometry because test data are usually obtained at low concentrations. The symptoms associated with excessive concentration include lower fluorescence intensity and a shortening or lengthening of the luminescence decay time. These can be caused by front-surface absorption or self-quenching.[22] High concentration may also result in reabsorption of light, which is re-emitted as secondary emission, and can thereby lengthen the observed lifetime.[5] A multiple decay function has been noted from a highly concentrated solution of acridine orange.[23]

Factors which may alter luminescence decay times in either direction are photodecomposition, or photolysis, due to prolonged exposure to exciting or ambient light; impurities*; and the selection of solvent or temperature.[22-26]

Some potential instrumental errors which may be experienced in decay-time fluorometry measurements are inadequate separation of exciting and emission wavelengths, photomultiplier saturation, erroneous readings due to excitation and viewing angle, and inadequate data processing. Most of these potential problem areas can be eliminated by careful instrument design and operation; however, some may not be resolved completely because certain components are presently unavailable.

A prime example of the latter is the separation of exciting and emission wavelengths. This problem is particularly acute when (a) excitation and emission wavelengths are in the ultraviolet or near ultraviolet region, (b) excitation and emission wavelengths are very close spectrally, and (c) for

---

*Because the luminescence lifetime of a contaminated solution is often different from that of a pure solution, this discernible difference can be used as a quick method for determining sample purity.

either of the preceding conditions as concentration becomes very low. These potential problems may result in the passage of scattered or reflected exciting light, together with luminescence to the detection system. This condition may influence the observed luminescence pulse by either increasing or decreasing the apparent lifetime, depending upon the time duration of the exciting light pulse relative to that of the luminescence pulse and the amount of exciting light which is passed through the emission filter. Problems of this type can virtually be eliminated by selection of excitation filters with sharp high-pass cutoff and emission filters with sharp low-pass cutoff, use of filters with very low transmission in the tail of the transmission curve, and selection of the excitation region to provide the greatest separation of excitation and luminescence wavelengths. The problem is most difficult to eliminate when excitation is in the ultraviolet (200–300 nm) and emission is in the near ultraviolet (300–400 nm), because presently available ultraviolet filters do not have sharp high-pass and low-pass cutoff characteristics.

Photomultiplier saturation, which is usually evidenced by a flattening of the luminescence pulse peak and lengthening of the decay curve, can be caused by overloading the photomultiplier with light or drawing too high an anode current. The first problem is easily remedied by reducing exciting or emission light intensity or sample concentration. To prevent overloading the photomultiplier current, a high-current-output photomultiplier should be used and system impedance should be as high as practical to preserve the electrical integrity of the luminescence pulse.[38]

Most luminescence-lifetime measuring instruments employ a 90° angle arrangement between excitation and emission optics.[5,6,12,14,18,22] At least one worker, however, has investigated other arrangements of excitation and viewing angles.[27]

Potential errors of lifetime measurements resulting from inadequate data processing are so important and complex that the next two sections, 2.2 and 2.3, are devoted exclusively to this subject. The major problems and associated potential errors are basically concerned with the detection and presentation of the true luminescence decay pulse.

2.2. Oscilloscope Techniques

The next four sections are concerned with techniques and equipment used primarily for obtaining, processing, and displaying accurate luminescence decay-time information. The oscilloscope techniques discussed in this section are, of course, used exclusively for the processing and display of luminescence pulses already produced and detected by some system. The process of producing and detecting the pulses can use any of several systems, but most common is an equipment arrangement employing a pulsed light source, an optical system, and a photomultiplier, as outlined in Figure 1.

Perhaps the most direct approach to recording and observing luminescence decay-time pulses is to display them on a standard, or sampling, oscilloscope for direct reading. For this purpose, various oscilloscopes have been used, including Tektronix 517 and 519; EG&G 707 (traveling-wave tube); Hewlett-Packard type 185A sampling oscilloscope; and the Tektronix 530; 540; 550, or 580 series with sampling plug-in units. While the instrumental approach is considerably simplified by using an oscilloscope, there are some serious limitations.

Probably the single most important limitation is that the oscilloscope presents what it sees. Thus, if the luminescence pulse is distorted due to the exciting light pulse and to detector-limited response time, the oscilloscope will present a distorted composite curve: the observed pulse shape on the oscilloscope tube face might not be the true luminescence decay pulse. Fortunately, with presently available components, this condition is usually not a problem if the pulse observed is 30 nsec or longer and the oscilloscope rise time is less than 3 nsec. Subsequent data processing of perturbed pulses by techniques described later can virtually eliminate this problem for all time ranges.

One advantage of using a fast, direct-reading oscilloscope such as the Tektronix model 519 is that a complete presentation can be obtained from one luminescence pulse. A sampling oscilloscope, although easier to use, requires repetitive inputs to obtain a complete presentation. A repetitive signal does not usually present a problem, however, since most decay-time fluorometers use repetitively pulsed stimulating light sources.

It may be difficult to determine precise data points from the limited observable luminescence pulse projected on the oscilloscope screen. Deciphering the data thus becomes a tedious task, and the information is of questionable accuracy. Several schemes have been used to resolve these problems. The most obvious are the direct photographing and storage of data on a storage-beam oscilloscope. Once the data have been frozen in this manner, the observer has sufficient opportunity to retrieve the information. An extension of the photographic approach is to make a transparency from the oscilloscope face, enlarge the plot by projecting the transparency onto a large quadrilleboard, draw the curve on the board, and thereby obtain higher-resolution data from this enlarged plot.

Some researchers have carried their data-reduction process further by automating the analysis of data obtained from oscilloscopes. A unique scheme described by Zarowin[14] images the curve displayed on the oscilloscope tube face through a slit onto a photomultiplier. The time density of imaged samples is then integrated to yield a decay curve on a chart recorder.

Others have obtained data from a sampling oscilloscope electrical output rather than from the tube face.[8,15,24] Subsequent processing is then performed by computers or chart recorders. Raw luminescence decay-time data for computer input need not be limited to sampling oscilloscope electrical signals but can also be derived from oscilloscope photographs.

2.3. Curve Normalization Techniques

When fluorescence lifetimes are excited by a pulse function which has finite time characteristics, the resulting fluorescence decay curve is a composite of the exciting pulse function and the true fluorescence decay curve which takes the form

$$F(t) = \int_0^t f(\tau) I(t - \tau) d\tau \qquad [1]$$

where $F(t)$ is the observed emission function, $I(t-\tau)$ is the combined response function of the exciting light and the detection system, and $f(\tau)$ is the true fluorescence time curve.[5] When the exciting light pulse function is short (perhaps ten times shorter) compared to the sample fluorescence decay curve, the amount of perturbation is small, and

$$F(t) \simeq f(\tau) \qquad [2]$$

While this is true for phosphorescence and some fluorescence decay curves, this is not, in general, the case for most fluorescence lifetimes. The obvious reason is that fluorescence lifetimes are often in the same time range as the exciting light pulse. That is, the fall time and pulse width of the light pulse may not be small compared to the time of fluorescence decay. To further complicate the situation, the detection system (photomultiplier and direct readout equipment beyond) has some finite response time which can introduce additional perturbation. For Eq. [2] to be true, the exciting light pulse function and combined equipment response time, R_e, should be approximately ten times faster than the fluorescence lifetime, τ, to be directly measured,* or

$$\tau \geq 10\, R_e \qquad [3]$$

The purpose of this section is to discuss curve normalization methods of removing the perturbations in composite fluorescence curves, a condition which exists whenever directly excited fluorescence is measured by direct

---

*Direct measurement refers to direct readout by such instruments as oscilloscopes, chart recorders, and plotters, without utilizing any method for resolving composite curve perturbations.

readout. The operation of all the instruments described in this section is based in some way on Eq. [1], so that the task of these instrumental techniques becomes one of reducing the composite curve, $F(t)$, into its components, and finally solving it for $f(\tau)$, the true fluorescence time curve.

2.3.1. Brody's Instrument

During the course of studying the fate of photons absorbed by photosynthesizing materials, Brody[5] found it desirable to observe the fluorescence decay curves of chlorophyll and other pigments. From the decay time, τ, an excited-state lifetime could be obtained which could be used to compute the extent of energy transfer between molecules. As an adjunct to investigation of energy transfer characteristics, Brody was also able to make measurements of quantum yield using

$$\tau = \tau_0 \, \phi \qquad [4]$$

where τ is the observed fluorescence lifetime, τ_0 is the computed "natural" lifetime when radiation emission is the only process by which the excited molecule loses its energy, and ϕ is the fluorescence or quantum yield. The advantages of this method of determining quantum yield are (a) light scattering does not interfere with the determination, and (b) other experimental

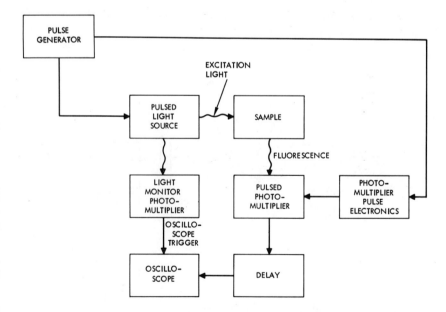

Figure 5. Author's block diagram of Brody's instrument for measuring fluorescence lifetimes.

difficulties involved in the measurement of absorbed and emitted quanta are eliminated.

One of the unique features of Brody's instrument is that the photomultiplier as well as the exciting light source is pulsed. A schematic block diagram of the instrument is shown in Figure 5, and an optical block diagram is shown in Figure 6. A pulse generator is used as trigger for the pulsed light source, a hydrogen flash lamp, and for the fluorescence-detecting photomultiplier. Two outputs are available from the pulsed light source. The primary light output is used to excite the sample, and a small fraction of the secondary light output is monitored by a photomultiplier and used for triggering the oscilloscope. The sample fluorescence is detected by the pulsed photomultiplier. By using a pulsed photomultiplier increased sensitivity, greater amplification, and shorter instrument response times are achieved. A delay line is used between the pulsed photomultiplier and the oscilloscope to facilitate oscilloscope triggering. Photographs of the exciting light pulse and the perturbed fluorescence decay curve were taken from the oscilloscope face and enlarged; points were extracted from curves and then either plotted by hand on semilogarithmic paper or fed to a computer for mathematical "normalization" into a true fluorescence time-course curve.

The arrangement of optical components shown in Figure 6 uses the standard 90° fluorescence arrangement. Filters are used to limit exciting light to the excitation region and to allow only fluorescence light to be detected by the pulsed photomultiplier.

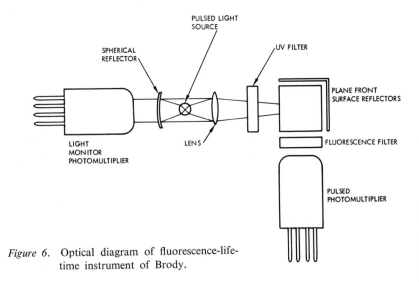

Figure 6. Optical diagram of fluorescence-lifetime instrument of Brody.

Brody's technique is significant, as it recognizes the perturbed nature of a displayed curve and provides a means for resolving true curve characteristics from a composite curve. The method, therefore, made possible very accurate lifetime measurements: ± 0.5 nsec in the 2 to 8 nsec region. As with most new techniques for obtaining this type of information, equipment is somewhat difficult to operate, and processing of data is tedious and expensive. The fact remains that a practical technique has been demonstrated which can be built upon, as will be seen.

2.3.2. Nanosecond Fluorometer and Instrument of Steingraber and Berlman

The basic curve normalization technique described by Brody[5] is used in a nanosecond fluorometer[15] and in an instrument designed by Steingraber and Berlman.[8] While both of these instruments employ the same basic technique, there are several differences between them and Brody's instrument. These three instruments can be compared by referring to Figures 7 and 8 and Table 1.

As can be seen, the major differences in these instruments are in light detection and data processing. The most automated of these instruments is undoubtedly the nanosecond fluorometer, which produces complete data in less than 3 min.[15] Of course, this system is also potentially the most expensive due to the on-line computers. Automated data processing by computer program has also been performed with the Steingraber and

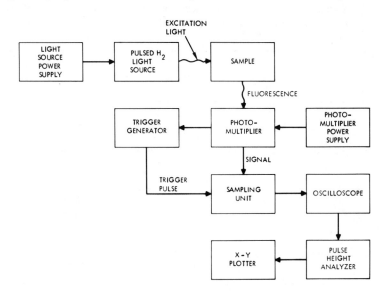

Figure 7. Block diagram of Steingraber and Berlman's instrument.

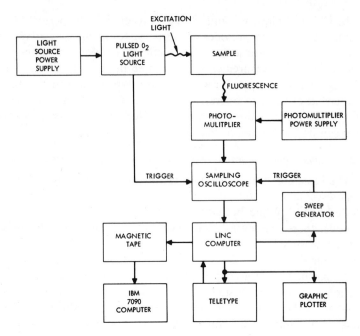

Figure 8. Block diagram of the nanosecond fluorometer.

Table 1. **Comparison of Major Components and Operation of Three Fluorescence-Lifetime Instruments—the Instrument of Brody, the Instrument of Steingraber and Berlman, and the Nanosecond Fluorometer**

Operation or component	Brody	Instrument of Steingraber and Berlman	Nanosecond fluorometer
Light source	Pulsed hydrogen flash lamp	Pulsed hydrogen flash lamp	Oxygen flash lamp
Fluorescence detector	Special pulsed photomultiplier	Photomultiplier	Photomultiplier
Readout device	Direct-reading oscilloscope	Oscilloscope with sampling input	Sampling oscilloscope and plotter from LINC computer
Data processing	Photograph from oscilloscope and plotting of normalized data	X–Y plotter via pulse-height analyzer, and plotting of normalized data	LINC computer, IBM 7090 computer with data-normalizing program

Direct Measurement of Fluorescence Lifetimes

Berlman instrument.[24] As also noted, the pulsed light source used in the nanosecond fluorometer was an oxygen lamp rather than a hydrogen discharge, as used by Brody and in the Berlman and Steingraber instrument.

2.3.3. TRW Model 75A Decay-Time Fluorometry System

As previously stated, instruments for directly measuring fluorescence lifetimes have been built ever since the work of Gaviola in 1926; however, the first commercial instrument for this task was not available until 1966, when TRW Instruments introduced the model 31A nanosecond spectral source system. The model 31A system consisted of the components shown in Figures 9 and 10 and was the predecessor to the present TRW model 75A decay-time fluorometry system.

The present model 75A decay-time fluorometry system (Figure 11) is basically the same as the model 31A system, the major difference being that the model 31A system utilized modular optical components while the model 75A decay-time fluorometry system is optically self-contained. The evolution of the model 75A system out of the 31A components is evident when the model 31A system block diagram, Figure 9, is compared to the model 75A system block and optical diagrams, Figures 12 and 13. In addition to optical packaging differences, the model 75A system has more flexibility in the sample chamber.

2.3.3(a). System Description. The TRW model 75A decay-time fluorometry system shown in Figures 11 and 13 consists of three major equipment groups. First is the power supply and control unit, which supplies high-voltage pulses to the gaseous-discharge lamp located in the optical

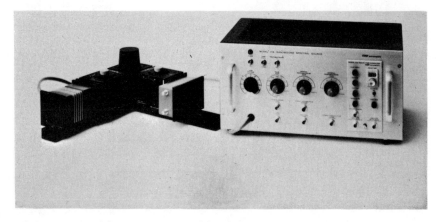

Figure 9. The TRW Instruments model 31A nanosecond spectral source system used for measuring luminescent lifetimes.

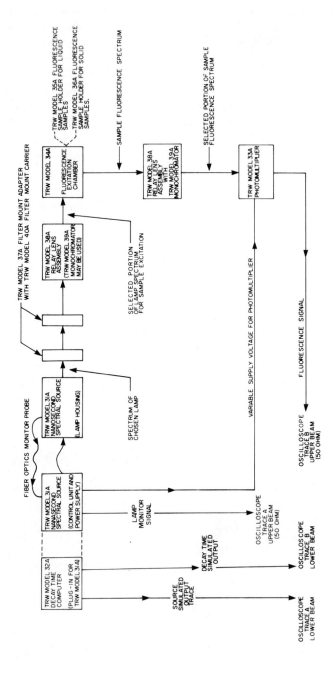

Figure 10. Block diagram of the TRW Instruments model 31A nanosecond spectral source system for fluorescence and phosphorescence decay-time measurement.

Direct Measurement of Fluorescence Lifetimes

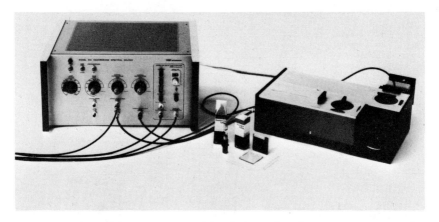

Figure 11. The TRW Instruments model 75A decay-time fluorometry system consisting of the model 75A decay-time fluorometer (optical system), model 31A nanosecond spectral source with model 32A decay-time computer, and lamps.

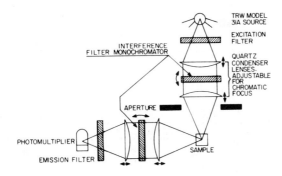

Figure 12. Optical schematic of the TRW Instruments model 75A decay-time fluorometry system.

system, provides for the model 32A decay-time computer plug-in module, and contains the electronics for photomultiplier voltages. The second equipment group consists of the optical system, and the third is comprised of the decay-time computer and oscilloscope, which provide data processing and presentation.

The exciting light source uses a gaseous-discharge lamp[12,17] similar to that of Malmberg,[16] except that (a) the gases employed are nitrogen, deuterium, carbon monoxide, or krypton,* and (b) the output power is considerably greater. The discharge is obtained from a 10 kv d-c power supply which can be pulsed at a variable rate from a few pulses per second to 5000

---

*Selected gases are used to yield different spectral output and time characteristics.

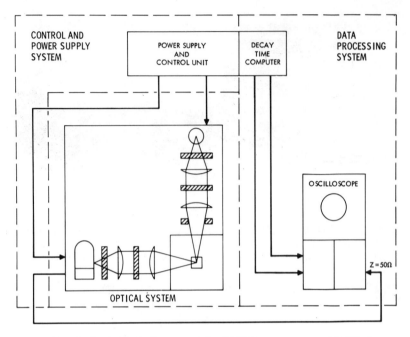

Figure 13. Author's block diagram of the TRW Instruments model 75A decay-time fluorometry system.

pulses per second. When the deuterium lamp is used, the light rise and fall time characteristics are approximately 2 nsec each with a half-width of 4 nsec. The decay-time computer obtains d-c voltage from the power supply and control unit. High voltage (-1050 v d-c) is also provided to the photomultiplier in the optical system.

The optical system diagrammed in Figure 12 uses an adjustable quartz condenser lens to collect, collimate, and focus light from the exciting light source onto the sample, as well as luminescence from the sample onto the detecting photomultiplier. A tunable interference filter monochromator and stationary filter positions are included in both the excitation and emission optical paths for precise wavelength selection and maximum separation of excitation and luminescence emission. The optical system is operable over a spectral range from 200 to 1000 nm and has a maximum aperture of $f/1.5$. An aperture control located between the exciting light source and the sample permits viewing the excitation at the same light intensity as the luminescence pulse when making short-lifetime measurements.[28] The sample chamber accepts liquid samples in various containers (e.g., test tubes or square cuvettes) and solid or cryogenic samples with special accessories.

A decay-time computer module located in the power supply and control unit is used in conjunction with a dual-beam independent time base oscilloscope to make the actual lifetime determination.

2.3.3(b). System Operation. The operating theory for the TRW model 75A decay-time fluorometry system is basically the same as for the instrument of Brody, the instrument of Steingraber and Berlman, and the nanosecond fluorometer. Thus, the system has (a) a fast exciting light source and power supply, (b) an optical system for holding samples, providing for separation of exciting light and luminescence emission, and detection of the luminescence pulse, and (c) a data-processing system for determining the true lifetime of luminescence. (Refer to Figure 13.)

The exciting light may be pulsed at any rate between a few pulses per second and 5000 pulses/sec; however, 1000 pulses/sec is customarily used to obtain a good presentation on the oscilloscope. Once filters have been selected and optics focused, the exciting light stimulates the sample. The sample time luminescence characteristics are then detected at 90° by a 931A photomultiplier. A 931A photomultiplier is used as it has the high current output capability necessary to develop maximum voltages across the electrical system's 50-ohm impedance. The 50-ohm system is necessary to preserve the fast time characteristics of the detected luminescence pulse. Output of the photomultiplier is connected to one beam of the dual-beam oscilloscope through a 50-ohm termination.

The other beam of the oscilloscope is used to display the decay-time computer generated functions—the computer-simulated light pulse and the computer-simulated decay pulse. A dual-beam oscilloscope with independent time bases is necessary since the time scales between the computer-simulated pulses and luminescence pulse often differ by 10^4, and the two traces need to be superimposed.

2.3.3(c). Operating Theory. As was previously indicated, the TRW decay-time fluorometry system operates on the principle of deciphering the true luminescence decay pulse from a detected light pulse which has been perturbed by the light source and detection system. In the instrument of Brody, the instrument of Steingraber and Berlman, and the nanosecond fluorometer, the complex detected luminescence pulse is solved for the true luminescence pulse-time characteristics by using a mathematical program in a computer either peripheral or external to the instrument system. In the TRW instrument, this operation becomes part of the measurement process.

The mathematical operation described by Eq. [1] to extract the true luminescence light pulse from the signal is performed in the TRW instruments decay-time fluorometry system in three steps. First, the curve $I(t-\tau)$, the combined response function of the exciting light and detection system,

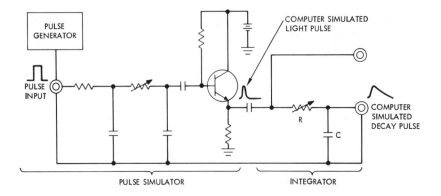

Figure 14. Schematic of model 32A decay-time computer.

is generated in the instrument's electrical equivalent circuit by using a special pulse-building signal generator shown in Figure 14. This pulse then becomes the input to a simple analog circuit whose RC time constant can be varied to simulate any luminescence decay constant. Finally, the composite output of this circuit appears as the computer-simulated decay pulse. In practice, the RC network is varied until the simulated decay pulse from the computer matches the actual luminescence decay pulse when both are displayed on the oscilloscope (see Figures 15, 16, and 17). By knowing the RC values in the analog circuit which created the simulated decay pulse, a true value of luminescence decay time, τ, can be read directly.[12,29]

One unique feature of this technique is that the decay-time computer

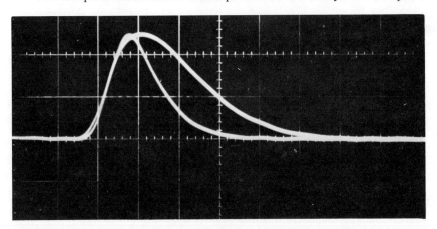

Figure 15. Quinine sulfate fluorescent decay pulse (longer pulse) and light pulse (short pulse). Tektronix Model 555 oscilloscope at 20 nsec/cm. (Note time distortion of pulses due to equipment pulse function and detection system limited time response.)

Direct Measurement of Fluorescence Lifetimes

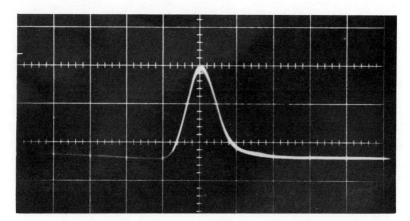

Figure 16. Actual and simulated source light pulses after matching by model 32A decay-time computer.

does not operate at the same speed as the luminescence pulse, thereby reducing substantially the cost and complexity of the computer. This, then, is the reason for the dual-beam independent time base oscilloscope. Another benefit of this approach is that a wide dynamic range is achieved simply by changing the scale factor (oscilloscope-sweep time bases) between the decay-time computer and the actual luminescence pulse. As a result, this instrument can be used for both fluorescence and some phosphorescence decay-time measurements.

Despite this fairly simple approach for rapidly determining lifetimes, the system has good accuracy and resolution, although not as good as some

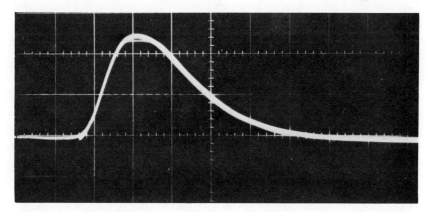

Figure 17. Decay pulse of quinine sulfate matched by model 32A decay-time computer simulated decay pulse. Fluorescence lifetime of 19.5 nsec.

reported.[8] Accuracy given by the manufacturer is 10% over the range of 5 nsec to 1 msec. In practice, greater accuracies have been achieved. As would be expected from the measurement technique employed, resolution and repeatability become a function of visual acuity in superimposing two traces on the face of an oscilloscope.

A potential limitation of the TRW technique, or for that matter any technique which assumes a single exponential decay, is in the measurement of multiple decay times. This is probably not as serious a limitation as first imagined because it has been found that (a) most pure compounds do exhibit almost true exponential decay even though some energy is lost before luminescence emission, and (b) these curves can be well matched with a single exponential decay.[5,8,15,22,24] If, however, significant difference from a single decay constant is suspected, a technique for measuring dual components of a complex lifetime has been described.[23]

2.4. Gated Photomultiplier Detection

In 1960, Bennett reported an equipment setup which has since become known as the gated photomultiplier technique.[6] Bennett described a stroboscopic technique employing a photomultiplier whose dynode string is "gated" on by a specially shaped pulse, which results in faster time resolution than is normally achieved with standard square-pulse techniques. Figure 18 shows Bennett's gated photomultiplier schematic. The feature of this design which permits it to have faster time resolution than previous techniques is that the photomultiplier time-resolution function is determined only by the duration of the gating pulse leading edge, T_1, and not by the entire width, T_2.

The gated photomultiplier is used with a fast gaseous-discharge lamp, other pulse-forming electronics, and a graphic chart recorder as shown in Figures 19 and 20. Operation of the apparatus begins with the 5 kHz free-

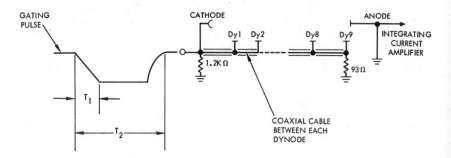

Figure 18. Schematic of gated photomultiplier.

Direct Measurement of Fluorescence Lifetimes

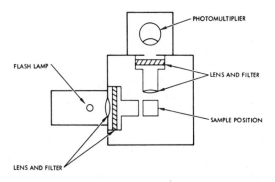

Figure 19. Schematic of Bennett's instrument arrangement.

running oscillator and trigger circuit, which provides a timing pulse for triggering the flash-lamp thyratron and the photomultiplier gate. The purpose of the fixed delay in the flash-lamp circuit is to allow the photomultiplier to be gated "on" before the flash lamp, and it also prevents interaction between the two circuits driven from the common oscillator and trigger. A motor drives the control of the variable delay line, which is used to determine the photomultiplier gating time with respect to the flash lamp. The gating pulse network generates the proper shaped photomultiplier gating pulse, as

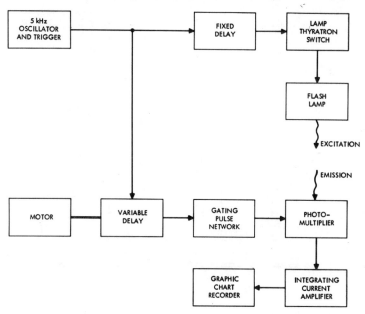

Figure 20. Author's block diagram of gated photomultiplier circuitry.

shown in Figure 18. The gated output from the photomultiplier is sent to an integrator/amplifier where several pulses are accumulated over time, amplified, and routed to a graphic recorder. The Y-axis of the recorder thus displays intensity and the X-axis displays time, which is varied by changing the variable delay line. A graphic display of the light pulse, using this technique, is shown in Figure 21 and can be compared to direct oscilloscope presentation of the same light pulse as shown in Figure 22.

One desirable feature of the gated photomultiplier technique is that the detected curve data are in plot form, an example of which is shown in Figure 23. As mentioned previously, the time-resolution function is determined by the gating-pulse leading edge and is therefore capable of providing faster system response-time characteristics than can be obtained with customary square gating pulses. This characteristic makes possible the stating of sample lifetime-measuring accuracy in terms of the variable delay line used, thus yielding an instrument whose accuracy can be checked.[30] The fact that optimum resolution is obtained by rather careful adjustment of the gating-pulse ramp voltage[6] could present some operational problems, especially for those not experienced in instrument building.

The shortcomings of the gated photomultiplier technique stem from

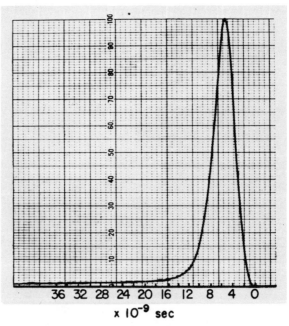

Figure 21. Response of gated photomultiplier system to the flash lamp.

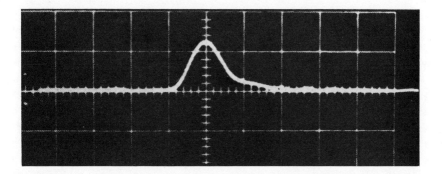

Figure 22. The excitation light as viewed by a gated 1P28 photomultiplier and presented on a Tektronix model 517 (direct reading) oscilloscope. Each major divison is 5 nsec.

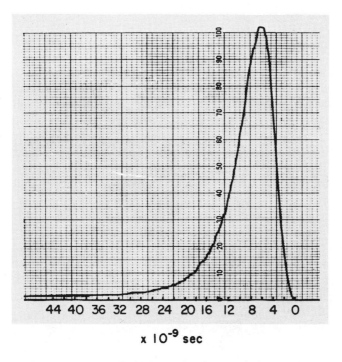

Figure 23. A typical chart plot from a gated photomultiplier showing the decay curve of 10^{-4} M fluorescein in water.

the electronics and data processing. The difficulty in adjusting the gating-pulse shape was described above. In addition to this, a rather high degree of sophistication is required in the triggering and timing functions, especially when short fluorescence lifetimes are involved.[31] However, some of the sophisticated electronics have since been replaced by more standard components, as described by Brown.[31] Even with the electronics somewhat simplified, there still remains the task of data processing. As shown in Figures 22 and 23, data are presented directly on a graphic plot. Sometimes this format is completely adequate; however, most data are usually processed further to yield a decay time, τ. This can be done by (a) making a rough calculation by measuring the time coordinates on the plot from the 100% to $1/e$ point on the decay slope, (b) making a logarithmic plot of the decay curve and determining the slope, or (c) inputting the curve data into a computer and performing a mathematical analysis of the curve.[32] In any of these methods, care needs to be exercised to determine when displayed lifetime data are perturbed by the shape of the light pulse, to what degree the accuracy of the data is affected, and what other data are needed to compensate for this source of error.[6,32]

2.5. Single-Photon Counting

The use of single-photon counting, also known as a specific application of delayed coincidence technique, has been associated primarily with nuclear physics and the measurement of scintillation materials where exciting sources are commonly gamma radiation.[10,33,34] However, the single-photon technique has recently been used to determine lifetimes of fluorescence materials.[35,36,39]

This basic technique can be employed where any source of repetitive luminescence is available. In this method, two separate detectors are employed, one to observe the exciting light source directly and the other to observe single photons from the sample. The signal from the first detector provides a zero-time reference and is used to gate the output from the second detector. The single photons from the second detector are then fed into a time-to-amplitude converter (or time-to-height converter), which converts the time difference of two pulses into a signal whose amplitude depends upon this difference. The output of the time-to-amplitude converter is displayed on a multichannel analyzer.

As would be suggested by the equipment employed, single-photon counting is a statistical technique. Equation [5] shows the basic relationships for single-photon counting where N is the number of photomultiplier output pulses per unit time, v is the light-source pulse rate per second, $\bar{n}$ is the average number of photons per pulse to the photomultiplier photocathode, and a is the quantum efficiency of the photocathode:[3]

$$N \simeq v\bar{n}a \qquad [5]$$

An important consideration of this technique is that the detecting photomultiplier must detect single photons from the light-emitting sample and then accumulate these over a period of time. It is necessary, therefore, that the sampling light intensity be reduced so that the average number of photons impinging on the sample photomultiplier per stimulating light pulse is less than 1.[10]

A typical single-photon equipment setup for measuring fluorescence decay-time characteristics is shown in Figure 24. When used to measure decay-time characteristics of scintillator materials, the pulsed light source is replaced by a pulse radiation source, such as gamma rays, neutrons, or alpha particles. The two photomultipliers shown in Figure 24 are used (a) to detect the light-source pulse and generate the zero-time reference pulse (light-source photomultiplier), and (b) to detect the single light photons emitting from the fluorescence sample (sample photomultiplier). The zero-time reference pulse is used to define the time origin and to gate the single-photon pulse from the sample photomultiplier. The light attenuator is adjusted so that the probability of a sample photomultiplier output pulse corresponding to more than one photon is extremely small. Single-photon pulses are counted when the gate to the time-to-amplitude converter is open. Counting of selected single-photon pulses is performed as a function of their delay after the zero-time reference pulse, using the time-to-amplitude converter and multichannel analyzer, and is continued until the numbers become statistically significant.[3]

As would be imagined from Figure 24, the equipment and operating techniques of single-photon counting are complex, and require a carefully designed system of electronics, even though the major components are commercial instruments. One instrument company has recently offered a photon-counting system.* One unique advantage of the single-photon technique is that the time resolution of the apparatus is established by the small statistical dispersion (usually about 0.3 nsec) and not by the photomultiplier response time, as is most common with other direct lifetime-measuring systems. In this instance, the time dispersion represents the uncertainty in the delay time between the emission of a photoelectron from the photomultiplier cathode and the arrival at the anode of the associated electron pulse. Decay-

---

*Model 1100 series photon counter, Solid State Radiations Inc., Los Angeles, Calif., 1968.

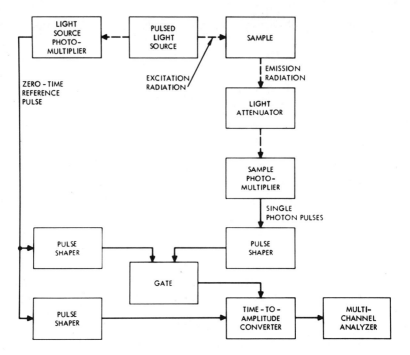

Figure 24. Typical equipment setup for single-photon counting.

time accuracies of 0.1 nsec for scintillator materials have been reported by Yates and Crandall.[34] Another unique feature of the single-photon technique is that it permits the observation of populating as well as depopulating radiation time characteristics.[33,37]

REFERENCES

1. E. Gaviola, Die Abklingungszeiten der Fluoreszenz von Farbstofflösungen, *Z. Physik.* **35**, 748–756 (1925–26).
2. E. Gaviola, Die Abklingungszeiten der Fluoreszenz von Farbstofflösungen, *Ann. Physik.* **81**, 681–710 (1926).
3. J. B. Birks and I. H. Munro, in "Progress in Reaction Kinetics" (G. Porter, ed.), Vol. 4, p. 257, Pergamon Press, London (1967).
4. M. F. Perrin, Polarisation de la lumière de fluorescence vie moyenne des molécules dans l'état excité, *J. Phys. Rad.* **7**, 390–401 (1926).
5. S. S. Brody, Instrument to measure fluorescence lifetimes in the millimicrosecond region, *Rev. Sci. Instr.* **28**, 1021–1026 (1957).
6. R. G. Bennett, Instrument to measure fluorescence lifetime in the millimicrosecond region, *Rev. Sci. Instr.* **31**, 1275–1279 (1960).

7. H. B. Phillips and R. K. Swank, Measurement of scintillation lifetimes, *Rev. Sci. Instr.* 24, 611–616 (1953).
8. O. J. Steingraber and I. B. Berlman, Versatile technique for measuring fluorescence decay times in the nanosecond region, *Rev. Sci. Instr.* 34, 524–529 (1963).
9. J. T. D'Alessio, D. K. Ludwig, and M. Burton, Ultraviolet lamp for the generation of intense, constant-shape pulses in the subnanosecond region, *Rev. Sci. Instr.* 35, 1015–1017 (1964).
10. L. M. Bollinger and G. E. Thomas, Measurement of the time dependence of scintillation intensity by a delayed coincidence method, *Rev. Sci. Instr.* 32, 1044–1050 (1961).
11. J. T. Dubois and R. L. Van Hemert, Lifetime of excited states in solution by the quenching method, *J. Chem. Phys.* 40, 923–925 (1964).
12. R. C. Mackey, S. A. Pollack, and R. S. Witte, Multiple watt submicrosecond high repetition rate light source and its application, *Rev. Sci. Instr.* 36, 1715–1718 (1965).
13. A Yariv, S. P. S. Porto, and K. Nassau, Optical maser emission from trivalent praseodymium in calcium tungstate, *J. Appl. Phys.* 33, 2519–2512 (1962).
14. C. B. Zarowin, New technique for the measurement of fluorescent lifetime, *Rev. Sci. Instr.* 34, 1051–1053 (1963).
15. L. Hundley, T. Coburn, E. Garwin, and L. Stryer, Nanosecond fluorimeter, *Rev. Sci. Instr.* 38, 488–492 (1967).
16. J. H. Malmberg, Millimicrosecond duration light source, *Rev. Sci. Instr.* 28, 1027–1029 (1957).
17. S. A. Pollack, Short duration light pulse during electrical breakdown in gases, *J. Appl. Phys.* 36, 3459–3465 (1965).
18. Principal characteristics of the TRW nanosecond spectral source fluorometry system, "TRW Fluorometry Handbook," TRW Instruments (1967).
19. R. G. Bennett and F. W. Dalby, Experimental determination of the oscillator strength of the first negative bands of N_2^+, *J. Chem. Phys.* 31, 434–441 (1959).
20. R. W. Swank and W. L. Buck, Decay times of some organic scintillators, *Rev. Sci. Instr.* 26, 15–16 (1955).
21. S. A. Pollack, Angular dependence of transmission characteristics of interference filters and application to a tunable fluorometer, *Appl. Optics* 5, 1749–1756 (1966).
22. S. Udenfriend, "Fluorescence Assay in Biology and Medicine," Academic Press, New York (1962).
23. R. F. Chen, G. G. Vurek, and N. Alexander, Fluorescence decay times: Proteins, coenzymes, and other compounds in water, *Science* 156, 949–951 (1967).
24. I. B. Berlman, "Handbook of Fluorescence Spectra of Aromatic Molecules," Academic Press, New York (1965).
25. D. W. Ellis, "Fluorescence and Phosphorescence Analysis" (D. M. Hercules, ed.), pp. 41–75, Interscience Publishers, New York (1967).
26. F. E. Lytle and D. M. Hercules, Luminescence of rethenium chelates, *J. Am. Chem. Soc.* 91, 253–257 (1969).
27. A. H. Kalantar, Isotropic rotational relaxation of photoselected emitters and systematic errors in emission decay times, *J. Phys. Chem.* 72, 2801–2805 (1968).
28. E. T. Meserve, Decay time fluorometry below ten nanoseconds with the TRW nanosecond spectral source system, *in* "TRW Fluorometry Handbook," TRW Instruments (1967).
29. E. T. Meserve, Measurement of short decay times with the TRW nanosecond spectral source system, *in* "TRW Fluorometry Handbook," TRW Instruments (1967).

30. W. R. Ware and B. A. Baldwin, Absorption intensity and fluorescence lifetimes of molecules, *J. Chem. Phys. 40*, 1703–1705 (1964).
31. G. C. Brown, Jr., Simplified system for the measurement of fluorescence litimes using the stroboscopic method, *Rev. Sci. Instr. 34*, 414–415 (1963).
32. W. R. Ware and P. T. Cunningham, Lifetime and quenching of anthracene fluorescence in the vapor phase, *J. Chem. Phys. 43*, 3826–3831 (1965).
33. A. Schwarzshild, A survey of the latest developments in delayed coincidence measurements, *Nucl. Instr. Methods 21*, 1–16 (1963).
34. E. C. Yates and D. G. Crandall, Decay times of commercial organic scintillators, *IEEE Trans. Nucl. Sci.*, pp. 153–158 (June 1966).
35. W. E. Blumberg, J. Eisinger, and G. Nanon, The lifetimes of excited states of some biological molecules, *Biophys. J. Soc. Abst. 12*, A106 (1968).
36. H. W. Offen and D. T. Phillips, Fluorescence lifetime of aromatic hydrocarbons under pressure, *J. Chem. Phys. 49*, 3995–3997 (1968).
37. J. Lowe, C. L. McClelland, and J. V. Kane, Lifetimes of the first excited states of C^{15} and B^{10}, *Phys. Rev. 126*, 1811–1819 (1962).
38. T. H. Bulpitt, Measurement of long decay times with the TRW nanosecond spectral source system, *in* "TRW Fluorometry Handbook," TRW Instruments (1967).
39. W. R. Ware, Fluorescence lifetime measurements by time correlated single photon counting, *Technical Report No. 3* under contract N00014–67–A–0113–006, Office of Naval Research, March 1, 1969.

Experimental Techniques: Chapter 2, Part C

PHOSPHORESCENCE INSTRUMENTATION AND TECHNIQUES

Eloise Kuntz

Biophysics Department
Michigan State University
East Lansing, Michigan

---

1. GENERAL INSTRUMENTATION

Three parameters of phosphorescence are most frequently measured. These are excitation and emission spectra, relative intensity or quantum yield of phosphorescence, and the lifetime of the triplet state. A fourth measurement—polarization of emission—would be very desirable but is less frequently performed due to technical problems.

The similarity in techniques for measuring low-temperature fluorescence and phosphorescence means that a single instrument may be designed for both types of measurements. Most of the designs and criteria discussed in Chapter 2A are applicable to phosphorescence, and the reader is referred to that section for such details. The essential features for phosphorescence studies will be stressed here. Further details may be obtained in a number of publications.[1-5]

Whether a commercial instrument or a custom-designed instrument is chosen, it is necessary that the sample area or compartment accommodate a cooling system, as well as accessories such as light choppers and shutters. While most investigators prefer an enclosed sample compartment to exclude room light, others leave this area open for maximum flexibility and either work in a darkened room or use light modulation and phase-sensitive detection which rejects unwanted light. Instruments designed for fluorescence

---

This work was supported by AEC Grant AT(11-1)1155 from the United States Atomic Energy Commission.

and phosphorescence studies of proteins and nucleic acids which utilize modulation and phase-sensitive detection are described by Longworth[4] and Eisinger.[5]

An instrument of high sensitivity should be chosen since the quantum yields of proteins and, in particular, nucleic acids may be very low. The sensitivity should be gained through optimizing the entire detector system, in preference to using an intense exciting light, since the quantum yields for photodecomposition are often large enough to cause appreciable decomposition during the observations. Fortunately, decomposition rates are much lower in frozen than in fluid systems.

For most purposes, xenon or mercury lamps of 150 to 200 watts are adequate when used with a single excitation monochromator. Eisinger[5] found that a 200-watt high-pressure mercury lamp provided sufficient light of high spectral purity after passage through two 0.25-m $f/3.6$ grating monochromators connected in series. One advantage of larger lamps is worth considering. The larger arcs may be positioned to minimize fluctuations due to arc wander. In either event, a beam splitter and photomultiplier between the emission monochromator and sample can be used to adjust automatically the sensitivity of the detector[4,5] to changes in the intensity of the exciting light.

A frequent source of difficulty is stray light.[6,7] This arises from several sources, including second-order bands from the excitation gratings, spurious wavelengths from the excitation monochromator due to dust or films on the mirrors and grating, multiple reflections from the walls of the dewar, and inaccurately placed sample capillary. When emissions of low intensity are being studied, the stray light should be measured using a blank in place of the sample. An example of such an evaluation is given by Eisinger,[5] who found that the integrated intensity of the stray light in his system, which used two excitation monochromators in series, was 10^{-5} times the intensity of the excitation band. This would permit good emission spectra from optically dense samples with quantum yields of 10^{-3}. If the ratio of scattered light to emission intensity is higher, a correction should be established by measuring a blank under identical conditions. This procedure may be used with either clear or turbid matrices since the observed emission intensity is relatively insensitive to turbidity.[8] If the sample capillaries are not accurately centered each time, the reproducibility is relatively low, with a standard deviation of 20% or more, as reported by Eisinger.[5] However, this can be improved to $\pm 7\%$ with careful centering of the sample capillary.

The relationship between sample size and sensitivity is dependent upon the optical characteristics of both the excitation and emission monochromators. If monochromators with large gratings of wide dispersion are employed, correspondingly large slits may be employed so that light may be gathered from and transmitted to relatively large areas. With well-designed systems, this results in substantial gains in sensitivity. However, it does place severe limitations on the design of sample holders for aqueous systems, such as are required for proteins and nucleic acids. The expansion and contraction of these matrices during the freezing and thawing cycles result in a high rate of breakage of most types of larger sample holders. Steen[12] has avoided this problem by using a shallow open aluminum dish for a sample holder. This is placed on top of a cold finger in a vacuum chamber. The sample is cooled to freezing before the chamber is evacuated.

The majority of workers have chosen to use quartz capillaries of 1 to 3 mm inside diameter for sample holders. This minimizes, but does not eliminate, breakage during freezing and thawing. With such small samples, worthwhile improvements in sensitivity can be achieved by using condensing and collecting lenses between the monochromators and the sample.[5] In the absence of such lenses, most of the advantages of large monochromators will be wasted, and smaller, less expensive instruments will do just as well when capillary sample holders are employed.

When capillary samples are used, maximum intensity of emission can be achieved by using optically dense samples so that observation geometry is essentially front face. Eisinger[5] has shown that, for a capillary with a mean pathlength of 1.3 mm, maximum fluorescence intensity will be observed if the effective optical density is at least 1.5. This situation also generally holds for phosphorescence. With macromolecules, quenching due to resonance transfer will not materially increase since the distance between chromophores within a macromolecule is fixed by the composition and conformation of the molecule. Practically, it may not be possible to achieve the desired optical thickness in many instances due to low solubility of the test compound.

1.1. Light Choppers

The long lifetimes of phosphorescence make it possible to isolate it from fluorescence or most types of stray light by means of light choppers. Two types of choppers are in common use. The older, more frequently used is a rotating can with slits spaced so as to block the emission path during the excitation period and then to block the excitation path during the emission period.

When rotating-can choppers are used, d-c electrometers are employed in the detector system. These are less stable than the phase-lock amplifiers discussed below. They also amplify all signals such as photomultiplier noise. Changes in photomultiplier noise and in amplifier performance can be monitored by periodically inserting a plastic disk containing 50 μc of C^{14}

together with a scintillator (New England Nuclear Corp.) into the emission path. Such a disk provides a constant low-level light source for calibration and checking.

Recently, tuning-fork light choppers have been described.[9] These have several advantages. They are small and can be used with a wide variety of cooling devices, in contrast to the rotating cans, which severely limit the dimensions of the cooling system. They are used to drive a phase-sensitive amplifier, which is inherently more stable than the d-c electrometers normally employed, since only those signals which are modulated in phase with the choppers will be amplified. They can be employed in three modes to measure either total emission, fluorescence alone, or phosphorescence alone. When total luminescence is to be recorded, only the emission chopper is employed. In this case, all the emitted light is amplified by the phase-sensitive amplifier. When fluorescence alone is to be observed, only the excitation chopper is employed, and the amplifier will amplify only the in-phase fluorescence from the short-lived singlet states. Phosphorescence alone is detected by operating the excitation and emission path choppers 180° out of phase with respect to each other. In this case, light which is at least 0.01 sec delayed from the excitation light will be detected.

1.2. Photomultipliers

Photomultipliers of high sensitivity and stability should be chosen. In addition, the region of high spectral sensitivity should cover the emission wavelengths of the molecules to be investigated as nearly as possible. Since unselected photomultiplier tubes vary widely in their sensitivity and noise levels, selected tubes should be used. Potted tubes further improve performance, especially in humid weather, by reducing ohmic leakage around the glass seals. Thermionic emission from the cathode, which is especially high in red-sensitive tubes, can be reduced by cooling. Photomultiplier tubes may be fatigued or damaged by exposure to room light even when no voltage is applied. If they are accidentally exposed, the noise level may be reduced by applying voltage to them in complete darkness.[10] Electrostatic charges on the photomultiplier tube may be reduced by wiping it with a cloth dampened with alcohol or acetone.[10]

1.3. Sample-Cooling Devices

A variety of devices for lowering and regulating the temperature of the sample have been described. These may be grouped into two general types—dewar and cold finger.

Dewars have the advantage of providing the most rapid cooling of the sample when filled with liquid N_2. Even this rate of cooling may not be rapid enough to prevent aggregation of monomers in low-viscosity media. Fortunately, the problem is less critical in polymers, where much of the chromophore spacings are determined by the polymer.

Well-evacuated dewars should remain frost-free for approximately 30 min when filled with liquid N_2. Poor seals between the glass and quartz or incomplete baking out of adsorbed gases before sealing will result in gradual loss of vacuum. In such cases, the viewing surface may be kept clear of frost by blowing dry nitrogen over the surface. Alternatively, dewars with sidearms may be evacuated each time they are used.

Two general types of dewars are used. The more widely used type (Figure 1A) is filled with a liquid coolant such as liquid N_2, and the sample capillary is positioned in this liquid. This is the simplest system to use but does have two disadvantages. It is often difficult to eliminate bubbling, which causes serious "noise" in the spectra. Only a very narrow range of temperatures is available with each coolant since these liquids, as commonly supplied, are close to their boiling points.

Practically speaking, liquid N_2 is the most widely useful coolant, providing temperatures close to 77°K. This temperature is sufficiently low for most purposes, but temperatures of about 4°K are desirable for certain studies and can be achieved with two-stage dewars utilizing precooling of an outer dewar with liquid N_2 and then with liquid helium in the inner compartment for final cooling. Other coolants such as liquid CO_2 are less useful since they are not cold enough for most phosphorescence investigations.

A recently described[5] tubular dewar (Figure 1B) utilizes gaseous N_2

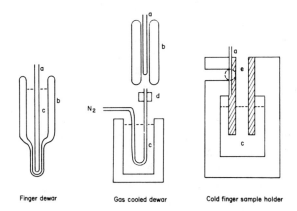

Figure 1. Low-temperature sample holders. (a) Sample capillary, (b) dewar, (c) liquid nitrogen, (d) heater, and (e) copper-tube cold finger.

which has been adjusted to any desired temperature at or above 77°K by passing through a coil immersed in liquid N_2 and then through a small heater before entering through the base of the dewar. Centering devices permit placing the sample in the intercept of the optical axes to obtain optimum intensity with minimum scatter.

A variety of cold-finger devices are adaptable for emission studies. The cold finger consists of a copper tube or rod which contains or dips into liquid nitrogen. The liquid N_2 reservoir and the insulation surrounding the cold finger may be polystyrene foam or preferably high-density polyurethane with suitably placed quartz ports for admitting and observing light. Dry gaseous nitrogen may be blown over the ports to prevent frosting. The cold finger described by Steiner[11] contains a groove into which the quartz capillary sample holder is clipped. A temperature close to 85°K can be maintained. More rapid cooling is obtained in the author's laboratory by substituting a copper tube filled with liquid N_2 for the rod (Figure 1C). Even so, the rate of cooling is much lower than is obtained with conventional dewars.

Steen[12] has described a cold-finger device in which the sample is placed in a small aluminum cup on top of the cold finger and viewed at right angles by means of a focusing mirror. Other ports allow excitation. A small heater near the top of the cold finger is used to maintain any temperature above 77°K. Evacuation of the sample chamber after the sample is frozen increases the efficiency of the system.

Several types of cryostats are available commercially and employ a variety of cooling devices. Some of these can be adapted to emission studies. However, they all require evacuation of the cold-finger sample-holder chamber to 10^{-5} to 10^{-6} torr in order to obtain reasonable efficiency. This severely restricts the design of the sample holder since very effective sealing is required to prevent evaporation and sublimation of the solvent.

2. MATRICES

The choice of a matrix for the sample should receive careful consideration since a variety of matrix effects are possible. In addition to emission from the solvent itself or an impurity in it, the solvent system can influence the rate of intersystem crossing, which will be apparent in the phosphorescence-to-fluorescence ratios, and it can cause collisional deactivations, which for long-lived triplets means that, with few exceptions, phosphorescence is completely quenched in liquids and even in ices until very low temperatures are reached. The tendency of the sample molecules to aggregate during freezing is greater with low-viscosity solvents. Fortunately, the spatial separation of chromophores is fixed by the polymer, so that this problem is less important than when monomers are being studied. However, different solvents do affect the conformation of the polymer, which in turn affects the phosphorescence by altering the microenvironment of the chromophores. This results from changes in the relative positions and spacings of surrounding residues and in the accessibility of the chromophores to the solvent system.

In order to eliminate collisional deactivations, the viscosity of the system must be approximately 10^{13} poise. This may be achieved by freezing fluids to very low temperatures or by imbedding in certain well-annealed plastics. For example, maximum phosphorescence can be observed at room temperature from compounds imbedded in degassed polymethyl methacrylate which has been annealed by heating to slightly below its melting temperature for several hours. Unfortunately, methacrylates strongly absorb in the UV region, and proteins and nucleic acids are not soluble in solvents for methacrylates. Polyvinyl alcohol (PVA) is water soluble and, with proper precautions, can be used for biopolymers. It does have a weak emission at 480 nm. This emission is lowest in PVA with a low content of residual carboxyl groups. Annealing by gentle heating while stretching PVA films decreases the rate of oxygen penetration so that at least some phosphorescence can be detected at room temperature, but it is usually necessary to lower the temperature to about 200°K to achieve maximum phosphorescence intensity.[13,14]

Proteins and nucleic acids are readily imbedded into polyvinyl alcohol. A 5% solution of a high molecular weight PVA with low carboxylic acid content is prepared by soaking the PVA in cold water and then heating until a clear homogeneous solution is formed. After cooling the PVA, the previously dissolved protein or nucleic acid is added and the mixture is poured onto clean glass coverslips to a sufficient thickness to achieve the desired optical density. The films are dried by evaporation and may be annealed by careful warming and stretching.

Since collision deactivation in water at room temperature occurs in 10^{-8} to 10^{-9} sec and the relaxation times of ices at temperatures moderately below freezing are still much shorter than the lifetimes of amino acid and nucleotide triplets, which are of the order of seconds, the viscosity must be increased to approximately 10^{13} poise. In the case of water or water plus moderate concentrations of additives such as glucose or salts, the viscosity necessary to achieve maximum emission is achieved at about 110°K for free amino acids, while 150°K appears to be low enough for many proteins[14] which impose their own constraints on the microenvironment surrounding the chromophores.

The temperature necessary to achieve maximum phosphorescence should be evaluated for each system. Simple alcohols such as methanol and ethanol do not achieve this even at 77°K. In some cases, it may be possible to find a temperature range in which maximum emission is observed before the cracking associated with low temperatures in many systems sets in. Fortunately, as pointed out elsewhere in this section, such cracking need not interfere with any measurements other than polarization.

Although the turbidity which develops at low temperatures when water is used as the matrix is not harmful, two other problems do arise. Simple monomers such as amino acids do not phosphoresce in ices of plain water. Also, the expansion during freezing and thawing causes frequent breaking of the capillary. Both of these effects can be overcome by addition of moderate concentrations of sugars or salts. The additives are believed to prevent quenching by preventing aggregation. However, they may also change the structure of the ice so as to decrease its interaction with the chromophore. Such additives are not necessary when biopolymers are used, since the polymer serves the same function.

Matrices which remain clear at 77°K are usually formed by 50% ethylene glycol or propylene glycol in water. These systems appear to be satisfactory for many studies on proteins and nucleic acids.[4,5] However, care must be exercised in selecting or purifying these polyalcohols, since even spectrograde lots may exhibit emission.

Other systems which remain clear at low temperatures include 4.5 M glucose, 5% PVA, and a mixture of dimethyl sulfoxide, ethylene glycol, and 1 M NaCl in the ratio of 4:5:1.

The probability of obtaining clear glasses is increased by using small-diameter capillaries since the cross-sectional area subject to expansion and contraction stresses is smaller.

3. POPULATION OF THE TRIPLET STATE

In the great majority of experiments, phosphorescence is observed following excitation to the lowest singlet state and intersystem crossing to the triplet state. It is also possible to observe phosphorescence following excitation to higher singlet states. In such instances, the phosphorescence is from the lowest triplet but the quantum yield is typically increased.

In some instances, the intersystem crossing rate is so low that phosphorescence is not observed. Excitation directly to the triplet is then required. Two techniques are available. The most direct technique, which is direct excitation at the absorption wavelengths of the ground state to triplet transition, requires intense flash lamps since the probability of this transition is very low in proteins and nucleic acids.

The second procedure is to populate the triplet state by photosensitization. This technique has proven valuable in the study of nucleotides and nucleic acids and is useful in a variety of energy-transfer problems. The donor should have its lowest excited singlet energetically lower than that of the acceptor, while its triplet must be higher than the triplet of the acceptor. In addition, the donor should have very high intersystem crossing and triplet to ground state transition probabilities. When such a donor is present in high concentrations, efficient energy transfer becomes possible. The experimental procedure is to excite the donor and observe the emission from the acceptor. Usually, there will be overlap of the donor and acceptor phosphorescence spectra which must be resolved. If the donor is properly chosen, it will have a much shorter lifetime than the acceptor. It then becomes possible to eliminate the emission from the donor by means of a rotating-can or tuning-fork chopper system operated at such a rate that the short-lived donor phosphorescence is eliminated. An example of this technique is reported by Eisinger,[5] who used 20% acetone as a sensitizer for protonated adenosine monophosphate. The lifetimes of the respective triplets are 3×10^{-4} and 1.75 sec, respectively. By operation of his tuning-fork choppers 180° out of phase and adjustment of the phase of the lock-in amplifier, the acetone signal could be eliminated entirely.

3.1. Spectra

Phosphorescence spectra from macromolecules may be characteristic of only one type of chromophore as a result of quenching or energy transfer from the other chromophores. The same macromolecule in a different configuration or other biopolymers may show complex spectra which are the sum of emissions from more than one type of chromophore.

Usually, the overlap of the spectra is such that the minor components appear as shoulders rather than as separate peaks. In such cases, several techniques may be required to assess the relative contributions of each chromophore. These techniques will be explained through their applications to the resolution of the tyrosine and tryptophan components in the phosphorescence spectra of a protein. Yeargers et al.[15] did this by reconstructing the tyrosine spectrum from the shoulder at 390 nm. They used the phosphorescence spectrum of ribonuclease A, which has its peak at 415 nm rather than at 390 nm, as in free tyrosine, to obtain the shape and position of the tyrosine peak. This is illustrated by curve c, which is constructed from the 390-nm shoulder of curve a in Figure 2.

Longworth[4] utilizes a different approach. He takes advantage of the fact that both tyrosine and tryptophan are excited at 275 nm while tryptophan, but not un-ionized tyrosine, absorbs at 295 nm. He records the

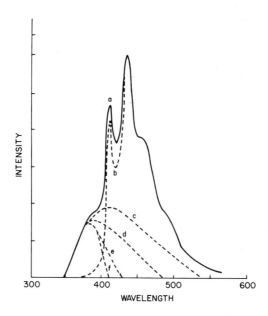

Figure 2. Alcohol dehydrogenase phosphorescence spectrum. (a) Excitation at 275 nm, (b) excitation at 295 nm, (c) tyrosine emission constructed from shape and position of tyrosine spectrum in ribonuclease A, (d) tyrosine emission reconstructed from shape of tyrosine peak in ribonuclease, but with peak at 390 nm, (e) tyrosine emission reconstructed from spectrum of free tyrosine, and (f) tyrosine emission reconstructed from fraction of emission having lifetime of tyrosine.

phosphorescence spectra resulting from each excitation and then subtracts the 295-nm emission spectrum from that excited at 275 nm to obtain the tyrosyl contribution. In the serum albumin studied by Longworth, the tyrosyl peak occurs in the 380–390 nm region. This technique assumes that the shape of the tryptophan spectrum remains constant when excited by 275 and 295 nm. This is not always quite accurate, as can be seen in the spectra of trypsin excited at 275 and 295 nm (Figure 3a,b). Therefore, in curve d of Figure 2 the tyrosyl spectrum is projected assuming that the peak is at 390 nm and the spectrum is broadened as in ribonuclease. A similar construction is made with the peak at 390 nm and the shape that of free tyrosine, (curve e of Figure 2.)

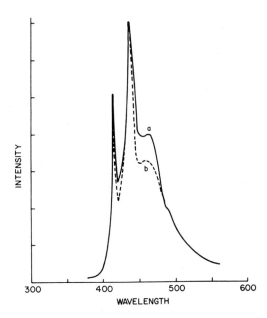

Figure 3. Trypsin phosphorescence spectrum. (a) Excitation at 295 nm, and (b) excitation at 275 nm.

An additional technique is available to determine which of these methods provides the best results in a given case. This is to determine the relative proportions of each chromophore by determining the proportion of the total emission at each emission wavelength exhibiting the lifetime characteristic of that chromophore (see Section 3.3). When this technique is applied to alcohol dehydrogenase, curve f of Figure 2 is obtained. This demonstrates the advisability of using two independent methods to arrive at the best estimate of the relative contributions of different chromophores.

An additional technique, which is employed in other types of spectroscopy, may prove helpful in certain cases for resolving the components of complex phosphorescence spectra. This is to convert the spectra to the first, or preferably the second, derivative. In the first derivative, the rate of change in intensity per unit wavelength or wavenumber is plotted. For the first derivative, this results in very sharp peaks with the highest point at the region of greatest rate of change and the true peak at the intercept with the baseline. The second derivative is generated by taking the derivative of the first derivative, resulting in even sharper spectra with the peak now corresponding in position to the true peak. Plotting such derivatives by hand or transferring the data to a computer is too laborious. However, amplifiers which directly plot the first derivative of signals from photomultipliers are available.

The use of the derivative technique should be limited to systems where the signal-to-noise ratio is very high. Otherwise, the artifacts will generate false peaks.

3.2. Quantum Yields

Either relative or absolute quantum yields may be determined, although relative yields are much easier to obtain and often suffice. The usual procedure is to record the fluorescence and phosphorescence spectra and then correct these spectra for wavelength sensitivity of the instrument.[16,17] These corrected spectra are then plotted on wavenumber paper to give equal areas for equal quanta. From a comparison of the areas of fluorescence and phosphorescence, the relative yields are obtained from the relationship

$$Q_{P\text{rel.}} = Q_{F\text{rel.}} \cdot \frac{A_P}{A_F}$$

If the absolute yield of fluorescence at room temperature is known, the absolute yield of phosphorescence may be obtained by comparing the areas of the fluorescence at room temperature and 77°K to obtain the 77°K fluorescence yield and from this the phosphorescence yield.

For proteins, free tryptophan fluorescence is often used as a standard as outlined in Chapter 2A, Section 1.3. Unfortunately, there is lack of agreement on the yield, with values of 0.20[18] and 0.14[19] being quoted. When these solutions are frozen, large differences in P/F ratios are found, suggesting that something is affecting the rate of intersystem crossing. In both cases, the total quantum yield at 77°K appears to be 0.9 to 1.0.

When low-temperature quantum yields are determined by comparison with room-temperature fluorescence yields, several corrections are required. Matrices change in volume upon freezing. This results in an increase or decrease in the concentration of the solute with a corresponding change in absorption and emission.[8] The correction factor for this is the ratio $(V_{\text{low temp.}}/V_{\text{room temp.}})$.

Volume changes also alter the density of the matrix. An increase in the index of refraction, n, reduces the fraction of the emitted light which falls within a narrow angle (perpendicular to the cell surface).[20] Borresen gives a correction factor of $(n + dn)^2/n^2$ for rectangular cells and approximately $(n + dn)/N$ for cylindrical cells.

The polarization of the emission results in unequal spatial distribution of the light[21] with phosphorescence preferential emission perpendicular to that of fluorescence. When polarization is positive, the magnitude of this factor is $(3 - P)/[3(1 + P)]$,[20] where the degree of polarization is $P = (I_V - I_H)/(I_V + I_H)$.

When turbid media or cracked glasses are employed as matrices, internal reflections depolarize the emission and alter the effective pathlength for absorption and emission and the cone of emission which enters the monochromator. The total effect of these factors has been estimated by varying the turbidity in a fluorescent solution and observing its effect on the intensity of the fluorescence.[8] Fluorescence intensity was found to be relatively insensitive to large changes in turbidity and light scatter.

The right-angle geometry of most spectrophosphorimeters requires accurate positioning of the sample capillary in the intercept of the excitation and emission monochromator lightpaths for quantitative work. The reproducibility of readings when either clear or turbid samples are positioned in the dewar by a sample holder such as that supplied with the Aminco spectrophosphorimeter is rather poor with a standard deviation of $\pm 20\%$.[8] This deviation can be lowered to 5–10% by accurate positioning of the capillary with a suitable holder. Alternatively, the reproducibility can be improved to about 7% by rotating the capillary to obtain averaged readings.[21]

3.3. Lifetimes

The transition rate of triplet to ground state is constant, which results in an exponential decrease in triplet population after the cessation of excitation. Since the phosphorescence intensity is proportional to the triplet population, there will be a corresponding exponential decrease in the intensity of the phosphorescence in accordance with the formula $I = I_0 e^{-kt}$, where I_0 is the initial intensity, I is the intensity at time t, and k is the rate constant.

The usual time-base instruments yield linear plots of intensity versus elapsed time (Figure 4). These must then be replotted on semilog paper (Figure 5). From this, one may determine the mean lifetime τ in accordance with the formula

$$I_\tau = I_0 e^{-1}$$

In this case, $I_\tau = 37\%$ of I_0.

Macromolecules frequently yield complex decay curves as a result of emission from chromophores which differ in their lifetimes. An example is the upper curve in Figure 5. If the differences in the lifetimes and the percent of each component are sufficiently high, the resulting semilog plot can be resolved into two or more straight-line regions in which each succeeding

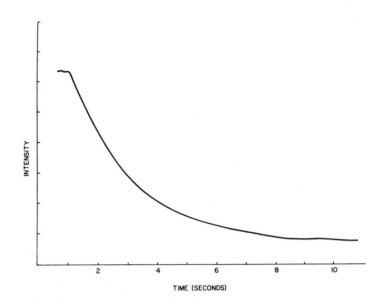

Figure 4. Lifetime of alcohol dehydrogenase decay of phosphorescence intensity.

region has a more gradual slope, indicative of a longer-lived component. These components are resolved by projecting the longer-lived component lines to their origins (B and C, Figure 5) and then subtracting region C from B and B from A to form lines A and B on the lower part of the graph. The mean lifetimes τ_a, τ_b, and τ_c are those times when $I_a = 37\% \ I_{a_0}$, $I_b = 37\% \ I_{b_0}$, and $I_c = 37\% \ I_{c_0}$.

Experimentally, the first step in determining the lifetime of the triplet state is to record the decrease in phosphorescence intensity as a function of time subsequent to excitation. Details and equipment for this procedure vary depending upon the range of lifetimes to be observed. For lifetimes in excess of 10^{-1} seconds, the normal excitation source may be interrupted with a rapid photographic shutter, while stray light is eliminated by the simultaneous use of a rapid rotating-can shutter or tuning-fork shutters 180° out of phase. For shorter lifetimes, an electronic flash lamp with a short pulse is the most convenient excitation source.

The decrease in phosphorescence intensity may be recorded using a variety of time-base instruments. Lifetimes in excess of 1 sec can be recorded on a conventional strip chart or an $X-Y$ recorder with time base. Shorter emission decay curves are usually displayed on an oscilloscope and photographed with high-speed Polaroid film. In either event, the recording may be initiated by hand or triggered by the flash lamp or shutter.

Phosphorescence Instrumentation and Techniques

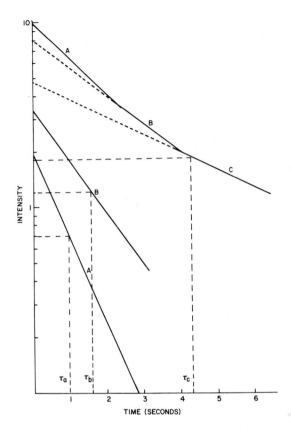

Figure 5. Upper curve: Semilog plot of Figure 4. Lower curves: Replotting of regions A and B by subtraction of B and C.

An alternate technique is to regulate the speed of a rotating-can shutter such that the period during which the slit is in the emission position somewhat exceeds the lifetime of the phosphorescence. This technique is limited to the range of speeds available with the driving motor but may be used in the range of 10^{-3} to 10^{-1} sec. For this procedure, a can-type rotating shutter with two oppositely placed slits with slightly less than 90° openings is preferable to the three-slit 60° opening type supplied with commercial instruments. The reason is that, with the 90° type, recording is initiated immediately after the exciting light is terminated, while with the 60° type there is a time interval equivalent to 30° rotation between illumination and observation, during which a short-lifetime component may be lost. It will also be impossible to determine the percent of the emission originating from each chromophore in multiple systems. In addition, the rate of build-up of triplets equals the rate of decay when the usual intensities of illumination are employed. If a 60° shutter is used, the population of triplets produced during illumination will be substantially less than with a 90° shutter operated at the same speed, and the observation period will be 30% less. A brush-type contact on the motor shaft of the rotating shutter can be employed to trigger the recorder or oscilloscope.

The above techniques are suitable when the emission contains either a single component or two components which differ markedly in lifetimes. However, nucleic acids and proteins may yield more complex curves. In such cases, it would be desirable to follow the decay over several orders of intensity in order to eliminate errors resulting from measuring only a portion of the complex decay curve. When this is done, several thousand determinations of each point are required to obtain sufficiently accurate curves when even moderately low noise levels are present.[22] The most practical method for accomplishing this is the method of moments proposed by Bay[23] for decay of short-lived radioactive species and applied to fluorescence lifetimes by Isenberg and Dyson.[22] In these procedures, each point on the curve is determined repetitively on a time-delay oscilloscope and averaged. The resulting curve is then compared with a series of computer-generated curves of the various lifetimes present in differing ratios.

3.4. Polarization of Phosphorescence

The polarization of phosphorescence from single chromophores is governed by the rules for photoselection,[24] while that from polymers may be modified by energy transfer between chromophores. By comparison of the degrees of depolarization of fluorescence and of phosphorescence, information on the extent to which energy transfer occurs in triplets versus singlets can be obtained.

Many spectrophosphorimeters can be adapted for polarization measurements by the incorporation of a polarizer in the excitation lightpath and an analyzer in the emission path. Either Glan–Thompson prisms may be used to pass the ordinary rays of the light, while refracting the extraordinary rays, or UV-transmitting Polacoat films may be used to pass the ordinary ray while absorbing the extraordinary rays. Whichever system is chosen, care must be taken to mount the units perpendicular to the optical axis with their vertical and horizontal axes accurately aligned. When Polacoat is used, the error can be held to less than 0.3% if the orientation is held within 2% of perpendicular to the optical axis and to less than 0.15% if the vertical and horizontal axis settings are correct within 2%.[25] A major advantage of Polacoat is that greater divergence and convergence of the light beams are permissible than with prisms.[26]

Polarization is measured by setting the polarizer with its electric vector in the vertical position and then measuring the relative intensities of emission when the analyzer is in the vertical position and in the horizontal position. Polarization would then be obtained from the expression

$$P = \frac{I_V - I_H}{I_V + I_H}$$

if there were no instrumental or sample artifacts. Two correction methods have been described. The first provides a correction factor for the difference in transmission of horizontal compared to vertical components of light by the emission monochromator.[27] To obtain this correction factor, an unpolarized source, such as a small lamp or a solution of small fluorescent molecules which become completely depolarized by rotation during the lifetime of the excited singlet, is placed in the sample position, and the ratio of emission intensity when the analyzer is in the vertical and the horizontal positions provides the correction factor, T, to yield the formula

$$P = \frac{I_V - TI_H}{I_V + TI_A}$$

An alternate and more readily obtained correction is to determine the ratio of light transmitted by the analyzer in its vertical and horizontal positions when the polarizer is in the horizontal position.[28] The correction for the relative transmission of the emission monochromator then becomes I_V'/I_H' where I_V' and I_H' are the intensities observed when the excitation polarizer is in the horizontal position. The expression for polarization then becomes

$$P = \frac{I_V - I_H(I_V'/I_H')}{I_V + I_H(I_V'/I_H')}$$

Another source of error is reflections from the surfaces of dewars and sample holders. Since reflections are partially polarized, they will alter the observed polarization.

Strains in the sample matrix or in the quartz glassware will also alter the observed polarization. Cracks or turbidity in the matrix depolarize the emitted light.

The technical problems listed above have discouraged many investigators from attempting polarization measurements of phosphorescence. However, the value of the method in studying energy transfer makes it worth attempting.

REFERENCES

1. C. A. Parker, "Photoluminescence of Solutions," Elsevier Publishing Co. (1968).
2. W. J. McCarthy and J. D. Winefordner, Phosphorimetry as a means of chemical analysis, *in* "Fluorescence Theory, Practice and Instrumentation" (G. G. Guilbault, ed.), p. 371, Marcel Dekker (1967).
3. J. D. Winefordner, W. J. McCarthy, and P. A. St. John, Phosphorimetry as an analytical approach in biochemistry, *in* "Methods of Biochemical Analysis," Vol. 15 (D. Glick, ed.), p. 369, Interscience (1967).
4. A. T. Longworth, Techniques for measuring fluorescence and phosphorescence of biological materials, *Photochem. Photobiol.* **8**, 589 (1968).
5. J. E. Eisinger, A variable temperature, U. V. luminescence spectrograph for small samples, *Photochem. Photobiol.* **9**, 247 (1969).
6. D. Redfield, Arc lamp intensity stabilizer, *Rev. Sci. Instr.* **32**, 557 (1961).
7. K. Schurer and J. Stoehorst, A simple device for stabilizing the output of a high-pressure xenon arc, *J. Sci. Instr.* **44**, 952 (1967).
8. E. Kuntz, F. Bishai, and L. Augenstein, Quantitative emission spectroscopy in media where appreciable light scattering occurs, *Nature* **212**, 980 (1966).
9. J. W. Longworth and F. A. Bovey, Conformation and interactions of excited states. I. Model compounds for polymers, *Biopolymers* **4**, 1113 (1966).
10. *Fluorescence News* **3** (2), 5 (1968). (Published by American Instrument Co.)
11. R. F. Steiner and R. E. Kolinski, The phosphorescence of oligopeptides containing tryptophan and tyrosine, *Biochemistry* **7**, 1014 (1968).
12. H. B. Steen, On the luminescence of L-tryptophane and L-tyrosine in aqueous solution at 77°K induced by X-rays and UV light, *Photochem. Photobiol.* **6**, 805 (1967).
13. E. Kuntz, Tryptophan emission from trypsin and polymer films, *Nature* **217**, 845 (1968).
14. E. Kuntz, R. Canada, R. Wagner, and L. Augenstein, Phosphorescence from tyrosine and tryptophan in different microenvironments, *in* "Molecular Luminescence" (E. C. Lim, ed.), p. 551, W. A. Benjamin (1969).
15. E. Yeargers, F. Bishai, and L. Augenstein, The nature of tyrosine phosphorescence from proteins, *Biochem. Biophys. Res. Commun.* **23**, 570 (1966).
16. W. H. Melhuish, The measurement of absolute quantum efficiencies of fluorescence, *New Zealand J. Sci. Technol.* **37**, 142 (1955).
17. R. Chen, Practical aspects of the calibration and use of the Aminco-Bowman spectrofluorimeter, *Anal. Biochem.* **20**, 339 (1967).
18. G. Weber and F. W. J. Teale, Determination of the absolute quantum yield of fluorescent solutions, *Trans. Faraday Soc.* **53**, 646 (1957).
19. R. F. Chen, Fluorescence quantum yields of tryptophan and tyrosine, *Anal. Letters* **1**, 35 (1967).
20. H. C. Borresen, The fluorescence of guanine and guanosine, *Acta Chem. Scand.* **21**, 920 (1967).
21. H. C. Hollifield and J. P. Winefordner, Rotating sample cell for low temperature phosphorescence measurements, *Anal. Chem.* **40**, 1759 (1968).
22. I. Isenberg and R. Dyson, The analysis of fluorescence decay data by a method of moments, abstract for the Symposium on Biological Molecules in Their Excited States, Arden House (1969).
23. Z. Bay, Calculation of decay times from coincidence experiments, *Phys. Rev.* **77**, 419 (1950).

24. A. C. Albrecht, Polarization and assignments of transitions: The method of photoselection, *J. Mol. Spectroscopy* 6, 84 (1961).
25. P. Johnson and E. G. Richards, A simple instrument for studying the polarization of fluorescence, *Arch. Biochem. Biophys.* 97, 250 (1962).
26. G. Weber, Fluorescence-polarization spectrum and electronic-energy transfer in tyrosine, tryptophan and related compounds, *Biochem. J.* 75, 335 (1960).
27. J. M. Price, M. Kaihara, and H. K. Howerton, Influence of scattering on fluorescence spectra of dilute solutions obtained with the Aminco-Bowman spectrophotofluorimeter, *Appl. Optics* 1, 521 (1962).
28. T. Azumi and S. P. McGlynn, Polarization of the luminescence of phenanthrene, *J. Chem. Phys.* 37, 2413 (1962).

Chapter 3

THE EXCITED STATES OF NUCLEIC ACIDS

J. Eisinger and A. A. Lamola

Bell Telephone Laboratories, Incorporated
Murray Hill, New Jersey

1. HISTORY AND INTRODUCTION

It is worthwhile to begin by recounting briefly the history of research into the fluorescence and phosphorescence properties of nucleic acids since it serves to illustrate several of the pitfalls which exist in this field.

In the 1930s there appeared a number of papers* describing the fluorescence properties of the nucleic acids and their bases. These observations were generally made with aqueous samples at room temperature using an excitation wavelength of 365 nm. The reported emissions were usually in the visible region. It has since become clear that the quantum yields of fluorescence of neutral bases and nucleic acids in aqueous solution at room temperature are vanishingly small. How is it possible therefore that these errors were made?

The reason seems to lie in the fact that 20 or 30 years ago scientists tended to use fluorescence properties as an empirical tool but had insufficient understanding of the physical processes entering into the absorption and emission phenomena. Without such an understanding, the observation of fluorescence from an impurity was often mistaken for that of the substance under study. Fluorescence and phosphorescence experiments are particularly prone to such errors since even a weakly absorbing impurity which does not manifest itself in absorption may be a suitable energy acceptor in energy transfer from the majority species. Because of their strong absorptions and weak emissions, nucleic acids require particular care in this respect, and the best method of guarding against this error is to ensure that the exci-

---

*Several of these references appear in Konev.[1]

tation spectrum of the fluorescence matches the absorption spectrum of the substance being investigated.[2,3] If this is not possible, one should at least make sure that there is no emission for excitation wavelengths at which the bases do not absorb (e.g., at 365 nm excitation employed by early workers in this field).

The common bases of DNA and RNA fluoresce weakly ($\Phi_f \leq 10^{-4}$) in aqueous solution at pH 7 at room temperature. Protonated purines, on the other hand, do fluoresce in room-temperature solutions,[4] as well as when adsorbed on chromatographic paper.[5] To gain some insight into the excited states of the neutral bases and nucleotides in polar solvents, one is, however, obliged to perform experiments at low temperatures.

The first low-temperature work on nucleic acids was reported by Agroskin *et al.* in 1960,[6] but the wavelengths of maximum fluorescence emission intensity given by these authors are generally higher than those given in 1964 by Bersohn and Isenberg,[7] who were the first to use a rigid glass as the medium. Konev[8] has ascribed this difference to improper calibration of the wavelength sensitivity of the emission monochromator used by the former authors, but a more plausible explanation is that their samples were dry powders in which the emitting molecules were aggregated and made intermolecular interactions (exciton shifts, excimer formation) likely. If it is desired to study the emission properties of isolated monomers, the use of a suitable glass matrix is imperative. Frozen-water solutions have also been used extensively for nucleic acid studies,[9] but the danger that solvent molecules may freeze out and form microcrystals or amorphous aggregates is acute, and such systems give results which are often difficult to interpret. Not only are energy levels apt to move in such aggregates, but energy transfer can occur easily and the rates of relaxation processes, such as intersystem crossing, can change in aggregates.

The phosphorescence of nucleic acids was first reported for adenine derivatives in snowy samples at low temperatures by Steele and Szent-Gyorgyi[10] in 1957, but the first results on isolated monomers were obtained by Longworth[11] in 1962 and Bersohn and Isenberg[7] in 1964.

Since then, the field has seen rapid clarification, particularly as the result of the careful characterization of the fluorescence and phosphorescence properties of the monomeric bases,[12,13] the discovery of exciplex formation between neighboring bases,[14] which explained the fluorescence spectrum of DNA, and the identification of the DNA triplet state as being that of thymine.[15,16] These developments will be recounted in a later section.

Since purines and pyrimidines are hardly ideal molecules for physical-chemical studies, it is clear that the considerable effort which has been expended in recent years in research on the excited states of these molecules is motivated by a desire to use these techniques to gain a deeper understanding of the biological role of these molecules. There are two broad areas in which the excited states of nucleic acids are of interest to the biophysicist. The first has to do with the lethal mutagenic damage centers induced in DNA by ultraviolet light and ionizing radiation and the role played by excited-state intermediates in these events. The second area concerns itself with the excited states of the nucleic acids as tools in determining ground-state properties of these molecules, primarily in connection with structural determinations which make use of energy transfer and emission spectroscopy, but also by development of analytical techniques using fluorescence.*

It is clear that for both of these applications an understanding of the excited states of nucleic acids under physiological conditions (i.e., in water at neutral pH and body or room temperature) is required. In view of the difficulty of obtaining such experimental data, most experiments have in fact been performed at or near liquid-nitrogen temperature, and the biophysicist is called upon to exercise deep insight in extrapolating results obtained for low-temperature glasses to room-temperature solutions. This extrapolation is complicated by the fact that many of the intramolecular processes which occur following excitation, such as nonradiative deexcitation, intersystem crossing, and excited-state interactions with neighboring molecules, including solvent molecules, occur at rates which are sensitive to temperature, solvent viscosity, and the state of aggregation. As a result, model systems must be used with greater caution, and only with a thorough understanding of the physical processes involved can a complicated molecular system like DNA in solution be approached.

The plan of this chapter is to preserve the biological emphasis and to stress those topics which appear to be most relevant for the biological applications outlined above. Consequently, we have omitted mention of the properties of the bases in organic solvents, which have been studied principally by Augenstein et al.[18-21] While we confine our attention mainly to the bases commonly found in DNA and RNA, we have included a section which deals with some of the "odd" bases found in transfer RNA.

2. STRUCTURES, NOMENCLATURE, AND ABBREVIATIONS

The chemical structures of the common and odd bases which occur naturally in nucleic acids are given below along with the conventional ring-numbering systems for purines and pyrimidines (Figure 1).

The following abbreviations are used for the ribonucleosides: A,

---

*Analytical techniques using fluorescence will not be discussed in this chapter, but excellent books on this subject are available.[17]

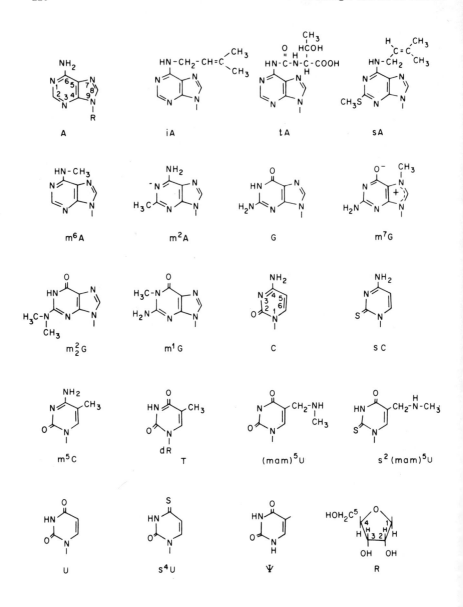

Figure 1. The chemical structures of the common bases bound in most DNAs and RNAs and of the odd bases which occur primarily in transfer RNA, along with the numbering scheme of purines and pyrimidines. The abbreviations correspond to the IUPAC–IUB convention.

adenosine; G, guanosine; T, thymidine (deoxyribonucleoside); C, cytidine; U, uridine; m$^{2'}$ G, 2'-O-methylguanosine; m^{7}G, 7-methylguanosine; m$_2^2$G, N(2)-dimethylguanosine, with analogous abbreviations for other methylated nucleosides. In addition, we use ψ, pseudouridine; I, inosine; s^{4}U, 4-thiouridine; iA, 6-(isopentenylamino)purine nucleoside; (mam)^{5}U, 5-(methylaminomethyl)uridine; s^2(mam)^{5}U, 5-(methylaminomethyl)-2-thiouridine; sA, 6-(isopentenylamino)-2-(methylthio)purine nucleoside; and tA, N-(purin-6-ylcarbamoyl)threonine.

The corresponding deoxyribonucleosides are indicated by the prefix d as in dA, deoxyadenosine, except for T, which already refers to the deoxyribonucleoside and for which rT represents the ribonucleoside. Unless otherwise stated, the mononucleotide indicated by the suffix MP, as in AMP, adenosine monophosphate, means that the phosphate ester linkage is at the 5-position of the sugar group. Dinucleotides represented by the formula XpY, as, for example, ApC, are 3'→5'; that is, the first nucleoside is connected at the 3'-position and the second at the 5'-position.

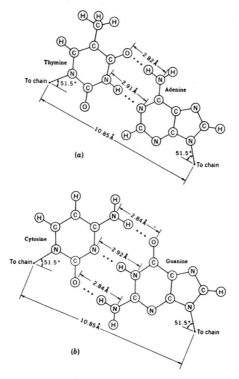

Figure 2. The structures of the hydrogen-bonded base pairs in DNA.

Homopolynucleotides are abbreviated poly rX or poly dX for the polyribonucleotide and polydeoxyribonucleotide, respectively, as, for example, poly dA is polydeoxyriboadenylic acid. The notation poly XY_3 represents the single-stranded polynucleotide containing X and Y randomly distributed and the proportion 1 to 3, respectively. Double-stranded polynucleotides are indicated by the use of a colon, as, for example, poly rG:rC represents the double-stranded polyribonucleotide in which one strand contains only guanine and the other strand cytosine. Poly dAT:dAT represents the double-stranded polymer in which each strand contains adenine and thymine in an alternating sequence.

The G-C and A-T hydrogen-bonded (Watson–Crick) base pairs are shown in Figure 2.

3. EXCITED STATES OF MONOMERS

3.1. Relevance of Low-Temperature Experiments

In this section we will consider the emission properties of isolated bases and nucleotides in neutral polar solvents. With the exception of protonated purines, these molecules have extremely weak emissions which have not yet been characterized in aqueous solutions at room temperature. The upper limit one can place on the fluorescence quantum yields is of the order of 5×10^{-4}, which means that their nonradiative de-excitation rates must be 10^{12} sec^{-1} or greater.* As a result of this severe quenching, most of our experimental information about nucleotides comes from experiments conducted at or near liquid-nitrogen temperature. The most desirable solvent for such experiments is one that approaches the polar properties of water, but water itself is not suitable since it "freezes out" solute molecules as microcrystals and other aggregates whose emission spectra, yields, and intersystem-crossing rates may be very different from those for isolated molecules. As a result, most investigators have turned to polar glasses such as ethylene glycol or propylene glycol, usually mixed with equal volumes of water.

Since the ultimate purpose of the investigations discussed in this chapter is to help in the understanding of nucleotides under physiological conditions, it is appropriate to review how the emission properties of molecules

---

*This estimate is based on the assumption that the absorbing and emitting levels are identical and that the relation between the Einstein coefficients of absorption and spontaneous emission holds (cf. "Introduction to Quantum Mechanics" by L. Pauling and E. B. Wilson, McGraw-Hill Book Company, 1935, p. 300).

depend on the temperature and viscosity of the solvent. Solvent effects connected with the dielectric constant of the solvent are, of course, well known[22] and will not be discussed here.

The energies of the excited states and the extinction coefficients are practically temperature independent, and the small changes in the absorption spectra which are observed between 77°K and room temperature generally have to do with narrowing of the vibronic levels at low temperature. The emission spectra undergo similar small changes with temperature as long as the molecule is in a rigid glass, but their intensities may change drastically if activated quenching occurs. Viscosity changes may have a profound effect on the shapes of the fluorescence spectra, and one commonly observes red shifts with increased temperature and lowered viscosity. These shifts, which may be as large as 3000 cm^{-1}, arise from two related mechanisms: The first is the formation of an excited-state complex ("exciplex") between the excited solute molecule and a single solvent molecule, and the second, which generally occurs when the viscosity is somewhat lower, is a reorientation of several molecules in the solvent shell of the excited solute molecule. Quite often, quenching sets in only after these excited-state interactions with the polar solvent have occurred.

Apart from the spectral changes and changes in nonradiative rates mentioned above, it has recently been shown that the intersystem-crossing rate may change drastically with the temperature. These effects will be discussed in section 5, in which the few room-temperature experiments dealing with excited nucleotides will be reviewed.

Phosphorescence is almost never observed in liquid media: many nonradiative de-excitation mechanisms, including quenching by dissolved oxygen molecules diffusing through the solvents, mitigate against it. Nucleotides in excited triplet states nevertheless play an important role in the photochemistry of nucleic acids (see section 6). The energies of the triplet levels of molecules in aqueous solutions can be determined by indirect means such as sensitization by triplet donors, and excited triplet states are important precursors for photochemical reactions, even though they are too short-lived for their phosphorescence to be observed. Triplet levels may be lowered as a result of solvent reorientation just like excited singlet levels. Such shifts would make no difference in their ability to be sensitized by an external triplet donor but would seriously hamper triplet energy migration among the nucleotides.

It should be mentioned that while the remarks we have made about the fluorescence quenching in nucleotides in aqueous solution apply to the common nucleotides found in DNA and messenger RNA, some rare nucleotides found in transfer RNA do fluoresce at room temperature (see section 8). There have also been recent reports of synthetically prepared nucleotides which fluoresce at room temperature which may be useful as structural probes when incorporated in biological polynucleotides.[23]

3.2. Emission Spectra and Other Experimental Parameters

The low-temperature fluorescence and phosphorescence spectra of the common mononucleotides are shown in Figure 3. All spectra were obtained[13,15,24,25] at 80°K in ethylene glycol–water glass (1:1), and their shapes are corrected for the wavelength dependence of the emission spectrometer used. These corrections[27] are less than 15% over the wavelength range of any of the spectra.

The thresholds of the fluorescence and phosphorescence spectra may be used to determine the energies of the excited singlet and triplet states. Since the efficiency of triplet energy transfer depends critically on the difference of the triplet energy levels of the donor and the acceptor,[28] the relative order of triplet energies is important in determining the likelihood of this mechanism. The quantum yields of fluorescence and phosphorescence (φ_P, φ_F) are determined from the integrated intensities of the suitably corrected spectra by comparison with the yield from some standard sample. It should be noted that φ_P is the same as the intersystem-crossing quantum yield given by $\varphi_{\rm isc}=k_{\rm isc}/(^1k_r+{}^1k_{\rm nr}+k_{\rm isc})$ only if the nonradiative de-excitation rate of the triplet state, $^3k_{\rm nr}$, is negligible compared to the radiative one, ^3k_r (^1k_r and $^1k_{\rm nr}$ are the radiative and nonradiative de-excitation rates of the excited singlet state). These mechanisms are shown schematically in Figure 4, which also contains typical values of the lifetimes of the various excited levels.

Fortunately, $\varphi_{\rm isc}$ can be determined directly from the intensity of the ESR signal,[29] and this has made it possible to estimate the fraction of nonradiative de-excitations from both the singlet and triplet states. These fractions are given by $1-\varphi_F-\varphi_{\rm isc}$ and $1-(\varphi_P/\varphi_{\rm isc})$, respectively, and the experimental values[13] for the common bases are listed in Table 1. It has already been pointed out that these bases do not fluoresce at room temperature and so that $^1k_{\rm nr}$ must exceed 10^{11} sec^{-1} under these conditions. It is interesting to note that the probability of nonradiative de-excitation is large (0.5 to 1) for singlet and triplet states even at low temperature so that the effective radiationless deactivation of the bases persists even in rigid media.

Apart from yielding values for $\varphi_{\rm isc}$, ESR experiments are also useful in supplying the dipolar interaction constants[30] D', D, and E for unpaired electrons in the triplet states.

Experimentally, D' is obtained from $H_{\rm min}$, the magnetic field for which the $\Delta m=2$ transition at a particular microwave frequency is observed. The

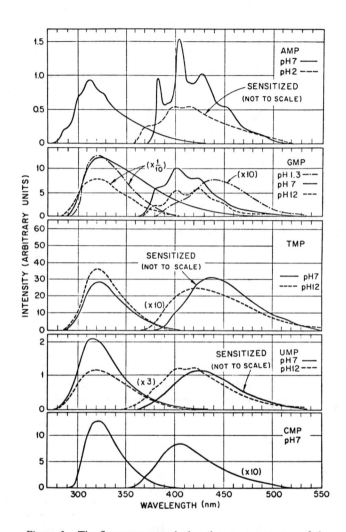

Figure 3. The fluorescence and phosphorescence spectra of the common nucleotides of DNA and RNA. The solvent was EGW glass and the temperature 80°K. The sensitized phosphorescence spectra are shown for the nucleotides which have negligible phosphorescence yields for low concentrations. Excitation wavelength was 265 nm.

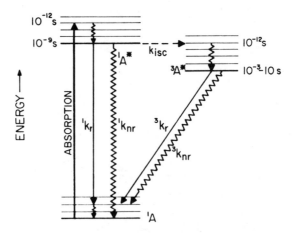

Figure 4. The Jablonski diagram showing energy levels of the ground state (^1A) and lowest excited singlet (^1A)* and triplet states (^3A*) of an aromatic molecule, along with the typical lifetimes of these states. Nonradiative processes are shown as wavy lines.

Table 1. **The Excited-State Properties of the Common**

	$^1E^b$ (10^3 cm^{-1})	$\lambda_{max}^{abs\ c}$ (nm)	λ_{max}^F	$^1\varphi_F$	$^1\tau_{calc}^d$ (nsec)	$^1\tau_{obs}^e$ (nsec)
AMP(pH 7)	35.2	258	313	0.01^i	0.02	2.8
AMP(pH 2)	34.8	257^c	—	~0	—	—
GMP(pH 7)	34.0	253, 274	323	0.13	1.6	~5
CMP(pH 7)	33.7	274	322	0.05	0.3	—
TMP(pH 7)	34.1	268	323	0.16^i	0.6	3.2
TMP(pH 12)	34.35	266	320	0.24	1.2	2.9
UMP(pH 7)	34.9	265	316	~0.01	0.03	—
UMP(pH 12)	35.0	263	320	0.005	0.01	—

^{a} The intersystem-crossing quantum yields (φ_{isc}), D', D, and E come from ESR spectra. ^1E and ^3E are the energies of the excited singlet and triplet levels and $^1\tau$ and $^3\tau$ are their lifetimes. Zero quantum yields generally imply an upper limit of 10^{-3}. Unless noted otherwise, all data come from Gueron et al.[13] and Lamola et al.[16] The quantum yields have experimental errors of the order of 30%.

^{b} ^1E is the energy of the 0–0 transition and is obtained by taking the midpoint between the thresholds of the absorption and fluorescence spectra plotted on an energy scale. For AMP at pH$_2$, no fluorescence is observed and ^1E comes from the threshold of the absorption spectrum.

^{c} These wavelengths of maximum absorption were obtained at 80°K, with the exception of the value for AMP at neutral pH, which refers to room temperature.

$\Delta m=1$ transitions, which are much more difficult to observe, yield values for D and E. Values for these parameters are listed in Table 1, and it is seen that they generally differ from one nucleotide to another. This makes it possible to gauge the relative triplet populations in a mixture of nucleotides which is being irradiated. In this, ESR spectra offer better resolution albeit lower sensitivity than phosphorescence spectra. Only TMP at pH 7 and pH 12 is seen to have almost identical values of D', and this degeneracy was responsible for the erroneous identification of the DNA triplet as T^- (see section 4.3.2).

Titration curves showing the pH dependence of the luminescence intensity at low temperature have been obtained for various mononucleotides and bases. It is important to bear in mind that this type of experiment reflects the ground-state pK, since re-equilibration in the excited state is precluded in a rigid medium. On the other hand, excited-state pK values may be determined from the differences in the fluorescence (or phosphorescence) thresholds of the conjugate acid and base forms. This is done by using a relationship derived by Forster and Weller,[33]

$$pK^* - pK = -\frac{0.625}{T}(\nu_{HB} - \nu_B) \quad [1]$$

where pK^* and pK are the excited-state and ground-state ionization constants, respectively, and ν_{HB} and ν_B are the frequencies of the emission spectral thresholds of the acid and base forms of the molecule, respectively.

Mononucleotides in EGW Glass at 80°K[a]

φ_{isc}	$1-\varphi_F-\varphi_{isc}$	$^3E^f$ $(10^3$ cm$^{-1})$	λ^P_{max}	φ_P	^3E (sec)	D' (cm^{-1})	D (cm^{-1})	E (cm^{-1})
0.02	0.97	26.7	404	0.015	2.4	0.126	0.121	0.027
~0	~1.0	27.35^g	398^g	~0	1.75^g	—	—	—
0.15	0.72	27.2	401	0.07	1.3	1.145	0.141	0.017
0.03	0.92	27.9	405	0.01	0.34	0.194	—	—
~0	0.84	26.3^g	432^g	~0	0.35^g	0.200	0.203	0.010
0.15	0.61	27.0	421	0.03	0.45	0.198	0.196	0.010
—	~0.99	27.4^{g,h}	423^{g,h}	~0	0.55^{g,h}	—	—	—
—	~0.99	28.4	404	0.002	—	—	—	—

[d] These singlet-state lifetimes (in nanoseconds) are obtained from the oscillator strengths and quantum yields (see Förster[40]).
[e] Fluorescence lifetime data are from Blumberg et al.[143]
[f] Obtained from the thresholds of the phosphorescence spectra.
[g] Obtained from sensitized phosphorescence spectra.
[h] Unpublished results.
[i] Hønnas and Steen[26] report $\varphi_F=0.005$ and $\varphi_F=0.30$ for AMP and TMP and $\varphi_P=0.01$ for AMP.

Table 2 gives values for $\Delta\nu = \nu_{HB} - \nu_B$ for the common mononucleotides.[13,16]

Values of $\Delta\nu$ are given rather than $pK^* - pK$ because pK is meaningless in the rigid glass at 77°K, for which equilibrium is probably not achieved even in the ground state. This is indicated by the fact that luminescence titration curves of the mononucleotides at 80°K in EGW as a function of the room-temperature pH give pK values about one unit higher than the room-temperature pK.[24] This probably corresponds to equilibrium near the freezing temperature of the glass.

Nearly all the $\Delta\nu$ values in Table 2 are negative, indicating that in most cases the proton is more firmly bound in the excited state. However, the values of $\Delta\nu$ are rather small. For comparison, β-naphthol has a $\Delta\nu$ of $+3300$ cm^{-1} corresponding to a pK shift of -7 at room temperature.[33]

The bases and the nucleotides have similar excited state properties as long as the 1-position in the pyrimidines and the 9-position in purines remain protonated. In certain cases, the position at which protonation occurs makes a profound difference in the emission properties of the bases. This is illustrated by the differences in the excited states of thymine, 1-methyl thymine, 3-methylthymine, and thymidylic acid, where the substituent affects the site of ionization at the high pH. Table 3 presents a collection of the relevant data.[34] Wierzchowski et al.[35,36] have shown that at pH 12 thymine exists as a mixture of the 1- and 3-anions. At pH values above 13, the di-anion is the most abundant species.[34] The room-temperature fluorescence[37] of thymine at high pH is due to the 1-anion. The room-temperature fluorescence[31,38] of protonated purines is not observed when the 9-position is substituted as in their nucleotides.

Hønnas and Steen[26] recently reported that they observed the fluo-

Table 2. The Change in Singlet (S) and Triplet (T) Excitation Energy $\Delta\nu$ upon Ionization (Loss of Proton)[a]

	Site of ionization	Room-temperature pK_a	$\Delta\nu(S)$ (cm^{-1})	$\Delta\nu(T)$ (cm^{-1})
TMP	N_3	10.0	-250	-700
AMP	N_1	3.8	-350	$+650$
GMP	N_7	2.4	-740	-1200
	N_1	9.4	-260	$+100$
CMP	N_3	4.5	-250	-300
UMP	N_3	9.5	-60	-1000

[a] The $\Delta\nu(S)$ values are average shifts for fluorescence and absorption edges; the $\Delta\nu(T)$ values are the shifts in the phosphorescence edges (Gueron et al.,[13] Lamola et al.,[16] and unpublished results of Lamola and Eisinger).

Table 3. The Excited-State Parameters of Thymine and Some of Its Derivatives at Neutral and High pH[a]

	φ_F (295°K)	λ^F_{max} (295°K) (nm)	φ_F	λ^F_{max} (nm)	φ_P	λ^P_{thr} (nm)	τ_P (sec)	D (cm^{-1})	E (cm^{-1})
Thymine (pH 7)	~0[c]	—	0.11	321	~0[d]	—	—	—	—
Thymine (pH 12)[b]	0.002	365	0.88	348	0.010	375	—	—	—
Thymine (pH > 13)	0.002	367	0.86	350	0.026	356	0.55	—	—
1-MeT (pH 7)	~0[c]	—	0.26	324	~0[d]	—	—	—	—
1-MeT (pH 12)	~0[c]	—	0.22	320	0.024	374	0.56	—	—
3-MeT (pH 7)	~0[c]	—	0.08	322	~0[d]	—	—	—	—
3-MeT (pH 12)	0.001	356	0.80	356	0.021	375	0.54	—	—
TMP (pH 7)	~0[c]	—	0.16	322	~0[d]	381[e]	0.35[e]	0.203[e]	0.010[f]
TMP (pH 12)	~0[c]	—	0.24	320	0.030	370	0.45	0.196	0.010
DMT (pH 7)	~0[c]	—	0.15	324	~0[f]	388[e]	0.63[e]	0.185	0.011

[a] The pK values for deprotonation of the 1 and 3 nitrogens are similar (~10). At pH greater than 13, both nitrogens are deprotonated. All data were obtained in EGW glasses at 80°K, except those in columns 1 and 2, which refer to aqueous solutions at room temperature. Data come from Gueron et al.,[13] Lamola et al.,[16] Berens and Wierzchowski,[35] and unpublished results by D. Patel, A. A. Lamola, and J. Eisinger; λ^P_{thr} is the wavelength of the blue threshold of the phosphorescence spectrum.
[b] λ_{ex} 290 nm.
[c] ≲10^{-4}.
[d] <10^{-3} for 1 mg/ml sample.
[e] Sensitized.
[f] For poly dAT : dAT.

rescence and phosphorescence spectra from adenine and thymine and their nucleoside derivatives in EGW glasses at 77°K when these were excited by X-rays. The observed spectra resembled those of the corresponding UV-excited bases, nucleosides, and nucleotides in all respects, except that the phosphorescence-to-fluorescence ratio was enhanced by approximately one order of magnitude. It is thought that the origin of this enhancement is that most of the excitations arise from ion recombinations, although direct excitation of the triplet manifold may occur as well.

3.3. Sensitized Phosphorescence Spectra

Three of the curves shown in Figure 3 are marked "sensitized (not to scale)." These refer to the phosphorescence spectra of protonated adenylic acid (AMP$^+$) and of neutral uridylic and thymidylic acids, none of which have an observable phosphorescence emission when directly excited. Such a condition can generally arise from one or both of the following possibilities: (1) The intersystem-crossing rate (k_{isc}) is too low to lead to a sufficient population of the triplet state. If ^1k_r and $^1k_{nr}$ are the radiative and nonradiative de-excitation rates of the excited singlet state, this generally means that $k_{isc} \ll {}^1k_r + {}^1k_{nr}$. (2) The k_{isc} is sufficiently great to populate the triplet state, but the nonradiative de-excitation rate of the triplet state is too great to make the phosphorescence observable, or $^3k_{nr} \gg {}^3k_r$. If only the first of these situations prevails, the triplet state of the molecule under study may be populated by providing for triplet–triplet transfer from a suitable donor molecule in its excited triplet state.[28,39] Ideally, this donor molecule has a lower singlet state than the acceptor, so that it may be excited at a wavelength at which the acceptor molecule does not absorb, and it has a higher-lying triplet state than the acceptor, so that the energy transfer may be efficient. These conditions were met by using acetone and acetophenone as triplet donors to sensitize the triplet states of TMP, UMP, and protonated AMP.[16] Under direct irradiation, none of these molecules has an observable phosphorescence.

Triplet sensitization is an extremely important technique, not only because it permits the study of molecules in their triplet states, which could not be populated by direct excitation, but it has also found an important application in selectively producing certain photolesions in DNA (see section 6.4).

Phosphorescence from thymine and its derivatives can also be induced by using solutions which are sufficiently concentrated to lead to the formation of aggregates.[16–18] Interactions between the aggregated thymines lead to an increase in the rate of intersystem crossing, but the mechanism for this is obscure in this case. The effect can be quite strong. For example,

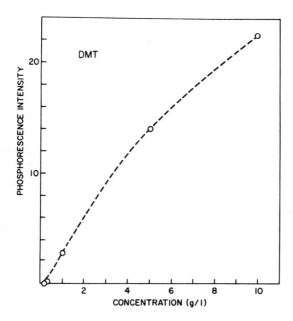

Figure 5. The relative phosphorescence intensity of dimethylthymine in EGW glass at 80°K as a function of concentration.

no phosphorescence is observed from a 10^{-3} M solution of 1,3-dimethylthymine (DMT) in EGW, but a 10^{-1} M solution phosphoresces with a quantum yield of a few percent[16] (see Figure 5).

3.4. Wavefunctions of the Excited States

3.4.1. Classification of Excited States

The low-lying excited electronic states of heterocyclic molecules like the bases being considered here may conveniently be classified as (π,π^*) states or (n,π^*) states, corresponding to the excitation of a π or n electron, respectively, to a previously empty π-orbital. Since the usual lowest absorption bands (260–270 nm) of the bases of the nucleic acids have extinction coefficients of the order of 10^4, they correspond to $\pi \to \pi^*$ transitions, but the possibility that weakly absorbing (n, π^*) states may have slightly lower energies and play an important part in the emission cannot be ruled out, since the weak bands corresponding to absorption by (n, π^*) states are difficult to recognize (see Chapter 1). For the pyrimidines, one may make a strong case for the (π, π^*) character of the emitting state since the emission spectrum, $f(\nu)$, and the absorption spectrum, $\varepsilon(\nu)$, plotted on a wavenumber scale are mirror images according to the relationship[40]

$$\frac{\varepsilon(\nu)}{\nu} \sim \frac{f(2\nu_0 - \nu)}{(2\nu_0 - \nu)^3} \qquad [2]$$

where ν_0 is the wavenumber of the 0–0 transition. When this relationship holds, one may conclude that the absorbing and emitting states are identical and the lowest-lying excited state is indeed (π, π^*). Figure 6 shows that the relation [2] provides a satisfactory fit for CMP and TMP but not for AMP (or GMP).[13] This suggests the presence of a low-lying (n, π^*) state in the purines, but not in the pyrimidines. The evidence for the existence of such a state is particularly strong in the case of adenine, where there is a large discrepancy between the observed singlet-state lifetime[41] (2.8 nsec) and the lifetime calculated from the fluorescence quantum yield, φ_F, and the radiative lifetime, τ_R, according to the relationship

$$\tau = \varphi_F \tau_R \qquad [3]$$

which is about 0.03 nsec.[13] The value of τ_R derived from the absorption properties of the molecule under the assumption of identity of the emitting and absorbing states (see Chapter 1) is given in Table 1 for the common nucleotides. The existence of an (n, π^*) band at approximately 290 nm has also been suggested on the basis of some solvent-shift experiments.[42] Fluorescence polarization experiments on AMP and GMP have been inter-

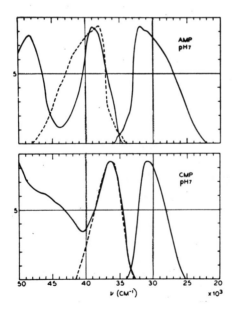

Figure 6. The absorption and fluorescence spectra of AMP and CMP in EGW at 80°K are shown in solid curves. The dashed curves are the "mirror images" of the emission spectra calculated according to Eq. [2].

preted[43] as proving that emission is from (π, π^*) levels in these nucleotides, but it is possible that emission from an (n, π^*) state reflects the polarization of the admixing state.

There are small changes in the absorption spectra of oligonucleotides and polynucleotides compared to the spectra of mononucleotides. The absorption peaks at 260 nm of ApA, oligo rA, and poly rA, for instance, are slightly blue-shifted compared to AMP, while their red absorption edges show slight red shifts.[44,45] While this is reminiscent of exciton splitting, this explanation is untenable since the concomitant rapid singlet energy migration in poly rA is not observed (see Section 7.4.1). The same spectral changes for adenosine can moreover be reproduced by changing the solvent from water to dioxane.[21] Gueron[44] has suggested that the 260 nm band consists of two (π, π^*) transitions, and the proximity of a neighboring base causes the stronger one to move to the blue and the weaker one to the red, producing considerable enhancement at the band edge.

Turning now to the triplet states of the common nucleotides, one may assign (π, π^*) character to all of them with confidence. The evidence for this comes from the fairly long lifetimes (~ 1 sec) observed for all the nucleotides (see Table 1). If the triplet states had appreciable (n, π^*) character, the transitions to the ground state would be enhanced as a result of spin–orbit coupling with (π, π^*) singlet states, which would result in much shorter lifetimes. Other evidence for the (π, π^*) nature of the triplet states is based on the observed values of the dipolar interaction constants D and E, which are also listed in Table 2. The experimental values agree quite well with estimates for (π, π^*) states.[16,25]

3.4.2. Molecular Orbital Calculations

There have been many calculations of ground-state and excited-state wavefunctions of the nucleic acid bases with the view of explaining their photochemical reactivities and abilities to form charge-transfer complexes and even of explaining such phenomena as spontaneous mutation and ultraviolet light–induced mutations. Unfortunately, the great majority of these calculations have been of little interest to the practicing spectroscopist, photochemist, or photobiologist, either because the calculations are too approximate or because the results have been in the form of correlations of little usefulness. In addition, the interpretations of calculations have often neglected basic notions of chemistry and molecular biology. A common error is to draw conclusions about reaction dynamics from calculated energies of reactants and products without considering the reaction mechanism. This is especially dangerous for photochemical reactions, which are nonadiabatic.

Theoretical calculations of molecular wavefunctions and experimentally observable parameters derived from them are helpful to the experimentalist only when they either explain apparent anomalies, assist in the interpretation of experimental data, or suggest new experiments. Quantitative predictions which meet these goals are not to be expected for the nucleotides at this time for a number of reasons. Even in the best calculations, electron correlation is not taken into account, and this severely limits the quality of the results for excited states. In addition, the calculations are for the gas-phase molecule, and therefore important solvation effects are neglected.

Extensive Hückel calculations of the wavefunctions and observable parameters such as ionization potentials, energy levels, and chemical reactivities have been performed by A. Pullman and B. Pullman et al.,[46-48] by P.O. Löwdin, and by others. An extensive review and bibliography of their work and that of others appeared recently.[46,47]

As an example of the most sophisticated molecular orbital calculations attempted for the bases of nucleic acids, we will mention the recent calculations of Snyder et al.[49] These authors performed all-electron all-integral LCAO-SCF calculations using gaussian basis functions for the ground and low-lying excited states of thymine (T) and the anions 1-HT$^-$ and 3-HT$^-$ derived by removing a proton from nitrogen atoms N_3 and N_1, respectively. Many ground-state properties, including the experimental carbonyl stretch frequencies, are accurately calculated. By making reasonable corrections for electron correlation energies and solvation effects, a very nice picture emerges which can rationalize the experimental data concerning the excited states. The calculations show that both the lowest-energy singlet and triplet $\pi\pi^*$ transitions of T and 1-HT$^-$ are very similar and are localized in the region of the C_5-C_6 bond. That is, they all involve the promotion of an electron from a π-orbital localized in the region of the C_5-C_6 double bond to a π^*-orbital involving the same atoms. These results fit the experimental observations that the ground states of T and 1-HT$^-$ are quite different, as judged by the differences in the carbonyl stretching frequencies, and yet the absorption, fluorescence, and phosphorescence spectra, as well as the electron spin resonance spectra, of the lowest triplet states in the two species are strikingly similar. On the other hand, the calculations show that the lowest-lying transitions in 3-HT$^-$ involve orbitals which are not localized in the region of the C_5-C_6 bond but include the nitrogens and the carbonyl groups. The absorption and fluorescence spectra of 3-HT$^-$ are greatly red-shifted from those of 1-HT$^-$ and T.

Unfortunately, the calculations cannot explain the difference in the intersystem-crossing efficiencies of T, 1-HT$^-$, and 3-HT$^-$ at 80°K (see Table 3). In all three cases, there are low-lying n,π triplet states which could facilitate intersystem crossing. An important prediction for the cases of T and 1-HT⁻ is that in the lowest excited singlet state the double-bond character of the C_5–C_6 bond is lost and the molecule may prefer to distort into a nonplanar geometry. This may be important in explaining in part the severe fluorescence quenching in these molecules and the temperature-dependent intersystem-crossing rate in thymine (see section 5).

4. EXCITED STATES OF OLIGONUCLEOTIDES AND POLY-NUCLEOTIDES AT LOW TEMPERATURE

4.1. Types of Interactions

When two or more chromophores lie sufficiently close to each other, interactions between them produce changes in the energy levels and the rates of the various radiative and nonradiative processes connecting the energy levels.[50] If the interaction is strong enough, new molecular complexes may be formed. Even at quite large separations (tens of angstroms) the dipolar interaction may be sufficient for energy transfer to occur. In order to unravel the possible interactions between the bases of the nucleic acids, the excited states of dinucleotides and other model systems were studied, and this work eventually led to an understanding of the excited states of polynucleotides and DNA in particular—at least at low temperature.

Nucleotides at the spacing which obtains in native DNA, 3.5 Å approximately, interact sufficiently weakly that the absorption spectrum of native DNA has virtually the same shape as the sum of the constituent mononucleotides. In other words, only a small exciton interaction is observed at this distance, although the stacking of the bases produces an appreciable hypochromicity.*[51]

4.2. Excited States of Dinucleotides

4.2.1. Fluorescence

Dinucleotides and other model systems in which two bases are joined by a short linkage are the simplest systems which permit the study of the excited-state interactions between the bases. As such, their emission prop-

---

*At the somewhat smaller separation (~2.8 Å) which obtains for a pair of thymine molecules in a rigid glass which originate from the splitting of a *cis-cis* cyclobutane dimer, there are appreciable changes in the shape of the 260 nm absorption band which are typical of exciton splitting.[52]

erties at low temperatures have been studied by groups of investigators in France,[53,54] Japan,[55] and the United States.[14,56–58] While there are some differences in the experimental approach of these investigators, such as the choice of glasses, their results are in overall agreement. These results have generally been presented by comparing the emission spectrum of a dinucleotide XpY and comparing it with that of an equimolar mixture of the mononucleotides XMP and YMP. Since the absorption spectra of these two samples differ only slightly, differences in the emission spectra can be ascribed to excited-state reactions between X and Y.

The origins of the small absorption changes (generally less than 5 nm) observed in dinucleotides (XpY) and in trimethylene-linked bases[59,60] (X-$(CH_2)_3$-Y) are not well understood. Whatever the detailed mechanism is,

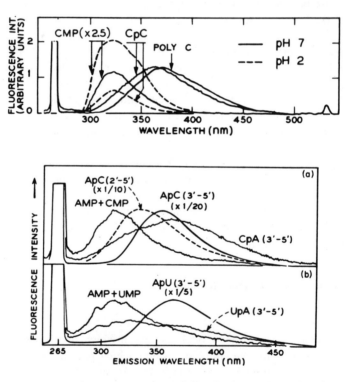

Figure 7. The fluorescence spectra of dinucleotides and a polynucleotide (poly C) contrasted with those of the corresponding mononucleotide or equimolar mixtures of mononucleotides. The red-shifted emissions of the dinucleotides and polynucleotides give evidence of exciplex formation since there are no corresponding changes in the absorption spectra. The solvent was EGW and the temperature 80°K.

the interactions producing these shifts are less than a few hundreds of wavenumbers in strength. This is the same order of magnitude as the exciton interaction[50,61] between two parallel dipoles separated by 3.5 Å and having magnitudes similar to the transition dipole moments of purines and pyrimidines.

It has been shown by optical rotary dispersion[62,63] and proton magnetic resonance[64,65] experiments that the bases in dinucleotides have a strong tendency to stack so that the relative positions of the chromophores can be expected to resemble those found in DNA.

In contrast to these relatively small changes in the absorption spectra, the emission spectra of dinucleotides and other linked bases are often quite different from those of the equimolar mixtures of the constituent bases. This is illustrated in Figures 7 and 8 and Table 4. It is seen from fluorescence spectra of dinucleotides and trimethylene-linked bases that in almost all cases the fluorescence is greatly red-shifted compared to the fluorescence of the constituent bases. This shift has been ascribed to the formation of an excited-state complex (exciplex) between the bases.[14,55,58] Exciplexes were first discovered in solution,[66] where they occur with a concentration dependence which proves that they are only formed during the collision of an excited molecule and a ground-state molecule. In molecules containing two

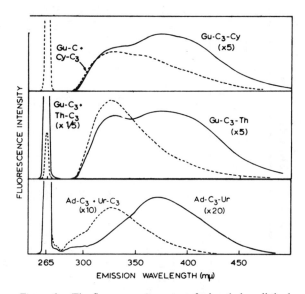

Figure 8. The fluorescence spectra of trimethylene-linked bases contrasted with those of equimolar mixtures of the propylated bases. The solvent was EGW and the temperature 80°K.

Table 4. A Comparison of the Emission Properties of Dinucleotides and Synthetic Models of Dinucleotides on the One Hand and of Equimolar Mixtures of the Corresponding Mononucleotides on the Other[a]

	Reference	Fluorescence			Phosphorescence		
		λ_{max} (nm)	φ_F	φ_F^E/φ_F^M	Type of emission	φ_P	φ_P/φ_F
AMP	13	313	0.010	—	A	0.015	1.50
ApA	56, 59	317	0.040	0	A	0.055	1.30
A—(CH$_2$)$_2$—A	59	324	0.010	0	A	0.010	1.00
A—(CH$_2$)$_3$—A	59	354	0.045	1.00	A	0.040	1.00
A—(CH$_2$)$_6$—A	59	331	0.010	0	A	0.010	1.00
GMP	13	323	0.130	—	G	0.070	0.60
G—(CH$_2$)$_3$—G	59	330, 360	0.010	0.10	G	0.060	6.00
CMP	13	322	0.050	—	C	0.010	0.20
CpC	58	356	0.090	1.00	C	0.010	0.10
C—(CH$_2$)$_8$—C	59	316	0.015	0	C	0.002	1.30
TMP	13	323	0.160	—	—	~0	~0
TpT	59	331$^\omega$	0.140	~0	T	0.010	0.06
T—(CH$_2$)$_3$—T	59	327	0.250	~0	T	0.005	0.02
AMP + CMP	58	319	0.020	—	A+C	0.010	0.50
ApC	58	358	0.060	1.00	A	0.070	1.20
ApC(2′-5′)	58	333	0.040	—	A	0.060	1.50
CpA	58	320, 380	0.010	0.10	A	0.010	1.00
A—(CH$_2$)$_3$—C	59	~330	0.010	—	A	0.030	3.00

AMP + TMP	58	322	0.060	—	A	0.016	0.10
ApT	58	355	0.060	1.00	T	0.020	0.30
TpA	58	330	0.050	—	T	0.010	0.20
A—(CH$_2$)$_3$—T	59	330, 365	0.010	0.01	T	0.030	0.30
AMP + UMP	58	312	0.010	—	A	0.006	0.60
ApU	58	362	0.030	1.00	A	0.040	1.30
UpA	58	360	0.010	1.00	A	0.030	3.00
AMP + GMP	58	327	0.040	—	A+G	0.030	0.80
ApG	58	335	0.080	—	A	0.110	1.30
GpA	58	360	0.030	1.00	A	0.110	3.00
GpA(2′-5′)	58	346	0.050	1.00	A	0.100	2.00
A—(CH$_2$)$_3$—G	59	362	0.120	1.00	A (97%)	0.010	0.10
GMP + UMP	56	325	0.050	—	G	0.030	0.70
GpU	58	323, 405	0.01, 0.008	0.30	∼G	0.009	1.00
GMP + CMP	55, 70	330	70(a.u.)	—	G	66(a.u.)	1.00
GpC	55, 70	335, 360	∼30(a.u.)	∼0.50	G	30(a.u.)	1.00
CpG	55, 56, 70	330, 360	27(a.u.)	∼0.50	G	27(a.u.)	1.00

a All data were obtained near 80°K in EGW glasses. Unless noted otherwise, XpY refers to the (3′–5′) dinucleotide of the bases X and Y (see section 2). Where no absolute quantum yields have been reported, the emission intensities are given in arbitrary units (a.u.). Where the fluorescence is shifted to longer wavelengths, exciplex formation is indicated. The column headed φ_F^E/φ_F^M gives a rough estimate of relative magnitudes of exciplex and monomeric fluorescence yields. The phosphorescence spectra and decay rates of the dinucleotides and trimethylene-linked bases generally have the characteristics of the phosphorescence of the nucleotide with the lower-lying triplet state, as can be seen in the column headed "type of (phosphorescence) emission." Data on some additional synthetic dinucleotide models involving odd bases are given in Leonard et al.[66] and Iwamura et al.[167] Note that exciplex formation between two adenines is favored when the chain linking them is three methylenes long. When the chain is (CH$_2$)$_2$, it appears to be too short for the adenine moieties to interact, and with (CH$_2$)$_6$ the number of alternative structures is too large.[59]

chromophores it is not unusual to find that every excitation of a chromophore leads to the formation of an exciplex.[66-68]

Figure 9 gives a schematic representation of how the energy levels of the bases depend on the separation between them: While there exists a potential minimum in the excited singlet level at an equilibrium separation, r_e, the ground-state potential is repulsive. An exciplex may lose its excitation energy radiatively or nonradiatively, but in either case the complex ceases to exist upon de-excitation. The exciplex red shift can be accounted for by the sum of the exciplex stabilization energy, E_e, and the ground-state repulsion energy, E_r. There appears to be no simple relationship between the monomer fluorescence quantum yields and the exciplex yields (Table 4). Since exciplex fluorescence has as its final state a repulsive ground state, it is not surprising that it does not have any vibrational structure. A broad featureless emission spectrum which has a considerable red shift compared to the sum of the fluorescence spectra of the constituent chromophores, along with an absorption spectrum similar to the sum of the absorption spectra of the constituent chromophores, is generally taken

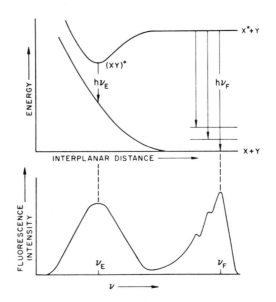

Figure 9. Schematic representation of the energy levels of an excited aromatic molecule (X*) as a function of its separation from a parallel aromatic molecule in a ground state (Y). The lower part of the figure shows the emission spectra of X*, which has vibrational structure, and of the exciplex (XY)*, which does not and whose emission is red-shifted by a considerable amount.

as evidence for the existence of an exciplex. The magnitude of the exciplex stabilization energy (ignoring E_r) is of the order of 5000 cm^{-1} for most of the exciplexes shown in Table 4, so that the excited-state interaction between the bases is seen to be much larger than the ground-state interactions reflected in the shifts of the absorption spectra.

It is not uncommon for samples of dinucleotides and trimethylene-linked bases to exhibit both monomer- and exciplex-like emission spectra (Figure 10).[58,59] It is thought that this reflects the heterogeneity of geometries for different pairs of bases.

Figure 7a shows that at neutral pH CpC (3′–5′) forms an exciplex or, better, an excimer. (An excimer is an exciplex formed between identical chromophores.) However, at pH 2, where both cytosine groups are protonated and strongly repulse each other, there is little difference between the fluorescence spectra of CMP and CpC.[14] Hélène and Garrestier-Montenay[69] have found that at intermediate pH values, where only one of the cytosine moieties is protonated, the fluorescence red shift is even larger than at neutral pH.

It is also worth noting that the particular linkage joining the bases has a profound effect upon the fluorescence one observes, and it is clearly meaningless to speak of an exciplex between two bases as a complex characterized by a single geometry. This point is well illustrated (see Table 4) by the different emission spectra reported[58,59,70] for ApC (3′–5′), ApC (2′–5′), CpA (3′–5′), and A-(CH$_2$)$_3$-C, the last compound representing the trimethylene-linked bases.

The sensitivity of the excited-state reaction leading to excimer formation is well illustrated by ApA (3′–5′) as well as poly rA, both of which have

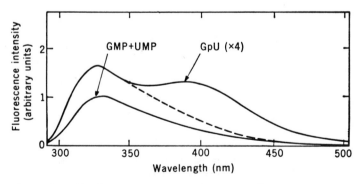

Figure 10. The fluorescence spectrum of GpU and of an equimolar mixture of GMP and UMP in EGW at 80°K. The emission of the dinucleotide appears to originate partly from monomers and partly from an exciplex.

fluorescence spectra which are not excimer-like in that they retain the vibrational structure of the monomer, AMP, but are red-shifted by a few hundred wavenumbers compared to it (see Figure 11).[27] In contrast to this, poly dA and A-$(CH_2)_3$-A have clearly excimer-like fluorescence spectra. The only safe conclusion one can draw from this result is that the ribo-oligoadenylic and polyadenylic acids have different structures than the deoxypolyadenylic acid and A-$(CH_2)_3$-A.

Beens and Weller[71] have shown that charge transfer plays an important role in the formation of many exciplexes. It is therefore expected that the magnitude of the stabilization energy of an exciplex depends on the stacking geometry of the two bases, as well as on their ionization potentials and electron affinities. While the latter parameters can be determined in principle, the geometry factor is sufficiently uncertain to make the theoretical prediction or interpretation of exciplex formation very uncertain.

No measurements of the lifetimes of base–base exciplexes have been published, but from the lifetimes of the excited monomers, which are of the order of a few nanoseconds, it is concluded that exciplex formation times must be of the order of 10^{-10} sec or less in many cases.

The question arises whether an exciplex (XY)* is as likely to be formed by an excited base X* interacting with base Y in the ground state as by Y* interacting with X. This was investigated[67] for the exciplex formed in ApC (3′–5′) by making use of the differences in the absorption properties of the two bases A and C. The intensity of the exciplex fluorescence was measured for different excitation wavelengths, and the results were found to be consistent with a model in which absorption by A would lead to exciplex for-

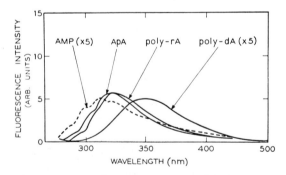

Figure 11. The fluorescence spectra of AMP, ApA, and ribo-, and deoxyriboadenylic acid in EGW at 80°K. Only in poly dA is there clear evidence for exciplex formation between neighboring adenine molecules.

mation with the same probability as does absorption by C. Similar results were obtained for the polynucleotide poly dAT:dAT.

4.2.2. Phosphorescence

In contrast to the excited singlet states of neighboring bases, which have been seen to be marked by excited-state complexes, the triplet states of almost all dinucleotides[53,54,68,70] and trimethylene-linked pairs of bases[59] are the same as that of one of the constituent monomers (see Table 4). This identification of the triplet state is made on the basis of the phosphorescence spectrum, and its decay rate as well as the ESR spectrum.

These experimental observations give in themselves no indication as to how the triplet state is populated. It is, on the one hand, conceivable that in the exciplex state the excitation "resides" principally on one of the bases which possesses the lower triplet level and that intersystem crossing gives the triplet state of this base. On the other hand, one could imagine that intersystem crossing in the exciplex can produce triplets of either base, but that immediately after the creation of the higher-lying triplet it would be transferred to the lower-lying triplet of the neighboring base. The fact that for all but two dinucleotides and trimethylene-linked bases which have been studied the triplet is that with the lower energy indicates that this triplet transfer is indeed effective, but it is still uncertain if intersystem crossing in these systems produces one or both triplets.

The first of the exceptions is that the triplet state of A-$(CH_2)_3$-iPA (iPA=isopentenyl adenine) resembles neither the A nor the iPA triplet and is considered to be a triplet exciplex.[60] The second exception[27] is the triplet state of the dinucleotide ApT (3'–5') at pH 11.5, where the thymine moiety is deprotonated. This dinucleotide has a monomer-like fluorescence spectrum very much like that of TMP at high pH (T$^-$), which has a much larger fluorescence quantum yield than does AMP (see Table 4.). This is an indication that the two bases have little tendency to enter into a stacked conformation, which would be a prerequisite for the formation of an exciplex and probably results from the negative charge on T. The phosphorescence of ApT (3 –5) at pH 11.5 shows both A and T$^-$ emissions in about the same proportions as are found in an equimolar mixture of AMP and TMP at pH 11.5 (Figure 12). It is concluded that the equilibrium conformation of this dinucleotide is such that A and T$^-$ are far enough apart so that the usual exchange mechanism for triplet–triplet transfer is inoperative. [27] Model studies show that A and T$^-$ could easily be 10 Å apart, and recent estimates[72] of triplet–triplet transfer rates at this distance show that they could indeed be slow compared to the phosphorescence decay rate of the higher-lying triplet state T$^-$ (3 sec^{-1}).

A striking demonstration of the occurrence of energy transfer in a dinucleotide is provided by the data reproduced in Figure 13. An equimolar mixture of AMP and GMP has a phosphorescence spectrum and phosphorescence decay rate which contain comparable intensities of AMP and GMP for excitation wavelengths at which they have similar molar absorptivities. At an excitation wavelength of 290 nm, on the other hand, practically all the light is absorbed by GMP and the phosphorescence component of AMP is negligible. Turning now to the dinucleotide ApG (3'-5'), which has an absorption spectrum like that of the equimolar mixture of AMP and GMP, it is seen that for both of these excitation wavelengths, the phosphorescence has the spectrum and decay rate characteristic of adenine, which proves that energy absorbed by the guanosine moiety has been transferred to adenine. Similar results are obtained in ApC.[67]

4.3. Excited States of Polynucleotides

4.3.1. Fluorescence

While the interaction between neighboring bases, one of which is in an excited singlet state, has been seen to be quite strong, it is not expected that this interaction, which is responsible for exciplex formation, would have a long range. It is therefore not surprising to find that the fluorescence properties of polynucleotides bear a great resemblance to those of the dinucleotides.

This point is perhaps best illustrated by comparing the dinucleotide ApT (3'-5') (Figure 12) with the alternating copolymer poly-d(A-T):d(A-T)—or poly dAT for short (Figure 14).[26] Both have exciplex singlet states with virtually the same spectral maximum (355 nm), and both have a thymine-like triplet state even though an equimolar mixture of AMP and

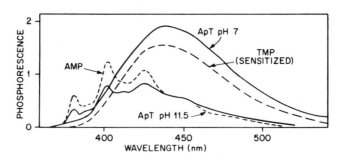

Figure 12. The phosphorescence spectra at 80°K of AMP, TMP, and ApT at neutral and basic *p*H. The TMP spectrum was obtained by sensitization and the solvent was EGW.

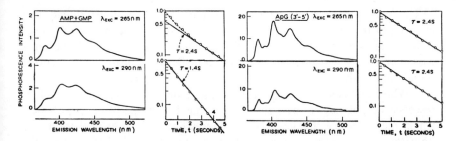

Figure 13. The phosphorescence spectra and decay rates of ApG and of an equimolar mixture of AMP and GMP, excited at 265 and 290 nm. At the shorter wavelength AMP and GMP have similar absorptivities, while at the longer wavelength only GMP has appreciable absorption. The solvent was EGW and the temperature 80°K.

TMP has an adenine-like phosphorescence with no discernible contribution from thymine. (This will be discussed in greater detail in Section 4.3.2.)

The copolymer poly dAT also offers a convenient model system for studying whether the exciplex (AT)* can be formed as a result of the excitation of either of the constituent bases.[68] The difference in the absorption spectra of A and T made it possible to use the excitation spectra of the exciplex singlet to derive the probability that the exciplex resulted from the excitation of T, and this probability was found to be 1/2 at the absorption peak. Similarly, the excitation spectrum of the T-like phosphorescence shows that the original excitation took place at A and T with equal probabilities. One can conclude that the reactions $A^* + T \rightarrow (AT)^*$ and $A + T^* \rightarrow$

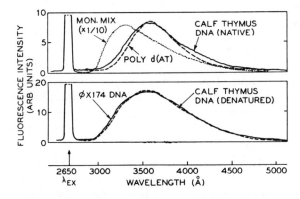

Figure 14. The fluorescence spectra of native and denatured calf thymus DNA, poly dAT, and single-stranded DNA from the phage ϕX174 contrasted with the fluorescence spectrum of an equimolar mixture of AMP, TMP, GMP, and CMP. Temperature was 80°K.

(AT)* have the same efficiency. The question of whether intersystem crossing of the exciplex produces both or only one of the A and T triplet states remains unanswered as it did for the analogous dinucleotide case discussed above, since triplet–triplet energy transfer would lead to the eventual population of the lower-lying triplet state (T) in any case.

While dinucleotides may be used as guides in anticipating the fluorescence of polynucleotides, there are some factors peculiar to polymers which may produce changes in their fluorescence. The first of these has to do with the secondary structure of the polynucleotide, which determines the relative geometry of neighboring chromophores and can therefore determine if exciplex formation is possible within the constraints of the molecule and which exciplexes may appear. We have already pointed out the differences between the fluorescence of poly dA and poly rA as illustrations of this point (see Section 4.2.1 and Figure 11).

Another effect of polymerization has to do with the formation of hydrogen bonds between bases belonging to separate strands. It has been shown that hydrogen bonding between G and C leads to almost complete quenching of the G and C fluorescence and the fluorescence yield from copolymers poly dG:dC and poly rG:rC is negligible.[14] The hydrogen bonds between A and T, on the other hand, appear to have little effect, as can be seen from the similarity of the emission spectra and yields of ApT and the double-stranded poly dAT, while the hydrogen bonds formed between A and U appear to lead to a quenching of A at the singlet levels leading to a quenching of both fluorescence and phosphorescence.[73,74]

The single-stranded polymer poly rAU has an exciplex fluorescence spectrum which is virtually the same as that of ApU (Figure 15). The random copolymer poly $r(A_{6.2}U)$, on the other hand, has a fluorescence spectrum[27] which appears to be a composite of the spectra of poly rA and of the ApU exciplex, while poly $r(A_2U)$ has a fluorescence intermediate between those of poly r AU and poly $r(A_{6.2}U)$ (Figure 15). This can be taken as evidence that although the A-U exciplex has a lower energy than the excited singlet state of poly rA, the exciplex is not an effective energy sink because of the short range of singlet energy transfer in poly rA (see Section 7.3.1).

Of particular interest is the fluorescence of DNA, which consists of stacked bases which in the case of native (double-stranded) DNA are hydrogen-bonded A:T and G:C base pairs. From what was said above, the G:C base pairs can be expected to contribute little. Their fluorescence was found to be efficiently quenched.[14,15] If we assume for the moment that this is indeed the case, we need only consider the fluorescence from A and T, and any exciplexes which may be formed among these bases. A T-T exciplex has never been observed, and the A-A and A-T exciplexes found in poly dA and poly dAT have fluorescence maxima at 350 and 355 nm, respectively. It is therefore gratifying that the fluorescence spectrum of DNA[14] is indeed an exciplex-like one with a maximum at 355 nm and slightly greater width than the fluorescence of poly dAT. The fluorescence quantum yield of DNA is only about one tenth as large as that of an equimolar mixture of the four constituent nucleotides or approximately 0.01 (see Figure 14).

The fluorescence spectrum of thermally denatured DNA or single-stranded DNA from the bacteriophage φX174[14] (see Figure 14) is also exciplex-like, although there is an appreciable monomer contribution at the short wavelengths. This is entirely consistent with the maintenance of base-stacking in denatured DNA.

4.3.2. Phosphorescence

It has already been pointed out that the phosphorescence of polynucleotides can be expected to be characteristic of the constituent chromophore with the lowest energy. This is indeed found to be the case in poly dAT[16,68], (Figure 16), where the lowest triplet state is that of T, and in poly rAU,[74,75] where the lowest triplet state is that of A. In homogeneous polymers, the triplet state is very similar to that of the mononucleotide, except that there is an appreciable red shift (about 10 nm) for the phosphorescence spectrum of poly rA compared to that of AMP[27]

DNA has a very weak phosphorescence (Figure 16) which has been the object of many investigations[7,12,13,16,75-78] since the problem of which

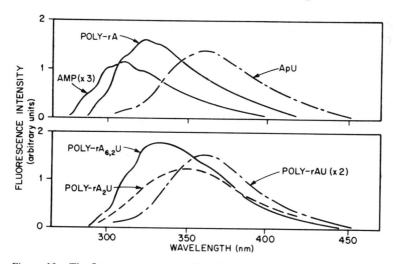

Figure 15. The fluorescence spectra of AMP, poly rA, and ApU contrasted with those of random copolymers containing A and U in different proportions.

base or bases are the source of the emission was first posed by Steele and Szent-Györgi.[10] While some of these workers used frozen aqueous media rather than glasses, this does not appear to cause as serious a complication for a polymer as it does for monomers, where aggregation can alter the spectroscopic properties of the solute molecules profoundly, and the suggestion of Douzou et al.[76] that the pyrimidines in DNA serve as triplet energy traps turned out to be correct. The earliest experiments using polar glasses, on the other hand, probably suffered from the presence of impurities or monomers in the sample and indicated that the purines were primarily responsible for the DNA phosphorescence.[7]

As the phosphorescence was characterized more and more carefully, it became clear that it consisted primarily of an emission which had a decay time of 0.3 sec and an unstructured spectrum with a peak at 450 nm.[75] The DNA triplet was also reported to have an ESR signal with an $H_{min}=$ 1054 gauss and the same decay rate.[75]

When these characteristics of the DNA triplet were compared with those of the triplet states of the constituent bases, it was clear that it did not resemble AMP, GMP, or CMP.[75] The emission and ESR properties of TMP were not available for comparison since the equilibrium triplet-state population, at least in isolated monomers in a glass, was too low. This led Rahn et al.[75] to propose that the DNA triplet state was that of thymine which has lost a proton at the N_3-position by transfer across the hydrogen bond to A. This model explained the absence of A phosphorescence (AH$^+$ had been shown to be quenched) and more importantly explained the striking similarity between the DNA and T$^-$ triplet states. An alternative suggestion was offered by Helene,[78] who suggested that T and T$^-$ have virtually identical triplet-state characteristics and that the DNA triplet was the thymine triplet, populated by triplet–triplet energy transfer.

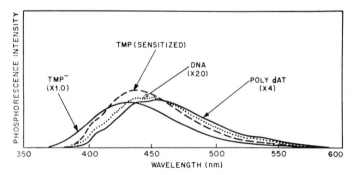

Figure 16. A comparison of the phosphorescence spectra (in EGW at 80°K) of DNA, poly dAT, TMP$^-$, and TMP. The last spectrum was obtained by sensitization using acetone as a triplet energy donor.

The problem was finally resolved when the TMP triplet was populated sufficiently by triplet–triplet energy transfer from acetophenone to permit its characterization by phosphorescence spectroscopy and ESR.[16] It emerged that Hélène's[78] assumption about the similarity of the T and T⁻ was correct and that the DNA triplet bears a closer resemblance to the triplet state of T than that of T⁻ (Figure 16). Lamola et al.[16] also showed that the T→A proton transfer postulated by Rahn et al.[75] is energetically unfavorable for any combination of singlet, triplet, or ground state of A or T with the ground state of the other base. Since T has the lowest-lying triplet state of all the bases,[16] the most reasonable model explaining the characteristics of the DNA triplet is one in which triplet–triplet transfer between neighboring bases populates the triplet state of thymine.

The question of where the intersystem crossing occurs remains unanswered. Lamola et al.[16] showed that while isolated TMP has a negligible intersystem-crossing rate, this rate becomes appreciable as the concentration is increased and aggregates are formed (see Section 3.3). It is therefore quite possible that thymine in DNA, where it is sandwiched between two other bases, has an appreciable intersystem-crossing rate so that some of the thymine triplets in DNA may have been produced there and need not have come there by energy transfer. In principle, this question could be resolved by measuring the excitation spectrum of the DNA triplet, but the differences in the absorption spectra between thymine and the other bases are so slight and the DNA triplet signal is so weak that the difficulties of such an experiment would be formidable.

5. EXCITED STATES AT ROOM TEMPERATURE

5.1. Energy Levels

Since there are only negligible changes in the absorption spectra of nucleic acids and their constituents on going from 80°K to room temperature, one may conclude that the energies necessary to populate the singlet states are essentially temperature independent. The same conclusion should obtain for the low-lying triplet states. On the other hand, various excited-state interactions and relaxation processes have temperature-dependent rates.* Thus, for example, the ability of these moieties to act as donors in excitation transfer processes could be seriously affected.

The excited-state interactions may be divided into those involving the

---

*The effect of temperature may be indirect—for example, in a case where the rate depends on the viscosity of the solvent.

excited group and either another chromophore or one or more solvent molecules (see Figure 17). The first of these interactions leads to the formation of an exciplex or excimer and has already been discussed in section 4. To this discussion must be added the possibility that, for those dinucleotides in which exciplex formation is precluded in rigid matrices (EGW at 80°K) because there is no ground-state stacking, there could be exciplex formation in a fluid solvent. The second kind of interaction leads to the familiar solvent reorientation red-shift and is said to arise from an exciplex as well if there is a particular solvent–solute complex formed with definitive stoichiometry and geometry.[79]

Large solvent reorientation shifts have been found for protonated purines in aqueous solution at room temperature (see Section 5.4). These spectral shifts are expected whenever solvent reorientation is rapid compared to the fluorescence lifetime. It is likely that solvent reorientation occurs for all the purines and pyrimidines in neutral aqueous solutions at room temperature, but, since they fluoresce so weakly under these conditions, there is meager experimental information.

No definitive information about the energies of the relaxed triplet states of the nucleotides in fluid solution is available, but there is no reason why the effects outlined above should not apply to triplet states. In fact, processes such as solvent reorientation are even more likely to occur for the longer-lived triplet states than for excited singlet states.

5.2. Nonradiative Rates in Aqueous Solution

Despite the handicap of not being able to use emission spectroscopy, a good deal has been learned about the lowest triplet and lowest excited singlet states of the nucleotides in fluid solution at room temperature by the use of two different approaches. The first involves the direct observation of molecules in their triplet states by means of their absorption using the flash photolysis technique. The second approach makes use of added compounds which act to quench excited states by collisional processes usually involving electronic energy transfer so that simple solution kinetics can be applied. In this approach, there must be some process of the donor or the acceptor (quencher) which can be easily monitored. Examples of this approach are the studies of the effects of various quenchers on the photoreactions of the pyrimidines (photohydration, photodimerization), which have been investigated in great detail (see Section 6).

A more generally applicable variation of the latter approach which has been developed by Eisinger and Lamola[80-82] makes use of the fact that the hexahydrated ion of europium (Eu^{3+}) can scavenge both the excited singlet states and triplet states of the nucleotides in water solution. Sub-

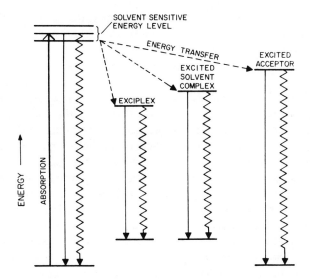

Figure 17. A schematic representation of the excited-state reaction which may occur. An excited molecule may form a complex with another solute molecule or a solvent molecule, or energy transfer to another solute molecule may take place.

sequently, the excited europium ion emits its characteristic line spectrum in the red. The simplest kinetic model based on collisional transfer seems to apply well. Using this kinetic model, Eisinger and Lamola interpret measurements of the intensity of the sensitized europium emission upon excitation of the donor at various europium ion concentrations. In this case, the quantum yield for producing molecules in the triplet state (φ_{isc}) was determined, and excited singlet-state and triplet-state lifetimes were estimated.

In its simplest form, the kinetic model used by Eisinger and Lamola[80,81] predicts that at a sufficiently high Eu^{3+} concentration molecules in excited singlet states are scavenged at a rate faster than the intersystem-crossing rate. The quantum yield of sensitized europium emission per photon absorbed by the donor, φ_f, is given by

$$\frac{\beta}{\varphi_f} = 1 + \left(\frac{1}{{}^1k_t\,{}^1\tau}\right)\frac{1}{[Eu^{3+}]} \qquad [4]$$

where β is the probability that an excited Eu^{3+} ion emits, ${}^1k_t$ is the bimolecular rate constant for singlet energy transfer from the donor to the europium, and ${}^1\tau$ is the lifetime of the donor singlet state. If, on the other hand, the Eu^{3+} concentration is sufficiently low and the intersystem-crossing yield is finite, triplet transfer is much more likely than singlet transfer, and

$$\frac{\beta}{\varphi_f} = \frac{1}{\varphi_{\text{isc}}}\left[1 + \left(\frac{1}{{}^3k\,{}^3\tau}\right)\frac{1}{[Eu^{3+}]}\right] \qquad [5]$$

where ^3k is the bimolecular rate constant for transfer of triplet excitation from the donor to the Eu^{3+}, and $^3\tau$ is the lifetime of the donor triplet state.

In developing this technique, it was convenient to use D_2O rather than H_2O as solvent because of the large increase in the Eu^{3+} fluorescence yield which results from this.[83] This isotope effect is about 20, which is much larger than the usual solvent isotope effects. Its magnitude can probably be explained by the extremely strong ligand binding of the hydration shell of the Eu^{3+} ion. The pH was kept near 5 to avoid association of the Eu^{3+} with the nucleotides. Typical sensitization curves are shown in Figure 18. The linear portions at high and low Eu^{3+} concentrations correspond, respectively, to the singlet transfer and triplet transfer limits discussed above. Extrapolation of the slope for triplet transfer to infinite Eu^{3+} concentration gives φ_{isc} directly as the inverse of the intercept [Eq. 5]. The values obtained[83] in this way for the five common nucleotides and orotic acid are given in Table 5.

According to Eqs. [4] and [5], the slopes of the sensitization curves are determined by the product of the bimolecular rate constant for transfer and the lifetime of the donor state. Unfortunately, the transfer rate constants are not known very well and so only estimates of the donor excited-state lifetimes could be obtained. Lower limits for the lifetimes can be determined by simply taking the largest possible value for the transfer rate constant, which is the diffusion-controlled value, about $5 \times 10^9\ \text{M}^{-1}\ \text{sec}^{-1}$ for aqueous

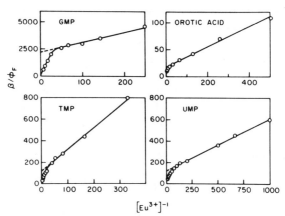

Figure 18. Sensitization of Eu^{3+} fluorescence by GMP, TMP, UMP, and orotic acid. The figures show experimental values obtained in D_2O at room temperature (see Section 5.2).

solution at room temperature. The rate constant for the quenching of tryptophan fluorescence by Eu^{3+} has been found to be exactly equal to this value. The transfer of triplet excitation from suitable donors to Eu^{3+}, on the other hand, occurs at rates much slower than the diffusion-controlled rate. In order to estimate the transfer rates, Eisinger and Lamola[80,81] performed competitive scavenging studies making use of the water-soluble diene (2,4-hexadienyl-)trimethylammonium chloride as a competitive triplet quencher since it could be reasonably assumed that the diene quenches the donor triplet at the diffusion-controlled rate. The estimated lifetimes of the excited singlet and triplet states of the nucleotides in D_2O solution saturated with air obtained by the Eu^{3+} method are given in Table 5.[82]

5.3. Triplet-State Molecules in Aqueous Solution

At sufficiently low monomer concentrations ($<10^{-3}$ M), photodimers of thymine, uracil, and orotic acid can be formed only as the result of the interaction of a ground-state monomer with a triplet-state molecule. This is so because collision between a ground-state molecule and an excited singlet-state pyrimidine is precluded because of the very short lifetime of the latter. Furthermore, the excited singlet states of the pyrimidines are too short-lived to be quenched by the oxygen in oxygen-saturated water solution. Thus Johns and his group were able to make use of the simple scheme illustrated in Figure 19 to interpret their elegant kinetic studies of the photodimerization of thymine,[84] uracil,[85] and orotic acid[86] in aqueous solution as a function of the pyrimidine concentration and the concentration of

Table 5. Room-Temperature Excited-State Parameters for Common Nucleotides and Orotic Acid in Water Saturated with Air, Obtained by the Europium Ion Method[a]

Donor[b,c]	φ_{isc}	$^1\tau(sec)^d$	$^3\tau(sec)^e$
TMP	8.0×10^{-3}	1.8×10^{-11}	5×10^{-7f}
UMP	7.3×10^{-3}	2.3×10^{-11}	1×10^{-6f}
CMP	1.5×10^{-3}	3.6×10^{-12}	2×10^{-6}
AMP	3.7×10^{-4}	1.3×10^{-12}	8×10^{-6}
GMP	4.6×10^{-4}	4.4×10^{-12}	5×10^{-6}
Orotic acid	6.4×10^{-2}	5.6×10^{-11}	7×10^{-8} (3×10^{-7})^g

[a] See Section 5.3.
[b] Donor concentration 0.5 g/liter as sodium salt in D_2O saturated with air; pD 4.5–5.5
[c] $\lambda_{ex} = 265$ nm.
[d] Using $^1k_t = 5 \times 10^9$ M^{-1} sec^{-1}.
[e] Using $^3k_t = 5 \times 10^7$ M^{-1} sec^{-1}.
[f] By competitive quenching method using (2,4-hexadienyl-)trimethylammonium chloride.
[g] Calculated for these conditions from data of Whillans and Johns[86] (see Table 6).

oxygen. The model leads to the following expression for the quantum yield of photodimer production ($\varphi_{\hat{pp}}$) at a monomer concentration [P]:

$$\varphi_{\hat{pp}} = \left(\frac{k_1}{k_1 + k_1'}\right) \varphi_{\text{isc}} \left[\frac{(k_1 + k_1')[P]}{(k_1 + k_1')[P] + k_2[O_2] + k_3}\right] \quad [6]$$

in which k_1 is the rate constant for dimer formation, k_1' is the rate constant for triplet self-quenching, k_2 is the rate constant for oxygen quenching, and k_3 is the rate at which the triplet decays in the absence of dimer production and oxygen quenching. It is clear from the rewritten form of the equation

$$\frac{1}{\varphi_{\hat{pp}}} = \left(\frac{k_1 + k_1'}{k_1}\right)\frac{1}{\varphi_{\text{isc}}}\left[1 + \frac{k_2[O_2]}{(k_1 + k_1')[P]} + \frac{k_3}{(k_1 + k_1')[P]}\right] \quad [7]$$

that measurements of $\varphi_{\hat{pp}}$ as a function of $[O_2]$ and $[P]$ yield

$$\left(\frac{k_1}{k_1 + k_1'}\right)\varphi_{\text{isc}}, \quad \left(\frac{k_2}{k_1 + k_1'}\right), \quad \text{and} \quad \left(\frac{k_3}{k_1 + k_1'}\right)$$

Johns, Hunt, and coworkers[86,87] have made successful use of flash photolysis to determine the kinetics of decay of the triplet states of the pyrimidines in solution. This was no easy undertaking because the triplet-state lifetimes are only barely resolvable by the conventional flash photolysis technique. As a result of these studies together with the photodimerization kinetics, the Canadian group was able to determine (k_1+k_1'), k_2, and k_3 for uracil and orotic acid in water solution. These are given in Table 6 along with their values for $[k_1/(k_1+k_1')]\varphi_{\text{isc}}$.

An important observation first made by Brown and Johns[85] is that the

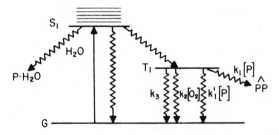

Figure 19. A model for the kinetics of excited pyrimidine molecules in aqueous solution. G, S_1, and T_1 are the ground-state and lowest excited singlet and triplet states, respectively. The meanings of the rate constants (k) are given in section 5, 3. The model was used by Johns et al.[84-87]

photodimer yield, but not the yield of photohydrate, in dilute solutions of uracil depends on the wavelength of the exciting light. A greater yield of dimers is obtained at the shorter wavelengths. As expected from the discussion above, the effect was attributed to a wavelength-dependent intersystem crossing, more triplets being formed at the shorter wavelengths. Similar observations were made for orotic acid[86] and thymine.[84] The data are shown in Figure 20 and 21. Later, Eisinger and Lamola found similar wavelength effects on the yield of sensitized Eu^{3+} emission for orotic acid[80] and TMP[82] as donors and in the range of Eu^{3+} concentration in which only the donor triplets are scavenged (Figures 20 and 21). These results confirmed that it is φ_{isc} which is wavelength dependent.

Two explanations have been suggested for this effect. The first involves the tautomeric forms of the bases. Suppose that a small fraction of the pyrimidine molecules exists as the enol tautomer, which absorbs at lower wavelengths than does the keto tautomer and which has a very high intersystem-crossing yield relative to the keto form. Then excitation at shorter wavelengths, at which the enol tautomer would absorb some of the light, would lead to an increase in the number of triplet molecules formed. This explanation has been ruled out in the case of orotic acid by Lamola and Eisinger,[82] who showed that 1,3-dimethylorotic acid and 1,3-dimethylthymine, which can exist only in the keto form, exhibits a similar wavelength-dependent intersystem-crossing yield.

The other rationalization is that intersystem crossing can occur from levels lying higher than the vibrationless level of the lowest excited singlet state so that exciting these levels by using shorter wavelengths leads to an

Table 6. **Kinetic Parameters for the Triplet States of Some Pyrimidines in Water at Room Temperature Obtained by Johns et al.**[84-87]a

	From dimerization rate			From flash photolysis		
	$\dfrac{k_1}{k_1 + k_1'} \varphi_{\text{isc}}$	$\dfrac{k_2}{k_1 + k_1'}$	$\dfrac{k_3}{k_1 + k_1'}$ (M)	k_3 (sec^{-1})	$k_1 + k_1'$ (M^{-1}sec^{-1})	k_2 (M^{-1}sec^{-1})
Thymine	4.7×10^{-4}	1.1	0.5×10^{-5}	0.8×10^4	2.3×10^9	3.4×10^9
TMP	5.6×10^{-4}	—	1×10^{-5}	—	—	—
Uracil	5×10^{-3}	1.1	5.9×10^{-5}	1.6×10^5	2.9×10^9	3.9×10^9
Orotic acid pH 1	0.12	1.1	0.5×10^{-5}	9.1×10^3	1.8×10^9	2.0×10^9
Orotic acid $pH > 3$	0.06	1.2	2×10^{-5}	1.25×10^4	6.8×10^8	8.1×10^8

a The rate constants are described in section 5.3.

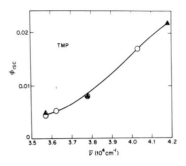

Figure 20. The intersystem-crossing yield of TMP and thymine in aqueous solution at room temperature, near-neutral pH versus the excitation wavenumber. The circles are values of φ_{isc} for TMP obtained by the Eu^{3+} method, while the triangles refer to relative dimerization yields of thymine, normalized to φ_{isc} of TMP at 265 nm.

increased intersystem crossing. In order for intersystem crossing to take place from these upper levels, it has to occur faster than vibrational relaxation if these upper levels are vibrational levels of the lowest excited singlet state, or faster than internal conversion if the upper levels belong to the second excited singlet state. The lifetimes of these upper levels of either kind are probably of the order of 10^{-12} sec. This is not very different from the lifetime of the lowest singlet state. Thus it is not unlikely that inter-

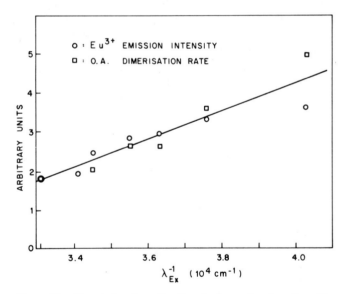

Figure 21. The circles show the dimerization rate of orotic acid in water at room temperature as a function of the wavenumber of the exciting light. This rate is seen to have virtually the same excitation energy dependence as the sensitized Eu^{3+} emission intensity (shown by squares) when Eu^{3+} is added to the solution.

system crossing could occur from these upper levels if it can occur at all. Since the intersystem-crossing efficiency in these compounds is small to begin with ($\sim$0.01), it is not surprising that the increase due to contributions by crossings from the upper levels is observable. Johns et al.[84] have suggested that the upper levels must be vibrational levels rather than another electronic state because the wavelength effect begins practically at the red edge of the lowest absorption bands of the compounds.

The wavelength dependence of the room-temperature intersystem-crossing efficiency in the pyrimidines provides one with a means for distinguishing between precursor states for photoproducts in polynucleotides: If the yield of a particular photoproduct has a similar wavelength dependence (after allowing for possible photoreversal of the product), then a triplet-state precursor is strongly indicated (see below). In the absence of such a wavelength effect, the precursor state is likely to be an excited singlet state.

For the purposes of comparison, selected values of the lifetimes of the excited singlet states and the intersystem-crossing yields at 80°K in EGW and in water at room temperature (RT) for the five common nucleotides were extracted from the available data and are given in Table 7. The ratios $^1k_{nr}$ (RT)/$^1k_{nr}$ (80°K) and k_{isc} (RT)/k_{isc} (80°K) calculated from these data are also listed. This analysis reveals that the ratio of internal conversion rates increases by one to three orders of magnitude on going from 80°K in EGW to room temperature in water. The intersystem-crossing rate appears to be almost temperature independent for GMP but increases by three orders of magnitude between 80°K and 300°K in TMP.

Thus the lack of fluorescence from the nucleotides at neutral pH at room temperature is apparently due to successful competition by internal

Table 7. Comparison of the Room-Temperature and Low-Temperature Values of Some Excited-State Parameters for the Common Mononucleotides

	φ_{isc}(80°K)	$^1\tau$(80°K) (nsec)	φ_{isc}(RT)	$^1\tau$(RT)^a (nsec)	$\dfrac{k_{isc}(RT)}{k_{isc}(80°K)}$	$\dfrac{^1k_{nr}(RT)}{^1k_{nr}(80°K)}$
TMP	$<3 \times 10^{-3}$	3	8×10^{-3}	0.015	$\geq$530	300
UMP	$<3 \times 10^{-3}$	0.2^b	7×10^{-3}	0.015	$\geq$30	13
CMP	$\sim 3 \times 10^{-2}$	0.3^b	1.5×10^{-3}	0.003	$\sim$5	100
AMP	2×10^{-2}	3	4×10^{-4}	$>0.002^c$	$<$30	$<$1500
GMP	1.5×10^{-1}	$\sim$5	4.6×10^{-4}	0.011	1.3	450

^{a} Average of the lower-limit value obtained by the Eu^{3+} method and the upper-limit value obtained from the limit of detectability of fluorescence. The two limits differed by less than a factor of 3 for all nucleotides except AMP.
^{b} Calculated value.
^{c} The value obtained by the Eu^{3+} is used since it differs by almost a factor of 100 from the estimated upper limit based on the lack of observable fluorescence.

conversion processes whose fast rates are temperature dependent and extremely fast at room temperature. For two of the compounds, TMP and UMP, the rate of intersystem crossing must increase dramatically in water at room temperature compared to EGW at 80°K, since higher yields of triplets are observed at room temperature despite the shorter excited singlet state lifetimes at the higher temperature.

It is important to point out that these increases in the rates of the radiationless decay processes may not be directly related to the temperature increase since the viscosity of the solvent has changed enormously between the temperature extremes. It is unfortunate that no data have been reported for the nucleotides in rigid matrices at room temperature. This is due to the difficulty of finding a suitable matrix.

5.4. Temperature Dependence of Fluorescence

Information which bears on the extremely fast internal conversion processes in the nucleotides can be found in the temperature dependences of their fluorescent emissions. The results for TMP in EGW are shown in Figure 22.[27] They can be analyzed in terms of a temperature-independent fluorescence yield, ^0I_F, which is approached at about 80°K, and a single temperature-dependent quenching process with a rate constant, $k_Q(T)$, which may be written

$$k_Q(T) = C[I_F^{-1}(T) - {}^0I_F^{-1}]$$
$$= C' \exp\left(-\frac{E_Q}{kT}\right)$$

where $I_F(T)$ is the fluorescence intensity at temperature T, E_Q is the "activation energy" for the quenching process, k is the Boltzman constant, and C and C' are constants. E_Q is then obtained from the slope of the curve obtained by plotting log k_Q versus $(1/T)$. A value of 3 kcal/mole is found for TMP in EGW. The wavelength of the maximum of the fluorescence spectrum is constant (321 nm) below 150°K, but shifts continuously to the red above that temperature until it reaches 326 nm at 250°K. Above this temperature the fluorescence is too weak to observe.

Radiationless decay could be facilitated in hydroxylic solvents because the nucleotide solute would be tightly coupled to the high-frequency O–H vibrations in the solvent because of the strong solvation and perhaps because of hydrogen bonding. Such a mechanism has been invoked for the case of fluorescence quenching of tryptophan in water. However, the results of two experiments indicate that such considerations are not important for thymine. Figure 22 shows the temperature dependence of TMP in deuterated EGW.[82]

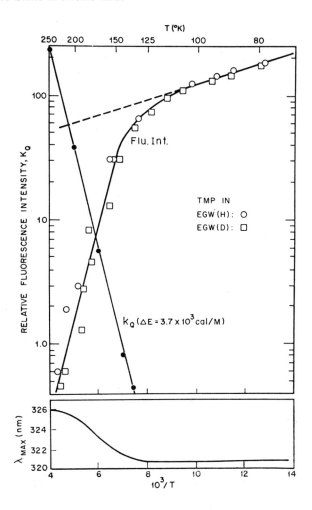

Figure 22. The temperature dependence of the fluorescence intensity and nonradiative quenching rate (k_Q) of TMP in EGW and in deuterated EGW. The lower part of the figure shows the maximum wavelength of the fluorescence as a function of the inverse absolute temperature.

The lack of an isotope effect argues against O–H vibrations or hydrogen bonding as being important. Figure 23 shows similar data[82] for 1,3-dimethylthymine in thoroughly dried 2-methyltetrahydrofuran. In this case, no hydrogen bonding with the solvent is possible. Nevertheless, the fluorescence quenching upon raising the temperature is more severe than for

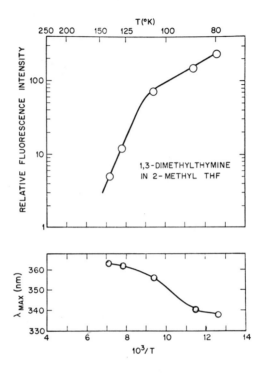

Figure 23. The temperature dependence of the fluorescence intensity and wavelength of the fluorescence maximum of dimethylthymine in 2-methyltetrahydrofuran.

TMP in EGW, and the red shift of the wavelength of maximum fluorescence (338 nm to more than 362 nm) is greater and occurs at a lower temperature.[82]

An even greater red shift (50 nm) is found for protonated adenine (AH^+) in EGW at *p*H 2 and is shown in Figure 24.[27,31,32,78,79] Equally large effects have been reported for other purines and their derivatives.[91]

It is important to note that the spectral shift for AH^+ (see Figure 24) as well as the other compounds mentioned is continuous with increasing temperature and does not appear to arise from a change in the relative populations of two different states since no isostylbic point is observed. The shifts seem to appear only above the softening point of the solvent. These observations are consistent with the mechanism suggested by Hercules and Rogers,[92] in which the shift arises from a lowering of the energy of the ground state as a result of the reorientation of surrounding solvent molecules,

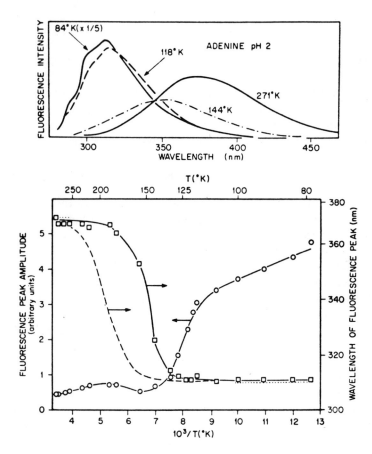

Figure 24. The upper part of the figure shows the fluorescence spectra of protonated adenine in EGW at various temperatures. The lower part of the figure shows the fluorescence yield and wavelength of maximum emission as a function of the inverse absolute temperature.

while the formation of a 1:1 exciplex with a solvent molecule would be accompanied by the occurrence of an isostylbic point. Such a mechanism is expected to be viscosity dependent, with the amount of shift depending on the extent of solvent reorientation achieved during the lifetime of the excited state. For example, in the case of AH^+ (Figure 24), if glycerin is used as the solvent instead of EGW, the red shift occurs at higher temperatures but is just as large. The same shift is obtained in water, but the liquid-water fluorescence disappears discontinuously as soon as the water freezes and is not observed again until much lower temperatures are reached.

While one expects a relationship between the quenching and spectral shifts, no consistent picture emerges. It is a striking feature of the results for AH$^+$ (Figure 24) that, as the sample warms up, quenching ceases as soon as the wavelength of the fluorescence maximum starts to shift. However, for TMP in EGW[27] as well as for tryptophan[93] the major temperature-dependent quenching process does not begin until the temperature at which solvent reorientation, as manifested by the spectral shift, has been exceeded. The results for DMT[82] in EGW fall somewhere in between.

5.5. Speculations about Fluorescence Quenching and Temperature Effects

The excited singlet states of the nucleotides lie more than 4 ev above their respective ground states. Thus there must be important ways in which they differ from compounds like the aromatic hydrocarbons, for which internal conversions between states with such large electronic energy gaps are usually slow and for which the sum of quantum yield of fluorescence and intersystem crossing is nearly unity.

It has been mentioned above that radiationless deactivation of the excited singlet state of dimethylthymine in an organic solvent in which hydrogen bonding is precluded is just as rapid as the deactivation of TMP in EGW.[82] Similar results have been obtained for purine derivatives by Augenstein et al.[18-21] Furthermore, the intensity of fluorescence at low temperatures from purine derivatives is not altered significantly on using solvents of differing polarity, although the temperature dependence of the fluorescence yield does differ from solvent to solvent. Thus the deactivation processes which persist at low temperatures seem to be essentially intramolecular. However, the nature of the solvent is no doubt an important factor for the temperature-dependent contribution to the radiationless deactivation, since the latter must be sensitive to solvent reorientation effects which depend on the polarity and viscosity of the solvent.

In general, the rate of radiationless intercombination between two states appears to increase as the difference in geometries of the two states becomes larger (see Chapter 1). This is due in part to more favorable Franck–Condon factors and in some cases to an increase in the electronic interaction between the states via vibronic coupling.

One kind of geometry change involves twisting about an essential single bond in the excited state. Stilbene (1,2-diphenylethylene) provides a nice example of this.[94] The latter is highly fluorescent at low temperatures in rigid matrices but poorly fluorescent at room temperature in fluid solutions. The quantum yield of isomerization (twist about the ethylenic double bond) is affected in exactly the opposite way, and excellent correlations can be made between fluorescence quenching and isomerization. The barrier to isomerization (and fluorescence quenching) is primarily internal (thermal), but the viscosity of the matrix also plays a part. These considerations also apply to the comparison of the fluorescence efficiencies of thymine and its two anions (see Table 3).

The molecular orbital calculations of Snyder et al.[49] show that for both thymine (T) and its anion formed by deprotonation at N_3 (1-HT$^-$) the lowest-energy singlet–singlet transition is approximately an ethylenic π,π^*-type localized at the C_5–C_6 bond, which is a double bond in the ground state and becomes essentially a single bond in the excited state. On the other hand, the lowest-energy transition of the anion formed by removing the proton at N_3 (3-HT$^-$) is not of this type, and the C_5–C_6 bonding is not severely disturbed by the excitation of one electron. One expects 3-HT$^-$ to remain planar in the excited singlet state, but the T and 1-HT$^-$ could undergo twisting about the C_5–C_6 bond if associated barriers can be surmounted.

Besides rationalizing the greater fluorescence quenching in T and 1-HT$^-$ compared to 3-HT$^-$ (see Table 3), this predicted geometry change in the lowest excited singlet state of T (and presumably U) might have something to do with the temperature-dependent intersystem crossing observed in the neutral pyrimidines (see Section 5.3).

A more general prediction, which has been discussed by Hochstrasser,[95] among others, is that planar molecules containing close-lying (n,π^*) and (π,π^*) excited states will tend to be distorted out of plane in the excited state due to interactions between these states (see Chapter 1). This can facilitate radiationless decay and has been invoked to explain the lack of luminescence from pyridine. Purines and pyrimidines possess a large number of close-lying (n,π^*) and (π,π^*) states in the energy region of interest. Thus this common feature of the nucleotides may well be the reason for the characteristically fast radiationless decay in these molecules. It should also be pointed out that these (n,π^*), (π,π^*) interactions and thus the radiative decay rates are expected to be temperature sensitive. Solvent sensitivity is also expected because the (n,π^*), (π,π^*) energy gap is usually solvent dependent.

It should be pointed out that the Stokes' shift for the fluorescent emissions from the nucleotides at 77°K in EGW is not unusually large, so that under those conditions geometry changes in the excited state do not appear to be very large.

The fast formation of short-lived photoproducts (photoisomers) is a remotely possible explanation for the efficient nonradiative decay in the nucleotides. Many such "chemical" decay paths are known. For example, benzene undergoes photoisomerization to give "Dewar benzene"[96] and "benzvalene,"[97] both of which revert to benzene in the dark. No such photoisomers of the nucleotides have been detected, however.

It should emerge from the preceding discussion that the physical reasons for the fluorescence quenching of nucleic acids, which is particularly severe at room temperature, remain unknown. It has been suggested that the short excited-state lifetimes and the consequent insensitivity to photodamage of nucleotides played a role in their selection as the primary carriers of genetic information.

6. EXCITED-STATE PRECURSORS OF PHOTOPRODUCTS

It is well known that exposure to ultraviolet radiation produces various kinds of stable photolesions in DNA and other nucleic acids. The photochemical reactions responsible for these changes may be written as

$$A_1 + h\nu \rightarrow A_1^*, \qquad A_2 + A_1^* \rightarrow A_3$$

where A_1 and A_1^* are the absorbing chromophore in its ground state and excited state, respectively. A_2 is a ground-state species which interacts with A_1^* and may be another chromophore or a solvent molecule, and A_3 is the stable photoproduct. In the present section, we will explore the nature of A_1^*, particularly whether A_1^* is in an excited singlet or triplet state under various conditions in which the photoreaction occurs.

6.1. Photohydrates

It has been clearly shown that the photohydration reactions of uracil and cytosine

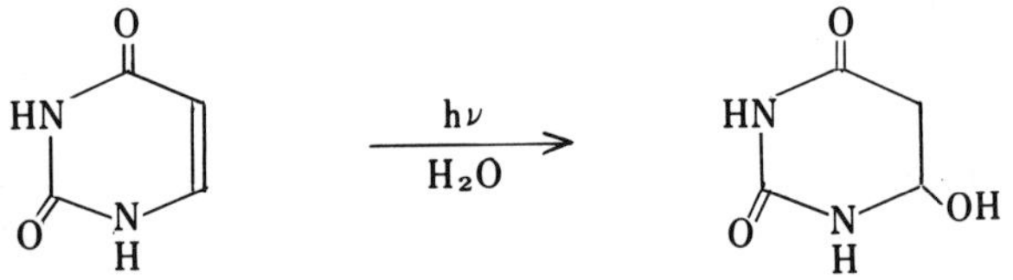

[8]

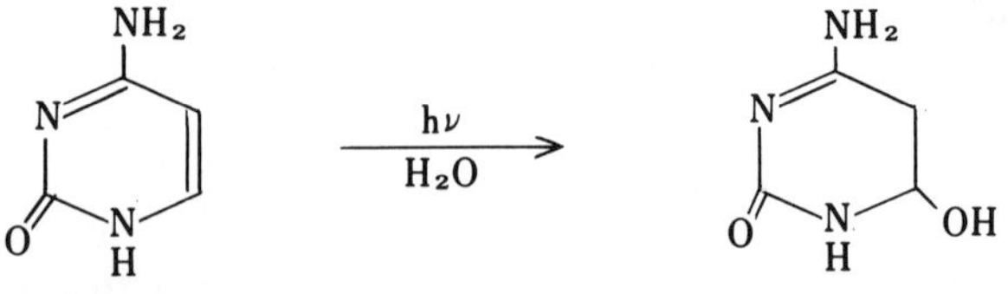

[9]

do not involve their triplet states. Two kinds of experiments have been performed in this regard. First, it was demonstrated by means of triplet sensitization that the triplet states of uracil and cytosine are inert toward the addition of water. That is, when the triplet states of uracil and cytosine as the free bases in water solution were populated by means of triplet energy transfer from sensitizer molecules having higher-lying triplet states, no photohydrates were produced, although dimers were formed.[98-101] The same results have been obtained for uracil and cytosine bases in polynucleotides.[100,101]

Triplet quenching experiments have led to the same conclusion in the case of uracil. The addition of triplet quenchers (2,4-hexadien-1-ol, or oxygen at concentrations too low to affect the excited singlet state) to water solutions of uracil does not affect the yield of photohydrate but does reduce the photodimer yield significantly.[102-104]

Burr et al.[104] have used oxygen quenching to completely suppress the production of uracil photodimers in studies on the photohydration reaction. The pH dependence of photohydration quantum yield has led to the suggestion that the primary process is the protonation of the excited singlet state, which has a pK near 4. The following scheme is a reasonable possibility:

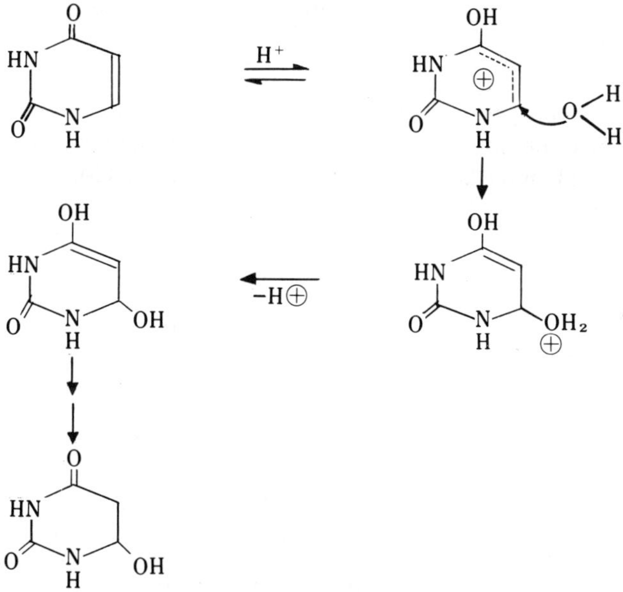

[10]

A photohydrate of thymine has not been observed. It is not clear whether the reason for this is that thymine does not form a photohydrate or that the photohydrate is very unstable.

6.2. The Cytosine–Thymine Adduct

An adduct derived from cytosine and thymine which has been isolated from irradiated DNA has been shown by Wang and Varghese[105,106] to have the following structure:

It is reasonable to assume that the compound isolated is the hydrolysis product of a photoadduct having the following structure:

An intriguing proposal for the mechanism of the formation of such an adduct involves phototautomerism in the G-C base pair, which has also been invoked to explain the lack of fluorescence from excited G-C base pairs:

Tautomeric base pairs

The imino double bond of the excited tautomeric form of cytosine could then add to the 5,6 double bond of a neighboring thymine, a reaction type which has precedent:

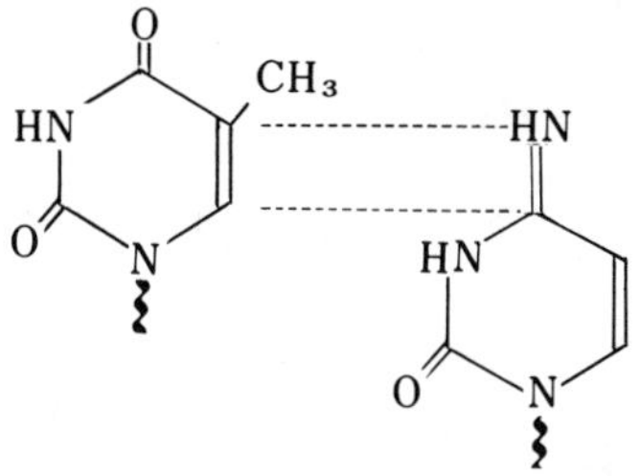

To the knowledge of these authors, the test of this mechanism involving the search for this product in single-stranded DNA has not been performed. The only fact about mechanism which is known so far is that the formation of the Wang adduct does not involve the triplet state of thymine, since this product is not formed in the acetophenone-sensitized photolysis of DNA.

6.3. Photodimers of Pyrimidines

Thymine and many of its derivatives undergo photodimerization under a variety of conditions.* In this section, the dimerization mode of concern is the cycloaddition at the 5,6 double bond to give

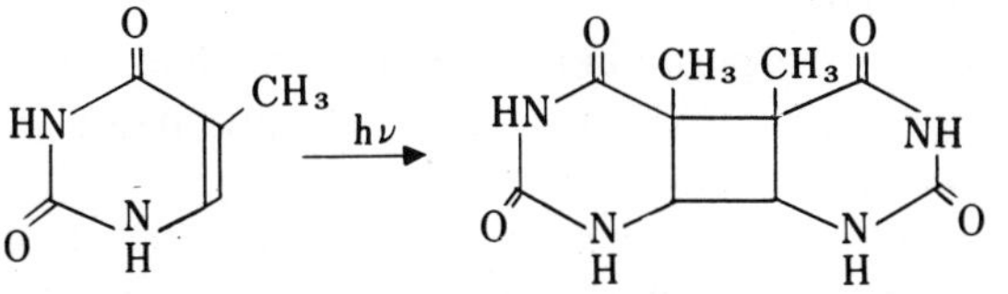

[11]

which is a cyclobutane derivative; this will be called *photodimerization*. Dimerization by means of addition at a carbonyl group or imino group will be called *photoaddition*.

The four possible *cis*-fused dimer configurations, two of which are enantiomers, are shown in Figure 25. All of the six isomers have been isolated from various photolysis mixtures.

There are many examples of this kind of photocycloaddition in the photochemical literature. For example, many simple olefins and cyclic enones form dimers and heterodimers in this manner. In all of these cases, including the photodimerization of thymine, there is no evidence for complicated multistep mechanisms involving free-radical or charged intermediates. On the contrary, in many cases the formation of the two new

---

*A recent review has been prepared by Wang.[107]

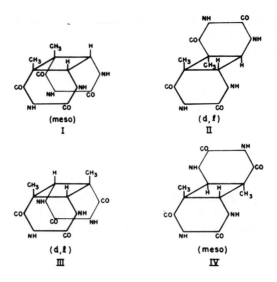

Figure 25. The molecular structures of the four *cis*-fused cyclobutane dimers of thymine.

carbon–carbon bonds appears to be concerted. Sequential formation of the two new bonds occurs in other cases, especially when the triplet state is the excited-state precursor.

With respect to the photodimerization of thymine and its derivatives, questions may be raised: Which excited-state precursor(s) are involved? What governs which isomers are formed?

It is convenient to discuss these questions with respect to five situations: (1) dilute solutions of thymine, (2) concentrated solutions, (3) aligned monomer pairs, (4) TpT, and (5) DNA.

It was pointed out in section 5.3 that the intermolecular photodimerization of thymine and its derivatives in sufficiently dilute solutions ($<10^{-2}$ M, depending on the solvent) where no ground-state association occurs demands the triplet state ($^3\tau \approx 10^{-6}$ sec) as an intermediate because the excited singlet state does not live long enough ($^1\tau \leq 10^{-11}$ sec) to take part in a bimolecular reaction. Triplet quenching experiments have shown that this is indeed the case.[102,103]

In general, all four configurational isomers are formed in solution. The ratio of isomers formed depends on the particular derivative and upon the solvent used. No definitive explanation of the observed product ratios exists. One complicating factor is that not all reactive encounters between a triplet thymine molecule and a ground-state molecule lead to a dimer. On the contrary, Wagner and Bucheck[108] have shown that self-quenching is the main result of such interactions for thymine in acetonitrile. The discrepancy between the infinite-concentration dimer yield obtained by Johns' group (see Section 5.3) for thymine in water ($\varphi_{\widehat{TT}} \sim 0.001$) and the triplet yield for TMP (see Section 5.3) ($\varphi_{isc} \sim 0.01$) found by Lamola and Eisinger is probably due to triplet self-quenching.

As the concentration of the thymine derivative is increased, a dramatic increase in the specific rate of dimer formation occurs when ground-state association (stacking) begins. A good deal of experimental data has been obtained for DMT. Lisewski and Wierzchowski[109] have shown that in the high-concentration range, triplet quenchers are ineffective at quenching the dimerization of DMT in water. They measured the concentration and temperature dependence of the quantum yield for dimer formation and analyzed the results in terms of the ground-state association model and found the association constant $K_a^{26°C} = 0.62$ lm^{-1}, in excellent agreement with values obtained by osmometry for pyrimidine bases (0.7–0.9 lm^{-1}). Extrapolation of their yield data to infinite DMT concentration gives $\varphi_{\widehat{TT}}$ [DMT]$_\infty$ = 0.125, in good agreement with the dimer yield in DMT crystals, $\varphi_{\widehat{TT}} = 0.165$. Smaller yields are found for concentrated solutions of DMT in organic solvents.[110] In this case, the dimerization can be partially quenched by triplet quenchers. This is consistent with the fact that organic solvents disrupt the association between the pyrimidines. The ratio of isomers in the quenched portion is the same as that obtained when the dimerization is sensitized by triplet energy donors. The unquenched portion has a different product ratio. These results indicate that different paths exist for dimer formation in aggregates and in dilute solution.

The situation which prevails in aggregated thymines can be understood from the results of studies on photodimerization of oriented monomers.

Wang[111] first suggested that the reason for the high efficiency ($\varphi_{\widehat{TT}} \sim 1$) of photodimerization of thymine in ice, where only the *chh* dimer is formed, is that, in freezing, microcrystals of thymine hydrate are formed in which neighboring thymines are parallel and suitably placed for dimerization. Such a crystal structure was found by Gerdil.[112] Eisinger and Shulman[113] examined frozen-water solutions of thymine at 80°K excited with 280 nm light and found that they do not emit. This is in contrast to thymine in EGW, where it is dispersed and a fluorescence yield of 0.2 is observed. On the other hand, thymidine and TMP fluoresce normally in frozen-water solutions, but neither of these compounds undergoes photodimerization very efficiently under these conditions. This striking correlation between the fluorescence quenching and efficient dimerization means, of course, that a process connected to dimer formation quenches the singlet state with high efficiency.

A very interesting aspect of the thymine-in-ice system is that, if irradiation at 280 nm is continued, monomer fluorescence does appear and grows with time. If the sample is subsequently irradiated with 239 nm light, the monomer fluorescence eventually disappears again. What is evidently happening is that, as dimerization proceeds in the crystal, monomers eventually become isolated in beds of dimers. Since these monomers cannot dimerize, they emit normal fluorescence. The short-wavelength light (236 nm) splits the dimers.

Eisinger and Lamola[114,115] were able to obtain a clear picture in their studies of oriented monomer pairs formed by breaking dimers in a rigid matrix. They dissolved pure dimers of ($\sim 10^{-3}$ M) DMT in EGW and cooled the sample to 80°K, where a clear glass is formed. Irradiation with 248 nm light was used to break some of the dimers into pairs of monomers, which remain suitably positioned to reform the dimers because of the rigidity of the matrix. The absorption spectra of "broken dimers", or monomer pairs, exhibit exciton splitting (Figure 26), and a distance between the transition dipoles in the broken *chh* dimer was calculated from the simplest exciton model to be 2.8 Å. The broken dimers do not fluoresce but redimerize with a quantum yield of 1.0 ± 0.1 (*chh* dimer). It is interesting that the quantum yield for breaking the dimers was also measured to be 1.0, showing that the paths for making and breaking the dimers do not cross.

The simplest interpretation of the results for thymine crystals and broken dimers is that the fluorescence quenching and efficient dimerization are directly coupled. That is, the excited singlet goes on to form dimer faster than it does anything else. One can imagine this occurring in two steps by

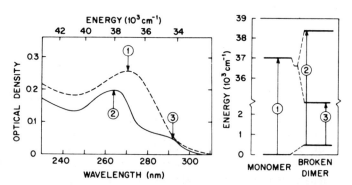

Figure 26. Left: The absorption spectrum in EGW at 80°K of dimethylthymine (DMT), shown as a dashed curve. The solid curve is the absorption spectrum of the "broken" *cis* head–to–head DMT dimer under the same conditions. The peaks labeled 2 and 3 are the exciton split peaks. Right: The energy levels of the monomer and the broken dimer consistent with these spectra. The energy scale refers to the absorption peaks, not to the 0–0 levels.

way of an excimer intermediate. Eisinger and Lamola constructed the model shown in Figure 27 to describe such a mechanism.

The alternative pathway would involve totally efficient intersystem crossing from the excited monomer pair followed by totally efficient dimerization from the triplet state. This would demand that the rate of intersystem crossing in the monomer pair be at least 10^4 times faster than it is in the isolated monomer.

It is important to note that, whether or not a triplet-state intermediate is involved, a strong interaction between the monomers in crystals, "broken dimers," or other aggregated forms of thymine is present at the excited singlet level. This interaction eventually leads to dimer formation either directly from a singlet state or by way of a triplet state which would not have been formed in the absence of the interaction at the singlet level. Interaction at the singlet level is precluded if diffusion or conformational changes requiring more than 10^{-10} sec are necessary to bring in the second monomer.

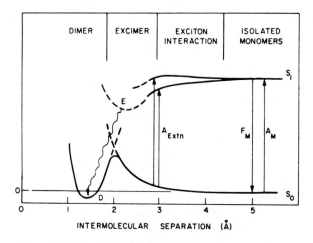

Figure 27. Potential energy profiles showing the interaction between two thymine molecules as a function of their separation. A photodimerization mechanism which proceeds by way of the excimer state is indicated by the arrow from E to D. The range of intermolecular separation is conveniently described by four regions—separated monomers, excitonic region, excimer region, and dimer. According to the observed exciton splitting in the absorption spectrum of the broken *chh* dimer, separation of the two chromophores is approximately 2.8 Å. Excitation (A_{Extn}) of the lower exciton state leads to the dimer (D) with unity quantum yield. The lower exciton state should be smoothly connected with the excimer state (E) so that the latter is most likely an intermediate in photodimerization.

The photodimerization of pyrimidines in nucleic acids can occur by way of the triplet state since the reaction can be sensitized with triplet donors. However, this tells us nothing of what the mechanism is when the light is absorbed by the nucleic acid.

Photodimerization of neighboring pyrimidines in dinucleotides (TpT, UpU), oligonucleotides, polynucleotides, and DNA occurs with a quantum yield on the order of 0.01.[85,116–118] Although several stereoisomers are produced in dinucleotides,[117,118] only *chh* dimers are produced in native DNA[119] or in well-stacked homopolynucleotides such as poly U.[118] Before discussing the experimental data concerning the excited-precursor question, it is interesting to consider what can be said about dimerization in the dinucleotides and polynucleotides from what has been presented above for the mononucleotides, free bases, and aggregated systems and "broken dimers." One quickly realizes that no strong conclusions can be drawn because none of these systems are proper models for two neighboring pyrimidines in a stacked polynucleotide or in DNA.

The Watson–Crick DNA structure provides for a 36° angle between the neighboring thymines, whose parallel molecular planes are spaced 3.4 Å apart. Thus neighboring thymines in TpT or in DNA are neither like TMP in dilute aqueous solution nor like the thymines in a broken dimer and in the thymine hydrate crystal, where they are very nicely juxtaposed for efficient dimer formation.

The quantum yield for dimers in TpT or DNA of about 0.01 is some 20 times larger than the limiting quantum yield in dilute thymine solutions and 100 times smaller than the yield in thymine crystals or "broken dimers." Photodimers are formed in concentrated aqueous solutions of thymine, where aggregates are present with a yield of 0.04.[84] The ratio of stereo-isomers formed is very similar to the ratio formed in TpT, so that concentrated solutions of thymine may be a suitable model for TpT. However, this provides no help in deciding whether the photodimerization is a singlet-state reaction controlled by the distribution of possible juxtapositions of neighboring thymine molecules, or whether the presence of a neighboring thymine induces intersystem crossing followed by dimerization from the triplet state.

Rahn[120] has measured the rate and extent of dimer formation in poly U in a frozen glass and observed that the number of dimers saturates with increasing dosage but that additional dimers can be formed if the glass is allowed to melt and is then refrozen. This observation is consistent with a "geometry-limited" dimerization mechanism. The polynucleotide in the frozen glass preserves its geometry, in which only a certain fraction of the pyrimidine pairs are in a suitable juxtaposition for dimerization. The same view has been taken of thymine dimerization in DNA and is supported by the observation that the photodimer yield is virtually constant below the DNA melting point but drops considerably at temperatures above it.[121]

Experiments with added quenchers have provided additional insight into the problem. The addition of specific triplet quenchers such as dienes and paramagnetic metal ions does not affect the rate of photodimer production in DNA[122,123] or in TpT.[87,124] This is consistent with a model in which dimerization is a singlet-state process. However, it may simply reflect a dimerization via the triplet state which occurs at a rate much faster than the quenching rate.

The Sutherlands[123] have shown that acridine orange, methyl green, ethidium bromide, and chloroquine reduce the dimer yield in DNA. They also observed DNA-sensitized dye fluorescence and have presented evidence that the sensitized dye fluorescence and dimer inhibition are due to Förster-type singlet–singlet transfer from the DNA to the dye. More recently, these investigators measured the dependences of the efficiencies of dimerization inhibition and dye fluorescence upon the wavelength of the exciting light.[125] They found that the efficiency of dimer inhibition by ethidium bromide is greater at shorter wavelengths of the exciting light. On the other hand, the efficiency of the sensitized dye fluorescence is wavelength independent. The data suggest that, at excitation energies higher than that necessary to excite the lowest vibrational levels of the DNA singlet state, singlet–singlet transfer cannot account for all the dimer quenching. Intersystem crossing is the only documented wavelength-dependent process in DNA constituents, and so the Sutherlands suggest that the model proposed by Brown and Johns[85] for poly U is applicable to DNA. In this model, dimers are formed directly from the excited singlet state, as well as via the triplet state, with triplets arising from intersystem crossing from higher vibrational levels of the excited singlet state.

6.4. Sensitized Pyrimidine Dimers in Polynucleotides

The energy levels for the lowest-lying excited singlet and triplet levels of the common nucleotides are shown schematically in Figure 28 together with those of acetone and acetophenone. On the basis of the order of these levels, it is possible to excite triplet states in DNA by means of triplet excitation transfer from acetone or acetophenone, which can be excited with light not absorbed by the DNA, e.g., 313 nm. The triplet of acetone can transfer to all the bases, while that of acetophenone can transfer only to thymine. Both acetone and acetophenone have high intersystem-crossing efficiencies ($\varphi_{\text{isc}} \sim 1.0$), but unfortunately the triplets of these molecules are short-lived ($\tau \leq 10^{-6}$) in water solution (even in the absence of oxygen), so that the efficiencies of triplet transfer to the bases in DNA are expected to be

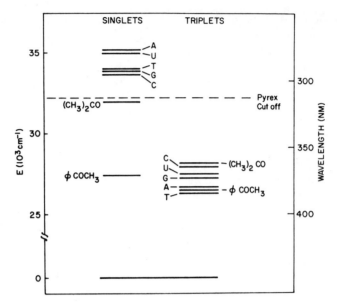

Figure 28. The energies of the lowest-lying singlet and triplet levels of the common nucleotides as well as the triplet sensitizers acetone and acetophenone. These values correspond to the thresholds of the low-temperature emission spectra.

low. One can nevertheless observe photochemical reactions in DNA which are sensitized by acetophenone[126] and acetone.[127] These occur at convenient rates and, what is more important, give pyrimidine dimers as the only major products. In contrast to this, the direct irradiation of DNA (254 nm) leads to a variety of products in comparable yields.

Since the acetophenone triplet can transfer only to thymine, it is expected that only products involving thymine or the sensitizer would be found. The most complete product analysis has been performed on *Escherichia coli* DNA irradiated at $\lambda \geq 313$ nm in water solution (0.1 M phosphate buffer) in the presence of acetophenone (10^{-2} M).[128] The results are presented in Table 8 and show a striking reduction in photoproducts containing cytosine compared to direct irradiation. Some results for acetone sensitization are also presented in the table and show an increase in cytosine dimers relative to acetophenone sensitization.

Meistrich *et al.*[129] have used two different derivatives of acetophenone to introduce thymine dimers into T4 phage. Using this technique, Meistrich[130,131] studied the inactivation of the phage due to the presence of TT, and Meistrich and Shulman[132] investigated the mutagenic effect of TT. Mennigmann and Wacker[133] have used acetone to sensitize the in-

Table 8. Comparision of the Initial Yields of Photoproducts Detected in *E. coli* DNA as a Result of Direct Irradiation at 254 nm and Sensitization by Acetophenone (10^{-2} M) and Acetone (1 M) at Wavelengths Greater Than 313 nm[a]

	Direct	Sensitized	
		Acetophenone	Acetone
TT̂	(1.0)	(1.0)	(1.0)
CT̂	0.8	0.03	0.14
CĈ	0.2	<0.0025	~0.01
5,6-Dihydrothymine	0.1	0.02	
Cytosine photohydrate	0.3	<0.003	
6,4'-[Pyrimidine-2'-one]-thymine	0.1	<0.0025	<0.002
Sensitizer addition		<0.02	

[a] The rate of production of TT̂ under direct irradiation is taken as unity.

activation of an *E. coli* mutant, and Chambers *et al.*[101] have used acetone sensitization to simplify the photochemistry of transfer RNA.

7. ENERGY TRANSFER IN POLYNUCLEOTIDES

7.1. General Considerations

In the preceding sections, we have described the singlet and triplet states of polynucleotides which are observed in emission at low temperature, but have said little about the mechanisms of energy transfer which may intervene between the absorption of the exciting photon and the creation of the excited states from which emission originates or the sites of photochemical reaction.

In the present section, we will first offer some general considerations about the nature and extent of singlet and triplet excitation energy transfer to be expected among the chromophores of a polymer. We will then offer brief theoretical discussions of the most important transfer mechanisms with estimates of the extent of transfer to be expected in polynucleotides and will compare these estimates with experimental results. Since it is difficult to obtain experimental evidence for transfer in these systems at room temperature, many of these findings apply to liquid-nitrogen temperature only. A basic understanding of the transfer mechanisms often makes it possible to at least place reasonable limits on the energy migration to be expected under physiological conditions in any particular system.

The two most important interchromophore interactions are the coulombic and exchange interactions. The coulombic interaction is dominated by the electric dipolar term and therefore has the form $(\mu_D \mu_A)/r^3$, where μ_A and μ_D are the transition dipole moments of the virtual transitions in the acceptor and donor which occur during transfer, and r is the separation between donor and acceptor. It will be seen later that the corresponding range of the transfer is proportional to the square of this interaction and will therefore drop off as r^{-6}. The exchange interaction is proportional to the overlap of the electronic wavefunctions of the acceptor and donor and therefore has an even more precipitous distance dependence. From these considerations, one may conclude that unless chromophores act as barriers or traps, the excitation migrations between nearest-neighbor chromophores will be much more likely than direct transfer to a distant chromophore.

Having determined that transfer between neighboring bases is the only one which needs to be considered, we may estimate the magnitude of the two interactions for the separation between stacked bases, which is usually taken to be 3.5 Å. For singlet–singlet transfer, the dipolar term involves allowed transitions both in the donor and the acceptor, and the interaction is on the order of a few hundred wavenumbers for all pairs of bases. The much smaller exchange interaction is estimated to be on the order of 10 wavenumbers. Since both of these interaction strengths are small compared to the widths of the absorption bands of the bases, no large changes in the absorption spectra of dinucleotides and polynucleotides compared to mononucleotides are expected or found.

While the exchange contribution is seen to be negligible for singlet–singlet transfer, it dominates triplet–triplet transfer, for which both μ_A and μ_D correspond to forbidden transitions. Coulombic triplet–singlet transfer, in which the triplet energy of a donor is used to excite the singlet state of an acceptor, is, on the other hand, quite efficient at low temperature since the small value of μ_D corresponding to a spin-forbidden transition is compensated for by the long donor lifetime.

The transfer of electronic energy from donor to acceptor may be so rapid that it precedes vibronic relaxation of the excited donor state. With smaller donor–acceptor interactions, the transfer follows vibronic relaxation, which is generally considered to have occurred approximately 10^{-13} to 10^{-12} sec following excitation.

All energy transfer, by whatever mechanisms, must of course be consistent with conservation of energy and the adiabatic approximation generally expressed by the Franck–Condon factors. In the case of transfer following vibrational relaxation, this means that the rate of energy transfer is proportional to the overlap between the normalized donor emission spectrum and the acceptor absorption spectrum.

At the time of this writing, the review on energy transfer in polynucleotides written by Gueron and Shulman[134] remains a valid presentation. In what follows we will therefore confine ourselves to a brief account of the theory and experimental background given in that review along with some recent experimental results and calculations.

7.2. Theory of Energy Transfer

We shall make no attempt to give a complete and formal theory of electronic energy transfer here but will summarize the basic results. More complete treatments of electronic energy transfer have been given by Förster[135,136] and others.[137–139]

Whenever chromophores interact, one may expect changes in the energy levels and energy transfer between the chromophores. The magnitude of these effects depends on the relative magnitudes of five energies—the coupling interaction, the width of the electronic bands, the width and spacing of the vibrational levels, and the electronic energy gap between the initial and final states. It has already been pointed out that the changes of energy levels as they manifest themselves in absorption are small for stacked bases, and these will not be considered here. We shall consider the energy transfer from an excited donor (D) to an acceptor (A) first for the idealized case in which the vibrational structures of D and A are neglected and then for three cases in which it is taken into account but the coupling matrix element $U = <\psi_{D^*}\psi_A | \mathbf{H}_C | \psi_D \psi_{A^*}>$ has different relations to the electronic and vibrational spectra of A and D. $\mathbf{H}_C$ is the coupling Hamiltonian, and ψ_D and ψ_A are the donor and acceptor electronic wavefunctions, with the asterisk denoting electronic excitation.

In the absence of vibrational structure of A and D, a good approximation for the rate of transfer between D and A is obtained by the following expressions:

$$n_D \sim \frac{2\pi |U|}{h} \quad \text{if} \quad \Delta E \leqslant 4|U| \qquad [12]$$

$$n_D = 0 \quad \text{if} \quad \Delta E \gg 4|U| \qquad [13]$$

7.2.1. Strong Coupling

The first case in which vibrational levels are considered is the so-called strong-coupling case, by which is meant that $|U|$ is larger than the widths of the electronic excited states of D and A, as well as their separation ΔE (the donor is taken to have the higher-lying state). In this case, excitation of any vibrational level in the D* manifold may be followed by transfer to any of the vibrational levels of A* so that the approximate equations given above apply, and

$$n \sim \frac{2\pi |U|}{h} \qquad [14]$$

7.2.1. Weak Coupling

In the weak-coupling limit, $|U|$ is smaller or comparable to the electronic bandwidths, and the vibrational levels are sufficiently narrow and resolved so that transfer from and to individual vibrational levels must be considered. The rate of transfer is derived from considerations similar to those employed for the strong-coupling case, but the interaction energy, U, must be replaced by a coupling energy, $V_{i,j}$ for the individual initial (D) and final (A) vibrational levels between which transfer can occur, and the rate is

$$n = \Sigma \frac{2\pi |V_{i,j}|}{h} \qquad [15]$$

where the summation is over all energetically allowed donor and acceptor transitions.

7.2.2. Very Weak Coupling

Since solvated nucleotides do not have spectra with resolved vibrational structure, the weak-coupling case is never applicable to them. As the coupling energy decreases, the strong-coupling case goes over directly to the very-weak-coupling case, which is characterized by U being smaller than the electronic absorption bandwidths and the density of vibrational levels being approximately constant and large on the scale of U. The number of available final states is therefore proportional to $|U|$, which results in a $|U|^2$ dependence for the transfer rate:

$$n \sim \frac{32|U|^2}{h} J \qquad [16]$$

where J is an overlap integral which satisfies the requirements of conservation of energy, the appropriate weighting of the vibrational levels as expressed by the Franck–Condon factors, and the density of levels. The exact form of J for the subcases of before-relaxation and after-relaxation transfer (J'' and J') will be given in the next section, which deals with the very-weak-coupling case, in which U arises from the electric dipole interaction.

Gueron[140] has derived* a useful approximate form of Eq. [16] in which the bandwidths of the electronic transitions of both the donor and acceptor are Δ and the energy band over which the donor emission spectrum and acceptor absorption spectrum have an appreciable overlap is δ. The rate of energy transfer is then given by

$$n = \frac{32}{h} |U|^2 \frac{\delta^3}{\Delta^4} \qquad [17]$$

It should be noted that all normal interactions between solvated purines and pyrimidines will fall into the very-weak-coupling category since (a) their spectra lack a resolved vibrational spectrum and (b) all interactions at the closest separations which they can normally attain (stacked bases) are much smaller than their electronic bandwidths.

7.3. Förster Energy Transfer

Singlet energy transfer in the very-weak-coupling limit resulting from the dipole–dipole interaction between donor and acceptor is often referred to as long-range or Förster transfer. The theory of this mechanism has been discussed by various authors,[136,139] and its predictions have been tested under a variety of experimental conditions.[138,141,142] The results of the theory may be presented either in terms of the energy transfer rate, k_t, or in terms of the Förster distance, R_0, which is defined as the separation between donor and acceptor at which k_t is equal to τ_D^{-1}, the decay rate of the donor in the absence of the acceptor:

$$k_t = \frac{8.8 \times 10^{-25} \varkappa^2 \varphi_D}{n^4 \tau_D r^6} J' \qquad [18]$$

$$R_{0(\text{cm})}^6 = \frac{8.8 \times 10^{-25} \varkappa^2 \varphi_D}{n^4} J' \qquad [19]$$

where $\varkappa^2$ is the dipolar orientation factor, φ_D is the fluorescence quantum yield of the donor, and n is the refractive index of the medium intervening between donor and acceptor. The overlap integral J' is given by

$$J' = \int_0^\infty F_D(\nu)\varepsilon_A(\nu)\nu^{-4}d\nu \qquad [20]$$

---

*To derive Eq. [17], Gueron assumed that the absorption and emission spectra have triangular shapes whose half-width is Δ. It follows that J will be of the order of δ^3/Δ^4.

where $F_D(\nu)$ is the emission spectrum of the donor normalized on the wavenumber scale, $\varepsilon_A(\nu)$ is the molar absorptivity spectrum of the acceptor, and ν is the wavenumber. It is assumed that vibrational relaxation takes place in the excited donor molecule before any transfer occurs.

The expressions given above apply in the so-called very-weak-coupling limit, which means that the donor and acceptor have absorption spectra with unresolved vibrational structure and that the (dipolar) interaction is small compared to the absorption bandwidths. This is the case which applies to solvated nucleotides.

It was pointed out in the previous section that for sufficiently great interaction between donor and acceptor, transfer may occur *before* vibrational relaxation.[13] This case may be treated in analogy with the after-relaxation transfer case by replacing the overlap integral J' by J'', defined by

$$J'' = \int_0^\infty F_B(\nu)\varepsilon_A(\nu)\nu^{-4}d\nu \qquad [21]$$

where $F_B(\nu)$ is the emission spectrum of the donor which would be observed in the absence of vibrational relaxation. $F_B(\nu)$ cannot be obtained by experiment but can be estimated to have the shape of $F(\nu)$ but displaced to the blue by a frequency interval which is about the same as the Stokes shift. It is clear that in certain cases J' can exceed J'' by several orders of magnitude. Nevertheless, before-relaxation transfer can actually occur only if its rate is comparable to or greater than the vibrational relaxation rate.

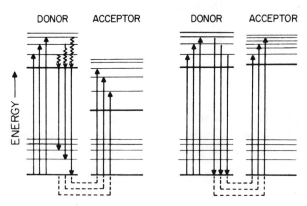

Figure 29. A schematic representation of the virtual, or coupled, transitions in the donor and the acceptor which are involved in (left) "after-relaxation" excitation transfer and (right) "before-relaxation" excitation transfer. Note that in latter, "after-relaxation" transfer is impossible for the donor–acceptor pair shown.

It is clear that unlike after-relaxation transfer, before-relaxation transfer will have a rate which will depend on excitation wavelength.

Figure 29 gives a schematic view of how transfer before and after relaxation differs. Note that for the case shown on the right side of the figure, after-relaxation transfer is energetically impossible.

7.4. Experiments and Calculations

7.4.1. Singlet Energy Transfer in Polyadenylic Acid

In section 3.5 it was pointed out that the absorption spectrum of poly rA differs from that of AMP in that the lowest band appears to be split in the former. If this spectral change were due to exciton splitting, the required interaction strength of several hundreds of wavenumbers would lead to a rate at which energy would be transferred between the bases on the order of 10^{13} sec.$^{-1}$

Since the fluorescence decay rate of poly rA at 80°K is four orders of magnitude smaller than this,[143] there should be practically complete depolarization of the emitted light from poly rA, following excitation by polarized light. This effect was looked for in different laboratories, and all investigators found very little depolarization of fluorescence,[140,144] which shows that the range of singlet energy transfer corresponds to only a few bases at most.

It was pointed out in section 4.3.1 that copolymers containing primarily adenosine residues and a small fraction of U or C residues, which form exciplex traps for the singlet energy, have fluorescence spectra which show again that the range of singlet energy transfer along the poly rA sections is short.[27]

How do these experimental findings compare with the predictions of the theory? If one assumes that transfer follows vibronic relaxation, the Förster theory indeed predicts transfer distances which are comparable to the interbase distance (see Section 4.7.3), and transfer only over one or two bases is expected.[13] Before-relaxation transfer rates depend on the number and strengths of the transitions in the 260 nm absorption band and probably lie between 10^{11} and 10^{12} sec^{-1}. It is questionable whether this is fast enough to compete with vibronic relaxation.

7.4.2. Triplet Energy Transfer in Polyriboadenylic Acid

The phosphorescence spectra of AMP, oligo As, and poly rA at 80°K are very similar in shape and decay time (2.5 sec, approximately), but the polymeric spectrum is blue-shifted by about 7 nm compared to the monomeric one, with the oligomeric spectra being shifted by intermediate amounts.[142] Bersohn and Isenberg[145] discovered that paramagnetic ions are capable of quenching the phosphorescence of poly rA nonstoichiometrically without affecting the fluorescence. Eisinger and Shulman[146] made a quantitative study of the effect of adding the paramagnetic transition metal ions Mn^{2+}, Ni^{2+}, and Co^{2+} to poly rA and found that all three ions are equally efficient quenchers of the phosphorescence, that each ion can quench the triplet excitation in a range of about 150 adenine bases with a small effect on the fluorescence, and that the decay time of the residual, unquenched phosphorescence remains 2.5 sec (see Figures 30 and 31). The diamagnetic Mg^{2+} and Zn^{2+} ions had no effect on the phosphorescence.[146]

These experimental results were interpreted as proving long-range triplet transfer along poly rA, with the triplet transfer proceeding in a stepwise fashion from base to base, until the proximity of a bound paramagnetic

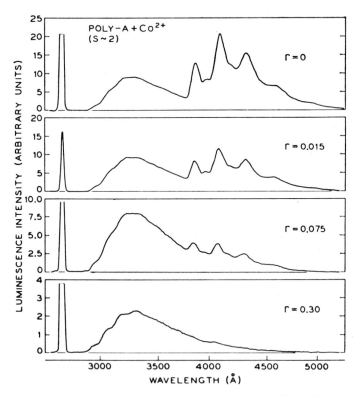

Figure 30. The luminescence spectra of poly rA in EGW at 80°K for various values of r, which is the ratio of bound Co^{2+} ions to phosphate groups. Note that Co^{2+} quenches primarily the phosphorescence emission.

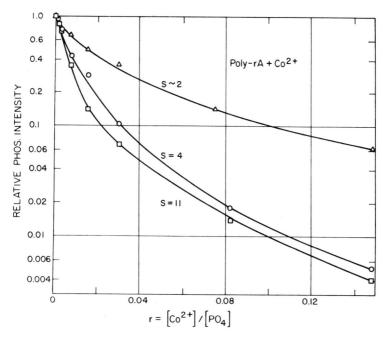

Figure 31. The relative phosphorescence intensities of poly rA of different lengths (characterized by its sedimentation coefficients) as a function of $r = [Co^{2+}]/[PO_4]$.

ion quenches the adenine triplet. However, in probing a little more deeply into this model, one encounters some difficulties. While the phosphorescence quenching data could be fitted by assuming that diffusion to a quenching site limits the range of triplet migration,[145,146] the same model would predict that the decay rate should increase with phosphorescence quenching, while it in fact remains constant. The reason for this could be that transfer barriers (kinks) limit the range of the triplets, but if a quenching ion and a triplet are between the same pair of barriers the triplet is quenched in a time less than 10^{-5} sec, which is the upper limit for a short-phosphorescence component.[147] Another reason, according to Gueron, could be that emission occurs not from the *bona fide* lowest triplet state of the polymer but from a trap (the trap depth would have to be about 1000 cm^{-1} at 77°K; see Section 7.4.4) to which triplet energy migrates in 10^{-5} sec or less. Such a trap might be provided by kinks in the polymer, but if this were the case the traps (which could not exceed 1% or so of the number of adenine residues) could be saturated by a sufficiently large triplet population. An alternative model for a trap is the lowered energy provided by solvent reorientation, which may persist to some extent in the glass at 80°K, the temperature at which these experiments were conducted. This could lower the triplet energy sufficiently to create a trap—the triplet would have dug itself in.[140]

Recently, there have been two reports of observations of delayed fluorescence from poly rA which is undoubtedly due to triplet–triplet annihilation.[148,149] While these experiments confirm the existence of triplet transfer, they do not resolve the problem of the constant decay rate of the quenched phosphorescence. The delayed fluorescence is second order with respect to the exciting light intensity, as it is predicted to be, but has an unexplained exponential decay time of 0.25 sec. Its intensity is two orders of magnitude smaller than the phosphorescence, so that it is unclear if triplet–triplet annihilation provides a major decay mechanism in poly rA.

It was pointed out above that the long-range triplet migration in poly rA occurs in a time less than 10^{-5} sec. A lower limit for this migration time comes from the observation that AMP and poly rA have $\Delta m = 2$ electron spin resonance lines of the same width; i.e., there is no indication of motional narrowing in the polymer, which would be expected if the triplet jump time between neighboring bases were less than 2×10^{-10} sec.[150] Somer and Jortner[151] recently made a theoretical study of the triplet exciton dynamics in poly rA and estimated the triplet jump time to be 10^{-9} sec, and we have estimated 10^{-7} sec (see Section 7.4.4). Both values fall between the two limits provided by experiment.

It is clear from the foregoing discussion that a detailed description of triplet excitation migration which takes place in poly rA eludes us at this time.

7.4.3. Singlet Energy Transfer in DNA

It was pointed out in Section 3.3.1 that virtually no monomer fluorescence is observed in DNA. The excitation spectrum of the exciplex emission in the somewhat simpler polynucleotide poly dAT showed that (AT)* was formed with the same probability whether the original excitation occurred in A or in T. If we ignore the extremely-short-range energy delocalization which occurs in exciplex formation, the only other type of singlet transfer which needs to be considered is the one among the bases before an exciplex it formed. Once an exciplex is created, its energy is too low to be transferred to any of the bases in DNA. In other words, the exciplex is an effective singlet excitation energy trap.

Gueron et al.[13] have calculated the rates for interbase energy transfer via the dipolar interaction in native DNA at liquid-nitrogen temperature. Before presenting their results, it is, however, important to state the assumptions which were made in the calculations. First of all, the emission and absorption spectra as well as the fluorescence quantum yields of the nucleotides in DNA were assumed to be identical to what they are in isolated mononucleotides. That is, the additional relaxation processes encountered in the polynucleotide, such as exciplex formation, were ignored. Secondly, the transition dipole orientations are those given in the literature for the principlal absorptions in the 260 nm bands, but the possibility of nearly degenerate states was ignored. Finally, since there is no solvent between stacked bases, n was taken as unity.

The results of these calculations are given in Table 9, which gives the ratio of the Förster distance, R_0, to the actual distance, d, between all pairs of bases in the Watson–Crick geometry of DNA. Here after-relaxation transfer is assumed. The table shows that transfer is likely to occur for about two thirds of the pairs of bases (values of R_0/d greater than unity), which means that for a random array of the four bases long-range migration of singlet excitation is not to be expected. It is interesting to note that transfer to A is very unlikely. This is the result of adenine having an appreciably higher excited singlet state than the other three bases. It must be remembered that these estimates represent upper limits for the transfer ranges, since exciplex formation and quenching processes would reduce them.

The rates of before-relaxation excitation transfers may be calculated by assuming that excitation occurs at the peak of the donor absorption band and that the before-relaxation emission has the same shape as the absorption spectrum. In this way, one obtains the transfer rates shown in Table 10, which again apply to the bases in native DNA. These estimates were obtained with the simplifying assumption that the four nucleotides have identical absorption and emission spectra and that the excitation wavelength is the same as the absorption maximum of the donor.[13] The rates are seen to be between 10^{-14} and 10^{-12} sec^{-1} in order of magnitude. The fastest

Table 9. Ratio of Förster Transfer Distance to Distance of the Neighboring Base in DNA for the Watson–Crick Geometry[a]

5'/3'	A	G	C	T
A	0.9	1.7 / —	1.3 / —	1.2 / —
G	— / 1.5	1.8	2.8 / 1.3	1.6 / 2.4
C	— / 1.1	0.8 / 1.4	1.8	— / 2.7
T	— / 1.0	1.7 / 1.1	2.2 / —	1.0

A — T base pair
A → T   0.87
T → A   —

G — C base pair
G → C   1.7
C → G   1.0

Direction of transfer

[a] Where no value is indicated, the ratio is smaller than 0.75. In each box, the upper left figure refers to transfer from row to column, the lower right figure to transfer from column to row.[13]

Table 10. The "Before-Relaxation" Transfer Rates, in Units of 10^{13} sec^{-1}, Between the Bases in the Watson–Crick Structure of DNA[a]

3'–5'→ ↓	A	G	C	T
A	2.5	2.0	0.6	1
G	1.0	0.9	2.2	4.0
C	0.15	0.15	2.3	6
T	0.3	0.4	1.5	0.35
A–T base pair: 0.17				
G–C base pair: 0.6				

[a] The assumptions underlying these estimates are given in Section 7.4.3.

transfers (e.g., 3'G→T5' or 3'C→T5') may well compete with vibrational relaxation, and before-relaxation transfer could actually operate.

In deciding if before-relaxation transfer is possible, one needs to know the rate of vibrational relaxation. At the present time, there exists little information on the magnitude of these rates. It is, however, possible to estimate by noting that the narrowest vibrational line in the AMP phosphorescence spectrum has a width, ΔE, of about 50 cm^{-1}. If this linewidth is indeed the width of the single vibrational level, its lifetime is of the order of $h/\Delta E$, or about 10^{-13} sec. In the other nucleotides, the vibrational relaxation rates may be even faster than this, since very little vibrational structure exists in their spectra.

A quantitative measurement of the intensity of DNA fluorescence as a function of A-T content and of the excitation wavelength would yield considerable information concerning singlet excitation transfer. Unfortunately, such measurements have not been carried out successfully because the fluorescence yield measurements give erratic results, probably due to the presence of mononucleotide impurities and single strands which have a much stronger fluorescence than does DNA. Nonetheless, a qualitative dependence of the excimer fluorescence intensity on the A-T content has been observed[152] which indicates that quenching in the G-C hydrogen-bonded pairs occurs at the singlet level and competes with energy transfer (see Section 4.3). This idea, together with the notion that exciplex formation also competes with energy transfer, forms the basis of the model which best explains the dependence of the DNA triplet yield on the A-T base-pair content, which has been measured.[16]

For low A-T content, the triplet yield, as judged by both measurements of the phosphorescence intensity and the ESR signal strength, is proportional to the A-T content. The explanation given was simply that light trapped by G-C pairs is quenched, whereas light trapped by A-T pairs gives excimers and triplets and there is little transfer of the trapped (after-relaxation) excitation between G-C and A-T pairs.

Gueron and Shulman[134] carried out a more detailed analysis of the data which resulted in an indication that before-relaxation transfer at the singlet level may occur. Their model is based on the following assumptions: (1) transfer (probably before-relaxation) is extensive enough to sample the base-pair distribution, (2) the transfer parameters are independent of A-T content, (3) processes internal to either base-pair, such as quenching in G-C, are independent of A-T content, and (4) only excitation trapped on A-T gives rise to the T triplet. For this model, they derived the expression

$$\frac{\varphi}{I} = 1 - \frac{q}{p} + \frac{q}{p}\left(\frac{1}{n}\right) \qquad [22]$$

where I is the triplet intensity, p and q are the probabilities of trapping the excitation energy on A-T and G-C base pairs, respectively, n is the fractional A-T content, and φ is the efficiency of forming a triplet in an excited A-T pair. Indeed, the plot of $1/I$ versus $1/n$ (excitation wavelength 265 nm) gives a straight line within experimental error (Figure 32). The value of q/p ob-

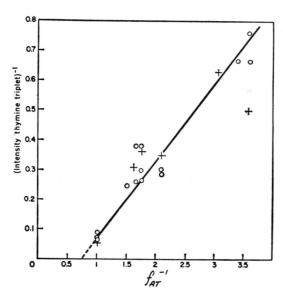

Figure 32. The inverse of the relative thymine phosphorescence intensity of various DNAs as a function of the inverse of fractional AT content of these DNAs, f_{AT}.

tained from the plot is 4; that is, 80% of the excitation is trapped at G-C sites and 20% is trapped at A-T sites. Since A-T and G-C pairs have nearly equal absorption coefficients at 265 nm, the ratio would be unity if there were no transfer.

Thus the following picture emerges: Light (265 nm) is absorbed equally well by both A-T and G-C pairs, and before vibrational relaxation can occur the excitation energy migrates over a sufficiently large range (more than four bases) to become localized preferentially at sites which have a higher trapping efficiency. In this way, excitation becomes localized at G-C and A-T pairs in the ratio of 4 to 1 multiplied by their respective relative concentrations. The excitation localized in G-C pairs is for the most part quenched at the singlet level. The excited A-T pairs go on to give exciplexes and the thymine triplet. The quenching and exciplex formation processes compete strongly with after-relaxation transfer.

7.4.4. Triplet Energy Transfer in DNA

7.4.4(a). Theory. From what was said in Sections 7.1 and 7.2 it is clear that triplet transfer among the bases of DNA will have to proceed by virtue of the exchange interactions between the various pairs of bases and that the very-weak-coupling limit of the general energy transfer theory applies. In the present section, we will derive numerical estimates of the transfer rates and the range of migration they lead to and present the results of experiments designed to determine this range.

Somer and Jortner[151] have estimated the strength of the exchange interaction between stacked adenine bases in DNA to be 10 cm^{-1}, and it is reasonable to expect this value to be approximately correct for all pairs of neighboring bases. Because of the much larger separation, the interaction between hydrogen-bonded base pairs should be smaller by at least a factor of 300 unless the hydrogen bonds play an extraordinary part in increasing the interaction.[153]

One is now in a position to use Eq. [17] to estimate the rate of triplet transfer between all possible pairs of bases; Δ is taken to be 5000 cm^{-1}, and δ for identical bases is assumed to be the same as the overlap between the absorption and fluorescence spectra or about 500 cm^{-1}. With $|U| \sim 10$ cm^{-1} for stacked bases, one obtains a triplet transfer rate of $k_t = 10^7$ sec^{-1} for identical bases in DNA. For those situations in which the acceptor triplet lies higher in energy than the donor triplet, this rate is reduced by the Boltzman factor. The spectral overlap increases by not more than a factor of 100 for the remaining cases in which the donor triplet state lies higher than that of the acceptor. The results are shown in Table 11.

These results also predict that before-relaxation triplet transfer should

Table 11. Theoretical Estimates of the Rates (in sec^{-1}) of Triplet–Triplet Transfer Between Stacked and Hydrogen-Bonded Base Pairs in DNA at 77°K[a]

Donor	Acceptor			
	T	A	G	C
T	10^7	10^4	10^{-1}	10^{-7}
A	10^8	10^7	10^3	10^{-2}
G	10^8	10^8	10^7	10^2
C	10^9	10^8	10^8	10^7
	Hydrogen-bonded base pairs			
T	—	10^{-1}	—	—
A	1	—	—	—
G	—	—	—	10^{-3}
C	—	—	10	—

[a] These estimates are proportional to the square of the exchange interaction (here assumed to be 10 cm^{-1} and 0.03 cm^{-1} for stacked and hydrogen-bonded bases, respectively) and may be used to estimate the rate of exchange-mitigated singlet energy transfer as well.

not occur in DNA since even the maximum rates of the before-relaxation transfer obtained by multiplying the values of Table 11 by 10^3 (that is, taking $\delta = 5000$ cm^{-1}) are much slower than vibrational relaxation times. The much slower rates for transfer within base pairs as opposed to rates of transfer along the same strand simply reflect the greater distance of separation in the former.

It is possible to make an estimate of the range of triplet migration in DNA from the rates listed in Table 11. Assume that 1 sec is the lifetime of the triplet state of each of the nucleotides in the absence of transfer and that there is no special quenching of the triplet in the G-C pair. Then it can be seen that with the exception of T→G, T→C, and A→C, triplet transfer between all other possible neighboring pairs along the same strand occurs at rates much faster than the decay rate. The T→G, T→C, and A→C rates, on the other hand, are much slower than the decay rate.

Since the rates for transfer along the same strand of DNA are much larger than those for interstrand jumps, the problem is reduced to considering the base sequence in one strand.

Quite generally, if the fraction of the 16 possible pairs of neighboring bases between which transfer occurs freely is x and if transfer is impossible in the remaining fraction $1 - x$, then the probability of finding a run of exactly n bases among which energy migrates freely is

$$P_n = (1-x)x^n \qquad [23]$$

It follows that the average number of neighboring bases in a random array of bases through which energy can migrate is

$$R_t = \sum_{n=0}^{\infty} np^n = (1-x)\sum_{n=0}^{\infty} nx^n = (1-x)^{-1} \qquad [24]$$

Let f_{AT}, f_{GC} be the normalized frequencies with which A-T and G-C base pairs occur in a random DNA molecule. Then there exist three types of pairs of neighboring bases which occur with different probabilities:

Type	Relative probability	Number of such pairs
1	$f_{AT}\,f_{GC}$	8
2	f_{AT}^2	4
3	f_{GC}^2	4

Table 11 shows that the three pairs of bases for which transfer is unlikely (T→G, T→C and A→C) all belong to type 1, so that

$$x = \frac{5f_{AT}f_{GC} + 4f_{AT}^2 + 4f_{GC}^2}{8f_{AT}f_{GC} + 4f_{AT}^2 + 4f_{GC}^2} = 1 - \frac{3}{4}f_{AT}f_{GC} \qquad [25]$$

Substituting Eq. [25] in Eq. [24], one obtains

$$R_t = 4(3f_{AT}f_{GC})^{-1} \qquad [26]$$

for the range of triplet transfer along random DNA strands. Figure 33 shows R_t as a function of f_{AT}, and it is seen that while the range is large for DNAs with extreme A-T or G-C content (e.g., poly dAT), it is almost insensitive to the AT content and equal to about five bases for DNA with comparable amounts of A-T and G-C base pairs (e.g., calf thymus DNA).

If the rates of Table 11 are too small by a factor of 100, the calculated range would be eight base pairs, and if the rates are too large by a factor of 100, the calculated range would be three base pairs along random DNA with $f_{AT}=f_{GC}=\frac{1}{2}$.

It should be pointed out that the prediction of a somewhat delocalized triplet state does not contradict the fact that the phosphorescence from DNA originates solely from the triplet state of thymine since the triplet transfer rates of Table 11 reflect, of course, the relative energies of the triplets and are such that the triplet excitation is expected to be at thymine an overwhelming fraction of the time.

The rates of transfer for the energetically less favorable cases in Table 11 will increase by a factor of 50 on going to room temperature because of the Boltzman factor. However, it is likely that the triplet decay rate increases by at least that factor, so that simply on the basis of this consideration

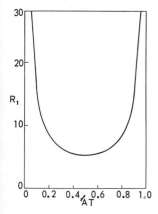

Figure 33. The triplet transfer range, R_t, as a function of the fractional AT content, f_{AT}, of DNA with a random array of bases, calculated according to Eq. [26].

the range of triplet migration in DNA at room temperature is expected to be smaller than the range at 77°K. Other effects of temperature which lead to spectral shifts should further narrow the range.

Somer and Jortner[151] assumed that the weak-coupling limit applies to transfer between the residues in polyadenylic acids. The transfer rate would then be given by $(2/h)\beta<\chi_0^*|\chi_0>^2$, where $<\chi_0^*|\chi_0>^2$ is the Franck–Condon factor for the 0–0 band. They calculated a jump time of 10^{-9} sec^{-1} for the triplet state in poly rA. However, the application of the weak-coupling limit is questionable even though there is some structure in the phosphorescence spectrum of A. The jump time of 10^{-7} sec given in Table 11 is based on the very-weak-coupling limit.

Lamola[100] calculated the transfer distances (distance at which the transfer rate equals the donor decay rate) for Förster transfer from the triplet states of TMP and AMP to the singlet state of proflavin and to the metal ions Mn^{2+}, Ni^{2+}, and Co^{2+}. The results are shown in Table 12.

7.4.4(b). Experimental Results. There have been several studies using dyes or paramagnetic metal ions bound to DNA which aimed at determining the range of triplet excitation migration in DNA at 77°K. The theory of the metal-ion quenching experiments may be summarized as follows. Let

Table 12. Förster Transfer Distances for Excitation Transfer from TMP and AMP in Their Triplet States at 80°K to Various Acceptors

Donor	Acceptor R_0, values (Å)			
	Proflavin	Mn^{2+}	Ni^{2+}	Co^{2+}
TMP	33	3.1	7.3	8.0
AMP	—	4.0	9.0	10.6

[M] and [P] be the molarities of bound quenching ions and of DNA phosphates, respectively. If $r = [M]/[P]$ and $r \ll 1$, virtually all divalent ions present in the solution will be bound in the absence of high concentrations of competing monovalent ions. It is now necessary to make two basic assumptions which have been implied but not stated in such studies: (1) The triplet migration range R_t is not larger than the average distance between bound quenchers, and (2) the binding of the quenchers is noncooperative. Under these conditions, an incremental increase in r, δr, produces a phosphorescence quenching of $\Delta I = -I\, R_t \delta r$, which upon integration gives

$$I(r) = I(0)e^{-rR_t} \qquad [27]$$

which under the usual experimental conditions can be simplified to

$$I(r) = I(0)\,(1 - rR_t) \qquad [28]$$

The most recent experiments of Isenberg et al.[155] on the quenching of DNA phosphorescence by metal ions gave 20, 17, 14, and 10 for the number of bases in the region quenched by one Cu^{2+}, Ni^{2+}, Co^{2+}, and Mn^{2+} ion, respectively. The DNA fluorescence is not quenched. The result for copper should be considered with caution since the ion is known to cause aggregation of DNA. The Förster distances for triplet transfer of excitation during the lifetime of the thymine triplet state to the metal ions given in Table 12 predict the quenched region to contain eight, eight, and three bases for Ni^{2+}, Co^{2+}, and Mn^{2+}, respectively. The differences between the experimental data and the predictions from Förster transfer theory could be due to triplet energy migration over five to 10 bases. Further evidence for triplet migration comes from the dependence of the range of quenching by Mn^{2+} on the A-T content of DNA. According to the transfer rates of Table 11, the range should increase with increasing A-T content. Indeed, Isenberg et al.[155] found a transfer range of five and 11 bases for M. lysodeikticus DNA (28% A-T) and salmon sperm DNA (56% A-T), respectively. However, they found a range of six bases for Mn^{2+} quenching of the phosphorescence of poly d(AT). The theoretical treatment given above would predict a large range for triplet migration in poly d(AT). Isenberg et al.[155] rationalized the results in terms of T→T transfer as being the only efficient mode of triplet migration in DNA. The number of T-T sequences increases with A-T content, but for poly d(AT) there are no such sequences. It should be pointed out that if the delocalization is only within thymine sequences, the average range in a DNA containing A-T and G-C base pairs in a random distribution is according to Eq. [24] only 1.3 bases.

In the light of these inconsistencies, the poly dAT quenching experiments of Isenberg et al. have recently been repeated. The results are given in Figure 34 and are virtually identical to those obtained by the above-mentioned authors. In interpreting these data, Isenberg et al. chose to ignore the portion of the curve below $r \sim 0.01$ and in this way obtained an R_t of about 6, which is in good agreement with the more recent results for $r > 0.01$. It must be remembered, however, that the region of the quenching curve most likely to satisfy the two assumptions stated at the beginning of this section is that for which $r < 0.01$. R_t estimated from this region ≈ 30 if the origin for shorter R_t at higher r values is a breakdown of assumption (2), i.e., if the metal binding is cooperative. If one assumes a model in which assumption (2) is satisfied but the poly dAT harbors regions of high and low R_t values, Eq. [27] must be modified to allow for two exponential terms and the low r region of the quenching curve leads to an R_t of about 100. Whichever model is assumed, it is clear that a long-range transfer with $R_t \gtrsim 30$ does occur in poly dAT, as is expected from Eq. [26] and Figure 33.

A sensitized delayed fluorescence from acridine orange and other acridine dyes bound to DNA at 77°K was observed by Isenberg et al.,[156] who showed that it is due to transfer from the thymine triplet state. They found that one dye molecule could be sensitized by any thymine among its 12 nearest neighbors, which corresponds to transfer within a distance of 11 Å. However, since the direct Förster distance for such a transfer is expected to be on the order of 30 Å (Table 12), this result provides no evidence for triplet migration in the DNA.

The most elegant study of triplet transfer in DNA to date is that of Galley.[157] He measured the dependence of the phosphorescence-to-fluorescence ratio of 9-aminoacridine bound to calf thymus DNA at 77°K as a

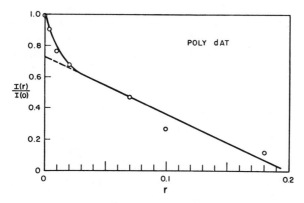

Figure 34. The ratio of the phosphorescence intensities $I(r)$ of poly dAT in the presence of Mn^{2+} ions, characterized by the parameter r, to the phosphorescence intensity in the absence of Mn^{2+} ions, $I(0)$. The solvent is EGW and the temperature 80°K ($r = [Mn^{2+}]/[PO_4]$).

function of the wavelength of the exciting light and as a function of the base-to-dye ratio. He was able to demonstrate triplet–triplet transfer from the DNA to the dye and obtained convincing data indicating that the triplet excitation can migrate over a short range in DNA. An average range of about 10 bases is indicated by these data, which, unfortunately, were obtained only at values of $r > 0.03$. He could detect no delay in the build-up of the dye phosphorescence upon exciting the DNA with a 15 msec flash compared to exciting the dye directly with the flash. This indicates that the triplet migration in DNA is not limited by diffusion and that the triplet jump rate is greater than 10^3 sec^{-1} (at least 10 jumps at a time faster than 0.015 sec). This is consistent with the observation that the unquenched portion of the DNA phosphorescence in the experiments with metal ions has the same decay time as that in the absence of quenchers.[155] Both the range of delocalization and the kinetics are consistent with the triplet transfer model presented above.

7.4.4(c). Excitation Migration in DNA at Room Temperature. The after-relaxation transfer efficiencies for both singlet and triplet excitation should be smaller at room temperature than at 77°K because the excited-state lifetimes are drastically shortened at the higher temperature. Relaxation processes which would reduce the donor energies are also more favorable at room temperature and would further reduce the transfer rates and efficiencies.

Weill and Calvin[158] studied the DNA-sensitized fluorescence of bound acridine dyes. The sensitization of dye fluorescence and concomitant quenching of thymine photodimerization were also investigated by the Sutherlands.[123] The latter investigators and Gueron and Shulman[134] have concluded that the dye sensitization is due to long-range Förster-type singlet–singlet transfer from the DNA to the dye before dimer formation. The data provide no indication of significant singlet excitation transfer in the DNA.

7.4.5. Energy Transfer in Other Systems

We have already discussed other instances of energy transfer in previous sections. Energy transfer in dinucleotides is described in Section 4.2 and in various polymeric systems (poly dAT, poly A_nU, poly rA, and DNA) in Sections 4.2, 7.1, and 7.4.1 to 7.4.4. Here we wish to mention some results on energy transfer in a few additional systems.

Excitation energy transfer in molecular aggregates of nucleic acid derivatives has been studied in frozen aqueous solutions at 77°K by Hélène et al.[77,78,159] Their findings may be summarized as follows: The formation of molecular aggregates of purine, adenosine, and guanosine leads to phosphorescence quenching when compared to the phosphorescence intensity in glassy solutions. The origin of this quenching is uncertain, but from the fact that the fluorescence is also quenched, one may conclude that it occurs at least partly at the singlet level. In mixtures of bases such as acetylcytosine and thymine, there is clear evidence for triplet migration to thymine,[159] which has the lowest-lying triplet state (see Sections 4.2.1 and 4.3.2).

In aggregates of adenosine and small amounts of thymidine, one finds that one thymidine molecule can quench the phosphorescence of 60 adenosine molecules, which proves that energy transfer occurs in adenosine aggregates.[159] In fact, when Co^{2+} ions were used as quenchers, the number of adenosine molecules quenched in aggregates by each ion was 110, which is almost the same as was found in poly rA[146] (see Section 7.4.1). Similar triplet energy transfer, albeit with a somewhat shorter range, was found in aggregates of guanosine in frozen aqueous solutions.[159] When thymidine acted as a quencher in these experiments, its phosphorescence was sensitized. The fact that 1,3-dimethylthymine could also be sensitized[159] showed once again that proton transfer was not necessary for the population of the triplet state (cf. Section 4.3.2).

8. TRANSFER RNA

8.1. The Role of Odd Bases

There have been numerous attempts to gain structural information about polypeptides by making use of the emission spectra of indigenous chromophores or of fluorescent labels; a recent review was written by Eisinger et al.[160] Energy transfer between chromophores offers another means of obtaining structural information.[161–163] In attempting to employ these techniques in polynucleotides, one is faced with several difficulties. First of all, as has been pointed out above, most bases do not emit under physiological conditions. Secondly, while proteins contain a few aromatic residues which play a special role, since they are the only amino acids capable of being excited by light of wavelengths greater than 2500 Å, DNA and most RNAs consist of just four bases with very similar absorption characteristics. The most important exceptions to this rule are found among the various transfer RNA molecules, which are richly endowed with "odd" bases, many of which absorb to the red of the common nucleotides, which permits one to excite them exclusively, and a few of which fluoresce in water at room temperature. It is likely that these odd bases, whose positions

Table 13. The Absorption and Emission Characteristics of Common and Odd Bases (or Their Derivatives) Which Occur in tRNA[a]

Nucleoside	Absorption		Fluorescence			Phosphorescence			Reference for emission data
	λ_{max} (nm)	ε (10^3)	λ_{max} (nm)	φ_F	λ_{thr} (nm)	λ_{max} (nm)	φ_P	τ (sec)	
A[b]	261	14.9	313	0.010	373	397	0.015	2.40	13
iA[c]	270	—	325	0.100	380	412	0.010	1.65	60
sA	247, 284	25.0, 17.8	335	0.350	390	425	0.250	0.36	
m⁶A	266	16.0	329	0.090	364	411	0.45	2.50	
m²A[d]	262	—	351	0.110	365	405	0.030	3.20	
tA	268, 277	—	321	0.370	360	409	0.030	1.95	
G[b]	252	13.7	323	0.130	365	402	0.070	1.30	13
m⁷G	257, 281	8.5, 7.4	337[g]	0.350[g]	357	448	0.005	1.80	i, 176
			381[e,g]	0.035[e,g]					
m²₂G[c]	259	16.3	338	0.270	373	414	0.140	1.40	167
m¹G[d]	248, 272	10.0, 7.9	326	0.150	381	423	0.010	1.60	
C[b]	271	8.9	322	0.050	352	405	0.010	0.30	13
sC[d]	281	—	—	<0.005	362	407	0.100	<0.04	
m⁵C	277	8.9	328	0.200	370	428	0.050	0.80	

Compound	λ_abs	ε	λ_fluor	φ_fluor	λ_phos	λ_phos	φ_phos	τ_phos	Ref
T^b	267	9.7	322	0.240	382[h]	438[h]	<0.005	0.30[h]	16
U^b	262	10.0	314	0.010	350[h]	405[h]	~0	0.55[h]	82
$(mam)^5U^a$	282	—	322	0.210	362	425	0.010	0.30	
$s^2(mam)^5U$	—	—	345	0.006	387	429	0.035	<0.04	
s^4U^f	244, 333	4.5, 21.0	—	<0.005	437	461	0.080	1.9×10^{-3j}	i, 177
Ψ	262	8.1	319	0.440	361	425	0.020	0.50	
Y^d	237, 315	30.3	405	0.330	402	450	0.010	2.40	170
			452[e]	0.030[e]	—	—	—	—	170

[a] The absorption data refer to room-temperature aqueous solutions and the emission data refer to EGW glasses at 80°K for all nucleosides with the exception of Y and m^7G, the only two bases to fluoresce at room temperature, for which room-temperature results are included. The abbreviations used in the first column are explained in section 2. Where no reference is given, the emission data represent previously unpublished results of B. Feuer and J. Eisinger, who are indebted to Dr. N. J. Leonard for his gift of sA, to Dr. G. B. Chheda for his gift of tA, and to Dr. J. Carbon for his gift of $(mam)^5U$ and $s^2(mam)^5U$; m^1s^4U was synthesized by B. Feuer.
[b] Nucleotide.
[c] $R_9 = (CH_2)_2CH_3$.
[d] Base.
[e] At room temperature.
[f] $R_1 = CH_3$.
[g] pH 5.2.
[h] Sensitized.
[i] Unpublished results of B. Feuer and J. Eisinger.
[j] Hélène et al. (177) give 4 msec for the s^4UMP phosphorescence lifetime. The same authors also give phosphorescence data for s^2U, s^2T, and s^6G, which have not been shown to occur in tRNA, however.

in the primary sequences of the tRNAs are in many cases known, will in the future play an important role in the determination of the three-dimensional structure of these important biological polymers.[164]

At this time, several of the odd bases found in tRNA have not been identified and the emission properties of others have not been characterized. Table 13 lists most of the known odd bases and their optical absorption and emission properties. With three known exceptions, these odd bases have observable emissions only at low temperatures. Low-temperature experiments may very well become useful tools in determining the three-dimensional structures of tRNAs, for instance by singlet or triplet transfer measurements. The assumption that tRNA molecules retain their solution structure in a frozen polar glass is at least as good as that the structure is unperturbed by crystallization.

The base 1-methyl-4-thiouracil, which is expected to resemble the nucleotide 4-thiouridine closely, has a negligible fluorescence quantum yield at room temperature both in water and in cyclohexane[165] 4-Thiouridine has nevertheless been reported to fluoresce[166] in tRNA^{Val} at room temperature. The explanation offered is that an exciplex emission is responsible for this fluorescence ($\lambda_{max} \sim 510$ nm). This base has also been implicated in a photochemical reaction between it and a cytosine five residues away from it.[166]

The nucleotide 7-methylguanosine (m^{7}G) has been found to fluoresce at room temperature in water at pH 5,[176] albeit with a quantum yield of only 0.02.[165] There is evidence both from absorption and from fluorescence titration studies that this molecule exists in neutral and protonated forms with a pK of about 7.2 and that only the protonated form emits at room temperature. 7-Methylguanosine occurs in several tRNA molecules including tRNA$^{Phe}_{yeast}$, where its fluorescence is apparently hidden by the

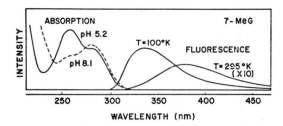

Figure 35. The absorption spectra at room temperature of m^{7}G at two pH values are shown on the left. On the right are shown the fluorescence spectra of m^{7}G in EGW at two temperatures.

much stronger emission from the base Y at about 440 nm (see below). When the Y base is removed by hydrolysis, tRNA^{phe} is found to fluoresce weakly with an intensity maximum at 370 nm, which is probably due to the m^{7}G. The nucleoside fluorescence peak is at 380 nm at room temperature but at 335 nm in a rigid glass with the temperature below 150°K (see Figure 35).[165] The base m^{7}G has been found only in the single-stranded regions of tRNA molecules, and the results given above indicate that it is well hydrated in tRNA$^{phe}_{yeast}$.

The position next to the 3'-end of the anticodon is frequently occupied by an adenine with a bulky substituent group.[164] Absorption and emission studies on model compounds have shown that this does not hamper the stacking with neighboring bases[60] which has been postulated to play an important role in determining the conformation of the anticodon loop.[168] The main function of the substituent groups appears to be to prevent spurious hydrogen bonding.

We will confine our attention for the remainder of this section to phenylalanyl tRNA of baker's yeast or tRNA^{phe}, in which the base Y, which fluoresces at room temperature and whose chemical structure is still unknown, is immediately adjacent to the A at the 3'-end of the anticodon m$^{2'}$G, A, A (m$^{2'}$G = 2'-O-methylguanosine).[169]

The position in the primary sequence occupied by Y makes this base a particularly valuable probe of a most interesting portion of tRNA^{phe}, the so-called anticodon loop. Since Y has an absorption peak at 315 nm, well to the red of the other bases in tRNA^{phe}, it can be excited by itself. The room-temperature emission spectrum is peaked at about 440 nm. In the sections which follow, we shall describe the results of several studies making use of the emission characteristics of Y and designed to shed light on the molecular structure of Y, its environment in tRNA^{phe}, the structure of the anticodon loop, and codon–anticodon binding.

8.2. tRNA^{phe} Studies

8.2.1. Mg^{2+}-Dependent Conformational Change

The fluorescence quantum yield of tRNA is 0.07 in water containing 10^{-2} M Mg^{2+} but decreases to half this value when the Mg^{2+} concentration is lowered to 10^{-4}.[170,171] Since the yield of Y, excised from tRNA^{phe} (denoted by Y$^+$) according to the method of Thiebe and Zachau,[172] shows no corresponding dependence on the Mg^{2+} concentration, one may conclude that the higher Mg^{2+} content brings about a conformation change in the tRNA^{phe} anticodon loop, which lowers the nonradiative de-excitation rate of Y, probably by altering its immediate environment. It is interesting to note that this conformational change occurs at about the same Mg^{2+} concentration as is needed for efficient protein synthesis *in vitro*.

8.2.2. The Environment of Base Y

Several experiments indicate that while the chromophore at the base Y in $tRNA^{phe}$ is not exposed to as many solvent molecules as is the free base Y^+, neither is it wholly shielded from them.[170] It is well known that the presence of polar solvent molecules with sufficient mobility to reorient themselves in the solvent shell of an excited molecule leads to a red shift of the fluorescence emission spectrum. Figure 37 shows that while the room-temperature fluorescence of $tRNA^{phe}$ is red-shifted compared to its spectrum at low temperature in a rigid polar glass, the room-temperature fluorescence of the base Y^+ is red-shifted by some 15 nm compared to the room-temperature spectrum of Y in $tRNA^{phe}$. These conclusions are confirmed by the existence of a sizable solvent isotope effect of the fluorescence yield, $\varphi(D_2O)/\varphi(H_2O)$. While the origin of this isotope effect (1.5 for $tRNA^{phe}$ and 1.9 for Y^+)[170] is not completely understood, it clearly reflects an isotope effect in a solvent-quenching mechanism.[93,173]

The position in the anticodon loop occupied by Y in yeast $tRNA^{phe}$ is, in most other $tRNAs$ whose anticodon begins with A, occupied by a substituted adenine.[164] It was therefore surmised the Y may have a similar structure. The phosphorescence spectrum of Y revealed itself to be devoid of the vibrational structure usually associated with adenine and its derivatives, but to bear a closer resemblance to the phosphorescence spectrum

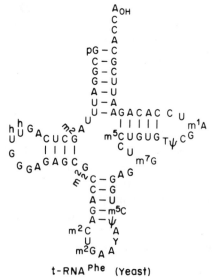

Figure 36. The primary sequence of bases in $tRNA^{phe}$ (yeast) in the so-called cloverleaf conformation.

and decay time of guanine, although red-shifted by some 30 nm from it.[170] It is noteworthy that the phosphorescence spectrum of Y^+ is further red-shifted when compared to Y in tRNA.[165] This shows that the energy of the triplet state of Y (as well as its excited singlet state) is lowered when Y is no longer surrounded by its neighboring bases in tRNA.

8.2.3. Separation Between Y and the 3'CpCpA End

Beardsley and Cantor[174] have conceived of an elegant method for determining the separation between the base Y and a dye molecule attached to the GGA end of tRNA^{phe}. Using Y as the donor and an acriflavine molecule as the acceptor, they calculate a Förster distance of 30 Å. This means that, if the separation between donor and acceptor is 30 Å and the donor and acceptor transition dipoles have random orientations with respect to each other, the fluorescence yield of Y will be halved while the fluorescence yield of acriflavine will be increased by a corresponding amount. In practice, various corrections have to be applied to the raw data, particularly because no wavelength could be found at which the acriflavine fluorescence (and that of two similar acceptors) could be monitored and at which the fluorescence of Y did not make an appreciable contribution. The small amount of singlet energy transfer which was observed made it possible to find a lower limit of 40 Å for the donor–acceptor separation with an upper limit near 60 Å. While the error is quite large, this result permits one to discard several models for three-dimensional tRNA^{phe} structures in which the folding of the usual "hairpin" model brings the anticodon loop and the 3'-end of the tRNA into close proximity.[174] The assumption that the transition dipoles of the donor and acceptor have random orientations with respect to each other was supported by Beardsley and Cantor,[174] who obtained comparable estimates for the donor–acceptor separations using three different acceptor molecules. Further support for this randomicity comes from fluorescence depolarization experiments which show that Y in tRNA^{phe} has a rotational relaxation time shorter than that of the whole tRNA molecule.[174]

8.2.4. Codon–Anticodon Binding

While there exists clear evidence for hydrogen bonding between the anticodon bases on tRNA and the codon bases on mRNA (UUU or UUC for phenylalanine) during protein synthesis on the ribosome, it has not been clear that this interaction is strong enough to cause codon–anticodon binding in a ribosome-free system. Since Y is immediately adjacent to the anticodon triplet, its emission properties are susceptible to small changes in environment, and in the presence of a large excess of pentauridylate a small blue shift in the emission spectrum of Y, corresponding to a slightly more hydrophobic environment, was indeed observed. While this shift is only of the order of 5 nm when pentauridylate is used as the codon molecule, it was possible to estimate the strength of this binding as 400 M^{-1} Table 14.[170] Polyuridylic acid was also observed to bind, but in the presence of polycytidylic acid no shift in the Y fluorescence was observed. It is interesting that this binding could only be observed in the presence of 10^{-2} M Mg^{2+}, which suggests that the tRNA conformation favored by high Mg^{2+} concentrations (see above) must obtain before the anticodon loop has a structure favorable for hydrogen bonding between the codon and anticodon triplets.

More recently the binding of various oligouridylic acids and poly rU which contain the wobble codon UUU, and of the codon UUC were determined at different temperatures by analyzing the fluorescence spectra as arising from linear combinations of the spectra of free and bound tRNA^{phe}.[175] Figure 38 shows experimental and theoretical binding curves for UUC at different temperatures and Table 14 gives the association constants for codons and wobble codons at different temperatures. The errors are about 20%. From the temperature dependence of K for UUC it is possible to derive values for the free energy, enthalpy, and entropy of binding. At 10°C these parameters are -4, -15, and -11 kcal/mole ($\Delta F = \Delta H - T\Delta S$). The fairly weak binding results from the large entropy term which arises from the need to select a conformation of UUC which permits the simultaneous formation of the base pairing hydrogen bonds. During protein synthesis the codon on the ribosome-held mRNA has

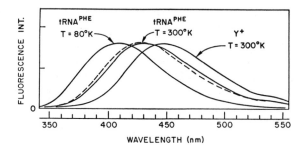

Figure 37. The fluorescence spectra of tRNA^{phe} in EGW at 80°K and in water at 300°K. The dashed curve corresponds to tRNA^{phe} in the presence of a twenty fold molar excess of pentauridylic acid in water at 300°K. Note that Y^+, the hydrolyzed fluorescent base of tRNA^{phe}, has a greatly red-shifted fluorescence in water. The concentration of tRNA was 0.7×10^{-3} M throughout, and the excitation wavelength was 313 nm. The Mg^{2+} concentration was 1.5×10^{-2} M.

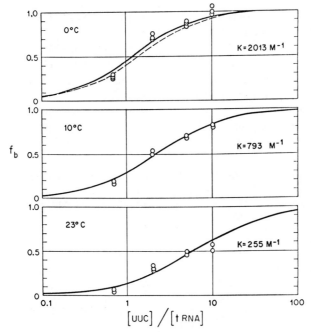

Figure 38. The fraction of anticodons of $tRNA^{phe}$ bound to their codons (UUC) at different temperatures as a function of the molar concentration ratio of UUC and of $tRNA$. $[tRNA] = 0.7 \times 10^{-3}$ M. The points are experimental and the curves are least-square-fitted binding curves which yield the binding constants shown.

presumably a conformation favoring this pairing. The entropy term would then be much smaller and the binding much stronger.

The fluorescence of the Y base in $tRNA^{phe}$ from yeast was also used to demonstrate strong binding between tRNAs with complementary

Table 14

Codon/wobble codon	K (0°C) M^{-1}	K (T°C) M^{-1}
U_3	373	
U_4	321	145 (8°C), 27 (22°C)
U_5	347	
poly U	395	
UUC	2013	1090 (5°C), 793 (10°C), 255 (23°C)

codons.[178] The association constant between tRNA^{phe} (anticodon GAA) and tRNA^{glu} (anticodon UUC) was found to be $5 \times 10^5 \text{M}^{-1}$ at 0°C, and since K had a very small temperature dependence one may conclude that the entropy term is small, as would be expected if the tRNA anticodons had similar, complementary structures of considerable rigidity (e.g., if they were portions of Watson and Crick helices).

REFERENCES

1. S. V. Konev, "Fluorescence and Phosphorescence of Proteins and Nucleic Acids," p. 141, Plenum Press, New York (1967).
2. G. Weber and F. W. J. Teale, *Trans. Faraday Soc.* 54, 640 (1958).
3. G. Weber, *Nature 190*, 27 (1961).
4. D. Duggan, R. Bowmann, B. B. Brodie, and S. Udenfriend, *Arch. Biochem. Biophys. 68*, 1 (1957).
5. J. D. Smith and R. Markham, *Biochem. J. 46*, 33 (1950).
6. L. S. Agroskin, N. V. Korolev, I. S. Kulaev, M. N. Me'sel, and N. A. Pomashchinkova, *Dokl. Akad. Nauk SSSR 131*, 1440 (1960).
7. R. Bersohn and I. Isenberg, *J. Chem. Phys. 40*, 3175 (1964).
8. S. V. Konev, "Fluorescence and Phosphorescence of Proteins and Nucleic Acids," Plenum Press, New York (1967).
9. C. Hélène, *Biochem. Biophys. Res. Commun. 22*, 237 (1966).
10. R. H. Steele and A. Szent-Györgyi, *Proc. Nat. Acad. Sci. 43*, 477 (1957).
11. J. W. Longworth, *Biochem. J. 84*, 104P (1962).
12. J. W. Longworth, R. O. Rahn, and R. G. Shulman, *J. Chem. Phys. 45*, 2930 (1966).
13. M. Gueron, J. Eisinger, and R. G. Shulman, *J. Chem. Phys. 47*, 4077 (1967).
14. J. Eisinger, M. Gueron, R. G. Shulman, and T. Yamane, *Proc. Nat. Acad. Sci. 55*, 1015 (1966).
15. C. Hélène, *Biochem. Biophys. Res. Commun. 22*, 237 (1966).
16. A. A. Lamola, M. Gueron, T. Yamane, J. Eisinger, and R. G. Shulman, *J. Chem. Phys. 47*, 2210 (1967).
17. S. Udenfriend, "Fluorescence Assay in Biology and Medicine," Vol. I (1962) and Vol. II (1969), Academic Press, New York.
18. J. Drobnik and L. Augenstein, *Photochem. Photobiol. 5*, 13 (1966).
19. J. Drobnik and L. Augenstein, *Photochem. Photobiol. 5*, 83 (1966).
20. V. Kleinwächter, J. Drobnik, and L. Augenstein, *Photochem. Photobiol. 5*, 579 (1966).
21. V. Kleinwächter, J. Drobnik, and L. Augenstein, *Photochem. Photobiol. 6*, 133 (1967).
22. E. Lippert, *Accounts Chem. Res. 3*, 74 (1970).
23. D. C. Ward, E. Reich, and L. Stryer, *J. Biol. Chem. 244*, 1228 (1969).
24. J. W. Longworth, R. O. Rahn, and R. G. Shulman, *J. Chem. Phys. 45*, 2930 (1966).
25. R. G. Shulman and R. O. Rahn, *J. Chem. Phys. 45*, 2940 (1966).
26. P. I. Hønnas and H. B. Steen, *Photochem. Photobiol. 11*, 67 (1970).
27. J. Eisinger, *Photochem. Photobiol. 9*, 247 (1969).
28. A. A. Lamola, *in* "Energy Transfer and Organic Photochemistry," Interscience Publishers, New York (1969).

29. M. Gueron, J. Eisinger, and R. G. Shulman, *Mol. Phys. 14*, 111 (1968).
30. S. P. McGlynn, T. Azumi, and M. Kinoshita, "Molecular Spectroscopy of the Triplet State," Prentice-Hall, Englewood Cliffs, N.J. (1969).
31. H. C. Borresen, *Acta Chem. Scand. 21*, 2463 (1967).
32. H. C. Borresen, *Acta Chem. Scand. 17*, 920 (1963).
33. A. Weller, *Prog. Reaction Kinetics 1*, 189 (1961).
34. J. Eisinger, A. A. Lamola, and D. J. Patel, unpublished results.
35. K. Berens and K. L. Wierzchowski, *Photochem. Photobiol. 9*, 433 (1969).
36. K. L. Wierzchowski, E. Litonska, and D. Shugar, *J. Am. Chem. Soc. 87*, 4621 (1965).
37. S. Udenfriend and P. Zaltman, *Anal. Biochem. 3*, 49 (1962).
38. J. Eisinger, *Photochem. Photobiol. 7*, 597 (1968).
39. A. A. Lamola, *Photochem. Photobiol. 8*, 601 (1968).
40. T. Förster, "Fluorescenz organischer Verbindungen," Vandenhoeck and Ruprecht, Göttingen (1951).
41. W. E. Blumberg, J. Eisinger, and G. Navon, *Biophys. J. 8*, 4106 (1968).
42. D. W. Miles, M. J. Robins, R. K. Robins, and H. Eyring, *Proc. Nat. Acad. Sci. 62*, 22 (1969).
43. P. R. Callis, E. J. Rosa, and W. T. Simpson, *J. Am. Chem. Soc. 86*, 2292 (1964).
44. M. Gueron, unpublished results.
45. J. Brahms, unpublished results.
46. E. D. Bergman and B. Pullman, eds., "Quantum Aspects of Heterocyclic Compounds in Chemistry and Biochemistry," Academic Press, New York (1970).
47. A Pullman, *in* "Theoretical Physics and Biology" (M. Marsis, ed.), Interscience, New York (1969).
48. B. Mely and A. Pullman, *Theoret. Chim. Acta 13*, 278 (1969).
49. L. C. Snyder, R. G. Shulman, and D. B. Neumann, *J. Chem. Phys. 53*, 256, (1970).
50. J. N. Murrell, "The Theory of the Electronic Spectra of Organic Molecules," Wiley, New York (1963).
51. R. F. Steiner and R. F. Beers, "Polynucleotides," Elsevier Publishing Company, Amsterdam (1961).
52. J. Eisinger and A. A. Lamola, *Mol. Photochem. 1*, 209 (1969).
53. C. Hélène, P. Douzou, and A. M. Michelson, *Biochim. Biophys. Acta 109*, 261 (1965).
54. C. Hélène, P. Douzou and A. M. Michelson, *Proc. Nat. Acad. Sci. 55*, 376 (1966).
55. K. Imakubo, *J. Phys. Soc. Japan 24*, 1124 (1968).
56. J. Koudelka and L. Augenstein, *Photochem. Photobiol. 7*, 613 (1968).
57. M. Gueron, R. G. Shulman, and J. Eisinger, *Proc. Nat. Acad. Sci. 56*, 814 (1966).
58. J. Eisinger and R. G. Shulman, *Science 161*, 1311 (1968).
59. D. T. Browne, J. Eisinger, and N. J. Leonard, *J. Am. Chem. Soc. 90*, 7302 (1968).
60. N. J. Leonard, H. Iwamura, and J. Eisinger, *Proc. Nat. Acad. Sci. 64*, 352 (1969).
61. M. Kasha, *Radiation Res. 20*, 55 (1963).
62. C. R. Cantor and I. Tinoco, *J. Mol. Biol. 13*, 65 (1965).
63. C. A. Bush and I. Tinoco, *J. Mol. Biol. 23*, 601 (1967).
64. S. I. Chan, B. W. Bangerter, and H. H. Peter, *Proc. Nat. Acad. Sci. 55*, 720 (1966).
65. P. O. P. Ts'o, N. Kondo, M. T. Schweizer, and D. P. Hollis, *Biochemistry 8*, 997 (1969).
66. T. Förster and K. Kasper, *Z. Elektrochem. 59*, 976 (1955).
67. M. Gueron, R. G. Shulman, and J. Eisinger, *Proc. Nat. Acad. Sci. 56*, 814 (1966).
68. J. Eisinger and R. G. Shulman, *J. Mol. Biol. 28*, 445 (1967).

69. C. Hélène and T. Garrestier Montenay, unpublished results.
70. C. Hélène and A. M. Michelson, *Biochim. Biophys. Acta 142*, 12 (1967).
71. H. Beens and A. Weller, *Acta Phys. Polon. 34*, (1968).
72. J. Eisinger, B. Feuer, and A. A. Lamola, *Biochemistry 8*, 3908 (1969).
73. R. O. Rahn, J. W. Longworth, J. Eisinger, and R. G. Shulman, *Proc. Nat. Acad. Sci. 51*, 1299 (1964).
74. R. O. Rahn, T. Yamane, J. Eisinger, J. W. Longworth, and R. G. Shulman, *J. Chem. Phys. 45*, 2947 (1966).
75. R. O. Rahn, R. G. Shulman, and J. W. Longworth, *J. Chem. Phys. 45*, 2955 (1966).
76. P. Douzou, J. Francq, M. Hausss, and M. Ptak, *J. Chim. Phys. 58*, 926 (1961).
77. C. Hélène, M. Ptak, and R. Santus, *J. Chim. Phys. 65*, 160 (1968).
78. C. Hélène, *Biochem. Biophys. Res. Commun. 22*, 237 (1966).
79. M. S. Walker, T. W. Bednar, and R. Lumry, *J. Chem. Phys. 47*, 1020 (1967).
80. A. A. Lamola and J. Eisinger, *in* "Molecular Luminescence" (E. C. Lim, ed.), p. 801, W. A. Benjamin, Inc., New York (1969).
81. J. Eisinger and A. A. Lamola, *Biochim. Biophys. Acta* (in press) (1971).
82. A. A. Lamola and J. Eisinger, *Biochim. Biophys. Acta* (in press) (1971).
83. J. L. Kropp and M. W. Windsor, *J. Chem. Phys. 42*, 1599 (1965).
84. G. J. Fisher and H. E. Johns, *Photochem. Photobiol.* (in press).
85. I. Brown and H. E. Johns, *Photochem. Photobiol. 8*, 273 (1968).
86. D. W. Whillans and H. E. Johns, *Photochem. Photobiol. 9*, 232 (1969).
87. D. W. Whillans, M. A. Herbert, J. W. Hunt, and H. E. Johns, *Biochem. Biophys. Res. Commun. 36*, 912 (1969).
88. H. C. Borresen, *Acta Chem. Scand. 21*, 920 (1967).
89. H. C. Borresen, *Acta Chem. Scand. 21*, 2463 (1967).
90. J. Drobnik and L. Augenstein, *Photochem. Photobiol. 5*, 83 (1967).
91. H. C. Borresen, *Acta Chem. Scand. 19*, 2100 (1965).
92. D. M. Hercules and L. G. Rogers, *J. Chem. Phys. 64*, 397 (1960).
93. J. Eisinger and G. Navon, *J. Chem. Phys. 50*, 2069 (1969).
94. J. Saltiel, O. C. Zafirou, E. D. Megarty, and A. A. Lamola, *J. Am. Chem. Soc. 90*, 4759 (1968); and the references given there.
95. R. M. Hochstrasser, *Accounts Chem. Res. 1*, 266 (1968).
96. H. R. Ward and J. S. Wishnok, *J. Am. Chem. Soc. 90*, 1085 (1968).
97. K. E. Welzbach, T. S. Ritscher, and L. Kaplan, *J. Am. Chem. Soc. 89*, 1031 (1967).
98. C. H. Krauch, D. M. Krämer, P. Chandra, P. Mildner, H. Feller, and A. Wacker, *Angew. Chem. Internat. Ed. Engl. 6*, 956 (1967).
99. C. L. Greenstock and H. E. Johns, *Biochem. Biophys. Res. Commun. 30*, 21 (1968).
100. A. A. Lamola, unpublished.
101. R. W. Chambers, H. P. Waits, and K. A. Freude, *J. Am. Chem. Soc. 91*, 7203 (1969).
102. A. A. Lamola and J. P. Mittal, *Science 154*, 1560 (1966).
103. C. L. Greenstock, I. H. Brown, J. W. Hunt, and H. E. Johns, *Biochem. Biophys. Res. Commun. 27*, 431 (1967).
104. J. G. Burr, B. R. Gordon, and E. H. Park, *Photochem. Photobiol. 8*, 73 (1968).
105. A. J. Varghese and S. Y. Wang, *Nature 213*, 909 (1967).
106. S. Y. Wang and A. J. Varghese, *Biochem. Biophys. Res. Commun. 29*, 543 (1967).
107. S. Y. Wang, *in* "The Basic Principles in Nucleic Acid Chemistry" (P. O. P. Ts'o, ed.), Academic Press, New York (in press).
108. P. Wagner and D. J. Bucheck, *J. Am. Chem. Soc. 92*, 181 (1970).
109. R. Lisewski and K. L. Wierzchowski, *Chem. Commun.*, p. 348 (1969).

110. H. Morrison, A. Feeley, and R. Kloepfer, *Chem. Commun.*, p. 358 (1964).
111. S. Y. Wang, *Nature 190*, 690 (1960); *Photochem. Photobiol. 3*, 395 (1964).
112. R. Gerdil, *Acta Crystallog. 14*, 333 (1961).
113. J. Eisinger and R. G. Shulman, *Proc. Nat. Acad. Sci. 58*, 895 (1967).
114. A. A. Lamola and J. Eisinger, *Proc. Nat. Acad. Sci. 59*, 46 (1968).
115. J. Eisinger and A. A. Lamola, *Mol. Photochem. 1*, 209 (1969).
116. H. E. Johns, M. L. Pearson, J. C. LeBlanc, and C. W. Helliner, *J. Mol. Biol. 9*, 503 (1964).
117. D. Wulff, *Biophys. J. 3*, 355 (1963).
118. H. E. Johns, S. A. Rappaport, and M. Delbruck, *J. Mol. Biol. 4*, 104 (1962).
119. D. Weinblum, *Biochem. Biophys. Res. Commun. 27*, 384 (1967).
120. R. O. Rahn, *Science 154*, 503 (1966).
121. J. L. Hosszu and R. O. Rahn, *Biochem. Biophys. Res. Commun. 29*, 327 (1967).
122. J. Eisinger, R. G. Shulman, and T. Yamane, "Proceedings of the Gatlinberg Conference" (W. Snipes, ed.), *Nuclear Science Set. 43, Nat. Acad. Sci. Nat. Res. Council* (1966).
123. B. M. Sutherland and J. C. Sutherland, *Biophys. J. 8*, 490 (1969); *9*, 1045 (1969).
124. J. Eisinger and A. A. Lamola, *Biochem. Biophys. Res. Commun. 28*, 558 (1967).
125. B. M. Sutherland and J. C. Sutherland, *Biophys. J.* (to be published).
126. A. A. Lamola and T. Yamane, *Proc. Nat. Acad. Sci. 58*, 443 (1967).
127. R. Ben-Ishai, E. Ben-Hur, and Y. Hornfeld, *Israel J. Chem. 6*, 769 (1968).
128. A. A. Lamola, *Photochem. Photobiol. 9*, 291 (1969).
129. M. Meistrich, A. A. Lamola, and E. Gabbay, *Photochem. Photobiol. 11*, 169 (1970).
130. M. Meistrich and A. A. Lamola, *Mut. Res.* (to be published)
131. M. Meistrich, *Mut. Res.* (to be published)
132. M. Meistrich and R. G. Shulman, *J. Mol. Biol. 46*, 157 (1969).
133. H.-D. Mennigmann and A. Wacker, *Photochem. Photobiol. 11*, 291 (1970).
134. M. Gueron and R. G. Shulman, *Ann. Rev. Biochem. 37*, 571 (1968).
135. T. Förster, *Dis. Faraday Soc. 27*, 7 (1959).
136. T. Förster, *in* "Comprehensive Biochemistry" (M. Florkin and E. H. Stotz, eds.), Vol. 22, Elsevier, Amsterdam (1967).
137. D. L. Dexter, *J. Chem. Phys. 21*, 836 (1953).
138. R. G. Bennett and R. E. Kellogg, Mechanisms and rates of radiationless energy transfer, *in* "Progress in Reaction Kinetics" (G. Porter, ed.), Vol. 4, Pergamon Press, London (1966).
139. A. A. Lamola, Electronic energy transfer in solution: Theory and applications, *in* "Energy Transfer and Organic Photochemistry" (P. A. Leermakers and A. Weissberger, eds.), Vol. XIV of "Techniques of Organic Chemistry," Interscience, New York (1969).
140. M. Gueron, unpublished results.
141. K. H. Drexhage, M. M. Zwick, and H. Kuhn, *Ber. Bunsen Physik. Chem. 67*, 62 (1963).
142. J. Eisinger, M. Gueron, and R. G. Shulman, *in* "Advances in Biological and Medical Physics," p. 219, Vol. 12, Academic Press, New York (1968).
143. W. E. Blumberg, J. Eisinger, and G. Navon, *Biophys. J. 8*, A-106 (1968).
144. J. W. Longworth, unpublished results.
145. R. Bersohn and I. Isenberg, *Biochem. Biophys. Res. Commun. 13*, 205 (1963).
146. J. Eisinger and R. G. Shulman, *Proc. Nat. Acad. Sci. 55*, 1387 (1966).
147. J. Eisinger, unpublished results.

148. J. Longworth, reported at Symposium on Biological Molecules in Their Excited States, Arden House, New York (October 1969). For a review of the meeting, see J. Eisinger, A. A. Lamola, J. Longworth, and W. Gratzer, *Nature* **226**, 113 (1970).
149. W. C. Galley, reported at Symposium on Biological Molecules in Their Excited States, Arden House, New York (October 1969). For a review of the meeting, see J. Eisinger, A. A. Lamola, J. Longworth, and W. Gratzer, *Nature* **26**, 113 1970.
150. M. Gueron and R. G. Shulman, unpublished results.
151. R. S. Somer and J. Jortner, *J. Chem. Phys.* **49**, 3919 (1968).
152. J. Eisinger, M. Gueron, and R. G. Shulman, unpublished results.
153. J. Eisinger, B. Feuer, and A. A. Lamola, *Biochemistry* **8**, 3908 (1969).
154. A. A. Lamola, unpublished results.
155. I. Isenberg, R. Rosenbluth, and S. L. Baird, Jr., *Biophys. J.* **7**, 365 (1967).
156. I. Isenberg, R. B. Leslie, S. L. Baird, R. Rosenbluth, and R. Bersohn, *Proc. Nat. Acad. Sci.* **52**, 379 (1964).
157. W. C. Galley, *Biopolymers* **6**, 1279 (1968).
158. G. Weill and M. Calvin, *Biopolymers* **1**, 401 (1963).
159. C. Hélène and T. Montenay-Garestier, *Chem. Phys. Letters* **2**, 25 (1968).
160. J. Eisinger, A. A. Lamola, J. W. Longworth, and W. Gratzer, *Nature* **126**, 113 (1970).
161. J. Eisinger, *Biochemistry* **8**, 8902 (1969).
162. W. C. Galley and L. Stryer, *Proc. Nat. Acad. Sci.* **60**, 108 (1968).
163. G. Weber and E. Daniel, *Biochemistry* **5**, 1900 (1966).
164. H. G. Zachau, *Angew. Chem. Internat. Ed.* **8**, 711 (1969).
165. J. Eisinger and B. Feuer, unpublished results.
166. A. Favre, M. Yaniv, and A. M. Michelson, *Biochem. Biophys. Res. Commun.* **37**, 266 (1969).
167. L. Iwamura, N. S. Leonard, and J. Eisinger, *Proc. Nat. Acad. Sci.* **65** 1025 (1970).
168. W. Fuller and A. Hodgson, *Nature* **215**, 817 (1967).
169. V. L. RajBhandary, S. H. Chang, A. Stuart, R. D. Faulkner, R. M. Hoskinson, and H. G. Khorana, *Proc. Nat. Acad. Sci.* **57**, 751 (1967).
170. J. Eisinger, B. Feuer, and T. Yamane, *Proc. Nat. Acad. Sci.* **65**, 638 (1970).
171. K. Beardsley and C. R. Cantor.
172. R. Thiebe and H. G. Zachau, *Europ. J. Biochem.* **5**, 546 (1968).
173. L. Stryer, *J. Am. Chem. Soc.* **88**, 5708 (1966).
174. K. Beardsley and C. R. Cantor, *Proc. Nat. Acad. Sci.* **65**, 39 (1970).
175. J. Eisinger, B. Feuer, and T. Yamane, *Nature* (in press) (1971).
176. M. Leng, F. Pochon, and A. M. Michelson, *Biochim. Biophys. Acta* **169**, 338 (1968).
177. C. Hélène, M. Yaniv, and J. W. Elder, *Biochem. Biophys. Res. Commun.* **31**, 660 (1968).
178. J. Eisinger (to be published).

Chapter 4

FLUORESCENT PROTEIN CONJUGATES*

W. B. Dandliker and A. J. Portmann

Department of Biochemistry
Scripps Clinic and Research Foundation
La Jolla, California

---

1. INTRODUCTION

This chapter deals with the methods for the preparation, purification, and characterization of fluorescent protein conjugates. The methods of preparation are approached from the viewpoint of the functionally reactive groups in proteins and in the molecule acting as a label. The purification of conjugates includes the important phases of removal of the free, unbound dye as well as methods to effect some degree of fractionation to yield materials of relatively uniform degrees of labeling. The methods of characterization are based upon a variety of chemical, optical, immunological, and miscellaneous physical properties such as sedimentation constant, electrophoretic behavior, and solubility. The application of fluorescent protein conjugates to a variety of chemical and biological problems is discussed.

Often the attachment of a fluorescent molecule to a macromolecule results in changes in decay time or emission characteristics. When changes in emission occur, so as to reduce either quenching or enhancement, this information may be used in an indirect way to deduce what changes have occurred in the environment of the fluorescent molecule, that is, what has happened to the macromolecule in terms of conformational changes, hydration, and the like.

The polarization of fluorescence has long been employed as a means

---

*Supported by The John A. Hartford Foundation, Inc., The National Science Foundation (GB 6887), The National Institute of Arthritis and Metabolic Diseases (AM 7508) of The National Institutes of Health, The American Cancer Society, California Division, Special Grant No. 480.

for determining absolute relaxation times of macromolecules. More recently, it has proved useful in the sensitive detection of changes in local brownian motion such as those accompanying folding, unfolding, or melting of macromolecular structures. Another area of application of polarization changes is in the monitoring of the kinetics and equilibria of macromolecular association or dissociation reactions in which one macromolecular component is labeled. Because the polarization changes can be followed rapidly, the course of even very fast reactions can be determined by this method.

Historically, one of the first applications of fluorescent labeling was made by Coons *et al.*, who labeled antibody so that its localization in tissue preparations containing antigen could be determined by fluorescence microscopy. This technique has become one of the most important and powerful now used in immunology.

Noncovalently bound fluorescent labels have been investigated in some cases from the viewpoint of dye binding to macromolecules, and in others as probes for some feature of macromolecular structure. For example, acridine dyes bind to helical nucleic acids and undergo large changes in absorption and emission. Certain anilinonaphthalenes bind to the hydrophobic areas of proteins with a large increase in quantum yield.

Steiner and Edelhoch[1] have given an excellent review on fluorescent protein conjugates up to 1962. Fluorescence polarization is dealt with in some detail, and shorter sections on the general features of fluorescence and energy transfer are included. Methods of preparing the conjugates and of investigating their properties are treated also in detail. The extensive literature on the use of fluorescent protein conjugates as histochemical stains is covered very thoroughly. The present review generally continues where the one by Steiner and Edelhoch left off, but in a few cases important papers earlier than 1962 are cited for one reason or another.

2. CHEMISTRY OF CONJUGATION

2.1. Functional Groups in Proteins and in the Label

The types of reactive groups used in labeling procedures obviously depend upon the reactive functional groups present in proteins. These include the reactive groups in the side-chains of the various amino acids as well as any terminal amino or carboxyl groups. While some discussion of this area can be rationally made from the viewpoint of functional groups on the two reactants, the details of the conformation of the macromolecule may modify the expected reactions drastically. This is especially well known in the case of —SH groups, which sometimes have to be "unmasked" before any reactivity can be detected at all. The same effects are true to varying degrees for other groups which may become accessible or inaccessible depending upon the nature of the solvent present, pH, and other factors. A summary of the known reactions of the side-chains of proteins with various reagents is given in Table 1, together with a rough, quantitative indication of the extent of reaction. Not all of these reactions have been used to produce fluorescent-labeled protein conjugates, but it is conceivable that any of them might be applicable in special cases, and so as many as possible have been included for the sake of completeness.

The variety of possible dye structures useful in producing fluorescent protein conjugates is very great. However, relatively few of the possible structures have found widespread application for one reason or another. The most widely employed dyes are probably the aminonaphthalenes, the fluoresceins and rhodamines. In order for a dye to be useful as a fluorescent labeling reagent, it must be possible to introduce a satisfactory functional group into the dye molecule for coupling to the protein. In addition, the final product must be fluorescent. Insofar as the protein is concerned, usually the α- or ε-amino groups are the sites of labeling through thiourea or sulfonamido linkages. Good methods are needed by which carboxyls, aliphatic hydroxyls, or phenolic rings could be labeled.

In addition to the data of Table 1, other special characteristics of the labeling reactions, as well as side-reactions, are further delineated individually below.

2.1.1. Anhydrides

$$\begin{matrix} R\text{-CO} \\ \phantom{R\text{-}}\diagdown \\ \phantom{R\text{-CO}}>O \\ \phantom{R\text{-}}\diagup \\ R'\text{-CO} \end{matrix} + \begin{cases} H_2N\text{———Prot} \\ HS\text{———Prot} \\ HO\text{-}\langle\bigcirc\rangle\text{-Prot} \end{cases} \rightarrow \begin{matrix} R\text{-CO-NH-Prot} \\ R\text{-CO-S-Prot} \\ R\text{-CO-O-}\langle\bigcirc\rangle\text{-Prot} \end{matrix} + \begin{matrix} R'\text{-COOH} \\ R'\text{-COOH} \\ R'\text{-COOH} \end{matrix} \quad [1]$$

This reaction gives amides with amino groups, thioesters (or acylmercaptans) with sulfhydryl groups,[2] and phenolic esters with the phenolic hydroxyl groups. However, the latter esters are easily hydrolyzed and are only stable if the pH is carefully maintained at 7.5 to 8 and if buffer systems containing nucleophiles (e.g., tris) are avoided. Otherwise, the specificity is limited to amino and sulfhydryl groups.[3] If a mixed anhydride is employed, it may be uncertain which part of the molecule will constitute the best leaving group and consequently with which group the protein becomes labeled. Unless some clear distinction can be made, a mixed product may be expected.

The anhydrides can be prepared in various ways, the most important ones being the interaction of an acyl halide with a carboxylic acid or its salt (in the former case, usually in the presence of pyridine):

Table 1. Reactions Between the Functional

		Carboxyl	Phenol	Aliphatic hydroxyl	Sulfhydryl	Thio-methyl
		—COOH	—⟨=⟩—OH	—OH[a]	—SH	—SCH$_3$
		Asp Glu C-terminal	Tyr	Ser Thr	Cys	Met
(1)	Anhydrides		±	?	2+	
(2)	Isocyanates ⎫		±		3+	
(3)	Isothiocyanates ⎭					
(4)	Arylsulfonyl chlorides		+		+	
(5)	Epoxides	3+	2+		2+	
(6)	Diazoacetyl derivatives	3+			2+	
(7)	Haloacetyl derivatives		±		3+	±
(8)	N-substituted maleimides				3+	
(9)	Quinones				3+	
(10)	Mustards	2+	+	?	3+	+
(11)	2-hydroxy-5-nitrobenzyl bromide		±		+	
(12)	Reactive aryl halides		+		3+	
(13)	Mercurials	?			3+	
(14)	Aromatic diazonium salts		3+		?	
(15)	N-bromosuccinimide					
(16)	Imidoesters					
(17)	Triazine derivatives		2+	2+	3+	

[a] It is interesting to note that the aliphatic hydroxyls of serine and threonine in proteins usually display a relatively low order of activity when contrasted to that of ordinary aliphatic hydroxyls of alcohols. It is not known whether this lack of reactivity is due to the fact that the hydroxyl is a part of the serine or threonine molecule or whether, as a result of the tertiary structure of proteins, the OH groups are hydrogen bonded or otherwise masked so that their reactivity does not appear.

Groups of Dyes and Proteins

Amino	Guanidyl	Indole	Imidazole	Amide	
$-NH_2$	$-NH-C-NH_2$ $\parallel$ NH	(indole structure)	(imidazole structure)	$-CONH_2$	
Lys N-terminal	Arg	Trp	His	Asn Gln (C-term)	References
3+ 3+	?		?		2 2
3+			+		7, 8
+ ±					2 2, 7, 11
±		±	±		2, 12, 13
?					2, 12, 14
3+ +		± 3+	+		12 18, 19, 20 21, 22, 23
2+			+		24, 25
±		?	3+	?	2, 14, 20, 26 2
		±	3+		27, 28, 29
3+ 3+	+	±	3+		30 32, 33

$$\text{R—COOH} + \text{Cl—CO—R'} \longrightarrow \text{R—CO—O—CO—R'} + \text{HCl} \qquad [2]$$

or the dehydration of carboxylic acids, which can be accomplished by agents such as acetyl chloride, acetic anhydride, or phosphorus oxychloride:

$$2 \text{ RCOOH} \xrightarrow{-\text{H}_2\text{O}} \text{R—CO—O—CO—R} \qquad [3]$$

2.1.2. Isocyanates

$$\text{R—N=C=O} + \begin{cases} \text{H}_2\text{N—Prot} \longrightarrow \text{R—NH—CO—NH—Prot} \\ \text{HS—Prot} \longrightarrow \text{R—NH—CO—S—Prot} \end{cases} \qquad [4]$$

In the reaction of isocyanates with amino groups a disubstituted urea is formed, while with the sulfhydryl group a thiocarbamate results. Despite the normal experience in organic chemistry that isocyanates react about equally well with alcohols or with amines, to our knowledge there is no evidence that such a similarity exists between the reactivity of amino groups of proteins and the aliphatic hydroxyls of serine or threonine residues in proteins.[2] In fact, it was shown long ago[4] that serine or threonine hydroxyls do not react with phenyl isocyanate.

There are several ways of preparing isocyanates; among them is the Curtius rearrangement of acyl azides:

$$\text{R—CO—N}_3 \xrightarrow{\text{heat}} \text{R—NCO} + \text{N}_2 \qquad [5]$$

The most widely employed method, however, utilizes the action of phosgene on primary amines,[5] and an intermediate carbamyl chloride is formed in the course of this reaction:

$$\text{R—NH}_2 + \text{COCl}_2 \longrightarrow \text{R—NH—CO—Cl} \longrightarrow \text{R—NCO} + \text{HCl} \qquad [6]$$

The product has to be prepared and stored with absolute exclusion of water; otherwise, some isocyanate will decompose (via the unstable carbamic acid) to the amine, which, in turn, reacts with more isocyanate to yield a disubstituted urea.

2.1.3. Isothiocyanates

$$\text{R—N=C=S} + \begin{cases} \text{H}_2\text{N—Prot} \longrightarrow \text{R—NH—CS—NH—Prot} \\ \text{HS—Prot} \longrightarrow \text{R—NH—CS—S—Prot} \end{cases} \qquad [7]$$

The reaction of isothiocyanates with the amino groups of proteins by analogy with simple organic compounds results in the production of a thiourea, while the reaction with sulfhydryl groups similarly should produce thiocarbamates. The reactivities of isocyanates and isothiocyanates with the groups in proteins seem to be similar, but the isothiocyanates present at least two distinct advantages—namely, their preparation employs the use of thiophosgene, which is considerably easier to manipulate than phosgene itself, and, in addition, the stability on storage is much better.[6]

2.1.4. Arylsulfonyl Chlorides

$$R\text{-}SO_2Cl + \begin{cases} H_2N\text{-}Prot \longrightarrow R\text{-}SO_2\text{-}NH\text{-}Prot \\ HO\text{-}\langle\!\!\!\bigcirc\!\!\!\rangle\text{-}Prot \longrightarrow R\text{-}SO_2\text{-}O\text{-}\langle\!\!\!\bigcirc\!\!\!\rangle\text{-}Prot \end{cases} \quad [8]$$

This reaction is perhaps the most frequently used in protein labeling in the form of 5-dimethylaminonaphthalene-1-sulfonyl chloride (dansyl chloride). The arylsulfonyl chlorides have many advantages: they are stable on storage and usually quite unreactive with water, while the reactivity with a variety of functional groups in proteins is high. The conjugates formed by the reaction with α- and ε-amino groups and with phenolic hydroxyls are sulfonamides and sulfonic esters, respectively; the sulfonamides are exceptionally stable. Derivatives of considerably diminished stability arise from the reaction with sulfhydryl and imidazole groups. The aliphatic hydroxyl of serine usually does not react with sulfonyl chlorides. In special cases, however, as for example in chymotrypsin, the neighboring groups change the reactivity of the serine so drastically that it becomes labeled by dansyl chloride.[7,8]

While the arylsulfonyl chlorides may be obtained by direct halosulfonation of aromatic compounds by chlorosulfonic acid, the usual method of preparation is the treatment of sulfonic acids with agents such as thionyl chloride, phosphorus pentachloride, or phosphorus oxychloride:

$$R\text{—}SO_3H + PCl_5 \longrightarrow R\text{—}SO_2Cl + POCl_3 + HCl \quad [9]$$

In many cases, it is sufficient to grind the sulfonic acid with PCl_5 without a solvent, then to extract the sulfonyl chloride with acetone, in which the unreacted acid is insoluble.[9,10]

2.1.5. Epoxides

$$R\text{-}\underset{O}{CH\text{-}CH_2} + \begin{cases} HOOC\text{-}Prot \longrightarrow R\text{-}CH(OH)\text{-}CH_2\text{-}O\text{-}CO\text{-}Prot \\ HO\text{-}\langle\!\!\!\bigcirc\!\!\!\rangle\text{-}Prot \longrightarrow R\text{-}CH(OH)\text{-}CH_2\text{-}O\text{-}\langle\!\!\!\bigcirc\!\!\!\rangle\text{-}Prot \\ HS\text{-}Prot \longrightarrow R\text{-}CH(OH)\text{-}CH_2\text{-}S\text{—}Prot \\ H_2N\text{-}Prot \longrightarrow R\text{-}CH(OH)\text{-}CH_2\text{-}NH\text{—}Prot \end{cases} \quad [10]$$

While the epoxides can react with a variety of groups in proteins, they are nevertheless of special potential interest because they are one of the few reagents which combine with carboxyl groups to yield esters.[2] In neutral or alkaline solution the reactions proceed as shown. However, in acidic solution epoxides are attacked at the tertiary carbon, and the product would presumably be a hydroxymethyl derivative. The reaction with phenols and sulfhydryl groups yields ethers or thioethers (sulfides), respectively. The normal course of the reaction with an amino group results in the formation of the secondary amine, but possibly tertiary amines might be formed with excess reagent.

One way to obtain epoxides is by the treatment of 1,2-chlorohydrins with alkali or strong organic bases such as sodium ethoxide:

$$\text{R—CH—CH—R'} \xrightarrow{\text{base}} \text{R—CH—CH—R'} \quad [11]$$
$$\phantom{\text{R—C}}\text{OH}\phantom{\text{—C}}\text{Cl} \phantom{\xrightarrow{\text{base}} \text{R—CH—CH—R}}\text{O}$$

The necessary chlorohydrins may be prepared in a number of ways, which include the action of chlorine and water on a double bond ("hypochlorous acid addition"), the action of chloroacetone on aromatic Grignard compounds, or the action of ethyl dichloroacetate on aldehydes or ketones. Starting from an alkene, there is also the possibility of direct epoxidation by means of a peroxy acid:

$$\text{R—CH=CH—R'} + \text{R''—COOOH} \longrightarrow \text{R—CH—CH—R'} + \text{R''COOH} \quad [12]$$
$$\phantom{\text{R—CH=CH—R'} + \text{R''—COOOH} \longrightarrow \text{R—CH—C}}\text{O}$$

2.1.6. Diazoacetyl Derivatives

$$\left.\begin{array}{c}\text{R—CO—CH=}\overset{+}{\text{N}}\text{=}\overset{-}{\text{N}} \\ \updownarrow \\ \text{R—CO—}\overset{-}{\text{CH}}\text{—}\overset{+}{\text{N}}\text{≡N}\end{array}\right\} + \left\{\begin{array}{l}\text{HS—Prot} \longrightarrow \text{R—CO—CH}_2\text{—S—Prot} + \text{N}_2 \\ \\ \text{HOOC—Prot} \longrightarrow \text{R—CO—CH}_2\text{—O—CO—Prot} + \text{N}_2\end{array}\right.$$

$$[13]$$

While most aliphatic diazo compounds are of low stability and are hence very difficult to work with, diazoacetyl derivatives are relatively stable, but still highly reactive as alkylating reagents for either —SH groups or carboxylic acid groups in proteins. By temporarily blocking the sulfhydryl groups with organic mercurials, these reagents are fully specific for the carboxyl groups.[2,7,11]

The preparation of these compounds is accomplished by the reaction of nitrous acid on a primary amino group alpha to a stabilizing group (e.g., keto or ester):

$$R-CO-CH_2-NH_2 + HONO \longrightarrow R-CO-CH=\overset{+}{N}=\overset{-}{N} + 2H_2O \quad [14]$$

or by the interaction of an acyl halide with an excess of diazomethane to remove HX formed:

$$R-CO-Cl + 2\ N_2CH_2 \longrightarrow R-CO-CHN_2 + CH_3Cl + N_2 \quad [15]$$

2.1.7. Haloacetyl Derivatives

$$R-CO-CH_2X + HS-Prot \longrightarrow R-CO-CH_2-S-Prot \quad [16]$$
X may be I or Br, sometimes Cl, but never F

With certain similarities to the diazoacetyl derivatives, haloacetyl derivatives also are powerful alkylating agents, but their specificity is completely different: the reaction with the carboxyl group seems to be totally lacking; they react instead easily with the thiol, thiomethyl, amino, and imidazole groups. Since the reaction with —SH is fastest, haloacetyl derivatives can be used specifically for the thiol groups by not applying an excess of the reagent. If cysteine is either absent or blocked, the reaction can be directed to some extent by the choice of pH. Anywhere above pH 2 the reaction with methionine is independent of pH; histidine and amino groups react in the unprotonated forms, i.e., above pH 5 and pH 7, respectively. The rarely observed reaction with phenolic groups takes place only above pH 9. The most commonly used reagents in this class are iodo- and bromoacetate. Somewhat interchangeably, haloacetamides or haloacetic esters are also used, although with some proteins reaction at specific sites may occur only with one but not the other. Higher homologues of α-haloacids, such as α-bromopropionate, are also reactive. The reactions with imidazole and with amino groups may lead to disubstituted products.[2, 12, 13]

The preparation of the α-haloacids and derivatives is accomplished by the reaction of bromine on the free acid in the presence of a trace of red phosphorus, thus converting the acid into the α-bromoacyl bromide (Hell–Volhard–Zelinsky reaction). This latter compound is then either hydrolyzed to the free α-brominated acid, alcoholyzed to the corresponding ester, or ammonolyzed to the amide:

$$R-CH_2-COOH + Br_2 \xrightarrow{P} R-CHBr-COBr \begin{cases} \xrightarrow{H_2O} R-CHBr-COOH \\ \xrightarrow{R'OH} R-CHBr-COOR' \\ \xrightarrow{NH_3} R-CHBr-CONH_2 \end{cases}$$

[17]

The same reaction is possible with chlorine instead of bromine, although there is some loss of specificity due to the tendency to chlorinate at additional locations. Iodo derivatives are not accessible by direct iodination, but are easily obtained by metathesis from the chloro or bromo derivatives:

$$R-CHCl-COOR' + KI \longrightarrow R-CHI-COOR' + KCl \qquad [18]$$

2.1.8. N-Substituted Maleimides

$$R-N\underset{O}{\overset{O}{\bigcirc}} + HS\text{-Prot} \longrightarrow R-N\underset{O}{\overset{O}{\bigcirc}}\overset{H}{\underset{H}{\bigvee}}\overset{S\text{-Prot}}{\underset{H}{}}$$

$$\downarrow \text{may hydrolyze}$$

$$R\text{-NH-CO-CH}_2\text{-CH-S-Prot}$$
$$|$$
$$COOH$$

[19]

At pH 7 the reaction is specific for sulfhydryl groups; below pH 6 the rate is very low, and above pH 7 the reagent begins to hydrolyze and groups other than sulfhydryl may also be attacked.[2, 12, 14]

N-substituted maleimides are synthesized by the reaction of a primary amine with maleic anhydride in the presence of polyphosphoric acid (PPA) according to the sequence[15-17]

$$R\text{-NH}_2 + O\underset{O}{\overset{O}{\bigcirc}} \longrightarrow \underset{HO\text{-C-CH}}{\overset{R\text{-NH-C-CH}}{\underset{\parallel}{\parallel}}} \xrightarrow{\text{PPA}}_{-H_2O} R\text{-N}\underset{O}{\overset{O}{\bigcirc}} \qquad [20]$$

2.1.9. Quinones

$$\underset{R'}{\overset{R}{\bigcirc}}\underset{O}{\overset{O}{\bigcirc}} + H_2N\text{-Prot} \longrightarrow \underset{R'}{\overset{R}{\bigcirc}}\underset{OH}{\overset{OH}{\bigcirc}}\text{NH-Prot} \qquad [21]$$

The reaction of quinones with sulfhydryl or amino groups proceeds by a 1,4-addition (only amino groups shown in Eq. [21]). The product is readily reoxidized by an unreacted quinone molecule which is then further unreactive provided R and R' are not hydrogens. If the substituents are hydrogens, the quinone molecule resulting by the oxidation step is able to react with another reactive protein group leading to crosslinking.[12] Inasmuch as the initial quinone probably will be a naphthoquinone or an anthraquinone, the crosslinking reaction will not ordinarily be of importance.

2.1.10. Mustards

The reaction of mustard gas and its analogues, the nitrogen mustards, with proteins occurs via a cyclic intermediate:

$$R-N\begin{matrix}CH_2-CH_2-Cl\\ \\ CH_2-CH_2-Cl\end{matrix} \longrightarrow R-\overset{+}{N}\begin{matrix}CH_2\\ \diagdown\\ CH_2\\ | \\ CH_2-CH_2-Cl\end{matrix} \quad Cl^- \qquad [22]$$

These intermediates are readily hydrolyzed in aqueous media, but they react even more readily with a variety of protein groups, the second reactive group on the open chain being hydrolyzed and not leading to crosslinking:[18-20]

$$R\text{-}\overset{+}{N}\begin{matrix}CH_2\\ \diagdown\\ CH_2\\ | \\ CH_2\text{-}CH_2\text{-}Cl\end{matrix} + \begin{cases}HOOC\text{——Prot} \longrightarrow HO\text{-}CH_2\text{-}CH_2\text{-}\overset{R}{N}\text{-}CH_2\text{-}CH_2\text{-}O\text{-}CO\text{-}Prot\\ HS\text{——Prot} \longrightarrow HO\text{-}CH_2\text{-}CH_2\text{-}\overset{R}{N}\text{-}CH_2\text{-}CH_2\text{-}S\text{——Prot}\\ HO\text{-}\bigcirc\text{-}Prot \longrightarrow HO\text{-}CH_2\text{-}CH_2\text{-}\overset{R}{N}\text{-}CH_2\text{-}CH_2\text{-}O\text{-}\bigcirc\text{-}Prot\\ H_2N\text{——Prot} \longrightarrow HO\text{-}CH_2\text{-}CH_2\text{-}\overset{R}{N}\text{-}CH_2\text{-}CH_2\text{-}NH\text{——Prot}\end{cases}$$

[23]

2.1.11. 2-Hydroxy-5-nitrobenzyl Bromide

The introduction of a hydroxyl group into the ortho position of benzyl bromide effects a great enhancement of the reactivity, generally, as well as a unique, almost exclusive specificity for the indole group of tryptophan. At neutral or acidic *p*H, the sulfhydryl group of cysteine is also attacked slowly, whereas in alkaline solutions the reaction is about equally fast with tryptophan, cysteine, and tyrosine. The structure of the coupled product is at present not elucidated. To what extent the parent compound could be substituted with a dye molecule and still retain its reactivity is also not yet known.[21-23]

2.1.12. Reactive Aryl Halides

$$\text{Ar-X} + \begin{cases} \text{HS} \longrightarrow \text{Prot} \longrightarrow \text{Ar-S} \longrightarrow \text{Prot} \\ \text{HN}\underset{}{\overset{=N}{\diagdown}}\text{Prot} \longrightarrow \text{Ar-N}\underset{}{\overset{=N}{\diagdown}}\text{Prot} \\ \text{H}_2\text{N} \longrightarrow \text{Prot} \longrightarrow \text{Ar-NH} \longrightarrow \text{Prot} \\ \text{HO}-\!\!\bigcirc\!\!-\text{Prot} \longrightarrow \text{Ar-O}-\!\!\bigcirc\!\!-\text{Prot} \end{cases} \quad [24]$$

These halides, e.g., 1-fluoro-2,4-dinitrobenzene (FDNB)[24] or 7-chloro-4-nitrobenzo-2-oxa-1,3-diazole (NBD chloride),[25] react with sulfhydryl, imidazole, α- and ε-amino, and phenolic hydroxyl groups. The most reactive are sulfhydryl groups.

2.1.13. Mercurials

$$\left.\begin{array}{c}\text{R}-\overset{+}{\text{Hg}}\text{X}^-\\ \updownarrow\\ \text{R}-\text{Hg}-\text{X}\end{array}\right\} + \text{HS}-\text{Prot} \longrightarrow \text{R}-\text{Hg}-\text{S}-\text{Prot} \quad [25]$$

These reagents have a high specificity for —SH alone. The complexes so formed can all be decomposed by excess free thiol, but are otherwise quite stable.[2, 14, 20, 26]

2.1.14. Aromatic Diazonium Salts

$$\text{Ar}-\overset{+}{\text{N}}\!\!\equiv\!\!\text{N}\ \text{Cl}^- + \begin{cases} \text{HO}-\!\!\bigcirc\!\!-\text{Prot} \longrightarrow \text{HO}-\!\!\bigcirc\!\!\overset{\text{Ar-N=N}}{-}\text{Prot} \\ \underset{\text{H}}{\overset{N}{\diagdown}}\!\!\text{Prot} \longrightarrow \text{Ar-N=N}-\!\!\underset{\text{H}}{\overset{N}{\diagdown}}\text{Prot} \end{cases} \quad [26]$$

The reaction of aromatic diazonium compounds with the phenolic ring was one of the earliest used reactions for attaching organic groups to proteins. However, the use in making fluorescent protein conjugates has been very limited, as in some cases the diazonium compound and the conjugate lack the fluorescence of the parent compound. Reaction can proceed not only with the phenolic ring but also with imidazole groups.[2] Large excesses of the diazonium compound might result in substituting both positions on the phenolic ring ortho to the hydroxyl group, as well as both free positions on the imidazole group; however, with the usually employed concentrations these disubstitutions should not be important.

The preparation of the diazonium compounds is accomplished by the reaction of an aromatic amine with nitrous acid in a cold, acidic medium:

$$\text{Ar}-\overset{+}{\text{NH}_3}\text{X}^- + \text{HONO} \longrightarrow \text{Ar}-\overset{+}{\text{N}}\equiv\text{N X}^- + 2\ \text{H}_2\text{O} \qquad [27]$$

The diazonium salts are in many cases rather unstable; usually they are not isolated but used immediately in solution. If a solid diazonium salt should be desired, the reaction is carried out in an acid solution of alcohol, using an alkyl nitrite as the diazotizing agent:

$$\text{Ar}-\overset{+}{\text{NH}_3}\text{X}^- + \text{R}-\text{ONO} \longrightarrow \text{Ar}-\overset{+}{\text{N}}\equiv\text{N X}^- + \text{ROH} + \text{H}_2\text{O} \qquad [28]$$

2.1.15. N-Bromosuccinimide

This reagent has been shown to modify only histidine under appropriate conditions and to convert it to a fluorescent group. Considerable study has been made of this reaction which shows that a nonfluorescent intermediate is first formed which then hydrolyzes to the final fluorescent compound. The exact nature of the chemical reactions is still unknown. Under some conditions, tryptophan may also react with this reagent.[27-29]

2.1.16. Imidoesters

$$\underset{\substack{\|\\ \text{R}-\text{C}-\text{OR}'}}{\overset{+}{\text{NH}_2}\text{Cl}^-} + \text{H}_2\text{N}-\text{Prot} \longrightarrow \underset{\substack{\|\\ \text{R}-\text{C}-\text{NH}-\text{Prot}}}{\overset{+}{\text{NH}_2}\text{Cl}^-} + \text{R}'\text{OH} \qquad [29]$$

This type of reagent seems to be highly specific for the conversion of an amino group into an amidino group, at least in the $p\text{H}$ range between 7 and 10. The resulting amidine is a much stronger base than the parent amine, exhibiting a pK_a of about 12.5.[30]

The preparation of the imidoesters (also called iminoesters, iminoethers, or alkyl imidates) is achieved by the Pinner synthesis, in which a nitrile and an alcohol are treated with dry HCl under anhydrous conditions; an imide chloride is formed as an intermediate:

$$\underset{\substack{\|\|\\ \text{R}-\text{C}}}{\text{N}} + \text{HCl} \longrightarrow \underset{\substack{\|\\ \text{R}-\text{C}-\text{Cl}}}{\text{NH}} \xrightarrow{\text{R}'\text{OH}} \underset{\substack{\|\\ \text{R}-\text{C}-\text{OR}'}}{\overset{+}{\text{NH}_2}\text{Cl}^-} \qquad [30]$$

The ease of this reaction diminishes with increasing size of R of the nitrile.

2.1.17. Triazine Derivatives

$$\text{R-NH-triazine-Cl} + \text{H}_2\text{N-Prot} \longrightarrow \text{R-NH-triazine-NH-Prot} \quad [31]$$

X may be an inactive substituent or a second reactive Cl which, however, does not participate in the labeling reaction.[31] The wide use of this type of reagent in commercial dyeing of cellulose fibers and wool ("reactive dyes") has led to an extensive study of its specificity: almost any functional group present in proteins was found to be attacked, at least to some extent, except disulfide, thiomethyl, and carboxyl groups.[32, 33]

In addition to the use of commercial dyes, the preparation of labeling agents from cyanuric chloride and a dye moiety containing an amino group has also been reported.[34, 35]

2.1.18. Carbodiimides

These reagents are not first introduced into the dye molecule as a functional group, but instead are present simultaneously with the dye and protein during the labeling procedure. As an example, the formation of an amide linkage is effected by the "activation" of a carboxyl group:[36, 37]

$$\text{R'}-\text{N}=\text{C}=\text{N}-\text{R''} + \text{RCOOH} \longrightarrow \begin{array}{c}\text{R}-\text{CO}-\text{O}\\ \diagdown\\ \text{C}=\text{N}-\text{R'}\\ \diagup\\ \text{R'}-\text{NH}\end{array} \quad [32]$$

$$\begin{array}{c}\text{R}-\text{CO}-\text{O}\\ \diagdown\\ \text{C}=\text{N}-\text{R''} + \text{H}_2\text{N}-\text{Prot}\\ \diagup\\ \text{R'}-\text{NH}\end{array} \longrightarrow \text{R}-\text{CO}-\text{NH}-\text{Prot} + \text{R'}-\text{NH}-\text{CO}-\text{NH}-\text{R''} \quad [33]$$

Inasmuch as both carboxyl and amino groups are present in proteins, the use of carbodiimides is likely to produce some undesirable intermolecular crosslinking. Presumably, this could be minimized by working at low protein concentrations.

2.1.19. γ-Irradiation

A novel method of coupling dyes to macromolecules has been described by Andersson.[38] He found that γ-irradiation of solutions of macromolecules with dyes resulted in some irreversible binding of the dye. Experiments were carried out with BSA, DNA, polyglutamic acid, and polymethacrylic acid. Dyes included fluorescein, methylene blue, *m*-dinitrobenzene, and N-dinitrophenyl glutathione. A free-radical mechanism was proposed to account for the reaction.

2.2. Dye Structures

Having considered the conceivable reactions which could be used to couple fluorescent dyes to protein groups, a discussion of the dye structures themselves is now in order. The general important considerations are whether or not an appropriate functional group can be attached to the dye molecule without destroying it and whether or not the final conjugate retains enough of the original fluorescence to be practically interesting.

2.2.1. Fluorescein

The most widely used fluorescent label is fluorescein, the structure of which is shown in formula [34] as the neutral molecule, carrying no charges; an extensive study of the photochemical decomposition and of the ionization forms was carried out by Lindqvist.[39]

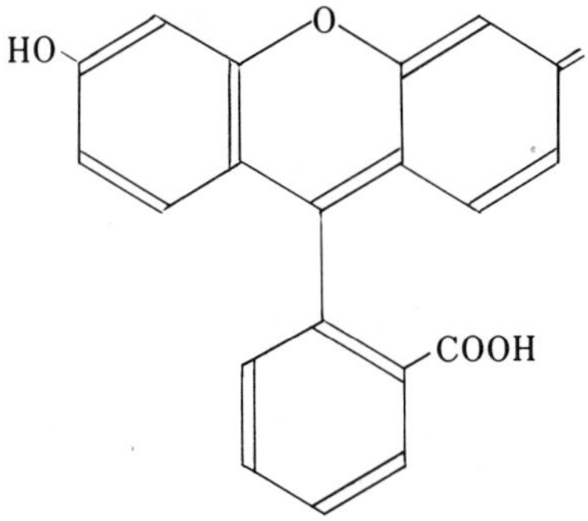

[34]

In order to provide a reactive group for coupling this dye to proteins, an additional substituent is needed. This can be achieved by using 4-nitrophthalic acid, instead of phthalic acid itself, in the synthetic process; the resulting nitrofluorescein can be reduced catalytically or with sulfide to aminofluorescein, which in turn is converted into the isocyanato[40] or isothiocyanato[6] derivative. More recently, the conversion to the dichlorotriazine derivative has also been reported:[35]

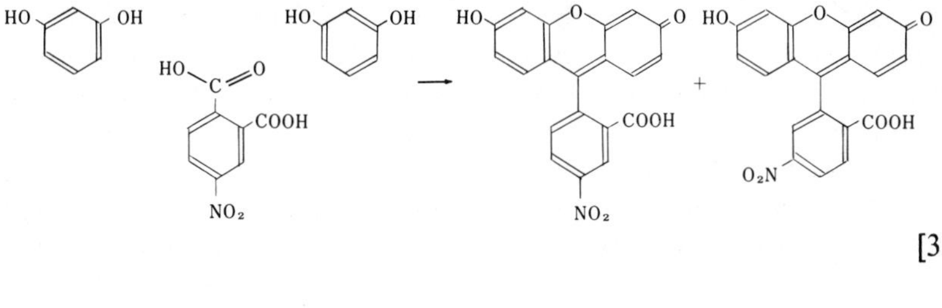

[35]

As indicated in formula [35], two isomers are possible reaction products, and the actual synthesis gives a mixture of both. The separation of the two isomers was originally achieved by fractional crystallization of the diacetate derivatives of the nitro compounds (Coons and Kaplan[40]), the so called isomers I and II resulting. Several chromatographic methods have since been developed for the isolation of the two isomers,[41-44] but their identity has been uncertain until recently. Borek[45] presented evidence from IR spectra that isomer I is 5'-substituted, i.e., para to the carboxylic acid group, whereas isomer II would be 4'-substituted, i.e., meta to the acid group. Corey and Churchill,[46] however, came to exactly the opposite conclusion by examining NMR spectra, and it appears certain now that isomer I is the meta isomer.

Attempts to introduce a fluorescein label into proteins by means of a reactive group other than isocyanate, isothiocyanate or dichlorotriazine, such as a diazonium salt,[47] an acid chloride,[34] or a sulfonyl chloride,[48] did not produce useful conjugates, mostly because the fluorescence intensity was sharply reduced after any of these modifications.

2.2.2. Rhodamines

With isocyanate or isothiocyanate as reactive groups, the rhodamine derivatives are mostly used in the form of their N-alkylated tetramethyl and tetraethyl derivatives. The synthesis is analogous to the one for fluorescein, the starting materials being m-dimethylaminophenol and 4-nitrophthalic anhydride:

[36]

Again, a mixture of two isomeric nitrorhodamines is obtained which are then reduced to the aminorhodamines and converted to isocyanates or isothiocyanates. A study of the equilibria of the various ionized forms of rhodamine B (the tetraethyl derivative), as well as of associations occurring at higher concentrations, was presented by Ramette and Sandell.[49]

A variant of the rhodamine dyes is aminorosamine B, made analogously from m-diethylaminophenol and p-nitrobenzaldehyde.[50] The purification of this dye should be less complicated because no isomers can be formed in the synthesis. However, the fluorescence of conjugates obtained from aminorosamine B isocyanate or isothiocyanate is of moderate intensity, and diazonium coupling produces conjugates with bright fluorescence below pH 3 only.

Another kind of reactive group, namely, a sulfochloride, can be introduced into the commercial dye lissamine rhodamine B200 (also called sulforhodamine B), which already contains sulfonic acid groups:[10, 51, 52]

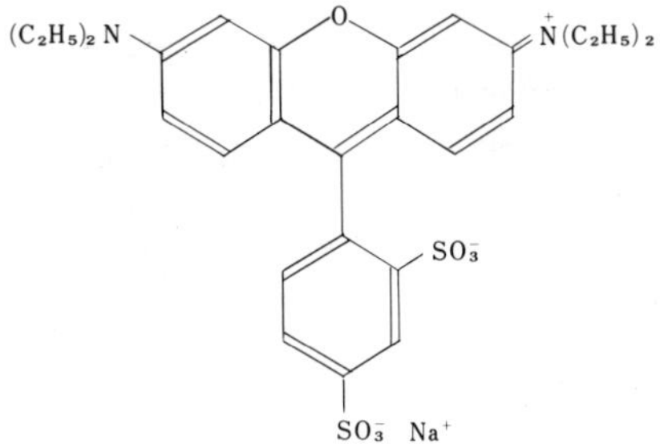

[37]

This dye has found widespread application, especially in the fluorescence microscopy field, despite the fact that recent measurements of its quantum yield showed that the fluorescence intensity is very low.[53]

2.2.3. 5-Dimethylaminonaphthalene-1-sulfonyl Chloride (Dansyl Chloride)

The synthesis of this dye starts with 5-aminonaphthalene-1-sulfonic acid, which is N-methylated with methyl iodide or dimethyl sulfate followed by conversion into the sulfonyl chloride:[9, 54]

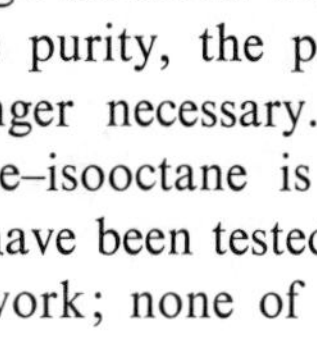

[38]

Since dansyl chloride is of high chemical stability and since it is commercially available in reasonable purity, the preparation of this dye in the laboratory usually is no longer necessary. For very demanding work, recrystallization from acetone–isooctane is indicated.[55] Several related aminonaphthalenesulfonates have been tested with respect to their possible use in fluorescent antibody work; none of the dyes tested except dansyl were useful for this purpose.[56]

2.2.4. 2-p-Toluidinonaphthalene-6-sulfonate (TNS)

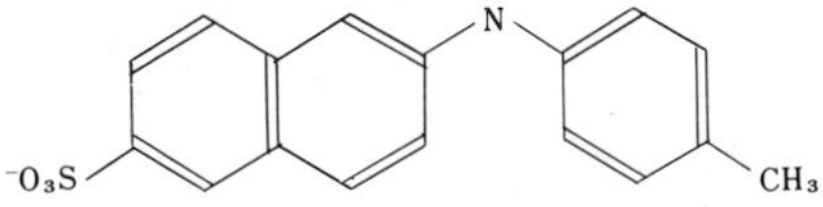

[39]

This compound, along with its anilino analogue (ANS), is extensively used as a fluorescent probe, i.e., not covalently bound to proteins (see section 10). Determination of bond length and angles by X-ray analysis was done by Camerman and Jensen,[57] and an explanation was given on this basis for the drastic solvent-dependent changes in fluorescent behavior (see section 5.2).

2.2.5. Miscellaneous Dyes

A number of other dyes have been investigated as labels in fluorescent protein conjugates. Although these dyes do not seem to have become very popular, some of them may have interesting properties for special applications. Examples are derivatives of anthracene,[58] acridine,* stilbene,[59-61] fluorene,[59] coumarin,[62, 63] pyrene,[62-64] benzimidazole,[65] and benzoxadiazole,[25] as well as N-substituted maleimides,[15-17] reactive dyes,[31-33] pyridoxal-5-phosphate,[66] and a substituted azobenzene[67] which fluoresces after chelation with Al^{3+}.

3. EXPERIMENTAL PROCEDURES FOR LABELING

3.1. Conditions of Labeling

For all common dyes the labeling procedures are very similar. A slightly alkaline to medium alkaline buffer, usually carbonate buffer, is added to the protein solution, the concentration of which is generally kept between 10 and 60 mg/ml. These concentration limits, however, seem relatively unimportant, and labeling at higher dilutions may be carried out with only slightly diminished efficiency. If a protein is to be labeled with an isocyanate or isothiocyanate, a buffer concentration of 0.05 M is adequate; if a sulfochloride is used as the labeling agent, a buffer with a rather large capacity, usually about 0.5 M, is recommended, first because one molecule of hydrochloric acid is liberated for every dye molecule attached to the protein, and second, in case the compound was prepared by the PCl_5 method, because the preparation may contain free acid unless the product has been extensively purified. The influence of pH has been investigated especially well in the case of labeling with fluorescein isothiocyanate.[68] The speed of the labeling reaction was found to increase with increasing pH within the range of pH 6 to 10. The optimal pH was considered to be

---

*A number of acridine mustards have been synthesized by Dr. H. J. Creech and coworkers at The Institute for Cancer Research, Philadelphia, Pa. The authors are indebted to Dr. Creech for gifts of several of these compounds.

at about 9.5, since hydrolysis of the isothiocyanate becomes pronounced above this value. The temperature at which the labeling is done was formerly always kept low, i.e., at 0°C or at refrigerator temperature. More recent evidence, however, points out that the labeling at room temperature is many times faster, without any disadvantages becoming apparent as long as organic solvents are absent.[68, 69]

The dye is then introduced into the buffered protein solution, either dissolved in a small volume of an organic solvent, mostly acetone or dioxane, or in the dry, finely divided form.[70, 71] With either method, some damage to the protein is likely to occur, possibly attributable to the organic solvent, if used, or to local overlabeling where the protein comes into contact with the concentrated dye solution or with the suspended dye particles. These drawbacks are avoided by employing dyes adsorbed to filter paper[72] or, according to Rinderknecht,[73, 74] to cellulose powder or Celite, from which they are released very slowly. Another method, introduced by Clark and Shepard,[75] to avoid these difficulties employs dialysis of the protein solution versus a solution of the dye. The amount of dye to be employed depends mainly on the intended degree of labeling, since invariably about 50–70% of the presented dye becomes bound to the protein.[68, 76] The formerly advocated amount of 50 mg dye for each gram of protein[40] results in rather heavily labeled proteins which generally are not desired. An amount of 6–8 mg/g protein, i.e., for gamma globulin a molar ratio of about 3:1, was found to give more satisfactory conjugates.[76] The degree of labeling is further influenced by the presence of ammonium salts which often remain in gamma globulin preparations from the isolation procedure. Since ammonium ions are competing with the protein for the dye, the efficiency of protein labeling is thereby reduced. For this reason, careful removal of all ammonium salts is important for obtaining reproducible labeling ratios.[77]

3.2. Isolation of the Labeled Conjugate

The older methods for removing low molecular weight substances from the labeling mixture include absorption with tissue powder or charcoal, which gives rise to large losses of protein, dialysis, which has to be performed for several days, and solvent extraction, which causes appreciable damage to the protein. These procedures have become unimportant since gel filtration became available. Chromatography on Sephadex or BioGel P columns gives an efficient separation of the high molecular weight protein and the unwanted products of low molecular weight, such as unreacted or hydrolyzed dye. This procedure is very gentle on the protein, it is fast, and the recovery of protein is practically complete.[78–82] In case

Sephadex, used for the separation of fluorescein isothiocyanate labeled conjugates, becomes contaminated with dye which cannot be removed by simple elution, the gel can be purified by a treatment with H_2O_2.[83]

After the isolation of the labeled protein, its dilution may be so high that at this point reconcentrating is desirable. This can be achieved by adding dry Sephadex to the solution of the conjugate, followed by filtration or centrifugation,[84] by dialysis versus a concentrated solution of a water-soluble, high molecular weight polymer, such as Carbowax,[81] or by pressure dialysis.[76]

3.3. Determination of the Degree of Labeling

Determination of the amount of label chemically combined to a protein or other macromolecule necessitates some accurate measure of both the amount of label and the amount of macromolecule present in the complex. The obvious choices for the determination of the amount of label are optical measurements quantifying either absorption or fluorescence. Alternatively, the fluorescent molecule might be marked with a second label, for example, a radioactive one, thus permitting quantification of the amount of fluorescent molecule in the complex. The method of radioactive labeling is obviously laborious, but the measurements should be quite unambiguous. Measurements of absorption or fluorescence, while more readily made, are subject to a number of uncertainties. For example, changes which were previously considered to be rather unimportant[85] may occur in the magnitude of the extinction coefficient, and in addition there may be a shift in the wavelength of maximum absorption when a fluorescent molecule is chemically coupled to a protein. The fluorescence spectrum is likely to undergo even greater changes than the absorption spectrum, so that methods employing these approaches must necessarily take such possible changes into account (see Figure 1, for example).

Determination of the protein concentration in a complex may be subject to some of these same difficulties when absorption measurements are used, but it appears that the effects on the protein absorption are usually a good deal less important than those on the dye. As a result, protein concentrations are usually measured from the known ultraviolet extinction coefficients of the proteins or by nitrogen analysis or some colorimetric method such as the Folin–Ciocalteu or the biuret method. When ultraviolet extinction coefficients are used to quantify protein concentration, in most cases a large correction is necessary to allow for the ultraviolet absorption of the dye molecule.

The average number, ν, of dye molecules bound per molecule of protein can be found from the equation

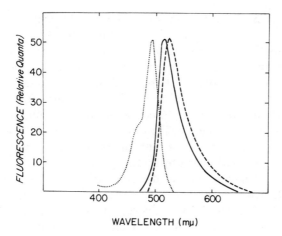

Figure 1. Excitation (······) and emission (- - - -) spectra of bovine γG, labeled with fluorescein isothiocyanate in a ratio of 1:1. Comparison with the emission spectrum of fluorescein (———) shows the shift of the emission which occurs upon conjugation. (Reproduced from *Arch. Biochem. Biophys. 133*, 265 (1969), with permission of R. F. Chen and Academic Press.)

$$\nu = \frac{\dfrac{A_1}{\varepsilon_1} \times \text{MW} \times E_2^{1\%}}{10\left[A_2 - \left(A_1 \dfrac{\varepsilon_2}{\varepsilon_1}\right)\right]} \qquad [40]$$

The subscripts 1 and 2 denote the wavelengths of maximum absorption of the bound dye and the protein, respectively. Equation [40] assumes that the protein has negligible absorption where the dye absorbs maximally. A is the absorbance of the solution in question, ε is the molar extinction coefficient of bound dye, $E^{1\%}$ is the absorbance of 1 cm of a 1% (1g/100 ml) solution of the protein, and MW is the protein molecular weight.

Creech and Peck[86] labeled BSA with aromatic isocyanates. These workers were well aware of the possible alteration of extinction coefficient of the dye when coupled and found correction factors for this effect from model compounds made by condensing the isocyanate with ε-aminocaproic acid. Molar extinction coefficients for the dye in the model compound were assumed to be the same as those in the protein. The protein concentration was determined by nitrogen analysis after suitable deduction for the nitrogen present in the dye. Under various conditions, from 13 to 62 dye molecules were combined to each BSA molecule.

A number of workers have investigated the labeling of proteins with fluorescein isocyanate. Emmart[87] investigated the absorption spectra at various pH for free fluorescein and fluorescein bound to gamma globulin. While no determinations of the degree of labeling were made, changes between the absorption spectra of free and bound fluorescein were noted. Sokol et al.[88] investigated the conjugation of fluorescein isothiocyanate with IgG. The quantum yield of fluorescence of the bound dye was found to decrease with increasing labeling ratio. Goldman and Carver[89] studied the labeling of serum globulins with fluorescein isothiocyanate. The degree of labeling was calculated by dialyzing the free dye out of the reaction mixture and determining the free dye concentration by comparison with a solution of fluorescein isothiocyanate of the same age as that of the reaction mixture. Both absorption and fluorescence intensity measurements were used to thus quantify the free dye. Under the conditions used, a fortyfold excess of isothiocyanate resulted in a labeling ratio of 20 (dye/protein, mole ratio), whereas a twentyfold excess gave a labeling ratio of about 10. From these measurements it was estimated that the fluorescence intensity of bound fluorescein is only about 10% of that of the free dye in neutral buffered solutions. McKinney et al.[90] recommended the use of fluorescein diacetate as a standard for absorption measurements. The advantages of the diacetate are ready purification and stability as a purified crystalline solid. The diacetate hydrolyzes rapidly when put into 0.1 M sodium hydroxide and gives the same extinction coefficient as that found for fluorescein amine isomer I, i.e., 87,000 liter mole^{-1} cm^{-1}. These workers were aware of the change in extinction coefficient upon binding and stated that this quantity is about 20% lower for bound than for the free dye. Dandliker et al.[91] labeled ovalbumin with fluorescein isothiocyanate and determined the difference in extinction coefficients between the bound and free forms of the dye. This was accomplished by dialysis of the reaction mixture to remove the free dye, which was then quantified by direct absorption measurement, assuming that the extinction coefficient of the reaction product was the same as that for free fluorescein. In this case, integrated areas under the absorption curves were compared. The integrated absorption area for a given amount of bound fluorescein was found to be about 12% less than that for the free dye. Direct measurement of sulfur content of the protein before and after labeling was also used to quantify the labeling ratio, but because of the small change in sulfur content, the results were not very accurate. Dedmon et al.[92] labeled rabbit or guinea pig γG with fluorescein isothiocyanate and removed excess free dye by chromatography on Sephadex G-25 followed by fractionation on DEAE-Sephadex. Labeling ratios from 1 to 4 moles of dye per mole of protein were obtained, assuming no change in extinction coefficient. Tengerdy[93] reported results on fluorescein-labeled gamma globulin purified on Sephadex G-25. Protein concentrations were determined by the Folin–Ciocalteu technique, and dye concentrations were measured by light absorption, assuming the same extinction coefficients for free and bound dye. Further results by Tengerdy and Chang[94] revealed that for fluorescein-labeled conalbumin the ratio of absorption for free to bound fluorescein was 1.18. The amount of bound dye was estimated by measuring the free dye separated from the reaction mixture by Sephadex G-25, taking into account the total amount of dye originally put into the system. Kodak disodium fluorescein was used as the standard. Wells et al.[95] determined the extinction coefficients of free fluorescein and fluorescein bound to IgG. A nomogram for both protein concentration and labeling ratio of IgG conjugates from absorption readings at two wavelengths was given. Jobbági and Király,[96] in studying the fluorescein labeling of gamma globulin, found that the wavelength of maximum absorption of the conjugate is at 496 mμ, while that of free fluorescein is at 490 mμ and that of the free isothiocyanate is at 491 mμ. In addition, the molar extinction coefficient of the bound dye is 24% smaller than that of the free dye.

Because of the widespread use of dansyl chloride, labeling with this dye has also been investigated in some detail. Hartley and Massey[97] investigated the labeling of chymotrypsin with dansyl chloride. The degree of labeling was determined by using a known amount of dansyl chloride at the beginning. After a reaction time sufficiently long to insure that any excess dansyl chloride had been hydrolyzed to dansylate, the free dansylate was dialyzed out and determined by an absorption measurement. Bound dansyl was then determined by difference and gave an extinction coefficient of 3.36×10^6 cm^2 mole^{-1} on chymotrypsin and 3.0×10^6 cm^2 mole^{-1} on ovalbumin; the wavelength of maximum absorption for these complexes varied from 335 to 340 mμ. The corresponding extinction coefficient for free dansylate is 4.55×10^6 cm^2 mole^{-1} and the maximum is at 312 mμ. Chen[55] applied a radioactive labeling technique for measuring the degree of labeling with dansyl chloride. Extinction coefficients of protein-bound dansyl were found to vary greatly with the nature of the labeled protein and to a slight extent with the labeling ratio. In all cases, the absorption of the bound dye is considerably less than that of the free dye. The radioactive method is recommended for any highly accurate work; for approximate purposes, an average extinction coefficient for protein-bound dansyl can be assumed to be 3400 liters mole^{-1} cm^{-1}. A shift in λ_{max} was also found to occur, that of free dansylate being at 315 mμ while if bound to ovalbumin appearing at 340 mμ. Other proteins investigated included BSA and chymotrypsinogen A. Chen[98] further studied the fluorescent properties of dansyl–BSA. The degree of labeling was found to affect the emission spectra, quantum yield, decay time, and polarization. In addition, apparently two types of sites exist in BSA, one having a characteristic emission at 500 mμ, and another at 540 mμ. At low labeling ratios, the sites emitting at 500 mμ seem to be occupied predominantly. The quantum yield was found to drop from about 0.7 for low labeling ratios down to about 0.3 for high ratios. Even with low labeling ratios, the decay curves were complex and nonexponential.

3.4. Fractionation According to the Degree of Labeling

The process of labeling a macromolecule leads to effects which may be divided into three categories. First, the conditions to which the macromolecule is necessarily exposed during the labeling procedure may *per se* result in some irreversible changes. This type of effect would be expected usually only if very low or very high pHs were utilized or if appreciable concentrations of organic solvents were present. Second, the mere presence of the label inserted in the macromolecular structure may significantly perturb the native structure and hence the properties of the macromolecule. *A priori* it might be supposed that there could be simple volume effects of the inserted dye molecule, as well as electrical effects depending upon the dipole moment or net charge of the dye. Third, except for very unusual circumstances, a labeling procedure leads to a heterogeneous population of labeled species because of the differing reactivity of the macromolecular functional groups. In a very few cases, as for example in the labeling of ribonuclease, only a single functional group may be involved.

What is important for this section is the existence of methods for fractionating the reaction mixture after the labeling process has been completed in order to obtain materials which have a uniform or nearly uniform degree of labeling. Even after this type of separation, it is very likely that the resulting purified products still consist of a large number of isomers corresponding to different points on the macromolecule where labeling has occurred.

Any fractionation method which has as its goal to separate the many products present in a normal reaction mixture after labeling obviously must focus upon some property which depends upon changes induced by the labeling. Of these, the net charge of the macromolecule is the most obvious choice since most dyes eliminate a protonated α- or ε-amino group, causing a shift in net charge of the macromolecule equal to -1. In addition, the dye molecule may carry a net charge of its own. For example, fluorescein at physiological pH would have a charge of about -1.5 of its own so that the total change in net charge involved in putting a fluorescein molecule into a protein would be about -2.5. Similarly, dansyl chloride also eliminates a protonated amino group. The net charge on the dansyl group itself at physiological pH is zero since the pK of the dimethylammonium group of dansyl in labeled proteins is about 1.9, while that for free dansylate is about 3.9.[99]

The alteration in net charge of the protein most naturally suggests ion exchange chromatography or electrophoretic methods as means of fractionation. Curtain[100] found that the fractions responsible for nonspecific staining in fluorescent antibody conjugates could be removed by chromatography on DEAE-cellulose using both an ionic strength and pH gradient. The materials coming off the column first were found to be lightly labeled, high antibody titer proteins. The later fractions were more highly labeled and gave higher nonspecific staining. Riggs et al.[101] also used chromatography on DEAE-cellulose but with a sodium chloride gradient at pH 7. From the eluents, fractions could be selected which gave high degrees of specific staining with low background. Goldstein et al.,[76] in studying the same problem, came to similar conclusions, that nonspecific staining is caused by excessively high labeling ratios. Labeled preparations were fractionated on DEAE-celluose using gradient elution, and fractions of increasing labeling ratio were obtained. Experimentation showed that if the ratio was higher than about 1.5, nonspecific staining became a problem. George and Walton,[80] in removing fluorescein or rhodamine from γG by gel filtration, found that the protein emerging after the void volume was repeatedly split into two peaks, the first being heavily labeled and the second only slightly labeled. Goldstein et al.,[102] in a further study of fluorescent antibody staining, found that unlabeled antibody molecules could also be removed from labeling mixtures by DEAE chromatography and that for successful fluorescent antibody staining this was just as important as removing the overlabeled material. McDevitt et al.[103] and Dedmon et al.,[92] using fluorescein labeling, Cebra and Goldstein,[104] using tetramethylrhodamine, and Takagi et al.,[105] using dansyl labeling, arrived at overall similar conclusions to those cited above. Wood et al.[71] emphasized the importance of prefractionation of γG on DEAE-cellulose before labeling in order to thoroughly remove other serum proteins. After labeling, the reaction mixture again was chromatographed on DEAE-cellulose to give a fractionation according to the degree of labeling, the lightly labeled fractions coming through the column first.

In a few cases, electrophoretic methods have been employed for fractionating labeled conjugates. Curtain[106] found that some of the fractions in fluorescent-labeled antibodies which are responsible for nonspecific staining could be removed by fractionation by electrophoresis convection; this procedure was later abandoned in favor of DEAE-cellulose chromatography. Kierszenbaum et al.[107] studied the fractionation of ovalbumin and BSA labeled either with dansyl or with fluorescein and found that fractionation according to the degree of labeling was readily obtained by acrylamide gel electrophoresis. With this method as conventionally used, the more highly labeled fractions emerge from the gel first. Although resulting fractions differed by as little as 0.05 to 0.1 in the average mole ratio of dye to protein, the nonintegral values obtained for labeling ratios make it clear that the preparations are still nonuniform with respect to the degree of labeling.

4. EFFECT OF THE LABEL ON THE PROPERTIES OF THE PROTEIN

As pointed out in the previous section, the types of alteration in a macromolecule include both those which are an inevitable result of the pH, ionic strength, solvent composition, etc., of the labeling medium and also those which are manifestations of the mere presence of the fluorescent label in the molecule. Usually, attempts are made to minimize alterations produced by the first set of circumstances, and in any case these could be evaluated to some extent by a control experiment in which the label is lacking. The second, more important type of effect arises from the inherent effects of having the fluorescent label present. In order to assess this factor, a variety of physical parameters might be examined such as, for example, molecular size and shape, electrophoretic properties, optical rotation, and functional properties such as antibody or enzymatic activity.

Well before the advent of the fluorescent antibody technique, the sensitivity of changes in immunological properties was appreciated and was used by Hopkins and Wormall[4] in characterizing proteins labeled with phenyl isocyanate. Coons et al.,[58] in labeling antibodies with β-anthrylisocyanate in 2:1 ratio, found that the specificity of antipneumococcus III was retained. Coons et al.,[108] using fluorescein isocyanate, found that the agglutinin titer of antibodies also was unchanged by labeling. Redetzki[109] labeled antienzyme antibodies with dansyl groups and found that the extent of inhibition of the enzyme by labeled or unlabeled antibody is the same. Albrecht and Sokol[110] found that the introduction of three to nine dansyl groups into antiencephalitis antibody produced no measurable change in the virus neutralization titer. George and Walton[80] and Killander et al.[79] reported negligible changes in antibody titer after fluorescein or rhodamine labeling. Nairn,[85] in discussing this general problem, quoted results on precipitin determinations which were used to illustrate the progressive loss of antibody activity as the degree of labeling was increased. The effect of light labeling was relatively small, but upon labeling with two to four times the optimal amount of dye, a considerable drop in the precipitable protein was found. He also noted that there was little change in solubility or stability of the labeled proteins, but that there were minor changes produced in the ultracentrifuge pattern. By way of comparison, Johnson et al.[111] investigated precipitin curves for iodinated antibody and found that iodination with seven atoms of iodine per molecule still allowed 70% of the antibody activity to be retained. In order to study antibody staining by both fluorescence microscopy and electron microscopy, Hsu et al.[112] labeled antibody both with fluorescein and with ferritin. The native immunological properties of the antibody were retained in all cases. Antibody labeled with fluorescein was found by Tengerdy[93] to behave quantitatively the same as I^{125}-labeled globulin, as measured by the inhibition of the γG/anti-γG precipitation, and it gave a line of identity with unlabeled antibody in the immunodiffusion test. Hebert et al.[113] found that the nonspecific staining produced by fluorescein-labeled globulins increased linearly with increase in labeling ratio and was also aggravated by impurities in the dyes.

A quantitative approach to the extent of immunochemical alteration in labeled molecules was reported by Kierszenbaum et al.[114] In this case the antigen was dansyl-labeled BSA and the antibody was unlabled. The association constant for the reaction between native BSA or dansyl–BSA and anti-native BSA was measured. The association constant for the reaction between the native antigen and its antibody was about two times as great as that for the labeled antigen and indicates that there is a measurable, although small, decrease in the immunochemical specificity of this antigen even after light labeling (see Figure 9).

Measurements of miscellaneous physical properties as well as enzymatic activity have also proven to be useful in detecting alterations due to labeling. Weber[9] noted that the enzymatic activities of dansyl-labeled ribonuclease and fumarase are comparable to those of the unlabeled enzymes, and that the solubility of ovalbumin at the isoelectric point is unchanged by labeling. Schiller et al.[115] made measurements of size, shape, and isoelectric point on fluorescein–BSA. The isoelectric point was found to be shifted about −0.13 units by the introduction of one to two dye molecules. The rate of the in vivo disappearance of fluorescein-labeled BSA was comparable to that obtained for isotope-labeled BSA. These workers noted that fluorescein-labeled guinea pig albumin did not give rise to antibodies in guinea pigs. This result might be caused either by the fact that this material is non-antigenic in guinea pigs or, conversely, by the possibility that antibodies against fluorescein albumin react strongly with native guinea pig albumin which is present in the serum in vast excesses. This would cause the antibodies to be automatically removed. Hartley and Massey[97] found that dansyl chloride labels chymotrypsinogen in the active site so that inhibition is complete when one molecule of dye has been bound. Competitive inhibitors were found to protect the enzyme during the labeling procedure. The labeling of insulin with fluorescein isothiocyanate was found by Halikis and Arquilla[116] to significantly decrease the biological activity of the hormone. Slayter and Hall[117] made the interesting discovery that the labeling of fibrinogen with dansyl protects fibrinogen against UV damage. Frommhagen[118] found a correlation between the pH dependence of solubility and the labeling ratio for fluorescein-labeled γG; fractionation according to the degree of labeling could be obtained by partial precipitation. Heavy labeling was found to increase the electrophoretic boundary spreading in the isoelectric zone. The ultracentrifuge showed no shift in the shape and position of the main sedimenting peak, but more aggregated material was detected at high labeling ratios.

5. EXCITATION AND EMISSION SPECTRA

5.1. Spectral Data

Several collections of spectral curves including absorption, excitation, and emission data are now available. Porro et al.[119] and Hansen[120] deal with dyes commonly used in biological work, while Berlman[121] gives fluorescence spectra of many aromatic molecules. A valuable collection of many types of spectral data is to be found in the new series of Landolt-Börnstein.[122]

5.2. Changes Due to Alterations in Environment of the Dye Molecule

The attachment of a fluorescent molecule to a macromolecule may alter the fluorescent properties of the dye in a variety of ways. First, there may be some effect from the change in chemical structure itself when the reactive functional group on the dye eliminates some group on the protein by combining with it. Second, and usually more important, is the fact that the nature of the surrounding environment in which the fluorescent molecule now finds itself may be considerably different from that of the free dye in terms of the number of neighboring water molecules, the number and type of ions present, and the presence of hydrophobic groups. An additional new feature of the environment also may be that a fluorescent molecule becomes fixed a short distance from another fluorescent molecule residing on the same macromolecule and hence the two can no longer move about independently by translational and rotational motion as molecules would in the unbound state.

To explain the manifold changes observed in individual dyes, either when these dyes are put into solvents other than water or when the dyes are coupled to macromolecules, several mechanisms have been invoked in individual instances enumerated below. These mechansims include (a) a change in ionization, either of the ground state or of an electronically excited state; (b) electronic energy transfer by a Förster[123] type of resonance mechanism; and (c) a change in the rate of intersystem crossing by which molecules in the first excited singlet state go over to an excited triplet which is nonfluorescent. The interpretations of these mechanisms have then been correlated with such macroscopic quantities as the dielectric constant or the Kosower Z parameter, which is an empirical measure of polarity of solvent.[124] Effects of binding a fluorescent molecule might also be described or explained in terms of changes in the amount of hydrophobic bonding and hydrogen bonding and the accessibility of large ions such as iodide, which may lead to deactivation of the excited state.

In addition to the various environmental factors mentioned above should be added at least a brief mention of the inner-filter effect and of the possibility of excimer formation. The inner-filter effect gives rise to an apparent decrease in quantum yield and can originate from two sources: namely, the absorption of an appreciable fraction of the exciting radiation as the incident beam traverses the cell and also reabsorption of the fluorescent radiation by the dye molecules, a process commonly termed *trivial reabsorption*. The inner-filter effect specifically does not include those processes which decrease the tendency of a molecule to fluoresce having been once excited. These processes are discussed below and give rise to quenching, that is, changes in quantum yield. The inner-filter effect can be circumvented by having sufficiently low concentrations of dye in the solution. A partial remedy for the inner-filter effect can be achieved by exciting and by viewing the fluorescence through the same face of the cell.[125]

In case a macromolecule is heavily labeled with a dye, it is conceivable that two other factors might be of importance in some instances. If many dye molecules are adjacent to one another, it is possible that a kind of dimer or polymer formation could occur with a subsequent drastic alteration in fluorescence properties. It is believed that this kind of interaction takes place between acridine dye molecules intercalated into the helix of nucleic acids. This type of interaction, which is probably a manifestation of the well-known tendency of dye molecules to aggregate in solution, should be carefully distinguished from excimer formation. Excimers are dimers which are stable only in the excited state and are known to account for some of the anomalies in the fluorescence properties of solutions of aromatic hydrocarbons at high concentrations. A discussion of excimer fluorescence can be found in Hercules.[126] Döller and Förster[127] investigated temperature effects on excimer fluorescence. At low temperatures, about 0°C, the emission of the excimer is observed; if the temperature is raised to 70°C, fluorescence of the excimer and the normal fluorescence are both very small (quenching), and it was concluded that if the temperature is raised further to about 100°C, dissociation occurs with reappearance of normal fluorescence. Lehrer and Fasman[128] have suggested that certain polyamino acids containing aromatic groups can exhibit conformation-dependent excimer fluorescence.

The early literature pertaining to the effect of environment on fluorescence of aromatic molecules has been reviewed extensively by Van Duuren.[129] Stryer[130] examined the fluorescent properties of a number of compounds in which proton donor groups were either absent or present; where such groups were present, either the un-ionized form only was fluorescent or both the unionized and ionized forms were fluorescent. If no proton-donating group was present in a molecule, then no appreciable isotope effect was noted upon measuring the fluorescence in D_2O as compared to that in water. With a proton-donating group present, if only the unionized form is fluorescent, then the fluorescent yield is higher in D_2O than in water. This effect was interpreted on the basis of equilibration of the protons of the compound with those of the solvent and a slower rate constant for the rate of ionization of the deuteron as compared to that of the proton. If a proton-donating group is present and both the un-ionized and ionized forms are fluorescent, then there is a shift in the shape of the emission curve as one goes from water to D_2O. Stryer suggested that the same type of mechanism might explain the marked fluorescence enhancement of the anilinonaphthalenes in media of low dielectric constant.

A quite different interpretation of the behavior of the anilinonaphthalenes was given by Turner and Brand.[131] These workers found that N-methyl-2-anilinonaphthalene-6-sulfonic acid, which has no proton-donating group, gave a large fluorescence enhancement in media of low dielectric constant similar to the effects observed with 2-anilinonaphthalene-6-sulfonic acid itself. They found, as Stryer[132] had noted previously, that there was a correlation between the fluorescence intensity and the dielectric constant in any one solvent, but that different solvents gave points falling on different lines. However, when correlation of the fluorescence properties was made versus the Kosower Z parameter, the best correlations were obtained, and a single line was found for all water-solvent systems investigated. On the basis of the anilinonaphthalenes noted above, these workers indicated that Stryer's mechanism[130] of proton donation from the excited state is not the correct explanation. Instead, they proposed that the difference in energy between the excited singlet and triplet states should decrease as solvent polarity is increased and in this way increase the rate of intersystem crossing so that excited molecules would terminate in a long-lived triplet state, thus decreasing the fluorescent yield. Weber[133] investigated the effect of the dielectric constant on the fluorescence spectrum of simple dansyl derivatives. A displacement toward longer wavelengths in media of high dielectric constant was found. This led to the conclusion that the dipole moment of the excited state is higher than that of the ground state by about 10 to 12 Debye units.

The present widespread interest in fluorescent probes seems to have originated with the work of Winkler.[134] He prepared antinaphthalenesulfonate antibodies by immunization with bovine gamma globulin coupled through an azo linkage to a naphthalenesulfonic acid. The striking effect which he noted was that combination of the resulting antibody with toluidinonaphthalenesulfonic (TNS) acid gave rise to a large fluorescence enhancement. Antibodies against fluorescent dyes had been prepared previously as, for example, against fluorescein,[135] but in these cases a quenching of fluorescence was noted upon combination with the antibody. Winkler's finding was rapidly investigated by other workers. McClure and Edelman[136] found that toluidinylnaphthalenes undergo a strong fluorescence enhancement in organic solvents or when bound to certain proteins. Proteins known from other data to have large hydrophobic regions gave the most pronounced enhancement effects. Concomitant with the enhancement effect, there is a shift in the wavelength of maximum emission toward shorter wavelengths in media of low dielectric constant. McClure and Edelman[137, 138] further applied these same ideas to probing the hydrophobic areas of α-chymotrypsin and for changes in these areas when chymotrypsinogen is activated to chymotrypsin. A variant of the well-known inhibition of carbonic anhydrase by sulfanilamide has been studied by Chen and Kernohan,[139] who investigated the inhibition of this enzyme with dansamide. Dansamide forms a highly fluorescent complex with the enzyme, and the extent of binding can be determined by fluorescent measurements or by quenching of the protein UV fluorescence. An additional interesting finding is that dansamide bound to the enzyme has very different fluorescent properties from the enzyme which has been simply conjugated with dansyl chloride. The correlation between fluorescence enhancement and the Kosower Z values mentioned above[131] would seem to supply a more preferable variable than the dielectric constant with which to discuss enhancement effects. A correlation with helical structure of polypeptides has been studied by Lynn and Fasman,[140] who found that only the beta helix but not the alpha-helical form or the random coil form of poly-L-lysine produced a fluorescent enhancement of toluidinylnaphthalenesulfonates. From this result it was inferred that only the beta helix binds this dye. Winkler[141] found that 6-aminonaphthalene-2-sulfonate fluorescence is quenched by bromate ions. The presence of either BSA or antibody against the dye protects against this quenching. However, in the case of BSA binding, there is a large blue shift of the excitation spectrum not found in the case of antibody binding. The conclusion was that the dye binding site on BSA was appreciably more hydrophobic than that on the antibody. Structural studies on TNS by Camerman and Jensen[57] have given insight as to why TNS is fluorescent in organic solvents. The fluorescence of the unhydrated crystal resembles that of the solution in organic solvents, while the water solution and the hydrated crystal are only very slightly fluorescent. Although the planes of the naphthalene and the benzene ring make an angle of about 50° with each other, the electron distribution around the nitrogen is such as to allow delocalization of the π electron systems throughout the molecule and to promote fluorescence. The very low fluorescence in water solution or in the hydrated crystal is attributable to H-bonding between the nitrogen atom and water which results in a blocking of the resonance.

Many workers have observed changes both in absorption and emission of dansyl groups when combined with proteins. Weber[9] found a variation in the absorption maxima for different dansyl protein conjugates. That for ovalbumin was at 344 mμ and for polylysine at 329 mμ. There is also a change in the magnitude of the extinction coefficient upon binding, as found by Hartley and Massey.[97] The extinction coefficient for the free acid was found to be 4.55×10^6 cm^2 mole^{-1}, while on chymotrypsin the value is 3.36×10^6 and on ovalbumin 3.0×10^6. In addition, there is a shift in the position of the maximum for different degrees of labeling; at a labeling ratio of 0.5 to 1, the maximum is at 340 mμ and at 3.8 : 1 it is at 335 mμ. The intensity of labeling also was found to affect the fluorescence emission, the color going from orange to yellow-green as the degree of labeling increased. These workers also pointed out the large differences in the wavelength of maximum absorption for different dansyl-labeled amino acids (Figure 2). The kind of shifts typically found for dansyl-labeled proteins is shown by Figure 3.[120] The change in the absorption spectrum of bound dansyl has been used by Klotz et al.[99] to estimate the pK of the dansyl group in proteins. The pK of the free dye is about 3.9, and this value drops to about 1.6 in dansyl–BSA. The effect was explained on the basis of apolar interactions between dye and protein; the lowering of the pK was abolished in urea solutions. Parker et al.[142] have studied the binding of the hapten, ε-dansyllysine, to antidansyl antibody. The quantum yield was found to increase by a factor of 25 to 30 upon binding and the maximum

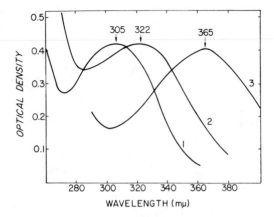

Figure 2. Absorption spectra of dansyl-labeled amino acids, showing the influence of conjugation on the wavelength of maximum absorption. (1) L-cysteine, labeled on —SH. (2) DL-lysine, labeled on ε-NH$_2$. (3) α-Benzoyl-L-histidine, labeled on the imidazole ring. (Reproduced from *Biochim. Biophys. Acta 21*, 67 (1956), with permission of B. S. Hartley, V. Massey, and Elsevier.)

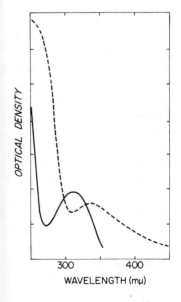

Figure 3. Absorption spectra for dansylate (———) and dansyl-labeled globulin (- - - -). The absorption maximum is shifted toward longer wavelength upon conjugation. (Reproduced from "Fluorescent Antibody Method," p. 135 (1968), with permission of M. Goldman and Academic Press.)

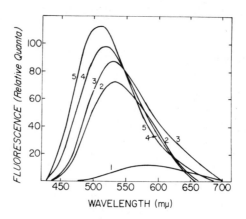

Figure 4. Emission spectra of dansyl-DL-tryptophan in various solvents: (1) water, (2) methanol, (3) ethanol, (4) *n*-butanol, (5) ethyl acetate. With decreasing dielectric constant, an increase in quantum yield, as well as a blue shift, takes place. (Reproduced from *Arch. Biochem. Biophys. 120*, 612 (1967), with permission of R. F. Chen and Academic Press.)

of emission to shift from 556 to about 500 mμ. The binding was found to be specific and limited to the *Fab* portion of the antibody molecule. The enhancement effect was explained conventionally in terms of the hydrocarbon environment of the bound hapten. A further publication by Parker *et al.*[143] utilized the fluorescence enhancement effect to study the variation of binding affinity of antibody during immunization. Chen[144] has studied the fluorescent effects of organic solvents on dansyl-labeled amino acids. In organic solvents there is a shift of the emission maximum toward shorter wavelengths and also a marked fluorescence enhancement as the dielectric constant is lowered (Figure 4). Similar shifts were observed when dansyl amino acids bind to the hydrophobic regions of proteins. Chen[98] investigated the fluorescent properties of dansyl–BSA. The degree of labeling was found to affect the emission spectrum (Figure 5), quantum yield, decay time, and polarization. The variations were interpreted in terms of two classes of sites, one of which is occupied first as the degree of labeling increases.

A very thorough study of the spectral and chemical properties of fluorescein was made by Lindqvist.[39] This paper also contains a discussion of the ionic equilibria undergone by the fluorescein molecule at various *p*H values. McDevitt *et al.*[103] found that fluorescein coupled through the isocyanate group to antibody gave a quantum yield of approximately twice

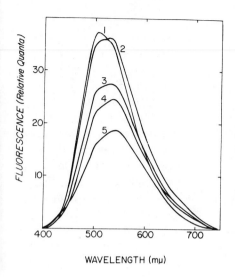

Figure 5. Emission spectra of BSA, labeled with dansyl to different degrees: (1) 1.2, (2) 2.5, (3) 3.3, (4) 4.3, (5) 16.2 dansyl groups per molecule. With increasing labeling ratio, there is a decrease in quantum yield and a second peak at a longer wavelength appears. (Reproduced from *Arch. Biochem. Biophys. 128*, 166 (1968), with permission of R. F. Chen and Academic Press.)

that of preparations made with the isothiocyanate. Klugerman[69] has examined the emission spectrum of fluorescein bound to bovine gamma globulin as a function of pH. Comparisons were made with the emission observed from fluorescein isothiocyanate, but these were complicated by hydrolytic reactions above pH 8.7. Jobbági and Király,[96] in investigating the degree of labeling by means of absorption measurements, pointed out the wavelength shift for fluorescein in different compounds. For free fluorescein the absorption maximum is at 490 mμ, for free fluorescein isothiocyanate at 491, for fluorescein coupled to a peptide at 492.5, and for fluorescein gamma globulin at 496. Tengerdy[145] investigated the reaction between fluorescein-labeled conalbumin and its antibody. He utilized the large quenching effect to obtain association constants and to follow the time course of the reaction.

In studying anti-DNP antibody, Amkraut[146] prepared a conjugate of ε-DNP–lysine with rhodamine isothiocyanate and found in the resulting complex an enhancement of the rhodamine fluorescence.

Churchich[66] has labeled several proteins with the coenzyme pyridoxal-5-phosphate. The changes in fluorescence properties on binding were found to be explicable on the basis of energy transfer from tryptophan within the framework of Förster's theory.

5.3. Electronic Mechanisms Responsible for Changes

In the previous section, the approach of the discussion was from the viewpoint of the alterations in atomic or molecular environment as being the significant factors in producing changes upon conjugation. In this section, a brief résumé of the electronic mechanisms by which these environmental factors produce effects is presented. A complete discussion of this same material can be found in Chapter 1. A review of the interaction of proteins with radiation has been given by Weber and Teale.[147]

Once a fluorescent molecule has been put into the first excited singlet state, vibrational relaxation normally follows very quickly, in about 10^{-12} sec, leaving the molecule in the lowest vibrational level of the first excited singlet state. At this point, several processes compete for the electronic excitation energy of the molecule. First, internal conversion may occur with the molecule returning to the ground state and all of the excitation energy appearing as heat. This process is usually favored in molecules with a large number of vibrational states, but may be either more or less favored in conjugates as evidenced by quenching in some cases and enhancement in others. A second kind of intramolecular process, termed *intersystem crossing*, involves a transition from an excited singlet state to a triplet state lying somewhat lower in energy. This transition, as Turner and Brand[131] have noted, is favored by an increase in polarity of the medium. Triplet states normally have long lifetimes and under favorable conditions of low temperature, which minimizes molecular vibration, may give rise to the long-lived emission termed *phosphorescence*. Collisions may be of a type which give so-called collisional quenching whereby the electronic excitation energy is imparted to another molecule which then falls to its ground state by nonradiative processes.

Another process which would not be expected to change quantum yield, but which could drastically alter the polarization of fluorescence, is noncollisional energy transfer. This nonradiative transfer of electronic excitation energy by a type of resonance mechanism was first analyzed by Förster. This process is not spin-forbidden and can occur over distances of 50 to 100 Å. Most likely, this is a primary reason that the polarization of fluorescence often decreases as the intensity of labeling is increased. The quantum yield would not be expected to be altered, since two molecules having the same energy levels simply can trade the energy back and forth until emission occurs from one of them.

Another type of quenching possibly of importance is that caused by the presence of oxygen. Oxygen is known to act powerfully as a quencher for solutions of many aromatic hydrocarbons in organic solvents, but this effect may be markedly less in aqueous solutions. For example, the fluorescence of pyrene in a hydrocarbon solvent is very sensitive to oxygen, but the fluorescence of an ionic derivative of pyrene, like the sulfonic acid in aqueous solution, shows no measurable oxygen effect.

5.4. Changes Due to Photochemical Reactions

The occurrence of photochemical reactions during the time of observation perhaps with or without the participation of molecular oxygen is another source of possible effects on fluorescence properties which have been only very poorly investigated. One of the most comprehensive pieces of work available in this area is the study of fluorescein decomposition by Lindqvist.[39] It is most likely that similar processes occur also in conjugates since a marked fading of fluorescence is usually observed for fluorescein-labeled ovalbumin unless a stabilizer such as normal gamma globulin is present. This fading is also a serious problem in fluorescent antibody staining where the sample is viewed while being illuminated by a strong UV source. The time of observation is frequently limited to a few minutes before the fluorescence has largely disappeared.

6. LIFETIME, DECAY TIME, AND QUANTUM YIELD

Once a molecule has been excited there is a certain probability, p_e, of emission which is also equal to the fraction of all molecules emitting per second. The reciprocal of p_e equals τ_0, the natural lifetime, which would be equal to the actually observable decay time if there were no other processes competing with emission. In the presence of competing processes, such as internal conversion, of probability, p_1, intersystem crossing, p_2, and energy transfer to another molecule which does not fluoresce, p_3, the probability of deactivation can be written as: $p = p_e + p_1 + p_2 + p_3$. The observed decay time, τ, is related to p by $p = 1/\tau$. The quantum yield, q, is simply the ratio τ/τ_0. Weber[165] pointed out that the last relationship may be expected to hold only if the probability of all competing processes remains constant with time after excitation.

The direct, accurate measurement of decay times has only recently become readily feasible. Ware[148] has described an excellent instrument utilizing a nanosecond flash lamp and a photomultiplier which can be gated before, during, or after the flash by the use of a delay line. Decay curves are plotted out by a recorder. Epple and Förster[149] have distinguished between classes of quenchers according to whether they exert changes on the absorption spectra or only upon the mean decay time of fluorescence. Experiments with rhodamine G, eosin, aminopyrenetrisulfonate, methylaminopyrenetrisulfonate, and hydroxypyrenetrisulfonate are reported. The quenchers studied included aniline, quinoline, and various phenols. Chadwick et al.[62] conjugated various dyes with BSA and measured the polarization of the complex. Assuming the decay time for the dansyl groups to be known as 11.8 nsec, the effective rotational relaxation time for BSA was then calculated by Perrin theory. Using the same relaxation time, the decay times for other dyes were then calculated by comparison of the polarizations. In this way, the decay time for hydroxypyrenetrisulfonate was found to be 90 nsec, and for 3-phenyl-7-isocyanatocoumarin a value of 2.5 nsec was obtained. More recent measurements in Dr. Johnson's laboratory (personal communication), as well as direct measurements of lifetime (Portmann and Dandliker, unpublished observations), have not confirmed the long lifetime for the pyrene derivative, and the correct value seems to be around 12 nsec. Frey and Wahl[150] carried out decay-time measurements on samples of BSA with varying amounts of dansyl combined per molecule. For lightly labeled material having on the order of one dansyl per BSA molecule, the decay was logarithmic and gave a decay time of 22 nsec. For heavily labeled material (up to 8.4 dansyl groups per BSA molecule), a nonlinear logarithmic decay curve resulted. It was possible to split this into two linear components, one with a time of about 6 nsec and the other of about 20 nsec. The significance of these findings was discussed in a second paper.[151] Two alternate possible explanations were offered for the complex decay curve of the heavily labeled material. One possibility is that the phenomenon is due to the energy transfer between dye molecules on the same protein molecule, and a second possible explanation is that there is more rotational freedom of the dye molecules with respect to the protein structure if the labeling is heavy. A discussion of the factors was given, but no clear-cut decision between the two possible explanations was possible. Chen et al.[152] have measured the decay curves for a large number of organic compounds using a recently available commercial instrument. In most cases, good agreement was found between the directly measured decay times and those computed by indirect methods such as by Perrin theory from polarization measurements. However, in a few cases, gross discrepancies were found. The decay times noted ranged from very short ones up to about 93 nsec for pyrene butyric acid. Chen[144] has measured decay times of a number of dansyl–amino acids in organic solvents and in protein solutions. The decay time for dansyl coupled to amino acids was found to be about 20 nsec in contrast to about 13 for the free dye. Wahl and Lami[153] have investigated the decay curves for dansyl-labeled lysozyme. Two decay times were necessary to explain the observed curves. The results were interpreted as indicating two different kinds of environment for the dansyl group, one giving rise to a characteristic decay time of 5 nsec and the other 15 nsec. Chen[98] has made a careful study of the fluorescent properties of dansyl BSA. The decay curves at all degrees of labeling were found to be complex and nonexponential. An average decay time was computed by fitting the curves as well as possible to two individual characteristic decay times. This average decay time decreases drastically at first with increase of labeling and then levels off. Anderson,[154] in applying Perrin theory to dansyl-labeled lactic dehydrogenase, measured the decay for different labeling ratios. For a ratio of 1.1, τ was found to be 16.5 nsec and for a ratio of 2.5 a decay time of 17.2 nsec was obtained.

An important new development in polarization theory and practice has been carried out by Wahl[155] and coworkers. This technique involves following the decay time for the horizontally polarized and vertically polarized components in the fluorescent light separately, by which data on the rotational relaxation time of the complex can be obtained. These papers are discussed in section 8.

7. ENERGY TRANSFER

By *energy transfer* is meant a process by which a molecule in an excited state loses energy to a molecule in the vicinity which is initially in the ground state. The newly excited molecule may subsequently either radiate or return to the ground state by a radiationless transition. If the donor and acceptor molecule are of different structures, the emitted radiation may be of a wavelength not appreciably absorbed by the acceptor molecule, in which case the process is termed *sensitized fluorescence*. If the acceptor molecule becomes deactivated by a radiationless transition, the process is termed *quenching*.

As will be discussed below, the occurrence of energy transfer between all-identical molecules decreases the polarization of fluorescence, and the decrease is a function of the concentration. This process, however, would not be expected to affect the quantum yield, and any decrease in quantum yield observed on increasing the dye concentration is due to the inner-filter effect mentioned in Section 5.2.

The actual relationship between the observed polarization of fluorescence and the concentration has been investigated both experimentally and theoretically by a number of workers. Weber[156] began with the empirical relationship of Feofilov and Sveshnikov, which is $1/p = 1/p'_0 - \alpha c\tau$, where p'_0 is the value of the polarization at zero concentration, τ is the decay time, which itself may be a function of concentration, and α is a constant. Weber pointed out that the theories of Vavilov and of Förster fail to predict this linear law except at low concentrations and attempted to improve the theory by calculating p after $\bar{n}$ transfers of electronic excitation energy and by calculating also the dependence of $\bar{n}$ on concentration. He assumed constant probabilities for either emission or transfer regardless of how many transfers had occurred. A linear law was obtained containing a parameter R, which is the distance at which the probability of emission equals the probability of transfer for oscillators aligned in parallel. The values of R for common dyes such as fluorescein were found to be in the range of 26 to 30 Å. Kawski[157] found that experimental results on the concentration dependence of polarization for fluorescein and for anthracene agree much better with Ore's extension of Förster's theory than with the theory in its original form. Ore's improvement was to consider that if molecule A is the nearest neighbor of molecule B, then the probability that B is the nearest neighbor of A is smaller than unity. Knox[158] has made an excellent critical evaluation of the several theories pertaining to energy transfer. Förster originally considered that transfer occurs only between an excited molecule A and its nearest neighbor B. The excitation energy was assumed to jump back and forth between A and B throughout the lifetime. Weber improved on this picture by assuming the excitation to be transferred from A to B to C, etc., with a loss of memory of the polarization at each stage. The most important omission in this theory is that the excitation is assumed never to return to the original site. Jablonski assumed that the excitation can be shared equally among all neighbors within a certain active sphere; this is a major approximation. The cluster theory of Knox seems to be an improvement on all the above. The excitation is assumed to stay within a certain cluster during the lifetime of the excited state; calculations were made for clusters of various sizes and the effects were summed. It may be noted that a cluster of two corresponds to Förster's theory. Knox's theory is also similar to Jablonski's, except that fewer physical assumptions are made. Knox's result fits experimental data even in regions or cases where Weber's theory fails to do so.

While earlier work was chiefly concerned with energy transfer between small molecules, the emphasis during the past few years has been on transfer between different parts of a single macromolecule. Shore and Pardee[159] have examined various biopolymers labeled with dansyl to determine whether or not there is energy transfer from other parts of the molecule to the dansyl group resulting in its sensitized fluorescence. No transfer was detected in dansyl-labeled DNA or RNA, but transfer was present in dansyl-labeled proteins when the excitation was between 250 and 285 mμ. Also, no transfer was detected in dansyl-labeled TMV when the RNA was excited. Velick et al.[160] have applied intramolecular energy transfer to a study of the antibody–hapten reaction. They found that the dinitrophenyl (DNP) hapten, whose absorption band overlaps the fluorescence emission band of tryptophan in proteins, exerts a strong quenching effect upon the fluorescence of antibody directed against DNP groups. The method could be quantified but is limited in scope to haptens having the desired absorption characteristics. Slayter and Hall[117] found that dansyl-labeling protects fibrinogen against UV radiation damage. The labeling was found to decrease the native UV fluorescence of the protein, but the dansyl fluorescence is enhanced, presumably by energy transfer from aromatic amino acids. Weber and Young[161] have studied the fragmentation of BSA by pepsin digestion. As an index of the extent of digestion the fluorescence enhancement of anilinonaphthalenesulfonate (ANS) was used. The polarization of fluorescence was found to be smaller when two or more ANS molecules were bound to the same BSA molecule. This effect was interpreted as being due to energy transfer.

Churchich[66] found energy transfer between tryptophan and pyridoxal-5-phosphate in proteins. The results were interpreted within the framework of Förster's theory. Chen and Kernohan[139] found that energy transfer between the seven tryptophan residues present in carbonic anhydrase occurred to the extent of 85% to dansamide presumably hydrophobically bonded. This efficiency in transfer is much greater than that observed if the same protein is simply labeled with dansyl chloride. Chen[98] investigated the fluorescent properties of BSA labeled to different extents with dansyl groups. With increase in labeling ratio, the polarization was found to slowly decrease from about 0.25 to about 0.17. In going from zero to about five dansyl groups per BSA molecule, both the lifetime and quantum yield decrease dramatically and thereafter change very much less as the labeling is increased. The effects were interpreted in terms of intramolecular energy transfer, but it would seem that the possibility of dye–dye interaction either in the ground state or as excimers has not been ruled out. Cheung and Morales[162] found a sensitized fluorescence imparted to 8-anilino-naphthalene-1-sulfonate from the aromatic amino acids in myosin.

8. POLARIZATION OF FLUORESCENCE

The theoretical interpretation of fluorescence polarization measurements has classically been made in terms of Perrin's equation in a form derived by Weber[163] and by Steiner and McAlister:[164]

$$\frac{1}{p} - \frac{1}{3} = \left(\frac{1}{p_0} - \frac{1}{3}\right)\left(1 + \frac{3\tau}{\varrho}\right) \quad [41]$$

This equation gives the relationship between the observed polarization, p, the polarization which would be observed in the absence of any rotary brownian motion, p_0, the decay time, τ, of the excited state, and the relaxation time, ϱ, of the rotation of the emission oscillator. This relaxation time may be defined as the time required for the average value of $\cos \theta$

to change from unity (at time zero) to $1/e$ (at time ϱ), where θ is the angle between the directions of the emission oscillator at time zero and time ϱ. If the particles are spherical, then ϱ becomes $\varrho_0 = 3\eta V/RT$, and Eq. [41] can be written

$$\frac{1}{p} - \frac{1}{3} = \left(\frac{1}{p_0} - \frac{1}{3}\right)\left(1 + \frac{RT\tau}{\eta V}\right) \qquad [42]$$

relating p also to the absolute temperature, T, and the viscosity, η, of the medium, and to the molar volume, V, of the spherical particle. Hence, a plot of $(1/p - 1/3)$ versus T/η is linear with a slope of $R\tau/V$ and an intercept of $(1/p_0 - 1/3)$. If the particles are not spherical, the slope of the T/η plot gives an average relaxation time which for ellipsoids of revolution is the harmonic mean of the two principal relaxation times of the ellipsoid.[163, 164]

The two most important circumstances under which Eq. [41] is valid are (a) when the molecule has spherical symmetry or (b) when the fluorescent residues are randomly oriented on the macromolecule. Linearity of the T/η plot is retained if the dye molecule is either rigidly attached to the macromolecule or rapidly rotating with respect to the macromolecule over the entire range of T/η investigated. In both cases, the slope of the T/η plot gives the relaxation time of the macromolecule, but in the latter case the intercept will give a polarization lower than the fundamental polarization, p_0. Nonlinearity in the Perrin plot can result from (a) A variation of τ with T or with the composition of the medium which is varied in order to vary η. A shortening of τ with an increase in T/η produces curvature concave to the T/η axis. (b) Polydispersity resulting in a spectrum of relaxation times over the entire T/η range. This situation results in curvature concave to the T/η axis. (c) Appearance of new relaxation times as T/η is varied. This condition can result, for example, from a dissociation into subunits or from thermally activated additional rotational motion and gives a curvature convex to the T/η axis. If a variation of η alone (by adding sucrose or glycerol) results in a different Perrin plot than that obtained when T is varied, a change in molecular association or internal flexibility may be the cause.

Fluorescence polarization literature, especially that dealing with protein studies, was covered in a review by Weber.[165] An excellent short review of the various aspects of fluorescence polarization has been given by Jozefonvicz.[166]

Details of the numerous applications of Perrin plots are discussed below. Weber[9] investigated dansyl conjugates of ovalbumin and BSA by the use of Perrin theory. Absolute relaxation times for the macromolecule were estimated from polarization data. The polarization of ovalbumin conjugates was found to be independent of pH between pH 1.5 and 14; however, with BSA conjugates the range of stability was only from pH 4 to 9. At other pHs, low values of p were obtained which were interpreted in terms of dissociation. Laurence[167] first utilized fluorescence polarization measurements to follow the binding of dyes to proteins and measured the association constant for the reaction between fluorescein and BSA. Harrington et al.[168] investigated the behavior of BSA in acid solution by means of fluorescence polarization. They coupled the protein both with dansyl chloride and with anthracene isocyanate. The results obtained by fluorescence polarization were correlated with those obtained by light scattering, the ultracentrifuge, and optical rotation. Steiner and Edelhoch[169] studied the effect of adding detergent or urea upon the fluorescence polarization of dansyl-labeled thyroglobulin. The relaxation times as computed from Perrin theory were found to be markedly decreased when either type of reagent was added, and the effects were reversible upon dilution. Chadwick and Johnson[63] applied Perrin theory to a number of proteins coupled to lissamine rhodamine, lissamine flavin, 3-phenyl-7-isocyanatocoumarin, and hydroxypyrenetrisulfonic acid. The proteins studied included BSA, gamma globulin, lysozyme, arachin, and thyroglobulin. Abnormal fluorescence polarization behavior was obtained with some dyes, and the effects were explained as being due either to participation of the macromolecule in the fluorescence mechanism or to internal flexibility of the protein molecule. Johnson and Richards[170] investigated the fluorescence polarization of legumin labeled either with dansyl or with anthracene isocyanate. The Perrin plots yielded a relaxation time of only about one-sixth that which might be expected from the ultracentrifuge molecular weight of 400,000. These results are possibly an instance in which a subunit upon which the label was bound rotates with a fair degree of independence from the main mass of the macromolecule. The review on fluorescent protein conjugates by Steiner and Edelhoch,[1] referred to earlier, also contains valuable sections on polarization. These same workers[171] were among the first to study the properties of gamma globulins by means of fluorescence polarization. Both rabbit antibody and bovine gamma globulin labeled with dansyl groups were investigated. The effects of alkali, acid, anionic and cationic detergents, urea, and guanidine were studied. Conditions were found by which unfolding to a state practically free of noncovalent bonds could be obtained and yet the native structure could be reformed upon removal of the unfolding agent. Chowdhury and Johnson[172] constructed Perrin plots for dansyl-labeled bovine gamma globulin by varying the temperature from 0 to 50°C. Straight-line plots were obtained, but the calculated relaxation time of 113 nsec was far lower than that to be expected (420 nsec) from hydrodynamic properties. Slight decreases in relaxation time were obtained upon the addition of urea. An important overall conclusion of this work is that the gamma globulin molecule possesses considerable internal flexibility. Young and Potts[173] have made a study of dansyl-labeled ribonuclease by means of Perrin plots. Absolute values of the relaxation time were compared with those from other methods and were in reasonable agreement. The rotational relaxation time deduced from the Perrin plots was used also to follow the reduction of ribonuclease and the influence of various added reagents on this reaction. Edelhoch and Steiner[174] have presented a careful discussion of how molecular parameters, the degree of association, and internal rigidity of molecules can be obtained by fluorescence polarization. Weber and Young[161] followed the pepsin degradation of BSA by measuring the polarization of dansyl-labeled BSA as well as the fluorescence intensity of noncovalently bound anilinonaphthalenesulfonate. With either method, a two-phase degradation process was observed. The polarization of the ANS–BSA complex was found to decrease when two or more ANS molecules were bound per BSA, probably indicating energy transfer. Gill[175] has studied crosslinked synthetic polypeptides as models for the tertiary structure of proteins. The Perrin plots were linear only in the lower portion and curved upward at higher T/η values. Winkler[176] made measurements on dansyl-labeled rabbit gamma globulin at various temperatures either alone or in the presence of unlabeled human gamma globulin or unlabeled ovalbumin. The polarization was found to be dependent upon the total protein concentration, which in this case was mostly the unlabeled gamma globulin or ovalbumin. The exact nature of this effect was not elucidated, but it is possible that part of it was caused by the additional viscosity of the added protein. The polarization was found also to depend upon the wavelength of excitation, but the interpretation of this effect also is not clear. Omenn and Gill[177] have measured Perrin plots for either the dansyl- or fluorescein-labeled copolymers of glutamic acid and lysine having an average molecular weight of 152,000. Different T/η plots were obtained depending upon whether T or η itself was varied, and the plots were always curved. Results were interpreted in terms of thermal transitions in the molecule leading to increased degrees of rotational freedom as well as to a slight rotation of the labeling dye molecule. Wahl and Frey[151] studied the fluorescence polarization of dansyl–BSA both under conditions where only the temperature was varied and also at constant temperature with the addition of sucrose to vary viscosity. The variation of temperature was found to yield linear Perrin plots while the addition of sucrose gave plots convex to the T/η axis. The effect was interpreted as being due to an alteration of structure by the sucrose. The effect of changing the degree of labeling was also studied on the Perrin plot. By varying the temperature with different degrees of labeling, several parallel plots extrapolating to different values of p_0 were obtained. The results were explained on the basis of increased ease of energy transfer and perhaps more rotational freedom of the dye molecule as the intensity of labeling was increased. Direct measurement of the decay time of fluorescence for each of these curves resulted in obtaining the same relaxation time of 130 nsec from any of the curves regardless of the degree of labeling. Weber and Daniel[178] studied the interaction of anilinonaphthalenes to BSA, and found a decrease in the polarization of fluorescence with increasing dye-to-protein ratio. The decrease in polarization was ascribed to energy transfer between dye molecules, and a simple theoretical treatment was developed based upon an assumed random distribution of the dye over the molecular surface and a single transfer of excitation energy. This treatment

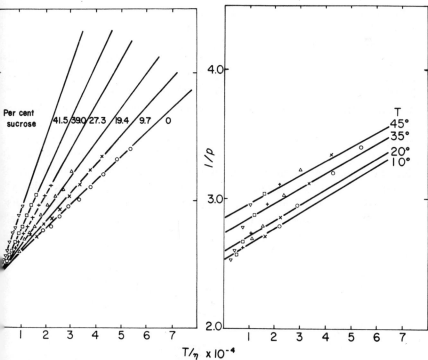

Figure 6. Perrin plots of dansyl-labeled human γG, obtained at various temperatures and various sucrose concentrations. Left: Points corresponding to equal sucrose concentrations are connected (temperature dependence), giving different slopes but approximately the same extrapolated p_0. Right: The lines are drawn through the points corresponding to equal temperatures (dependence upon sucrose concentration); in this way, the slopes are nearly the same, but the extrapolation to p_0 gives different values. (Reproduced from *Biochemistry* 6, 1440 (1967), with permission of J. K. Weltman, G. M. Edelman, and American Chemical Society.)

indicated an average distance between binding sites of about 21 Å and an average angle between oscillators of about 33°. Essentially similar conclusions were reached by measurement of the tryptophan fluorescence. Winkler et al.[179] compared the polarization of dansyl-labeled native gamma globulin together with the reduced and alkylated protein. The two molecules seemed to have the same relaxation time and same internal rigidity. Gill et al.[180] have measured the fluorescence properties of either dansyl- or fluorescein-labeled polypeptides and compared these with the properties of the free dye in solutions of high viscosity. They concluded that dansyl could be used between pH 2.5 and 14, but fluorescein only from pH 6 to 8 without artifacts being produced. Calculated values of the rotational relaxation time were much shorter than were to be expected for a rigid molecule of the size of the polypeptide. Moreover, the relaxation times were twice as long for dansyl labeling as those for fluorescein labeling. This difference was ascribed to tighter binding of the fluorescein by the polypeptide chain, but the extrapolations to $T/\eta = 0$ were quite long and no independent data on the decay times were available. Weltman and Edelman[181] found, as Wahl and Frey[151] had for BSA, that different relaxation times were obtained from Perrin plots depending upon whether or not the temperature or the viscosity was changed (Figure 6). These results were explained on the basis of a certain amount of independent rotation of the dye about the bond axis. These workers also confirmed Winkler's[176] finding that p is a function of the wavelength of excitation. They concluded that the gamma globulin molecule is very rigid and that the results were incompatible with flexible models. Wahl and Weber[182] also studied dansyl conjugates of gamma globulin and noted the differences in Perrin plots obtained upon varying viscosity versus those upon varying the temperature.

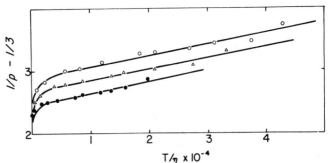

Figure 7. Perrin plots of dansyl-labeled rabbit γG (two dansyl groups per molecule): ●, 7.5°C; △, 25°C; ○, 35°C. The viscosity was varied by adding sucrose. (Reproduced from *J. Mol. Biol. 30*, 375 (1967), with permission of P. Wahl, G. Weber, and Academic Press.)

The difference between temperature and viscosity variation was attributed to the thermal activation of rotational motion of the attached dye. The relaxation time at neutral pH was found to correspond to a rigid molecule of the size of 7S γG (Figure 7). Knopp and Weber[64] introduced the use of pyrenebutyric acid as a fluorescent label. From a Perrin plot on BSA labeled with this dye they deduced that the decay time was of the order of 100 nsec. This result checked closely with preliminary direct measurements of decay time (80–110 nsec).

Jablonski[183, 184] and Wahl[155] have developed a powerful new method for investigating rotary brownian motion. The measurement consists in following the decay of the two polarized components in the fluorescent light after a pulse of exciting light. Analysis of the data results in the determination of one or more relaxation times characteristic of the rotational relaxation of the fluorescent molecule.

Consider the excitation of an isotropic fluorescent solution with incident light propagated in the y direction and polarized in the z direction. Let the observer be on the x-axis. The emission anisotropy, r (instead of the polarization), is a convenient parameter[185] in the Jablonski–Wahl theory:

$$r = \frac{E_z - E_x}{E_x + E_y + E_z} = \frac{E_z - E_y}{E_x + E_y + E_z} \qquad [43]$$

The quantities E are rates of energy emission through an entire closed surface, e.g., a sphere, surrounding a given sample under a given intensity of illumination; the subscripts indicate the direction of the electric vector in the emission. The total rate of energy emission, E, is just the sum of $E_x + E_y + E_z$. In measuring the fluorescent intensity, $I_{z,x}$, polarized in the z direction and propagated in the x direction, a magnitude proportional to E_z results. Similarly, $I_{y,x}$ is proportional to $E_y = E_x$. Hence, we can formally write $I = I_{z,x} + 2I_{y,x}$ where I is related to E by the same proportionality constant relating, e.g., $I_{z,x}$ to E_z. The emission anisotropy can now be expressed in terms of readily measurable quantities:

$$r = \frac{I_{z,x} - I_{y,x}}{I_{z,x} + 2I_{y,x}} = \frac{I_{z,x} - I_{y,x}}{I} \qquad [44]$$

or

$$r = \frac{I_\| - I_\perp}{I_\| + 2I_\perp} \qquad [45]$$

where $I_\|$ and $I_\perp$ are the intensities of the components of the fluorescent light polarized parallel and perpendicular, respectively, to the direction of polarization of the incident radiation.

The decay of I with time, t, follows a simple exponential relationship:

$$I(t) = I_0 e^{-t/\tau} \qquad [46]$$

in which I_0 is the value of I at the moment of excitation, and τ is the decay time of the fluorescence. From the definition of r [Eq. 45], it follows that

$$I_\parallel = \frac{I}{3}(1 + 2r) \qquad [47]$$

and

$$I_\perp = \frac{I}{3}(1 - r) \qquad [48]$$

Since r is a function of time, the decays of $I_\parallel$ and $I_\perp$ follow the equations

$$I_\parallel(t) = \frac{I_0}{3}[1 + 2r(t)]e^{-t/\tau} \qquad [49]$$

and

$$I_\perp(t) = \frac{I_0}{3}[1 - r(t)]e^{-t/\tau} \qquad [50]$$

which are not simple exponentials since r is a function of t. The function $r(t)$ is related to rotary brownian motion. Subtraction of Eq. [50] from Eq. [49] and taking logarithms results in

$$\log r(t) = \log[I_\parallel(t) - I_\perp(t)] - \log I_0 + \frac{t}{\tau} \log e \qquad [51]$$

If the molecules are spherical,

$$r = r_0 e^{-3t/\rho_0} \qquad [52]$$

where r_0 is the fundamental emission anisotropy, i.e., the anisotropy in the absence of rotary brownian motion. For spherical molecules, $\log r$ is linear in time and the slope of the plot is $-3/\rho_0$, where ρ_0 is the relaxation time of the sphere. If the molecules are ellipsoids of revolution, $r(t)$ involves two relaxation times and $\log r$ is no longer linear in time. The initial tangent determines a mean relaxation time. If all orientations of the fluorescent group with respect to the axes of the molecule are equally probable, then the harmonic mean, ρ_h, results. If the dye molecule is freely rotating (during nanosecond times), the following approximation in which $A < r_0$ holds: $r(t) \approx A \exp(-3t/\rho)$. Here, ρ is simply the experimentally determined relaxation time of the macromolecule.

The pulse method of Jablonski and of Wahl has several apparent advantages over the steady-state polarization method originated by Perrin.

Relaxation times can be determined from a single pulse type experiment at a given temperature and viscosity and do not require an extrapolation to $T/\eta \to 0$. No materials such as sucrose or glycerol need be added. The pulse method more easily detects the presence of more than one relaxation time. Since the experiments are rapid, the pulse method is convenient and can be used to study solvent effects, conformational transitions, and the like, as Stryer[186] has done. Wahl[187] has employed his theory, referred to earlier, in a study of dansyl–BSA. At pH 7.2 only one relaxation time of 150 nsec was found. At pH 2, however, several exponential decays appear. Within the limit of experimental error the curves could be fitted satisfactorily by assuming two time constants, one of 31 nsec and the other of 181 nsec, yielding a mean relaxation time, ϱ_h, of 58 nsec. This value of ϱ_h is in good agreement with that obtained from a Perrin plot on the same material.

Fayet and Wahl[188] have studied the polarized fluorescence of fluorescein-labeled gamma globulin and interpreted the results by means of Wahl's theory. The decay of the difference $(I_{\parallel} - I_{\perp})$ changes with labeling ratio, indicating energy transfer, which is to be expected. However, the relaxation time of γG extrapolated to a labeling ratio of zero is only 57 nsec, in agreement with the value previously obtained by Wahl[187] with dansyl–γG. These measurements confirmed his contention of extensive free internal rotation in the gamma globulin molecule and are in contrast to the results obtained by Weltman and Edelman[181] and by Wahl and Weber.[182] In both of these latter papers, the technique was to construct conventional

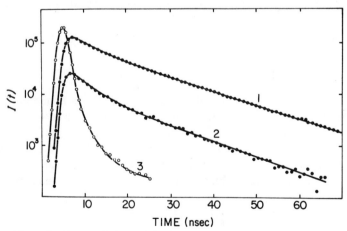

Figure 8. Decay of the polarized components of the emitted fluorescence of dansyl-labeled γG: (1) $s(t) = I_{\parallel}(t) + 2\,I_{\perp}(t)$; (2) $d(t) = I_{\parallel}(t) - I_{\perp}(t)$; (3) $g(t) =$ apparatus response. (Reproduced from *Biochim. Biophys. Acta* 175, 60 (1969), with permission of P. Wahl and Elsevier.)

Perrin plots. Wahl[189] has investigated fluorescence polarization of dansyl gamma globulin by following the decay curves of the V and H components separately (Figure 8). Two relaxation processes were found, one with a time constant of 370 nsec and one with a time constant of 23 nsec. Wahl describes the short relaxation time as arising from internal brownian motion of a rotating globular part of the γG molecule. This subunit would have a molecular weight of about 15,000 and leads to the conclusion that the γG molecule has a considerable amount of flexibility. Wahl and Timasheff[190] have applied Wahl's method and theory to the molecular association of β-lactoglobulin A. The molecule was labeled noncovalently with dansylate ion either with 0.2 or 0.7 mole of dansylate per mole of protein. Relaxation times obtained from fluorescence measurements and the derived molecular dimensions were in fair agreement with X-ray determinations of β-lactoglobulin.

Another important application of fluorescence polarization measurements utilizes the changes in polarization caused by changes in rotary brownian motion which accompany the reaction of one molecule with another. This idea was first applied to reactions between macromolecules by Steiner,[191] who used it to study the equilibria between lysozyme, nucleic acids, and BSA. Complexes were found to form when the charges on the two partners were opposite in sign and dissociation could be produced by increasing ionic strength. The conclusion was that the major forces involved were electrostatic in nature. Extensions of this concept were developed by Dandliker et al.[91] for equilibrium measurements and by Dandliker and Levison[192] for kinetic measurements, especially in antigen/antibody systems.

The general approach for interpreting results on macromolecular equilibria follows the pattern

$$\text{Ligand} + \text{Receptor} \rightleftarrows \text{Complex} \qquad [53]$$

The ligand can be present in a fluorescent form, F, or a nonfluorescent form, N, i.e., without added label. The receptor, R, is always present unlabeled. Molar concentrations, F, of the fluorescent conjugate in either the bound (subscript b) or free (subscript f) form, are related to polarization, p, and to molar fluorescence, Q (fluorescence intensity in arbitrary units divided by concentration of fluorescent conjugate), by the equations

$$\frac{F_b}{F_f} = \frac{Q_f}{Q_b}\left(\frac{p - p_f}{p_b - p}\right) \qquad [54]$$

and

$$\frac{F_b}{F_f} = \frac{Q_f - Q}{Q - Q_b} \qquad [55]$$

If the binding sites are nonuniform and can be characterized by a Sips distribution of binding free energies, the mass law can be expressed, if only F and R are present, by

$$\log F_f = \frac{1}{a_F} \log\left(\frac{F_b}{F_{b,\max} - F_b}\right) - \log K_{0F} \qquad [56]$$

If, instead, F, N, and R are simultaneously present, then

$$\log F_f = \frac{1}{a_F} \log\left(\frac{F_b}{F_{b,\max} - F_b - N_b}\right) - \log K_{0F} \qquad [57]$$

and

$$\log N_f = \frac{1}{a_N} \log\left(\frac{N_b}{F_{b,\max} - F_b - N_b}\right) - \log K_{0N} \qquad [58]$$

in which N is the molar concentration of N, a is the heterogeneity constant, $F_{b,\max}$ is the equivalent concentration of binding sites on the receptor, R, and K_0 is the association constant. The interpretation of fluorescence polarization kinetics can be carried out either by initial rate or integrated rate equations:

$$\left(\frac{dF_b}{dt}\right)_0 = -\left(\frac{dF_f}{dt}\right)_0 = \frac{Q_f}{Q_b}\left(\frac{F_{f0}}{p_b - p_f}\right)\left(\frac{dp}{dt}\right)_0 \qquad [59]$$

and

$$\log(p_e - p) = \log(p_e - p_f) - (\log e)(k_1[AB]^{N_1} - k_{-1})t \qquad [60]$$

In the kinetic equations, p is the instantaneous value of polarization and p_e is the equilibrium value. The rate constants k_1 and k_{-1} refer to the forward and backward rates, [AB] is antibody concentration, and N_1 is the order of the reaction with respect to antibody.

The equations above have now been extensively applied to antigen–antibody and to hapten–antibody reactions by Dandliker and Feigen[135] and Dandliker et al.[91, 193] An attempt by Haber and Bennett[194] to interpret antigen–antibody equilibria by use of Perrin's theory was apparently subject to errors from stray light in the instrument used. Deranleau and Neurath[195] labeled a variety of amino acids with dansyl groups and studied the combination of these materials with chymotrypsin and chymotrypsinogen. The extent of interaction of these materials, which were either substrates or inhibitors, was followed by fluorescence polarization measurements. Tengerdy[196] utilized polarization measurements on fluorescein-labeled conalbumin and its antibody for the quantification of antibody. Anderson[154] has used fluorescence polarization to resolve conflicting evidence about the dissociation of lactic dehydrogenase into subunits.

LDH was found to maintain a constant size and shape over a wide range of concentrations and at various ionic strengths. The rotational relaxation time of 220 nsec was found to be independent of labeling ratio from 1.1 to 3.6 moles of dansyl per mole of protein. Identical Perrin plots were obtained upon varying the temperature or the viscosity, and no thermally activated rotations were found. Equilibrium measurements by Kierszenbaum et al.,[114] in which both labeled and unlabeled BSA were simultaneously reacting with anti-BSA, showed that the presence of 2.7 dansyl groups in the BSA molecule lowers the association constant for combination with BSA to one-half the value for unlabeled BSA (Figures 9 and 10). Kinetic results on antigen–antibody systems obtained by Dandliker and Levison,[192, 198] and Levison et al. [197, 199] have revealed that the antigen–antibody reaction is not diffusion controlled and that the mechanism of the reaction depends dramatically on the nature of the environment. In the presence of chaotropic ions, the rate law becomes first order with re-

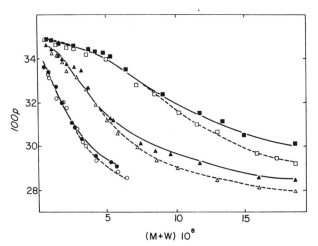

Figure 9. Titration of rabbit anti-BSA with dansyl-labeled BSA (closed symbols) and with a mixture of native and dansyl-labeled BSA (open symbols), followed by fluorescence polarization. $M = F_b + F_f$ and $W = N_b + N_f$ [see Eqs. 57 and 58]. Relative amounts of antibody are $\bigcirc$, 1; $\triangle$, 2; $\square$, 4. The titrations obtained with the mixed antigens show the competition between labeled and native antigen for the antibody, the labeled one having less affinity than the unlabeled. The association constants for the reaction between the antibody and the labeled antigen (K_{0F}) and the unlabeled antigen (K_{0N}) are 2.0×10^8 and 3.4×10^8 liters mole^{-1}, respectively. (Reproduced from *Immunochemistry* 6, 132 (1969), with permission of F. Kierszenbaum, J. Dandliker, W. B. Dandliker, and Pergamon Press.)

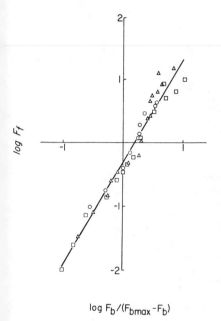

Figure 10. Sips plot of fluorescence polarization data according to Eq. [56]. The data points were derived from those shown as closed symbols in Figure 9. These data are for the combination of antibody with labeled BSA alone (no unlabeled BSA present). The straight line shown represents the best least-squares fit of the data and corresponds to $a_F = 0.61$ and $K_{0F} = 2.04 \times 10^8$ liters mole^{-1}. (Reproduced from *Immunochemistry* 6, 133 (1969) with permission of F. Kierszenbaum, J. Dandliker, W. B. Dandliker, and Pergamon Press.)

spect to both antigen and antibody, while in less chaotropic media the order with respect to antibody becomes half order (Figures 11 and 12). These findings together with the large heat of activation found (~ 12 kcal/mole) suggest extensive structural changes possibly involving the macromolecules themselves, the solvent, or surrounding ions prior to or during reaction.[197]

Despite a great deal of research activity in the area of fluorescence polarization, there are still, at the date of writing, no very satisfactory instruments for measuring polarization of fluorescence (fluorescence polarometers*) commercially available. However, a wide variety of instruments varying greatly in complexity and performance have been described in the literature. One of the earliest photoelectric instruments was described by Weber.[200] A basic difficulty with this instrument was that very great demands of constancy on the instrument are imposed by having first to set the instrument to a reference signal and then subsequently to make the measurement on the fluorescent compound. An improved visual type of polarometer was described by Harrington et al.[201] A good discussion on many of the factors to be considered in polarometer design has been given by Laurence.[54] Memming[202] described an instrument utilizing

---

*For some time we have been using *polarometer* to denote an instrument for measuring the degree of polarization as contrasted to optical rotation.

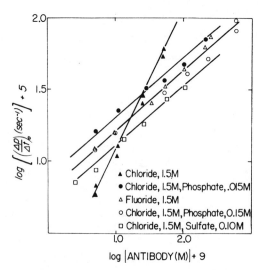

Figure 11. Effect of nonchaotropic ions on the order of reaction with respect to antibody concentration for the fluorescein-labeled ovalbumin divalent antiovalbumin system: ▲, 1.5M KCl, 0.01M tris; □, 1.5M KCl, 0.1M K_2SO_4, 0.01M tris; △, 1.5M KF, 0.01M tris; ●, 1.5M KCl, 0.01M K_2HPO_4; 0.005M KH_2PO_4; ○, 1.5M KCl, 0.1M, K_2HPO_4, 0.05M KH_2PO_4. (Reproduced from *Biochemistry* 9, 324 (1970), with permission of S. A. Levison, F. Kierszenbaum, W. B. Dandliker, and American Chemical Society.)

an optical null with a rotating polarizer and a polarization compensator consisting of glass plates. The null readout was made photoelectrically. A very simply designed and constructed polarometer was described by Johnson and Richards.[203] Bromberg et al.[204] have described the design and use of a polarization fluorometer. A fluorescence polarometer utilizing an optical null and separate measurements of the V and H components manually was described by Dandliker et al.[205] Ainsworth and Winter[206] gave a very detailed description of the construction of an automatic recording polarization spectrofluorometer. Chen and Bowman[207] described a fluorescence-polarization modification of a commercial spectrofluorometer. The main difficulty with this instrument seemed to be obtaining adequate rejection of the incident wavelength in making the fluorescence measurement. In addition, emprical corrections had to be made for unequal reflection of the polarized components at the gratings. Aurich[208] has described a luminescence polarometer utilizing a correction plate to make the sensi-

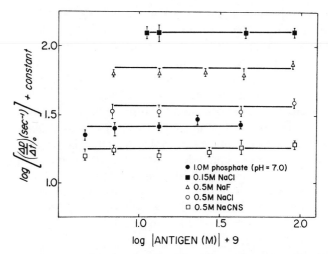

Figure 12. Determination of the order of reaction with respect to antigen for the dansyl-labeled bovine serum albumin (BSA) divalent anti-BSA system in various ionic media: ●, 1.0M phosphate (pH 7.0); ■, 0.15M NaCl, 0.01M neutral tris; △, 0.5M NaF, 0.01M neutral tris; ○, 0.5M NaCl, 0.01M neutral tris; □, 0.5M NaCNS, 0.01M neutral tris. (Reproduced from *Biochemistry 9*, 324 (1970), with permission of S. A. Levison, F. Kierszenbaum, W. B. Dandliker, and American Chemical Society.)

tivity of the detecting system the same to the two polarized components. Deranleau and Neurath[195] have described a phototube instrument in which both outputs are fed into an analog computer which then gives a direct readout of polarization and total intensity. Witholt and Brand[209] have described a polarization spectrofluorometer capable in different modes of measuring corrected excitation, emission, and polarization spectra and also conventional absorption and transmission spectra. Polarization could be measured to within 1%, and some empirical correction factors could be used to further improve this value. A direct-reading fluorescence polarometer with a response time of about 1 sec (White *et al.*[210]) has been used extensively by Dandliker and Levison[192, 197-199] to measure the rates of antigen–antibody reactions.

9. VISIBLE TRACING

9.1. Coons Fluorescent Antibody Technique

The Coons fluorescent antibody technique has become an immensely important tool in immunology and has hence given rise to an enormous volume of literature. The basis of the method was laid approximately 30 years ago when Fieser and Creech[5] investigated the fluorescent labeling of amino acids with isocyanates of anthracene and benzanthracenes. On the basis of this work, Fieser and Creech suggested that probably proteins could likewise be coupled to fluorescent dyes. Coons et al.[58] applied this idea and labeled antibody with β-anthryl isocyanate and found that the antibody retained its immunological specificity against type II pneumococcus. In a subsequent paper, Coons et al.[108] labeled antipneumococcus antibody with fluorescein isocyanate and demonstrated that pneumococci in infected tissue could be specifically stained with this labeled antibody. An interesting account of the early history of this story has been given by Coons.[211]

The methodology and literature on the fluorescent antibody technique were reviewed some years ago by Steiner and Edelhoch[1] and by Nairn.[85] A more recent very comprehensive coverage of the entire subject is given by Goldman.[212] Additional short reviews of general interest are those of Coons,[47] Beutner,[213] Smith et al.,[214] and Miller et al.[215] The latter especially serves well as a very short introduction to the whole technique.

In the review that follows below, no attempt has been made to completely cover the fluorescent antibody literature, but, instead, a few papers have been singled out which contain within them some unusual feature such as improvement in the method, an analysis of the factors involved, or some feature different than just a routine application of the fluorescent antibody technique. Clayton[216] has used multiple antibody staining all on the same tissue section. One antibody was labeled with fluorescein, the second with dansyl, and a third with nuclear fast red. In this way, the distribution of antigens to all three antibodies could be obtained simultaneously by examining the green, yellow, and red fluorescence, respectively. Curtain[106] described a method employing electrophoresis convection by which labeled antibody preparations can be fractionated in order to remove materials responsible for nonspecific staining. One of the earlier standard methods for removal of these materials was by adsorption on liver powder. Johnson[217] examined some of the factors which are important in this procedure. He found that the preliminary dialysis extending over a several-day period normally seems unnecessary and that premoistening of the liver powder gives good conjugates and decreases the losses, which are otherwise rather high. Smith et al.[218] described the use of fluorescent counterstains. The antibody was labeled with fluorescein, and this labeled material was applied admixed with normal gamma globulin labeled with lissamine rhodamine. The latter provides a general background of red fluorescence against which the green fluorescence of fluorescein is readily visible. Curtain[100] developed a considerably improved method for fractionation of fluorescent antibody preparations. A preliminary Sephadex G-25 column was used to remove soluble low molecular weight fluorescein derivatives. This separation was followed by one on DEAE-cellulose using gradient elution. He made the important observation that lightly labeled high-titer proteins were eluted from the column first, and that later fractions were more highly labeled and gave rise to higher degrees of nonspecific staining. Fothergill and Nairn[219] compared the purification of fluorescent antibody preparations on charcoal and on Sephadex G-25. Charcoal was found to be satisfactory for rhodamine conjugates, but for fluorescein-labeled antibodies the losses were excessively high, going up to 30 or 40%. Sephadex G-25 was found to be as efficient as charcoal for this separation, with practically no losses. It is difficult to see how the Sephadex purification alone could serve to do more than to remove unbound fluorescein. George and Walton[80] studied the purification of labeled antibody on Sephadex. They found that this process is more rapid, simple, and efficient than older methods such as dialysis, adsorption, or extraction. On Sephadex the emerging protein peak was split into two fractions. The first one was rather heavily labeled and the second only lightly labeled, indicating the possibility of separation according to the degree of labeling using gel filtration. Both fluorescein- and rhodamine-conjugated gamma globulins were used in this study. The physical basis for the separation according to the degree of labeling is not immediately obvious, but possibly the process is essentially adsorption chromatography rather than gel filtration. Griffin et al.[220] pointed out that more highly specific staining materials could be obtained by using less dye during the labeling procedure than had been originally employed by Coons. Goldstein et al.[76] further studied some of the factors involved in nonspecific staining. They found that the presence of other serum proteins in the γG preparations being labeled, the presence of low molecular weight fluorescent by-products, or the presence of impurities in the fluorescein isothiocyanate were not of appreciable importance. As previous work had already indicated, they confirmed that the most important variable in nonspecific staining is the molar labeling ratio and that this should be less than about 1.5 to 1 in order to eliminate nonspecific staining. Goldstein et al.[102] pointed out that, while it is important to eliminate heavily labeled molecules from fluorescent antibody preparations, it is equally essential to also remove unlabeled materials, the presence of which obviously must decrease the sensitivity of the method. They found that DEAE chromatography was also capable of removing unlabeled molecules. Ideally labeled antibody preparations lie somewhere in the intermediate fractions between the very heavily labeled materials and the unlabeled. Frommhagen and Spendlove[221] have contended that degradation products of fluorescein isothiocyanate which can still label proteins are partly responsible for the origin of nonspecific staining. They recognized overlabeling as being an important source of nonspecific staining but also believed that the presence of nonglobulin components of serum could contribute to this factor. Hsu et al.[112] prepared antibodies labeled both with ferritin and with fluorescein so that studies could be made both by electron microscopy and fluorescence microscopy. Antibodies labeled in this way still show good immunochemical reactivity. McDevitt et al.[103] compared fluorescent antibody labeling with fluorescein isocyanate and with fluorescein isothiocyanate. Complexes with both these dyes were purified on DEAE-cellulose and the properties were then examined. In both cases, materials giving rise to nonspecific staining were successfully removed on DEAE-cellulose. Although the isocyanate is somewhat more difficult to work with chemically, conjugates made with this material have approximately twice the quantum yield of those preparations made with the isothiocyanate. Lewis and Brooks[222] examined visually the fluorescence intensity of sections specifically stained with antibody labeled with different dyes. They found that the staining titers decreased in the order fluorescein > lissamine > rhodamine > tetramethylrhodamine > dansyl. Bernier and Cebra[223] have studied antibody formation in cells by utilizing antibody to either heavy or light chains labeled with either fluorescein or tetramethylrhodamine. These workers found that a single cell can produce both heavy and light chains, but only one type of light chain is produced in an individual cell. Cebra and Goldstein[104] found that purification on DEAE-cellulose columns was effective both for tetramethylrhodamine and for fluorescein conjugates. Wolf et al.[224] have pointed out that in making fluorescent antibody preparations best results are obtained if only 7S gamma globulin is used. Mansberg and Kusnetz[225] have described a method and instrumentation for microscopic automatic scanning of preparations stained with fluorescent antibody. Scanning could be done at two wavelengths for doubly stained preparations, and the necessary intensities were obtained by a laser source. Hebert et al.[113] found that nonspecific staining increased monotonically with increasing labeling ratio. In addition, impure dyes were found to give higher degrees of nonspecific staining. These workers favored the dialysis method of labeling carried out for a short time as being the best procedure for making fluorescent antibody conjugates. Evans et al.[226] have studied the immunochemical reactivity of heavy and light chains of antistreptococcal antibody by the labeling method. The labeled chains were tested for reactivity with streptococci. The heavy chain was found to contain 12% of the activity of intact γG, whereas the light chain had only about 0.3% of the activity.

There are some interesting variations of fluorescent antibody staining in which objects other than tissue slices are stained specifically. Paronetto[227] incorporated antigens into agar or adsorbed them onto cellulose acetate. He then found these materials could be specifically stained with labeled antibody. The staining of agar slices was also used by Allen,[228] who employed the method for both the identification and quantification of soluble antigens. Toussaint and Anderson[229] have developed the cellulose acetate technique into a quantitative method. As in earlier work, the soluble antigens were adsorbed onto disks of cellulose acetate and then challenged with fluorescent antibody. The resulting stained disks were then quantitatively assayed for fluorescence in a fluorometer. Further improvements were reported by Toussaint.[230] Danielsson[231, 232] collected bacteria directly on a nonfluorescent membrane filter, stained the bacteria specifically *in situ*, and examined the filter by fluorescence microscopy.

An interesting aspect of the properties of fluorescent-labeled conjugates has to do with their metabolic fates in intact animals. This subject has not received wide attention, but some data were reported by Weiser and Laxson.[233] The soluble complex of BSA and fluorescein-tagged anti-BSA was injected into mice. The fluorescent complex disappeared rapidly from most of the sites where it became early deposited except in the glomerulus where deposits persisted for more than a month. In contrast, fluorescein-tagged BSA alone did not persist at any site more than a few days. Mayersbach and Pearse[234] studied the fate of injected ovalbumin. The ovalbumin was detected either by having it labeled with fluorescein or by staining it with fluorescent-labeled antibody. With either method, the protein was found to be localized in the kidney tubules.

9.2. Quantitative Precipitation Test

Redetzki[109] studied the effect of labeling on the quantitative precipitin test. Antibody was made against alcohol dehydrogenase and labeled with dansyl chloride. Quantitative precipitin tests with the labeled or with the unlabeled antibody gave the same quantity of precipitated complex, and also no difference was noted between the labeled and unlabeled material in ability to inhibit the enzyme. Fluorescent-labeled antibodies against diphtheria toxin and toxoid were used by Head[235] in a modified quantitative preciptin test. The washed flocs containing the fluorescent antibody were dissolved in DMF and measured fluorometrically. Tengerdy[236] carried out a variant of the quantitative precipitin test in which the fluorescence of the supernatant, due to the presence of excess antibody, was measured fluorometrically. Bennett and Haber[237] have employed fluorescent tagging of the antigen to detect trace amounts of antigen left in immunospecifically purified antibody. Neurath[238] has devised a modification of the quantitative precipitin test using fluorescence tagging. The insoluble antigen–antibody complex was collected on a millipore filter and, after washing, fluorescence of the filter pad was measured directly in a fluorometer. Tengerdy[239] has described a means for determining gamma globulin concentrations in human serum. The test depends upon the inhibition of the precipitation of fluorescein-labeled human gamma globulin in the presence of anti-human gamma globulin serum. The amount of inhibition is a function of the amount of gamma globulin present in the serum sample. Hochberg et al.[240] used fluorescent-labeled antibody to detect and to quantitate group A streptococci in broth cultures. After washing, the cells are stained with fluorescent-labeled antistreptococci antibody. After removal of the excess antibody, the bound antibody is put in solution by dissociating with distilled water. The amount of fluorescence in solution is related to the number of streptococcal cells present in the original sample. Cathou[241] has devised a means by which considerable labor may be saved in performing quantitative precipitin determinations by means of fluorescent-labeled antigens. After the precipitates have been formed, the supernatants of all those tubes on the antibody-excess side of the equivalence point are devoid of fluorescence, and fluorescence first appears in that tube following the equivalence point. With this information, just a few tubes in the neighborhood of the equivalence point can then be quantitated by nitrogen analysis or other means to yield the amount of precipitable antibody.

9.3. N-Terminal Analysis

Labeling with dansyl chloride has also found application in determining certain functional groups in polypeptides and proteins, as described by Gray.[242] Under appropriate conditions, dansyl chloride can be made to form stable derivatives only with the ε-amino groups of lysine, N-terminal amino groups, and the phenolic OH group of tyrosine.

10. NONCOVALENTLY BOUND LABELS

Noncovalent binding of fluorescent dyes to macromolecules arises presumably by virtue of electrostatic forces, hydrophobic interactions, and H-bonding, possibly with additional small contributions from London dispersion forces. Early work in this area was concerned with the binding of dyes such as fluorescein, for example, to BSA and was an outgrowth of a very extensive older literature on the binding of nonfluorescent dyes to proteins. A second area which drew the attention of biochemists grew out of the observation that certain anilinonaphthalenes bind presumably by hydrophobic interactions and in doing so undergo large fluorescence enhancements. Still a third area is concerned with the intercalation of dyes, especially acridines, into nucleic acids. This latter phenomenon has been used extensively in detecting changes in folding and conformation of nucleic acids in solution. Berns and Singer[243] did a comparison of specific and nonspecific dye–protein interactions. They found that the binding of an acridine to BSA leads to a twenty-fold greater fluorescence enhancement than that observed when bound to antibodies specific for the dye. Weber and Young[161] followed the peptic digestion of the BSA molecule by monitoring the fluorescence intensity and polarization of noncovalently bound anilinonaphthalenesulfonate ion. Alexander and Edelman[244] found a change in the quantum yield of adsorbed TNS upon the denaturation of proteins. The phenomenon was used to study the transition temperature of several enzymes and to note the influence of stabilizers. Stryer[132] found that ANS binds stoichiometrically both to apomyoglobin and to apohemoglobin. Binding does not occur if a heme is present in either case. The binding results in a large fluorescence enhancement which can be mimicked by placing ANS into nonpolar solvents. Daniel and Weber[245] measured the binding of anilinonaphthalenes to BSA over a range of pH, temperature, and ionic strength. Between pH 5 and 10, 5 moles of dye is bound per BSA molecule. In a subsequent paper,[178] fluorescence polarization measurements on the same system were reported. McClure and Edelman[136] have investigated the fluorescence enhancement of TNS which occurs on being bound to certain proteins and have compared this dye with ANS, which fluoresces both in water and organic solvents. The fluorescence of TNS is enhanced by low dielectric constant and by high viscosities to a lesser extent. The TNS enhancement observed on binding to BSA, chymotrypsin, and β-lactoglobulin is quite large, while a much smaller enhancement occurs on binding to lysozyme, IgG, or ovalbumin. The former proteins are known to possess large hydrophobic areas. McClure and Edelman,[137, 138] in a series of two papers, have studied the TNS fluorescence in organic solvents and upon being bound to α-chymotrypsin. The change in enhancement could be used also for following the chymotrypsinogen-to-chymotrypsin conversion. Chen and Kernohan[139] have investigated the fluorescence enhancement which occurs when dansamide is bound to carbonic anhydrase. A 1 : 1 mole ratio of dye to protein was found, and the binding apparently occurs only at the active site since the activity is totally inhibited. The fluorescence enhancement was explained on the basis of the hydrophobic character of the binding site and also attributed partly to the loss of a sulfonamide proton on binding. Edelman and McClure[246] have presented a brief review of the fluorescence enhancement of TNS and related compounds (so-called fluorescent probes), which show large increases in fluorescence intensity when they are placed in nonpolar environments. Flanagan and Ainsworth[247] have studied the partition of several aminonaphthalenesulfonate derivatives between water and dodecylamine in hexane and between water and detergent micelles. They came to the conclusion that electrostatic as well as nonpolar forces are responsible for binding of these dyes to proteins. Since these dyes are also adsorbed by positively charged micelles but not by negatively charged ones, they concluded that the sulfonate group of the dye provides the main electrostatic contribution to the binding. Binding to uncharged detergents occurs only if the hydrophobic part of the dye molecule is sufficiently large. Approximately 2.4 moles of dansylate is bound per BSA molecule. The pH dependence of this binding suggests two different classes of binding sites. Franke[248] has made equilibrium measurements of the binding of polycyclic aromatic hydrocarbons to human serum albumin. Lynn and Fasman[140] showed that binding of TNS occurs to the beta form of poly-L-lysine, but not to the alpha form or to the random-coil form. Stryer[186] has given a brief review of the properties of the so-called fluorescent probes such as dansamide, ANS, and TNS and the use of these dyes in obtaining information about proteins concerning polarity, interatomic distances, rotational motion, and flexibility. Turner and Brand[131] have shown that the fluorescence enhancement of compounds such as ANS, TNS, dansyl, methylaminonaphthalenesulfonate, and aminonaphthalenesulfonate correlates better with the empirical Z values of solvents (Kosower[124]) rather than simply with low dielectric constant. Cheung and Morales[162] found that the protein fluorescence of myosin A is quenched by its interaction with ANS. The quenching was attributed to energy transfer as shown by the fluorescence excitation spectrum. Kotaki et al.[61] have discovered an additional hydrophobic probe which is the only one known, thus far, that is not a naphthalene derivative. The compound is a monobenzoyl derivative of 4,4'-diaminostilbene-2,2'-disulfonate (MBAS). A 1:1 complex forms between MBAS and HSA and results in a 150-fold increase in fluorescence yield. Like the other fluorescent probes, it also shows a large enhancement when placed in nonpolar solvents. Winkler[141] studied the fluorescence properties of 6-aminonaphthalene-2-sulfonate, which is fluorescent either in aqueous solution or bound to BSA or to antibody specific for this dye. However, the fluorescence in water is strongly quenched by bromate ions, and this quenching is lacking when the dye is bound either nonspecifically to BSA or specifically to antibody. An interesting application of fluorescence enhancement has been described by Hartman and Udenfriend.[249] These workers treated acrylamide gels with 8-anilinonaphthalene-1-sulfonate and were then able to directly visualize protein bands in the gel. Staining is very rapid (a few minutes) and destaining is unnecessary. The stained protein fractions can be used directly as antigens without removal of the gel or of the dye.

The interaction of haptenic dyes with their antibodies may be regarded as a type of noncovalent labeling. Using antibodies against diazonium-coupled aminonaphthalenesulfonic acid, Winkler[134] was able to show that the fluorescence enhancement obtainable upon binding TNS to hydrophobic regions also resulted upon binding to the antibody site. Parker et al.[142] showed an increase in fluorescence and a shift in the emission maximum from 556 to about 500 mμ when ε-dansyl–lysine reacts with anti-dansyl antibodies. Yoo and Parker[250] found that antiaminonaphthalenesulfonate antibodies give a large fluorescence enhancement when ANS is added, whereas ANS is practically unaffected by normal γG.

As pointed out above, the binding of dyes, especially the acridines, to DNA and RNA has excited considerable interest because of the information on molecular conformation which can be obtained by a study of the optical properties of the dye. This work has previously been reviewed by Van Duuren.[251] Stone and Bradley[252] studied the absorption properties of acridine orange bound to DNA and were able to follow the helix-coil transition by the change in the optical absorption. Faddeeva[253] found a quenching of fluorescence upon binding acridine orange to DNA. This quenching is reversed by heating the complex to 100°C for 20 min, showing that the quenching is dependent upon the double helix configuration of DNA. Boyle et al.[254] further studied the interaction of DNA and acridine orange. Weill and Calvin[255] investigated the optical properties of proflavine and acridine orange upon being bound to polyphosphates and DNA. Changes were found both in the absorption and fluorescence spectra and in the quantum yield and fluorescence polarization. The findings supported the intercalation theory of Lerman,[256] who had employed fluorescence polarization measurements as a means of inferring the structure of the DNA–acridine complex. Churchich[257] studied the binding of acriflavine to sRNA. Borisova and Tumerman[258] investigated the differences between native and denatured DNA by examination of the optical properties of bound acridine orange. Only one fluorescent band was observed with native DNA, while two were found with the denatured form. Both with the native and denatured forms, increases in decay time and fluorescence yield of the dye were noted. The two fluorescence bands of the denatured form apparently are connected with two quite different types of binding since the decay time for the 530 mμ band is 5 nsec compared to 21 nsec for the 640 mμ band. Tomita[259] measured the absorption, excitation, and emission spectra of acridine orange bound to both DNA and RNA. He found evidence for two types of complexing. In one, the dye is bound to phosphate groups and in so doing becomes nonfluorescent, while in the second type of binding, by intercalation between base pairs, a greatly enhanced fluorescence results. Tomita[260] compared the behavior of acridine orange with some other fluorescent dyes bound to nucleosides or nucleic acids. The dyes thionine, methylene blue, and proflavine give moderate-to-strong quenching upon being bound, while acridine orange undergoes a strong enhancement. Löber[261] has measured binding constants with DNA for several of the acridine dyes. He found that the binding tendency increases with the basicity of the dyes. Thus, acridine binds less than 3-aminoacridine < 3,6-diaminoacridine < 3,6-diamino-10-methylacridine. The highest association constants were noted with those dyes which contain the ring nitrogen alkylated. He explained the results on the basis of both electrostatic and nonelectrostatic contributions to the binding process.

ACKNOWLEDGMENT

The authors are indebted to G. K. Turner Associates for the generous use of their library facilities.

REFERENCES

1. R. F. Steiner and H. Edelhoch, Fluorescent protein conjugates, *Chem. Rev.* **62**, 457–483 (1962).
2. H. Fraenkel-Conrat, *in* "Methods in Enzymology" (S. P. Colowick and N. O. Kaplan, eds.), Vol. 4, pp. 247–269, Academic Press, New York (1957).
3. J. F. Riordan and B. L. Vallee, *in* "Methods in Enzymology" (C. H. W. Hirs, ed.), Vol. 11, pp. 565–570, Academic Press, New York (1967).
4. S. J. Hopkins and A. Wormall, Phenylisocyanate protein compounds and their immunological reactions, *Biochem. J.* **27**, 740–753 (1933).
5. L. F. Fieser and H. J. Creech, The conjugation of amino acids with isocyanates of the anthracene and 1,2-benzanthracene series, *J. Am. Chem. Soc.* **61**, 3502–3506 (1939).
6. J. L. Riggs, R. J. Seiwald, J. H. Burckhalter, C. M. Downs, and T. G. Metcalf, Isothiocyanate compounds as fluorescent labeling agents for immune serum, *Am. J. Pathol.* **34**, 1081–1097 (1958).
7. F. W. Putnam, *in* "The Proteins" (H. Neurath and K. Bailey, eds.), Vol. 1, part B, pp. 893–972, Academic Press, New York (1953).
8. H. R. Horton and D. E. Koshland, *in* "Methods in Enzymology" (C. H. W. Hirs, ed.), Vol. 11, pp. 856–866, Academic Press, New York (1967).
9. G. Weber, Polarization of the fluorescence of macromolecules. II. Fluorescent conjugates of ovalbumin and bovine serum albumin, *Biochem. J.* **51**, 155–167 (1952).
10. H. Uehleke, Neue Möglichkeiten zur Herstellung fluoreszenzmarkierter Proteine, *Z. Naturforsch. B.* **13**, 722–724 (1958).
11. P. E. Wilcox, *in* "Methods in Enzymology" (C. H. W. Hirs, ed.), Vol. 11, pp. 605–617, Academic Press, New York (1967).

12. J. L. Webb, "Enzyme and Metabolic Inhibitors," Vol. 3, Academic Press, New York (1966).
13. F. R. N. Gurd, in "Methods in Enzymology" (C. H. W. Hirs, ed.), Vol. 11, pp. 532–541, Academic Press, New York (1967).
14. J. F. Riordan and B. L. Vallee, in "Methods in Enzymology" (C. H. W. Hirs, ed.), Vol. 11, pp. 541–548, Academic Press, New York (1967).
15. Y. Kanaoka, T. Sekine, M. Machida, Y. Soma, K. Tanizawa, and Y. Ban, Protein sulfhydryl reagents. I. Synthesis of benzimidazole; derivatives of maleimide; fluorescent labeling of maleimide, *Chem. Pharm. Bull. (Tokyo)* 12, 127–134 (1964).
16. Y. Kanaoka, M. Machida, Y. Ban, and T. Sekine, Fluorescence and structure of proteins as measured by incorporation of fluorophore. II. Synthesis of maleimide derivatives as fluorescence–labeled protein–sulfhydryl reagents, *Chem. Pharm. Bull. (Tokyo)* 15, 1738–1743 (1967).
17. Y. Kanaoka, M. Machida, H. Kokubun, and T. Sekine, Fluorescence and structure of proteins as measured by incorporation of fluorophore. III. Fluorescence charateristics of N-[p-(2-benzoxazolyl)phenyl]maleimide and the derivatives, *Chem. Pharm. Bull. (Tokyo)* 16, 1747–1753 (1968).
18. R. M. Herriott, M. L. Anson, and J. H. Northrop, Reaction of enzymes and proteins with mustard gas (bis(β-chloroethyl)sulfide), *J. Gen. Physiol.* 30, 185–210 (1946).
19. H. Fraenkel-Conrat, in "The Enzymes" (P. D. Boyer, H. Lardy, and K. Myrbäck, eds.), Vol. 1, pp. 589–618, Academic Press, New York (1959).
20. J. L. Webb, "Enzyme and Metabolic Inhibitors," Vol. 2, Academic Press, New York (1966).
21. D. E. Koshland, Y. D. Karkhanis, and H. G. Latham, An environmentally-sensitive reagent with selectivity for the tryptophan residue in proteins, *J. Am. Chem. Soc.* 86, 1448–1450 (1964).
22. H. R. Horton and D. E. Koshland, A highly reactive colored reagent with selectivity for the tryptophan residue in proteins. 2–Hydroxy–5–nitrobenzyl bromide, *J. Am. Chem. Soc.* 87, 1126–1132 (1965).
23. H. R. Horton and D. E. Koshland, in "Methods in Enzymology" (C. H. W. Hirs, ed.), Vol. 11, pp. 556–565, Academic Press, New York (1967).
24. C. H. W. Hirs, in "Methods in Enzymology" (C. H. W. Hirs, ed.), Vol. 11, pp. 548–555, Academic Press, New York (1967).
25. P. B. Ghosh and M. W. Whitehouse, 7-Chloro-4-nitrobenzo-2-oxa-1,3-diazole: A new fluorigenic reagent for amino acids and other amines, *Biochem. J.* 108, 155–156 (1968).
26. F. Karush, N. R. Klinman, and R. Marks, An assay method for disulfide groups by fluorescence quenching, *Anal. Biochem.* 9, 100–114 (1964).
27. L. Brand and S. Shaltiel, Appearance of fluorescence on treatment of histidine residues with N-bromosuccinimide, *Biochim. Biophys. Acta* 75, 145–148 (1963).
28. L. Brand and S. Shaltiel, Introduction of fluorescence into proteins by treatment with N-bromosuccinimide, *Israel J. Chem.* 1, 51–52 (1963).
29. L. Brand and S. Shaltiel, Modification of histidine residues leading to the appearance of visible fluorescence, *Biochim. Biophys. Acta* 88, 338–351 (1964).
30. M. L. Ludwig and M. J. Hunter, in "Methods in Enzymology" (C. H. W. Hirs, ed.), Vol. 11, pp. 595–604, Academic Press, New York (1967).
31. R. Hess and A. G. E. Pearse, Labeling of proteins with cellulose-reactive dyes, *Nature* 183, 260–261 (1959).

32. J. Shore, Mechanism of reaction of proteins with reactive dyes. I. Literature survey, *J. Soc. Dyers Colour.* **84**, 408–412 (1968).
33. J. Shore, Mechanism of reaction of proteins with reactive dyes. II. Reactivity of simple model compounds with chlorotriazine dyes, *J. Soc. Dyers Colour.* **84**, 413–422 (1968).
34. C. S. Chadwick and R. C. Nairn, Fluorescent protein tracers; The unreacted fluorescent material in fluorescein conjugates and studies of conjugates with other green fluorochromes, *Immunology* **3**, 363–370 (1960).
35. V. E. Barsky, V. B. Ivanov, Y. E. Skliar, and G. I. Mikhailov, Dichlorotriazinylaminofluorescein—a new fluorochrome for cyto- and histochemical detection of proteins, *Izv. Akad. Nauk SSSR Ser. Biol.* 744–747 (1968) (in Russian).
36. D. G. Hoare and D. E. Koshland, A method for the quantitative modification and estimation of carboxylic acid groups in proteins, *J. Biol. Chem.* **242**, 2447–2453 (1967).
37. F. Kurzer and K. Douraghi–Zadeh, Advances in the chemistry of carbodiimides, *Chem. Rev.* **67**, 107–152 (1967).
38. L. O. Andersson, Coupling of dyes to various macromolecules by means of gamma-irradiation, *Nature* **222**, 374–375 (1969).
39. L. Lindqvist, A flash photolysis study of fluorescein, *Ark. Kemi* **16**(8), 79–138 (1960).
40. A. H. Coons and M. H. Kaplan, Localization of antigen in tissue cells. II. Improvements in a method for the detection of antigen by means of fluorescent antibody, *J. Exp. Med.* **91**, 1–13 (1950).
41. J. De Repentigny and A. T. James, A chromatographic separation of the aminofluorescein isomers, *Nature* **174**, 927–928 (1954).
42. L. C. Felton and C. R. McMillion, Chromatographically pure fluorescein and tetramethylrhodamine isothiocyanates, *Anal. Biochem.* **2**, 178–180 (1961).
43. H. S. Corey and R. M. McKinney, Paper chromatographic system for 5- and 6-substituted fluoresceins, *Anal. Biochem.* **10**, 387–394 (1965).
44. W. B. Dandliker and R. Alonso, Purification of fluorescein and fluorescein derivatives by cellulose ion exchange chromatography, *Immunochemistry* **4**, 191–196 (1967).
45. F. Borek, Spectral evidence for the structures of the nitrofluorescein isomers, *J. Org. Chem.* **26**, 1292–1294 (1961).
46. H. S. Corey and F. C. Churchill, Immunochemically significant fluoresceins: Structure determination by nuclear magnetic resonance spectroscopy, *Nature* **212**, 1040–1042 (1966).
47. A. H. Coons, Histochemistry with labeled antibodies, *Internat. Rev. Cytol.* **5**, 1–23 (1956).
48. H. Uehleke, Untersuchungen mit fluoreszenz-markierten Antikörpern. IV. Die Markierung von Antikörpern mit Sulfochloriden fluoreszierender Farbstoffe, *Schweiz. Z. Allgem. Pathol. Bakteriol.* **22**, 724–729 (1959).
49. R. W. Ramette and E. B. Sandell, Rhodamine B equilibria, *J. Am. Chem. Soc.* **78**, 4872–4878 (1956).
50. F. Borek and A. M. Silverstein, A new fluorescent label for antibody proteins, *Arch. Biochem. Biophys.* **87**, 293–297 (1960).
51. C. S. Chadwick, M. G. McEntegart, and R. C. Nairn, Fluorescent protein tracers; a trial of new fluorochromes and the development of an alternative to fluorescein, *Immunology* **1**, 315–327 (1958).

52. C. S. Chadwick, M. G. McEntegart, and R. C. Nairn, Fluorescent protein tracers. A simple alternative to fluorescein, *Lancet 1*, 412–414 (1958).
53. R. F. Chen, Fluorescent protein–dye conjugates. II. Gamma globulin conjugated with various dyes, *Arch. Biochem. Biophys. 133*, 263–276 (1969).
54. D. J. R. Laurence, *in* "Methods in Enzymology" (S. P. Colowick and N. O. Kaplan, eds.), Vol. 4, pp. 174–212, Academic Press, New York (1957).
55. R. F. Chen, Dansyl labeled proteins: Determination of extinction coefficient and number of bound residues with radioactive dansyl chloride, *Anal. Biochem. 25*, 412–416 (1968).
56. H. Wolochow, Fluorescent labels for antibody proteins. Application to bacterial identification, *J. Bacteriol. 77*, 164–166 (1959).
57. A. Camerman and L. H. Jensen, 2-*p*-Toluidinyl-6-naphthalene sulfonate: Relation of structure to fluorescence properties in different media, *Science 165*, 493–495 (1969).
58. A. H. Coons, H. J. Creech, and R. N. Jones, Immunological properties of an antibody containing a fluorescent group, *Proc. Soc. Exp. Biol. Med. 47*, 200–202 (1941).
59. R. M. Peck and H. J. Creech, Isocyanates of dimethylaminostilbenes and acetylaminofluorene, *J. Am. Chem. Soc. 74*, 468–470 (1952).
60. J. E. Sinsheimer, J. T. Stewart, and J. H. Burckhalter, Stilbene isothiocyanates as potential fluorescent tagging agents, *J. Pharm. Sci. 57*, 1938–1945 (1968).
61. A. Kotaki, M. Naoi, M. Harada, and K. Yagi, A new hydrophobic probe for protein, *J. Biochem. (Tokyo) 65*, 835–837 (1969).
62. C. S. Chadwick, P. Johnson, and E. G. Richards, Depolarization of the fluorescence of proteins labelled with various fluorescent dyes, *Nature 186*, 239–240 (1960).
63. C. S. Chadwick and P. Johnson, Depolarization of the fluorescence of proteins labelled with various fluorescent dyes, *Biochim. Biophys. Acta 53*, 482–489 (1961).
64. J. Knopp and G. Weber, Fluorescence depolarization measurements on pyrene butyric–bovine serum albumin conjugates, *J. Biol. Chem. 242*, 1353–1354 (1967).
65. K. C. Tsou, D. J. Rabiger, and B. Sobel, Fluorescent alkylating agents. 1-(β-Chloroethyl)bisbenzimidazoles, *J. Med. Chem. 12*, 818–822 (1969).
66. J. E. Churchich, Energy transfer in protein pyridoxamine-5-phosphate conjugates, *Biochemistry 4*, 1405–1410 (1965).
67. W. R. Dowdle and P. A. Hansen, Labeling of antibodies with fluorescent azo dyes, *J. Bacteriol. 77*, 669–670 (1959).
68. R. M. McKinney, J. T. Spillane, and G. W. Pearce, Factors affecting the rate of reaction of fluorescein isothiocyanate with serum proteins, *J. Immunol. 93*, 232–242 (1964).
69. M. R. Klugerman, Chemical and physical variables affecting the properties of fluorescein isothiocyanate and its protein conjugates, *J. Immunol. 95*, 1165–1173 (1965).
70. J. D. Marshall, W. C. Eveland, and C. W. Smith, Superiority of fluorescein isothiocyanate (Riggs) for fluorescent-antibody technic with a modification of its application, *Proc. Soc. Exp. Biol. Med. 98*, 898–900 (1958).
71. B. T. Wood, S. H. Thompson, and G. Goldstein, Fluorescent antibody staining. III. Preparation of fluorescein-isothiocyanate-labeled antibodies, *J. Immunol. 95*, 225–229 (1965).
72. M. Goldman and R. K. Carver, Preserving fluorescein isocyanate for simplified preparation of fluorescent antibody, *Science 126*, 839–840 (1957).

73. H. Rinderknecht, A new technique for fluorescent labeling of proteins, *Experientia* 16, 430 (1960).
74. H. Rinderknecht, Ultra-rapid fluorescent labeling of proteins, *Nature* 193, 167–168 (1962).
75. H. F. Clark and C. C. Shepard, A dialysis technique for preparing fluorescent antibody, *Virology* 20, 642–644 (1963).
76. G. Goldstein, I. S. Slizys, and M. W. Chase, Studies on fluorescent antibody staining. I. Non-specific fluorescence with fluorescein-coupled sheep anti-rabbit globulins, *J. Exp. Med.* 114, 89–109 (1961).
77. L. Kaufman and W. B. Cherry, Technical factors affecting the preparation of fluorescent antibody reagents, *J. Immunol.* 87, 72–79 (1961).
78. J. Zwaan and A. F. Van Dam, Rapid separation of fluorescent antisera and unconjugated dye, *Acta Histochem.* 11, 306–308 (1961).
79. J. Killander, J. Pontén, and L. Rodén, Rapid preparation of fluorescent antibodies using gel-filtration, *Nature* 192, 182–183 (1961).
80. W. George and K. W. Walton, Purification and concentration of dye–protein conjugates by gel filtration, *Nature* 192, 1188–1189 (1961).
81. W. Lipp, Use of gel filtration and polyethylene glycol in the preparation of fluorochrome-labelled proteins, *J. Histochem. Cytochem.* 9, 458–459 (1961).
82. T. Tokumaru, A kinetic study on the labeling of serum globulin with fluorescein isothiocyanate by means of the gel filtration technique, *J. Immunol.* 89, 195–203 (1962).
83. J. I. Reisher and H. C. Orr, Removal of fluorescein isothiocyanate from Sephadex after filtration of conjugated proteins, *Anal. Biochem.* 26, 178–179 (1968).
84. P. Flodin, B. Gelotte, and J. Porath, A method for concentrating solutes of high molecular weight, *Nature* 188, 493–494 (1960).
85. R. C. Nairn, "Fluorescent Protein Tracing," E. and S. Livingstone Ltd., Edinburgh (1962).
86. H. J. Creech and R. M. Peck, Conjugates synthesized from proteins and the isocyanates of certain systemic carcinogens, *J. Am. Chem. Soc.* 74, 463–468 (1952).
87. E. W. Emmart, Observations on the absorption spectra of fluorescein, fluorescein derivatives and conjugates, *Arch. Biochem. Biophys.* 73, 1–8 (1958).
88. F. Sokol, A. Hulka, and P. Albrecht, Fluorescent antibody method. Conjugation of fluorescein isothiocyanate with immune γ-globulin, *Folia Microbiol.* 7, 155–161 (1961).
89. M. Goldman and R. K. Carver, Microfluorimetry of cells stained with fluorescent antibody, *Exp. Cell Res.* 23, 265–280 (1961).
90. R. M. McKinney, J. T. Spillane, and G. W. Pearce, Fluorescein diacetate as reference color standard in fluorescent antibody studies, *Anal. Biochem.* 9, 474–476 (1964).
91. W. B. Dandliker, H. C. Schapiro, J. W. Meduski, R. Alonso, G. A. Feigen, and J. R. Hamrick, Application of fluorescence polarization to the antigen–antibody reaction, *Immunochemistry* 1, 165–191 (1964).
92. R. E. Dedmon, A. W. Holmes, and F. Deinhardt, Preparation of fluorescein isothiocyanate-labeled γ-globulin by dialysis, gel filtration, and ion-exchange chromatography in combination, *J. Bacteriol.* 89, 734–739 (1965).
93. R. P. Tengerdy, Properties of fluorescein-labeled human gamma globulin used in quantitative immunofluorescence tests, *Anal. Biochem.* 11, 272–280 (1965).
94. R. P. Tengerdy and C. A. Chang, Optical absorption and fluorescence of fluorescent protein conjugates, *Anal. Biochem.* 16, 377–383 (1966).

95. A. F. Wells, C. E. Miller, and M. K. Nadel, Rapid fluorescein and protein assay method for fluorescent-antibody conjugates, *Appl. Microbiol. 14*, 271–275 (1966).
96. A. Jobbági and K. Király, Chemical characterization of fluorescein isothiocyanate–protein conjugates, *Biochim. Biophys. Acta 124*, 166–175 (1966).
97. B. S. Hartley and V. Massey, The active centre of chymotrypsin. 1. Labelling with a fluorescent dye, *Biochim. Biophys. Acta 21*, 58–70 (1956).
98. R. F. Chen, Fluorescent protein–dye conjugates. I. Heterogeneity of sites on serum albumin labeled by dansyl chloride, *Arch. Biochem. Biophys. 128*, 163–175 (1968).
99. I. M. Klotz, E. C. Stellwagen, and V. H. Stryker, Ionic equilibria in protein conjugates: Comparison of proteins, *Biochim. Biophys. Acta 86*, 122–129 (1964).
100. C. C. Curtain, The chromatographic purification of fluorescein-antibody, *J. Histochem. Cytochem. 9*, 484–486 (1961).
101. J. L. Riggs, P. C. Loh, and W. C. Eveland, A simple fractionation method for preparation of fluorescein-labeled gamma globulin, *Proc. Soc. Exp. Biol. Med. 105*, 655–658 (1960).
102. G. Goldstein, B. H. Spalding, and W. B. Hunt, Studies on fluorescent antibody staining. II. Inhibition by sub-optimally conjugated antibody globulins, *Proc. Soc. Exp. Biol. Med. 111*, 416–421 (1962).
103. H. O. McDevitt, J. H. Peters, L. W. Pollard, J. G. Harter, and A. H. Coons, Purification and analysis of fluorescein-labeled antisera by column chromatography, *J. Immunol. 90*, 634–642 (1963).
104. J. J. Cebra and G. Goldstein, Chromatographic purification of tetramethylrhodamine-immune globulin conjugates and their use in the cellular localization of rabbit γ-globulin polypeptide chains, *J. Immunol. 95*, 230–245 (1965).
105. T. Takagi, Y. Nakanishi, N. Okabe, and T. Isemura, Hydrophobic fluorescent group coupled to taka-amylase A. Change of its environment accompanying splitting of disulfide bonds, *Biopolymers 5*, 627–638 (1967).
106. C. C. Curtain, Electrophoresis of fluorescent antibody, *Nature 182*, 1305–1306 (1958).
107. F. Kierszenbaum, S. A. Levison, and W. B. Dandliker, Fractionation of fluorescent-labeled proteins according to the degree of labeling, *Anal. Biochem. 28*, 563–572 (1969).
108. A. H. Coons, H. J. Creech, R. N. Jones, and E. Berliner, The demonstration of pneumococcal antigen in tissues by the use of fluorescent antibody, *J. Immunol. 45*, 159–170 (1942).
109. H. M. Redetzki, Labelling of antibodies by 5-dimethylamino-1-naphthalene sulfonyl chloride, its effect on antigen–antibody reaction, *Proc. Soc. Exp. Biol. Med. 98*, 120–122 (1958).
110. P. Albrecht and F. Sokol, Fluorescent antibody method. Optimal conditions for conjugation of 1-dimethylaminonaphthalene-5-sulfonyl chloride with γ-globulin, *Folia Microbiol. 6*, 49–63 (1961).
111. A. Johnson, E. D. Day, and D. Pressman, The effect of iodination on antibody activity, *J. Immunol. 84*, 213–220 (1960).
112. K. C. Hsu, R. A. Rifkind, and J. B. Zabriskie, Fluorescent, electron microscopic and immunoelectrophoretic studies of labelled antibodies, *Science 142*, 1471–1473 (1963).
113. G. A. Hebert, B. Pittman, and W. B. Cherry, Factors affecting the degree of nonspecific staining given by fluorescein isothiocyanate labeled globulins, *J. Immunol. 98*, 1204–1212 (1967).

114. F. Kierszenbaum, J. Dandliker, and W. B. Dandliker, Investigation of the antigen–antibody reaction by fluorescence polarization. Measurement of the effect of the fluorescent label upon the bovine serum albumin (BSA) anti-BSA equilibrium, *Immunochemistry* 6, 125–137 (1969).
115. A. A. Schiller, R. W. Schayer, and E. L. Hess, Fluorescein-conjugated bovine albumin. Physical and biological properties, *J. Gen. Physiol.* 36, 489–505 (1953).
116. D. N. Halikis and E. R. Arquilla, Studies on the physical, immunological and biological properties of insulin conjugated with fluorescein isothiocyanate, *Diabetes* 10, 142–147 (1961).
117. H. S. Slayter and C. E. Hall, Protection of proteins against ultraviolet damage using the resonance transfer mechanism, *J. Mol. Biol.* 8, 593–601 (1964).
118. L. H. Frommhagen, The solubility and other physicochemical properties of human γ-globulin labeled with fluorescein isothiocyanate, *J. Immunol.* 95, 442–445 (1965).
119. T. J. Porro, S. P. Dadik, M. Green, and H. T. Morse, Fluorescence and absorption spectra of biological dyes, *Stain Technol.* 38, 37–48 (1963).
120. P. A. Hansen, "Fluorescent Compounds Used in Protein Tracing. Absorption and Emission Data," University of Maryland (1964).
121. I. B. Berlman, "Handbook of Fluorescence Spectra of Aromatic Molecules," Academic Press, New York (1965).
122. A. Schmillen and R. Legler, in "Landolt-Börnstein, Numerical Data and Functional Relationships in Science and Technology, New Series" (K. H. Hellwege, ed.), Group II, Vol. 3, Springer-Verlag, Berlin (1967).
123. T. Förster, "Fluoreszenz organischer Verbindungen," Vandenhoeck and Ruprecht, Göttingen (1951).
124. E. M. Kosower, The effect of solvent on spectra. I. A new empirical measure of solvent polarity: Z values, *J. Am. Chem. Soc.* 80, 3253–3260 (1958).
125. S. Udenfriend, "Fluorescence Assay in Biology and Medicine," Academic Press, New York (1962).
126. D. M. Hercules, "Fluorescence and Phosphorescence Analysis. Principles and Applications," Interscience Publishers, New York (1966).
127. E. Döller and T. Förster, Der Konzentrationsumschlag der Fluoreszenz des Pyrens, *Z. Phys. Chem. (Frankfurt)* 34, 132–150 (1962).
128. S. S. Lehrer and G. D. Fasman, Fluorescence studies on poly-α-amino acids. II. Conformation-dependent excimer emission band in poly-L-tyrosine and poly-L-tryptophan, *Biopolymers* 2, 199–203 (1964).
129. B. L. Van Duuren, Effects of environment on the fluorescence of aromatic compounds in solution, *Chem. Rev.* 63, 325–354 (1963).
130. L. Stryer, Excited-state proton-transfer reactions. A deuterium isotope effect on fluorescence, *J. Am. Chem. Soc.* 88, 5708–5712 (1966).
131. D. C. Turner and L. Brand, Quantitative estimation of protein binding site polarity. Fluorescence of N-arylaminonaphthalenesulfonates, *Biochemistry* 7, 3381–3390 (1968).
132. L. Stryer, The interaction of a naphthalene dye with apomyoglobin and apohemoglobin. A fluorescent probe of non-polar binding sites, *J. Mol. Biol.* 13, 482–495 (1965).
133. G. Weber, in "A Symposium on Light and Life" (W. D. McElroy and B. Glass, eds.), pp. 82–107, The Johns Hopkins Press, Baltimore (1961).
134. M. Winkler, A molecular probe for the antibody site, *J. Mol. Biol.* 4, 118–120 (1962).

135. W. B. Dandliker and G. A. Feigen, Quantification of the antigen–antibody reaction by the polarization of fluorescence, *Biochem. Biophys. Res. Commun.* 5, 299–304 (1961).
136. W. O. McClure and G. M. Edelman, Fluorescent probes for conformational states of proteins. I. Mechanism of fluorescence of 2-*p*-toluidinylnapthalene-6-sulfonate, a hydrophobic probe, *Biochemistry* 5, 1908–1919 (1966).
137. W. O. McClure and G. M. Edelman, Fluorescent probes for conformational states of proteins. II. The binding of 2-*p*-toluidinylnapthalene-6-sulfonate to α-chymotrypsin, *Biochemistry* 6, 559–566 (1967).
138. W. O. McClure and G. M. Edelman, Fluorescent probes for conformational states of proteins. III. The activation of chymotrypsinogen, *Biochemistry* 6, 567–572 (1967).
139. R. F. Chen and J. C. Kernohan, Combination of bovine carbonic anhydrase with a fluorescent sulfonamide, *J. Biol. Chem.* 242, 5813–5823 (1967).
140. J. Lynn and G. D. Fasman, A fluorescent probe, toluidinylnaphthalene-sulfonate, specific for the β-structure of poly-L-lysine, *Biochem. Biophys. Res. Commun.* 33, 327–334 (1968).
141. M. H. Winkler, A fluorescence quenching technique for the investigation of the configurations of binding sites for small molecules, *Biochemistry* 8, 2586–2590 (1969).
142. C. W. Parker, T. J. Yoo, M. C. Johnson, and S. M. Godt, Fluorescent probes for the study of the antibody–hapten reaction. I. Binding of the 5-dimethylaminonaphthalene-1-sulfonamido group by homologous rabbit antibody, *Biochemistry* 6, 3408–3416 (1967).
143. C. W. Parker, S. M. Godt, and M. C. Johnson, Fluorescent probes for the study of the antibody–hapten reaction. II. Variation in the antibody combining site during the immune response, *Biochemistry* 6, 3417–3427 (1967).
144. R. F. Chen, Fluorescence of dansyl amino acids in organic solvents and protein solutions, *Arch. Biochem. Biophys.* 120, 609–620 (1967).
145. R. P. Tengerdy, Equilibrium and kinetic studies of the reaction between conalbumin and anticonalbumin, *Immunochemistry* 3, 463–477 (1966).
146. A. A. Amkraut, Determination of antibody–hapten binding constant by fluorescence enhancement, *Immunochemistry* 1, 231–235 (1964).
147. G. Weber and F. W. J. Teale, in "The Proteins" (H. Neurath, ed.), 2nd ed., Vol. 3, pp. 445–521, Academic Press, New York (1965).
148. W. R. Ware, Time resolved emission spectroscopy, in "Symposium on Biological Molecules in Their Excited States," Arden House, New York (October 6–9, 1969), Abstract B-3.
149. R. Epple and T. Förster, Untersuchungen über Löschung und Polarisationsgrad der Fluoreszenz in Lösung, *Z. Elektrochem.* 58, 783–787 (1954).
150. M. Frey and P. Wahl, Durée de vie de fluorescence de la sérum-albumine de boeuf conjuguée avec le 1-diméthylaminonaphtalène 5-sulfonyl, *Compt. Rend. Acad. Sci. Paris Ser. D.* 262, 2653–2656 (1966).
151. P. Wahl and M. Frey, Dépolarization de fluorescence de la sérum-albumine de boeuf conjuguée avec le 1-diméthylaminonaphtalène 5-sulfonyl, *Compt. Rend. Acad. Sci. Paris Ser. D.* 262, 2521–2524 (1966).
152. R. F. Chen, G. G. Vurek, and N. Alexander, Fluorescence decay times: Proteins, coenzymes, and other compounds in water, *Science* 156, 949–951 (1967).
153. P. Wahl and H. Lami, Etude du déclin de la fluorescence du lysozyme-1-diméthylaminonaphtalène-5-sulfonyl, *Biochim. Biophys. Acta* 133, 233–242 (1967).

154. S. R. Anderson, Fluorescence polarization studies of conjugates of beef heart lactic dehydrogenase with 1-dimethylaminonaphthalene-5-sulfonyl chloride, *Biochemistry* 8, 1394–1396 (1969).
155. P. Wahl, Sur l'étude des solutions de macromolécule par la décroissance de la fluorescence polarisée, *Compt. Rend. Acad Sci. Paris* 260, 6891–6893 (1965).
156. G. Weber, Dependence of the polarization of the fluorescence on the concentration, *Trans. Faraday Soc.* 50, 552–555 (1954).
157. A. Kawski, Zwischenmolekulare Energiewanderung und Konzentrationsdepolarisation der Fluoreszenz, *Ann. Phys. (Leipzig)* 8, 116–119 (1961).
158. R. S. Knox, Theory of polarization quenching by excitation transfer, *Physica* 39, 361–386 (1968).
159. V. G. Shore and A. B. Pardee, Energy transfer in conjugated proteins and nucleic acids, *Arch. Biochem. Biophys.* 62, 355–368 (1956).
160. S. F. Velick, C. W. Parker, and H. N. Eisen, Excitation energy transfer and the quantitative study of the antibody–hapten reaction, *Proc. Nat. Acad. Sci.* 46, 1470–1482 (1960).
161. G. Weber and L. B. Young, Fragmentation of bovine serum albumin by pepsin. I. The origin of the acid expansion of the albumin molecule, *J. Biol. Chem.* 239, 1415–1423 (1964).
162. H. C. Cheung and M. F. Morales, Studies of myosin conformation by fluorescent techniques, *Biochemistry* 8, 2177–2182 (1969).
163. G. Weber, Polarization of the fluorescence of macromolecules. I. Theory and experimental method, *Biochem. J.* 51, 145–155 (1952).
164. R. F. Steiner and A. J. McAlister, Use of the fluorescence techniques as an absolute method for obtaining main relaxation times of globular proteins, *J. Polymer Sci.* 24, 105–123 (1957).
165. G. Weber, in "Advances in Protein Chemistry" (M. L. Anson, K. Bailey, and J. T. Edsall, eds.), Vol. 8, pp. 415–459, Academic Press, New York (1953).
166. J. Jozefonvicz, Polarisation de fluorescence des solutions de polymère, *Ann. Chim. (Paris)* 3, 37–47 (1968).
167. D. J. R. Laurence, A study of the adsorption of dyes on bovine serum albumin by the method of polarization of fluorescence, *Biochem. J.* 51, 168–180 (1952).
168. W. F. Harrington, P. Johnson, and R. H. Ottewill, Bovine serum albumin and its behavior in acid solution, *Biochem. J.* 62, 569–582 (1956).
169. R. F. Steiner and H. Edelhoch, The properties of thyroglobulin. VI. The internal rigidity of native and denatured thyroglobulin, *J. Am. Chem. Soc.* 83, 1435–1444 (1961).
170. P. Johnson and E. G. Richards, The study of legumin by depolarization of fluorescence and other physicochemical methods, *Arch. Biochem. Biophys.* 97, 260–276 (1962).
171. R. F. Steiner and H. Edelhoch, Structural transitions in antibody and normal γ-globulins. II. Fluorescence polarization studies, *J. Am. Chem. Soc.* 84, 2139–2148 (1962).
172. F. H. Chowdhury and P. Johnson, Physico-chemical studies on bovine γ-globulin, *Biochim. Biophys. Acta* 66, 218–228 (1963).
173. D. M. Young and J. T. Potts, Structural transformations of bovine pancreatic ribonuclease in solution: A study of polarization of fluorescence, *J. Biol. Chem.* 238, 1995–2002 (1963).
174. H. Edelhoch and R. F. Steiner, in "Electronic Aspects of Biochemistry. Proceed-

ings of the International Symposium held at Ravello, Italy, September 16–18, 1963" (B. Pullman, ed.), pp. 7–22, Academic Press, New York (1964).
175. T. J. Gill, Crosslinked synthetic polypeptides. II. Evaluation of the internal structure of intramolecularly crosslinked polymers by polarization of fluorescence measurements, *Biopolymers 3*, 43–55 (1965).
176. M. H. Winkler, Fluorescence depolarization: A study of the influence of varying excitation wavelength and solution concentration, *Biochim. Biophys. Acta 102*, 459–466 (1965).
177. G. S. Omenn and T. J. Gill, Studies of polypeptide structure by fluorescence techniques, *J. Biol. Chem. 241*, 4899–4906 (1966).
178. G. Weber and E. Daniel, Cooperative effects in binding by bovine serum albumin. II. The binding of 1-anilino-8-naphthalenesulfonate. Polarization of the ligand fluorescence and quenching of the protein fluorescence, *Biochemistry 5*, 1900–1907 (1966).
179. M. H. Winkler, M. B. Goldman, and E. A. Sweeney, Fluorescence depolarization. II. The intramolecular relationships of the constituent fragments of γ-globulin, *Biochim. Biophys. Acta 112*, 559–564 (1966).
180. T. J. Gill, E. M. McLaughlin, and G. S. Omenn, Studies of polypeptide structure by fluorescence techniques. III. Interaction between dye and macromolecule in fluorescent conjugates, *Biopolymers 5*, 297–311 (1967).
181. J. K. Weltman and G. M. Edelman, Fluorescence polarization of human γG-immunoglobulins, *Biochemistry 6*, 1437–1447 (1967).
182. P. Wahl and G. Weber, Fluorescence depolarization of rabbit gamma globulin conjugates, *J. Mol. Biol. 30*, 371–382 (1967).
183. A. Jablonski, Über die Abklingungsvorgänge polarisierter Photolumineszenz, *Z. Naturforsch. A16*, 1–4 (1961).
184. A. Jablonski, *in* "Luminescence of Organic and Inorganic Materials, International Conference, New York, October 9–13, 1961," pp. 110–114, John Wiley and Sons, New York (1962).
185. A. Jablonski, On the notion of emission anisotropy, *Bull. Acad. Pol. Sci., Ser. Sci. Math. Astr. Phys. 8*, 259–264 (1960).
186. L. Stryer, Fluorescence spectroscopy of proteins, *Science 162*, 526–533 (1968).
187. P. Wahl, Détermination du temps de relaxation brownienne de la sérum-albumine en solution par la mesure de la décroissance de la fluorescence polarisée, *Compt. Rend. Acad. Sci. Paris Ser. D, 263*, 1525–1528 (1966).
188. M. Fayet and P. Wahl, Etude du déclin de la fluorescence polarisée de la γ-globuline de lapin conjugée avec l'isothiocyanate de fluoresceine, *Biochim. Biophys. Acta 181*, 373–380 (1969).
189. P. Wahl, Mesure de la décroissance de la fluorescence polarisée de la γ-globuline-1-sulfonyl-5-diméthylaminonaphthalène, *Biochim. Biophys. Acta 175*, 55–64 (1969).
190. P. Wahl and S. N. Timasheff, Polarized fluorescence decay curves for β-lactoglobulin A in various states of association, *Biochemistry 8*, 2945–2949 (1969).
191. R. F. Steiner, Reversible association processes of globular proteins. IV. Fluorescence methods in studying protein interactions, *Arch. Biochem. Biophys. 46*, 291–311 (1953).
192. W. B. Dandliker and S. A. Levison, Investigation of antigen–antibody kinetics by fluorescence polarization, *Immunochemistry 5*, 171–183 (1968).
193. W. B. Dandliker, S. P. Halbert, M. C. Florin, R. Alonso, and H. C. Schapiro,

Study of penicillin antibodies by fluorescence polarization and immunodiffusion, *J. Exp. Med. 122*, 1029–1048 (1965).
194. E. Haber and J. C. Bennett, Polarization of fluorescence as a measure of antigen–antibody interaction, *Proc. Nat. Acad. Sci. 48*, 1935–1942 (1962).
195. D. A. Deranleau and H. Neurath, The combination of chymotrypsin and chymotrypsinogen with fluorescent substrates and inhibitors for chymotrypsin, *Biochemistry 5*, 1413–1425 (1966).
196. R. P. Tengerdy, Quantitative determination of antibody by fluorescence polarization, *J. Lab. Clin. Med. 70*, 707–714 (1967).
197. S. A. Levison, A. N. Jancsi, and W. B. Dandliker, Temperature effects on the kinetics of the primary antigen–antibody combination, *Biochem. Biophys. Res. Commun. 33*, 942–948 (1968).
198. S. A. Levison and W. B. Dandliker, Effect of phosphate ion on ovalbumin–antiovalbumin kinetics, as measured by fluorescence polarization and quenching techniques, *Immunochemistry 6*, 253–267 (1969).
199. S. A. Levison, F. Kierzenbaum, and W. B. Dandliker, Salt effects on antigen–antibody kinetics, *Biochemistry 9*, 322–331 (1970).
200. G. Weber, Photoelectric method for the measurement of the polarization of the fluorescence of solutions, *J. Opt. Soc. Am. 46*, 962–970 (1956).
201. W. F. Harrington, P. Johnson, and R. H. Ottewill, An improved method for measuring degree of polarization, *Biochem. J. 63*, 349–352 (1956).
202. R. Memming, Fluoreszenzpolarisation von einigen Derivaten des Aminopyrens, Dissertation, Technische Hochschule, Stuttgart (1958).
203. P. Johnson and E. G. Richards, A simple instrument for studying the polarization of fluorescence, *Arch. Biochem. Biophys. 97*, 250–259 (1962).
204. N. S. Bromberg, A. Pesce, and N. O. Kaplan, A polarization fluorometer and its use, *Appl. Spectroscopy 17*, 132 (1963).
205. W. B. Dandliker, H. C. Schapiro, R. Alonso, and D. E. Williamson, Instrumentation for the Measurement of the Polarization of Fluorescence, in "Proceedings of the San Diego Symposium for Biomedical Engineering (1963)," pp. 127–132.
206. S. Ainsworth and E. Winter, An automatic recording polarization spectrofluorimeter, *Appl. Optics 3*, 371–383 (1964).
207. R. F. Chen and R. L. Bowman, Fluorescence polarization: Measurements with ultraviolet-polarizing filters in a spectrophotofluorometer, *Science 147*, 729–732 (1965).
208. F. Aurich, Eine Apparatur zur Messung der Lumineszenzpolarisation, *Spectrochim. Acta 22*, 1073–1080 (1966).
209. B. Witholt and L. Brand, Versatile spectrophotofluorometer—polarization fluorometer, *Rev. Sci. Instr. 39*, 1271–1278 (1968).
210. J. U. White, D. E. Williamson, S. A. Levison, and W. B. Dandliker, A rapid, direct reading fluorescence polarometer (in preparation).
211. A. H. Coons, The beginnings of immunofluorescence, *J. Immunol. 87*, 499–503 (1961).
212. M. Goldman, "Fluorescent Antibody Methods," Academic Press, New York (1968).
213. E. H. Beutner, Immunofluorescent staining: The fluorescent antibody method, *Bacteriol. Rev. 25*, 49–76 (1961).
214. C. W. Smith, J. F. Metzger, and M. D. Hoggan, Immunofluorescence as applied to pathology, *Am. J. Clin. Pathol. 38*, 26–42 (1962).

215. J. N. Miller, R. A. Boak, C. M. Carpenter, and F. Fazzan, Immuno-fluorescent methods in the diagnosis of infectious diseases, *Am. J. Med. Technol. 29*, 25–32 (1963).
216. R. M. Clayton, Localization of embryonic antigens by antisera labelled with fluorescent dyes, *Nature 174*, 1059 (1954).
217. G. D. Johnson, Simplified procedure for removing non-specific staining components from fluorescein-labelled conjugates, *Nature 191*, 70–71 (1961).
218. C. W. Smith, J. D. Marshall, and W. C. Eveland, Use of contrasting fluorescent dye as counterstain in fixed tissue preparations, *Proc. Soc. Exp. Biol. Med. 102*, 179–181 (1959).
219. J. E. Fothergill and R. C. Nairn, Purification of fluorescent labeled protein conjugates; comparison of charcoal and Sepadex, *Nature 192*, 1073–1074 (1961).
220. C. W. Griffin, T. R. Carski, and G. S. Warner, Labeling procedures employing crystalline fluorescein isothiocyanate, *J. Bacteriol. 82*, 534–537 (1961).
221. L. H. Frommhagen and R. S. Spendlove, The staining properties of human serum proteins conjugated with purified fluorescein isothiocyanate, *J. Immunol. 89*, 124–131 (1962).
222. V. J. Lewis and J. B. Brooks, Comparison of fluorochromes for the preparation of fluorescent-antibody reagents, *J. Bacteriol. 88*, 1520–1521 (1964).
223. G. M. Bernier and J. J. Cebra, Polypeptide chains of human gamma-globulin: Cellular localization by fluorescent antibody, *Science 144*, 1590–1591 (1964).
224. P. L. Wolf, B. Pearson, M. Rosenblatt, J. Vazquez, and T. Jarkowski, A rapid method to diminish nonspecific reactions in immunofluorescence, with the use of tagged 7S gamma-globulin immune serums, *Am. J. Clin. Pathol. 43*, 47–52 (1965).
225. H. P. Mansberg and J. Kusnetz, Quantitative fluorescence microscopy: Fluorescent antibody automatic scanning techniques, *J. Histochem. Cytochem. 14*, 260–273 (1966).
226. C. H. Evans, S. B. Herron, and G. Goldstein, Antibody activity associated with the fluorescent polypeptide chains of human immunoglobulin G, *J. Immunol. 101*, 915–923 (1968).
227. F. Paronetto, The fluorescent antibody technique applied to titration and identification of antigens in solution or antisera, *Proc. Soc. Exp. Biol. Med. 113*, 394–397 (1963).
228. J. C. Allen, Immunofluorescence applied to protein solutions, *J. Lab. Clin. Med. 62*, 517–524 (1963).
229. A. J. Toussaint and R. I. Anderson, Soluble antigen fluorescent-antibody technique, *Appl. Microbiol. 13*, 552–558 (1965).
230. A. J. Toussaint, Improvement of the soluble antigen fluorescent-antibody procedure, *Exp. Parasitol. 19*, 71–76 (1966).
231. D. Danielsson, A membrane filter method for the demonstration of bacteria by the fluorescent antibody technique. 1. A methodological study, *Acta Pathol. Microbiol. Scand. 63*, 597–603 (1965).
232. D. Danielsson and G. Laurell, A membrane filter method for the demonstration of bacteria by the fluorescent antibody technique. 2. The application of the method for detection of small numbers of bacteria in water, *Acta Pathol. Microbiol. Scand. 63*, 604–608 (1965).
233. R. S. Weiser and C. Laxson, The fate of fluorescein-labeled soluble antigen–antibody complex in the mouse, *J. Infect. Dis. 111*, 55–58 (1962).

234. H. Mayersbach and A. G. E. Pearse, The metabolism of fluorescein-labelled and unlabelled egg-white in the renal tubules of the mouse, *Brit. J. Exp. Pathol. 37*, 81–89 (1965).
235. W. F. Head, Immunochemical study of the diphtheria toxin–fluorescent antitoxin system, *J. Pharm. Sci. 51*, 662–665 (1962).
236. R. P. Tengerdy, Quantitative immunofluorescein titration of human and bovine gamma globulins, *Anal. Chem. 35*, 1084–1086 (1963).
237. J. C. Bennett and E. Haber, Studies on antigen conformation during antibody purification, *J. Biol. Chem. 238*, 1362–1366 (1963).
238. A. R. Neurath, Fluorometric estimation of antigens (antibodies), *Z. Naturforsch. B20*, 974–976 (1965).
239. R. P. Tengerdy, Gamma globulin determination in human sera by the inhibition of the precipitation of fluorescein-labeled gamma globulin, *J. Lab. Clin. Med. 65*, 859–868 (1965).
240. H. M. Hochberg, J. K. Cooper, J. J. Redys, and C. A. Caceres, Identification of fluorescent-antibody labeled Group A streptococci by fluorometry, *Appl. Microbiol. 14*, 386–390 (1966).
241. R. E. Cathou, A rapid fluorescence method for the determination of the equivalence zone in the precipitin reaction, *Immunochemistry 5*, 508–511 (1968).
242. W. R. Gray, in "Methods in Enzymology" (C. H. W. Hirs, ed.), Vol. 11, pp. 139–151, Academic Press, New York (1967).
243. D. S. Berns and S. J. Singer, A fluorescence study of specific and non-specific dye–protein interactions, *Immunochemistry 1*, 209–217 (1964).
244. B. Alexander and G. M. Edelman, A fluorescent probe for conformational changes of proteins, *Fed. Proc. 24*, 413, Abstract No. 1570 (1965).
245. E. Daniel and G. Weber, Cooperative effects in binding by bovine serum albumin. I. The binding of 1-anilino-8-naphthalenesulfonate. Fluorimetric titrations, *Biochemistry 5*, 1893–1900 (1966).
246. G. M. Edelman and W. O. McClure, Fluorescent probes and the conformation of proteins, *Accounts Chem. Res. 1*, 65–70 (1968).
247. M. T. Flanagan and S. Ainsworth, The binding of aromatic sulphonic acids to bovine serum albumin, *Biochim. Biophys. Acta 168*, 16–26 (1968).
248. R. Franke, Die hydrophobe Wechselwirkung von polycyclischen aromatischen Kohlenwasserstoffen mit Humanserumalbumin, *Biochim. Biophys. Acta 160*, 378–395 (1968).
249. B. K. Hartman and S. Udenfriend, A method for immediate visualization of proteins in acrylamide gels and its use for preparation of antibodies to enzymes, *Anal. Biochem. 30*, 391–394 (1969).
250. T. J. Yoo and C. W. Parker, Fluorescent enhancement in antibody–hapten interaction. 1-Anilinonaphthalene-8-sulfonate as a fluorescent molecular probe for anti-azonaphthalene sulfonate antibody, *Immunochemistry 5*, 143–153 (1968).
251. B. L. Van Duuren, in "Fluorescence and Phosphorescence Analysis" (D. M. Hercules, ed.), pp. 195–205, Interscience, New York (1966).
252. A. L. Stone and D. F. Bradley, Aggregation of acridine orange bound to polyanions: The stacking tendency of deoxyribonucleic acids, *J. Am. Chem. Soc. 83*, 3627–3634 (1961).
253. M. D. Faddeeva, Spectral properties of a complex of DNA with acridine orange, *Tsitologiya 4*, 231 (1962) (in Russian).
254. R. E. Boyle, S. S. Nelson, F. R. Dollish, and M. J. Olsen, Interaction of deoxyribonucleic acid and acridine orange, *Arch. Biochem. Biophys. 96*, 47–50 (1962).

255. G. Weill and M. Calvin, Optical properties of chromophore–macromolecule complexes: Absorption and fluorescence of acridine dyes bound to polyphosphates and DNA, *Biopolymers 1*, 401–417 (1963).
256. L. S. Lerman, The structure of the DNA–acridine complex, *Proc. Nat. Acad. Sci. 49*, 94–102 (1963).
257. J. E. Churchich, Fluorescence studies on soluble ribonucleic acid labelled with acriflavin, *Biochim. Biophys. Acta 75*, 274–276 (1963).
258. O. F. Borisova and L. A. Tumerman, Luminescence of complexes of acridine orange with nucleic acids, *Biophysics 9*, 585–595 (1964).
259. G. Tomita, Fluorescence-excitation spectra of acridine orange–DNA and –RNA systems, *Biophysik 4*, 23–29 (1967).
260. G. Tomita, Absorption and fluorescence properties of some basic dyes complexing with nucleosides or nucleic acids, *Z. Naturforsch. B23*, 922–925 (1968).
261. G. Löber, Complex formation of acridine dyes with DNA. IV. Equilibrium constants of substituted proflavine and acridine orange derivatives, *Photochem. Photobiol. 8*, 23–30 (1968).

Chapter 5

THE LUMINESCENCE OF THE AROMATIC AMINO ACIDS

I. Weinryb

Department of Biochemical Pharmacology
Squibb Institute for Medical Research
New Brunswick, New Jersey

and

R. F. Steiner

Department of Chemistry
University of Maryland Baltimore County
Baltimore, Maryland

1. INTRODUCTION

The fluorescence and phosphorescence of natural proteins arise from emission by the fluorogenic ring moieties of the aromatic amino acids. Such macromolecular luminescence may be visualized as corresponding to the intrinsic luminescence representing the summed contribution of the amino acid fluorogens, as perturbed by incorporation into a polypeptide and by the influence of the secondary and tertiary structure of the protein. Accordingly, an understanding of the luminescence properties of tryptophan, tyrosine, and phenylalanine is basic to an interpretation of the fluorescence and phosphorescence of the proteins.

The systematic investigation of the fluorescence properties of the aromatic amino acids dates from scarcely more than a decade ago. The earliest studies dealt primarily with the determination of the excitation and emission maxima[1-4] in connection with the development of fluorescence assays. Early measurements of the quantum yields of fluorescence include those of Shore and Pardee,[5] as well as the subsequent work of Teale and Weber.[6]

Serious interest in the low-temperature phosphorescence of the aromatic amino acids probably dates from the consideration of the relationship between the phosphorescence of these amino acids and that of proteins by Debye and Edwards,[7] as well as from the later discussion and experiments of Steele and Szent-Gyorgyi[8] on the possibilities of emission from the triplet states of biological molecules.

This chapter will be concerned with the available information about the luminescence of tryptophan, tyrosine, phenylalanine, and their derivatives, together with a discussion of the excited-state characteristics and relaxation mechanisms suggested by these data. The luminescence properties of indole, phenol, and benzene will also be discussed when deemed pertinent.

2. EXCITATION OF THE AROMATIC AMINO ACIDS

2.1. Excitation by Near-Ultraviolet Radiation: Ultraviolet Absorption Spectra

Luminescence is usually induced by irradiation at the long-wavelength maximum of the optical absorption spectrum of the compound to be studied, for reasons of sensitivity, selectivity, and minimization of photodecomposition. An appropriate assignment of excitation wavelength hence requires a knowledge of the absorption spectrum in solution of the compound. The longest-wavelength electronic transitions for tryptophan, tyrosine, and phenylalanine occur in the near ultraviolet; a brief description of the major features of their spectra at wavelengths greater than 200 nm will be presented here. Readers desiring further information on the ultraviolet absorption spectra of the aromatic amino acids may consult the compendia of Hershenson,[9, 10] the reviews of Wetlaufer[11] and Beaven and Holiday,[12] as well as additional selected references.[13–18]

2.1.1. Benzene and Phenylalanine

The spectrum of benzene in liquid solution at wavelengths greater than 200 nm shows a weak absorption band centered about 255 nm with molecular extinction coefficient, ε, equal to 200 and a stronger band near 205 nm with ε equal to 7400.[19, 20] Vibrational fine structure is very evident, even in aqueous solution. These electronic transitions result in low absorption intensities because of their forbidden character, a consequence of the high symmetry of the benzene molecule; only vibrational interactions with lower-symmetry, higher-energy electronic states enable them to occur

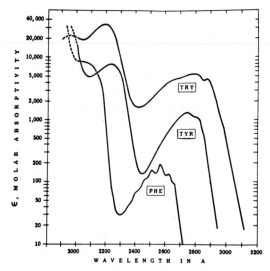

Figure 1. Ultraviolet absorption spectra of tryptophan, tyrosine, and phenylalanine at pH 6 (from Wetlaufer[11]).

at all.[21-23] Some extremely weak transitions, which are only about 10^{-5} times as intense as the 255 nm absorption, have been reported for benzene at wavelengths near 340 nm[24-26] and may represent direct singlet–triplet excitation processes.

The spectrum of phenylalanine parallels that of benzene (Figure 1). The absorption maximum at 258 nm is shifted to slightly longer wavelengths, but its intensity ($\varepsilon \cong 190$) is hardly changed. Very little shift of the 205 nm band occurs, but its intensity ($\varepsilon = 9600$) is somewhat increased, perhaps because of a slight lowering of molecular symmetry upon ring substitution. Variation of the pH gives rise to small changes in the spectrum of phenylalanine, an indication that the ionization of amino and carboxyl groups perturbs the ring absorption.[12]

2.1.2. Phenol and Tyrosine

The substitution of benzene by a hydroxyl group markedly changes its absorption spectrum. The absorption maxima shift to 210 and 270 nm, the intensity of the latter transition increasing about sevenfold. Formation of the anion results in a further red shift and intensification of the spectrum, peaks appearing at 235 ($\varepsilon = 9400$) and 287 ($\varepsilon = 2600$) nm.[20] Such effects are due (1) to the influence of the hydroxyl substituent in lowering the ring symmetry (and hence increasing transition allowedness) through inductive distortion of ring charge distribution as well as direct ring-substituent charge interaction,[22] and (2), in the case of the anion, to the resonance stabilization of additional structures.[27, 28]

The spectrum of tyrosine is very similar to that of phenol (Figure 1). At pH 6, electronic transitions occur at 275 nm ($\varepsilon = 1400$) and 222 nm ($\varepsilon = 9000$). Upon ionization of the phenolic side-chain, the maxima shift to 240 nm and 293 nm, and the corresponding extinction coefficients increase to about 11,000 and 2300, respectively.[12]

2.1.3. Indole and Tryptophan

Indole and tryptophan behave analogously to the preceding two examples and show essentially similar spectra. For indole in a number of solvents, major peaks occur at about 225 nm, with a value of ε equal to 25,000, and in the vicinity of 270 nm, where ε is close to 6000.[29, 30] While structure is evident in the long-wavelength band for all solvents, the degree of resolution and the exact wavelength positions of the subpeaks are solvent dependent. For indole in the nonpolar solvent cyclohexane, secondary maxima can be observed near 280 and 288 nm.[29]

The fine structure of the primary absorption band of tryptophan in water is somewhat blurred. The spectral maxima (Figure 1) occur near 220 nm, with ε equal to 36,000, and at 280 nm, with ε equal to 5500.[11, 12] In addition, a secondary maximum is found at 288 nm and a shoulder at 265 nm, with values of ε close to 4500 in each case.[11, 12] A weak long-wavelength absorption near 300 nm, with ε near 700, has also been reported; it appears as a shoulder in aqueous solution.[11]

The fluorescence polarization studies of Weber[28] indicate that at least two independent electronic transitions are responsible for the spectra of indole and tryptophan in the 260–310 nm region. Recent quantum mechanical (self-consistent field) calculations for tryptophan predict two transitions in this region.[31] The benzenoid character of the indole chromophore and the wavelength of absorption, as well as considerations dealing with the vibrational fine structure, all have suggested that at least one of the transitions is of the π-π* type, similar to that of the long-wavelength absorption bands discussed above for phenylalanine and tyrosine.[11, 28]

2.1.4. The Peptide Bond

The peptide bond shows a peak absorption at about 190 nm for peptides and proteins in the random coil form (for example, see Rosenheck and Doty[32]). This is well removed from the wavelengths of excitation usually employed for aromatic amino acids and peptide derivatives (250–290 nm). Setlow and Guild[33] have pointed out that the peptide bond tail absorption persists out to 270–280 nm, although extremely weak ($\varepsilon \cong 5$) at wavelengths greater than 250 nm. Because absorption is so weak above 250 nm, emission by the peptide bond *per se* can usually be ignored. However, even weak absorption by the peptide bond in this wavelength region may be of photochemical interest, since, as McLaren[34] has observed, 1 einstein of light quanta of 253.7 nm possesses an energy of 112 kcal, or over twice the molar energy necessary to rupture a peptide bond under appropriate conditions. The possibility of peptide photodecomposition at relatively long wavelengths of excitation thus exists.

2.2. Excitation by Higher-Energy Radiation

Luminescence studies involving excitation with vacuum ultraviolet radiation, fast electrons, and X-rays have been reported on aromatic amino acids, both in solution and in powder form. As the energy of the exciting radiation increases, higher excited singlet states tend to be populated to an increasing degree. Photochemical processes resulting in the formation of radical species and in irreversible alteration of the absorbing molecules become more likely. These reactions are competitive with respect to direct emission of fluorescence.

Some major conclusions of these investigations may be summarized as follows:

(1) In general, high-energy radiation is less efficient in inducing luminescence than excitation at the usual near-ultraviolet wavelengths (250–280 nm).[35, 36] In particular, the excitation of dilute aqueous solutions of aromatic amino acids by soft (7 kev) X-rays results in fluorescence yields much lower (0.4 to 0.02%) than those resulting from ultraviolet excitation.[35]

(2) The fluorescence emission spectrum induced by X-ray irradiation of amino acid powders is similar to that elicited by ultraviolet light.[38]

(3) With increasing energy of the exciting radiation, the phosphorescence/fluorescence ratio for tryptophan powders increases markedly.[36] This may reflect the stronger vibrational coupling of the higher excited singlet and triplet states, with the consequent facilitation of intersystem crossing. For tryptophan solutions at low temperatures ($<140°K$), ion–electron recombination may be an important factor in the enhancement of phosphorescence observed.[37]

(4) The thermal quenching of the X-ray induced fluorescence of powders is less pronounced than that of the ultraviolet-induced fluorescence of solutions.[39]

3. ENVIRONMENTAL EFFECTS UPON THE FLUORESCENCE OF THE AROMATIC AMINO ACIDS

3.1. Temperature

The variation of quantum yield with temperature for an aromatic amino acid is governed primarily by the energies of activation of the various radiationless processes which deactivate the excited state, in competition with the direct emission of fluorescence. Since these have finite energies of activation and increase in importance with increasing temperature, the quantum yield normally falls with increasing temperature.

In general, the quantum yield of any fluorescent species may be represented by

$$Q = \frac{k_f}{k_f + \sum_i k_i} \qquad [1]$$

where k_f is the rate constant for emission of fluorescent radiation, and the k_i are the rate constants for the various competitive nonradiative processes which deactivate the excited state.

Equation [1] may be rewritten

$$Q^{-1} - 1 = k_f^{-1} \sum_i k_i \qquad [2]$$

If, as is probable, k_f is independent of temperature, the temperature dependence of Q may be expressed by

$$Q^{-1} - 1 = k_f^{-1} k_0 + k_f^{-1} \sum_i f_i \exp - \frac{E_i}{RT} \qquad [3]$$

Here, k_0 represents the summed rate constants of any deactivation processes which are independent of temperature. These may include intersystem crossing to the triplet state. The summation in Eq. [3] corresponds to the contribution of the temperature-dependent processes; f_i and E_i are the frequency factor and activation energy, respectively, for the ith process.

If only one temperature-dependent deactivation process is present, then

$$Q^{-1} - 1 = k_f^{-1} k_0 + k_f^{-1} f \exp - \frac{E}{RT} \qquad [4]$$

Upon differentiating Eq. [4], we have

$$\frac{dQ^{-1}}{d(1/T)} = - \frac{fE}{k_f R} \exp - \frac{E}{RT} \qquad [5]$$

and

$$\ln\left[-\frac{dQ^{-1}}{d(1/T)}\right] = \ln\frac{fE}{k_fR} - \frac{E}{RT}$$

In this case, a plot of $\ln[-dQ^{-1}/d(1/T)]$ versus $1/T$ yields a straight line of slope $-E/R$.[40] However, the graphical differentiations required in this formulation demand data of relatively high precision for successful analysis. In practice, if $Q < 0.1$ over the temperature region of interest, the term $k_0 k_f^{-1}$ in Eq. [4] is small compared with $Q^{-1}-1$ and may be neglected. In such cases, a plot of $Q^{-1}-1$ versus $1/T$ yields a straight line of slope $-E/R$.

Equations [3] to [5] describe the temperature dependence of quantum yield. Gally and Edelman[40, 41] have found that the fluorescence quantum yields of tryptophan and tyrosine in aqueous solution indeed decrease with increasing temperature. Tryptophan fluorescence appears to be somewhat more temperature sensitive than that of tyrosine. Oligopeptides containing either tryptophan or tyrosine likewise exhibit this kind of behavior. A discussion of the probable nature of the radiationless deactivation processes, which are responsible for the thermal quenching of tryptophan and tyrosine fluorescence, will be postponed until section 6.

3.2. Physical State

The physical state of an aromatic amino acid in liquid solution may be altered simply by cooling the solution to liquid-nitrogen temperatures or below, where a rigid glass matrix is formed for many solvents, such as 50% ethylene glycol/water. Aside from the enhancement of fluorescence to be expected from the decreased temperature *per se*, at least two other factors serve to increase the fluorescence emission. First, quenching by interaction with solvent molecules is greatly diminished in the rigid glass, and, second, any internal conversion mechanisms dependent upon the rotational and vibrational freedom of the solute molecules are curtailed. Thus Bishai *et al.*[42] found that in a rigid glass (0.5% glucose) at 77°K the total emission yield (fluorescence plus phosphorescence) from tyrosine, phenylalanine, and tryptophan approaches unity. In addition, the wavelength of maximum fluorescence for tryptophan is "blue-shifted" by about 25 nm or more from its value at room temperature. The wavelength shifts for tyrosine and phenylalanine are small.

Fluorescence emission from other solid states has been observed. Studies at liquid-nitrogen temperatures by Nag-Chaudhuri and Augenstein[43] revealed that the wavelength of maximal fluorescence was somewhat "red-shifted" for amino acid powders, as compared with rigid aqueous or organic solvent glasses. The fluorescence yield of phenylalanine powders at 300°K is enhanced more than tenfold relative to the value in aqueous solution[44] (see also Augenstein et al.[36]). The authors suggest that the increased yield is due to a lessening of collisional self-quenching, the elimination of quenching by interaction with solvent molecules, and restraints on the ease of intersystem crossing introduced by the crystalline structure.

The fluorescence of tryptophan in solid polyvinyl alcohol (PVA) has also been examined. Konev et al.[45] found that the fluorescence lifetime for tryptophan in PVA plates was some 50% higher than the value in aqueous solution. Accordingly, Kuntz[46] has reported that the fluorescence quantum yield of tryptophan in a PVA film is doubled and the wavelength of maximal emission "blue-shifted" 15 nm relative to tryptophan in water.

3.3. Solvent

The fluorescence emission spectrum of a molecule is, in general, more sensitive to the nature of the solvent than is its absorption spectrum. The reason for this lies in the differences between the absorption and emission processes. Absorption is a transition from an equilibrium ground state to a "Franck–Condon" excited state which is not at equilibrium with respect to interactions with solvent, while emission occurs from an equilibrium excited state and results in a nonequilibrium ground state; hence, the two processes are not exactly complementary. Solvent reorientation during the equilibration after absorption and before emission has a significant influence.

Solvents may affect the fluorescence of a molecule in one or both of the following ways: (1) via a nonspecific interaction which can be related to the polarity or dielectric constant of the solvent, and (2) through a specific solute–solvent interaction resulting in the formation of a stoichiometric complex. The first kind of interaction expends energy as the solvent rearranges to an equilibrium orientation about the excited solute molecule and is of a dipole–dipole nature, depending upon the polarities of the solvent and excited solute molecules. Lippert[47] has shown that the excited-state dipole moment of the solute may be determined, in principle, from a knowledge of the ground-state polarity and of the wavelength shifts in fluorescence emission induced by a series of solvents of varying polarity, provided that only mechanism (1) is important. Excited aromatic molecules are often more polar than their ground-state counterparts; this is another reason for greater solvent effects on luminescence emission than on absorption.[47]

The influence of solvent upon the fluorescence properties of indole and tryptophan derivatives has received considerable attention. Table 1

Table 1. Effect of Solvent on Fluorescence Emission Maxima of Indole and
N-Acetyltryptophan Methyl Ester[a]

Solvent	Dielectric constant	Fluorescence emission maximum (nm)	
		Indole	N-Ac-Trp-Me
n-Pentane	1.8	296	
Cyclohexane	2.0	297	
n-Hexane	1.9	300	310
Diethyl ether	4.3	303	
Benzene	2.2	305	
Butyl ether	—	310	330
Dioxane	2.2	310, 315	330
Acetonitrile	37.5	320	
n-Butyl alcohol	17.1	326, 330	340
Methyl alcohol	32.6	328, 330	340
n-Propyl alcohol	20.1	330	340
Ethyl alcohol	24.3	330	
Water	78.5	350	350

[a] Data taken from Van Duuren,[29] Cowgill,[48] and Walker et al.[49] In cases of differing values for the same system, both values are listed.

summarizes the effects of solvents of varying polarity upon the wavelengths of maximum fluorescence intensity for indole and N-acetyltryptophan methyl ester; the data are from Van Duuren,[29] Cowgill,[48] and Walker et al.[49] The general pattern of behavior is unambiguous. The emission wavelength is clearly shifted to the red with increasing solvent polarity, whereas solvent effects on the absorption spectrum of indole are small.[29] The quantum yield for these compounds decreases in polar solvents.[48]

The shift of emission spectrum to longer wavelengths in polar solvents is to be expected if indole and N-acetyltryptophan methyl ester undergo a π-π* transition upon excitation with an increase in dipole moment.[47] A calculation of the dipole moments of indole and tryptophan in the excited state has yielded values of 5.6 and 8.5 D, respectively.[50] According to Mataga et al.,[51] the value for indole represents an increase of 5 D upon excitation. A ground-state (π-electron) dipole moment for tryptophan of 2.3 D has been computed by Yeargers.[31] Van Duuren[52] has pointed out the likelihood of significant contributions to the indole excited state by ionic mesomeric structures (which would be a factor in the elevation of dipole moment).

However, dipole–dipole interactions of the excited indole with solvent do not suffice to explain all of the observations. Low concentrations of polar solvents have a dramatic and disproportionate effect upon the fluorescence of indole in cyclohexane. Van Duuren[29] observed that the addition of 1% ethanol to indole in cyclohexane caused a red shift of 18 nm in the emission spectrum. Since the bulk dielectric constant of the solvent is almost unchanged, this finding is not easy to explain on the basis of dipole–dipole interactions with the solvent.

Walker et al.[49, 53] have made similar observations for indole in several mixtures of polar and nonpolar solvents and have proposed that a specific and stoichiometric complex, or "exciplex," is formed by the excited indole molecule and the polar solvent [an example of mechanism (2)]. This model has been applied with some success to observations of the fluorescence of indole in pentane in the presence of low levels of several alcohols, including methanol and butanol. A 1 : 2 indole–alcohol complex* has been proposed for these systems.[49]

The alternative explanation exists that selective solvation of the indole by the polar solvent results in a *local* dielectric constant which is much higher than the bulk value, permitting dipole–dipole interactions to assume disproportionate importance. A choice between the two mechanisms can be made from the spectral distribution of fluorescence. The "exciplex" model predicts that an isoemissive wavelength exists for a range of concentrations of polar solvent, while the alternative model predicts only a gradual shift of emission spectrum. Walker et al.[49] have reported that an equivalent analysis for indole in pentane–alcohol mixtures confirms the exciplex model.†

Bobrovich et al.[54] have concluded from measurements of the absorption and fluorescence polarization spectra of indole in polar and nonpolar solvents that emission occurs from the second excited singlet (^1L_a) state in polar solvents and from both the first (^1L_b) and second states in nonpolar solvents.‡ As Walker et al.[55] have pointed out, the results in the paraffin nonpolar solvent used by the Russian workers may have been complicated by the presence of polar impurities in the solvent. Walker et al.[55] have suggested that an inversion of the energies of the two states occurs upon exciplex formation in polar solvents so that emission occurs from the ^1L_b state in nonpolar solvents and from the ^1L_a state in polar solvents.

However, recent fluorescence polarization studies of indole and several indole carboxylic acids have led Kurtin and Song[56] to conclude that emission occurs from both the ^1L_a and ^1L_b states in both polar (glycerol–

---

*Longworth[114] has reported a stoichiometry of 1:1 for the indole–isopropanol exciplex.

†Use of this kind of analysis had led Eisinger and Navon[90] to conclude that exciplexes between tryptophan and ethylene glycol–water solvent are not formed.

‡The use of L_a and L_b notation to describe the near-ultraviolet transitions of tryptophan is discussed by Weber.[28]

methanol) and nonpolar (diethyl ether–isopentane–ethanol) media, with a possible exception in the case of indole-N-acetic acid. They suggest that the contribution from ^1L_a predominates in polar solvents. Several other workers have reached similar conclusions.[51, 57, 58] The issue must be regarded as unsettled at the present time.

The presence of dissolved oxygen in a solution of an aromatic compound generally results in fluorescence quenching and is often a practical problem in fluorescence studies. The quenching is not serious for indole, being of the order of 5–10%.[29] A study of a series of indole derivatives (including tryptophan) by Bridges and Williams[59] revealed that only indole and 5-hydroxyindole showed enhanced fluorescence upon deoxygenation of solutions. Quenching of the X-ray excited fluorescence of tryptophan (and tyrosine) by oxygen and nitric oxide has also been observed.[60, 61] Both of the gases are paramagnetic (i.e., possess triplet ground states) and probably enhance intersystem crossing in the excited species.

Contrary to the results noted above for X-ray excited fluorescence of tyrosine, Feitelson[62] has found that dissolved oxygen has no effect on the ultraviolet-excited fluorescence of several tyrosine and phenylalanine derivatives. The effects of certain other added solutes were also investigated.

Fluorescence quenching by oxygen is a dynamic process dependent upon the ability of dissolved oxygen molecules to diffuse to the excited fluorogen and interact within the lifetime of the excited state.[63] The quenching efficiency decreases in general with decreasing excited lifetime and is small for molecules of lifetime less than about 5 nsec, including the aromatic amino acids.

The effect of solvent viscosity upon the fluorescence of tryptophan derivatives has been examined by Weinryb.[64] A comparison of the emission from a series of tryptophan compounds in two amphiprotic solvents of similar dielectric constant but widely differing viscosity (methanol and propylene glycol) reveals varying extents of fluorescence enhancement in the high-viscosity solvent, which may be related to the ionic nature of the particular derivative.

Several observations on the effect of solvent upon phenol, tyrosine, and phenylalanine emission are available. According to Weber and Rosenheck,[65] the fluorescence quantum yield and wavelengths of maximum emission of phenol are relatively insensitive to the polarity of solvent. However, Cuatrecasas et al.[66] report a marked enhancement of N-acetyltyrosinamide fluorescence upon addition of ethylene glycol, ethanol, or dioxane to aqueous solution. Cowgill[48] has likewise observed major increases in the quantum yields of phenol and acetyltyrosine ethyl ester with decreasing solvent dielectric constant. Feitelson[67] has found that tyrosine in aqueous solution is quenched by amides such as dimethylacetamide and dimethylformamide. The latter compound is a strong quencher of both tyrosine and O-methyltyrosine fluorescence.

The fluorescence properties of phenylalanine have been reported to show little or no dependence upon solvent.[68]

4. FLUORESCENCE QUANTUM YIELDS AND LIFETIMES FOR THE AROMATIC AMINO ACIDS

Literature values for the fluorescence quantum yields and lifetime of tryptophan, tyrosine, and phenylalanine, as well as their parent aromatic compounds in aqueous and nonaqueous solvents, are listed in Table 2. There is substantial uncertainty at present as to the correct quantum yields for the aromatic amino acids. The first determinations of these quantities were those of Shore and Pardee.[5] Their results were challenged by Teale and Weber,[6] who determined the quantum yields of tryptophan, tyrosine, and phenylalanine in aqueous solution with considerable care. However, they neglected to mention the pH and temperature conditions of measurements, although later papers by White[69] and Weber and Rosenheck[65] indicate that measurements were made at neutral pH and at temperatures near 26°C. These measurements and those of Weber and Teale[70] for phenol and indole have been widely accepted and have been routinely used to compute absolute quantum yields of fluorescence for proteins and aromatic model compounds by the comparative method. More recently, these values have been corroborated by Bishai et al.,[42] although no discussion of the associated experimental uncertainty was included.

Chen[71] has reinvestigated the quantum yields in water of the aromatic amino acids by comparison with the quantum yield of quinine, a commonly employed reference standard. In contrast, Teale and Weber[6] used various glycogen preparations as scattering standards of unit quantum yield. Chen finds values which are uniformly lower by one third than the results of Teale and Weber. He points out the instability and likelihood of contamination of glycogen preparations. However, the use of quinine as a standard may present problems, also.[72-75] Furthermore, Bishai et al.[42] used the same method as Chen[71] and arrived at different conclusions.

The most recent indications are that Chen's values may be the most reliable. These include the recent study on the fluorescence of substituted indoles by Bridges and Williams,[59] in which a quinine standard was employed and in which the quantum yield of tryptophan obtained (at neutral pH) was in good agreement with that of Chen. Also, Dawson and Windsor[73] employed the same basic method as Weber and Teale[70] (the use of a nonabsorbing, colloidal reference solution to scatter excitation light),

Table 2. Fluorescence Quantum Yields (Q_F) and Lifetimes (τ_F) for the Aromatic Amino Acids and Parent Compounds

Compound	Solvent	pH or state of ionization	Temperature (°C)	Q_F	τ_F (ns)	Reference
Indole	H_2O			0.45		70
	H_2O	7			5.0	106
	H_2O	7		0.23		90
	H_2O		25	0.24		101
	H_2O		25	0.4	4.1	78
	H_2O	7	23		2.7	77
	H_2O	Neutral (3.3–11.0)		0.46		59
	H_2O	Anion ($H_0 = 17.4$)		0.145		59
	n-Butanol			0.58^a		48
	Dioxane			0.62^a		48
Tryptophan	H_2O	Neutral		0.20^b		6
	H_2O	$-COO^-$, $-NH_2$		0.51^c		80
	H_2O	$-COOH$, $-NH_3^+$		0.085^c		80
	H_2O				3.0	78
	H_2O	7	23		2.6	77
	H_2O	7			2.5	106
	H_2O	Cation (pH 1)		0.091		59
	H_2O	Zwitterion (4.1–8.6)		0.149		59
	H_2O	Anion (10.7–11.1)		0.289		59
	H_2O	Dianion ($H_0 = 17$)		0.072		59
	H_2O	8.9		0.51^c		68
	H_2O	10.2			6.1	106
	H_2O		27	0.14		90
	H_2O			0.12		104
	H_2O		Room		2.8	90, 102

Table 2. (Continued)

Compound	Solvent	pH or state of ionization	Temperature (°C)	Q_F	τ_F (ns)	Reference
	H_2O	6	23	0.13		71
	H_2O + 0.055% glucose	0	27	0.05		42
	H_2O + 0.055% glucose	6	27	0.20		42
	H_2O + 0.055% glucose	12	27	0.17		42
	Ethylene glycol–water (1:1)	4.7	25		2.9	79
	Solid polyvinyl alcohol				4.3–4.7	45
	Polyvinyl alcohol film			0.43^a		46
	Tetrahydrofuran			0.52^c		68
	N,N'-dimethylformamide			0.39^c		68
	Dimethyl sulfoxide			0.81^c		68
Phenol	H_2O			0.19		104
				0.22		70
	H_2O		25	0.13		101
	H_2O	6	23	0.14		71
	Ethanol (O_2-free)			0.16		103
				0.19		104
	Methanol		26	0.25		65
	Isopropyl alcohol		26	0.27		65
	Acetonitrile		26	0.28		65
	Trimethylpentane		26	0.20		65
Tyrosine	H_2O	Neutral		0.21		6

	Solvent	pH	Temp (°C)	ϕ	τ (ns)	Ref.
	H$_2$O					78
	H$_2$O	6				71
	H$_2$O		23			102
	H$_2$O		Room			77
	H$_2$O	7	23	0.14	3.2	104
	H$_2$O	8.9				68
	Tetrahydrofuran			0.09	3.6	68
	N,N'-dimethylformamide			0.12^c	2.6	68
	Dimethyl sulfoxide			0.01^c		68
				0.22^c		
				0.27^c		
Phenylalanine	H$_2$O	Neutral		0.04^b		6
	H$_2$O	6	23	0.024		71
	H$_2$O				6.4	78
	H$_2$O + 0.55% glucose	0	27	0.02		42
	H$_2$O + 0.55% glucose	6	27	0.03		42
	Trifluoroethanol			0.02^c		68
	Dimethyl sulfoxide			0.02^c		68
	Tetrahydrofuran			0.02^c		68
	Diglyme [bis(2-methoxyethyl)ester]			0.02^c		68
Benzene	Ethanol			0.10		105
	Hexane			0.053		73
	Cyclohexane			0.07		107

^{a} Determined relative to a value of 0.40 for indole in water.
^{b} Determined by comparative method from value for tyrosine given in Teale and Weber.[6]
^{c} Determined by comparative method from appropriate value given in Teale and Weber.[6]
^{d} Determined relative to value of 0.20 for tryptophan in water.

but nevertheless failed to reproduce Weber and Teale's yields for various compounds. Other investigators[76] have found the tryptophan quantum yield obtained by Teale and Weber[6] inconsistent with their data. Although Weisstuch and Testa[76] add perhaps a final measure of confusion to the situation in suggesting that the early values of Shore and Pardee[5] are most consistent with their results, on balance the evidence to date appears to favor the values reported by Chen[71] (cf., however, pp. 368-369).

The confusion about absolute quantum yields does not invalidate conclusions based on *relative* yield. As Table 2 shows, the fluorescence yield of indole in water is reduced when it is incorporated into tryptophan. The quantum yield of tryptophan is dependent upon its state of ionization. Among the various ionized species, the order of increasing quantum yield is NH_3^+-Trp-COOH (acid form, pH 2) < NH_3^+-Trp-COO$^-$ (zwitterionic form, neutral pH) < NH_2-Trp-COO$^-$ (basic form, pH 9). Below pH 2, the yield decreases further because of collisional quenching by hydrogen ion. In strongly alkaline solution (pH > 11), the yield also decreases, probably as a result of proton dissociation by the excited indole.

The incorporation of phenol into tyrosine hardly affects its quantum yield. The yield is reduced by protonation of the carboxylate group and greatly reduced by ionization of the phenolic hydroxyl.[42, 69] In strong acid, hydrogen ion quenching becomes important. There is some indication of a similar effect for phenylalanine.[42]

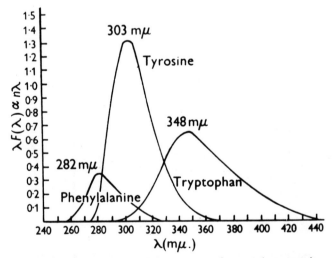

Figure 2. Fluorescence emission spectra of tryptophan, tyrosine, and phenylalanine in water (from Teale and Weber[6]). The ordinate is proportional to relative quanta emitted.

Teale and Weber[6] found that the fluorescence excitation spectra of the aromatic amino acids over the 200–320 nm range closely corresponded to their respective absorption spectra. This indicates that the quantum yields are independent of the excitation wavelength over the same range. This in turn implies that fluorescence occurs from the same (lowest-singlet) excited energy level in the particular solvent (H_2O), regardless of the extent to which higher excited states may be populated by absorption.[73]

Fluorescence emission spectra of tryptophan, tyrosine, and phenylalanine, as obtained by Teale and Weber,[6] agree closely in shape, emission maxima, and relative areas under the curves (or relative quantum yields) with those of Chen,[71] and are reproduced in Figure 2.

The short fluorescence lifetime encountered for aqueous solutions of aromatic amino acids lead to relatively large uncertainties in the values obtained. Within these limits, the emission lifetimes for tryptophan and tyrosine from different laboratories are in satisfactory agreement. There is a discrepancy in the indole lifetimes reported by Chen et al.[77] and by Gladchenko et al.[78]

5. FLUORESCENCE OF DERIVATIVES OF THE AROMATIC AMINO ACIDS

Much attention has been directed toward the question of the relationship between molecular structure and fluorescence properties for the aromatic amino acid derivatives.[48, 69, 79-90] Only the results will be summarized here; a detailed discussion of the mechanisms involved will be postponed to the following section.

5.1. Tryptophan Derivatives

Probably the greater part of the accumulated data has concerned the fluorescence of variously substituted tryptophan model compounds, including tryptophan peptides. The following relationships have been observed:

(1) Acetylation or benzyloxycarbonylation of the primary amino group abolishes its ionic character and increases the quantum yield and excited lifetime relative to the parent molecule with a protonated α-amino group.[48, 69, 79, 80] This is the case irrespective of the state of ionization or substitution of the α-carboxyl group.

(2) Esterification or amidation of the α-carboxyl group reduces the quantum yield and lifetime relative to unmodified tryptophan, irrespective of the state of ionization of the α-amino group.[48, 69, 79, 80] The effect

persists, although to a diminished extent, if the amino group is acetylated or benzyloxycarbonylated or if the α-carboxyl is separated by another residue from the tryptophan, as in Trp-Gly.[79]

(3) Formation of a peptide bond at either the α-carboxyl or α-amino position quenches the fluorescence.[69, 79-82, 87]

(4) Increasing the separation between a protonated α-amino group and the tryptophan residue, as in the series NH_3^+-$(Gly)_n$-Trp-COO^-, raises the quantum yield and lifetime.[79] In the above series the quantum yield increases as n increases from 1 to 3, while the degree of quenching resulting from protonating the α-amino group decreases.[83] In contrast, the series NH_3^+-Trp-$(Gly)_n$-COO^- shows only progressive quenching with increasing n.[79]

(5) The fluorescence efficiency of histidyl tryptophan is sharply reduced by protonation of the imidazole ring.[84]

(6) The proximity of a disulfide group, as in bis-indole-3-methylene disulfide, quenches indole fluorescence.[89] The quenching of fluorescence persists in the reduced, sulfhydryl form of this derivative.[89]

(7) For substitutions of the types listed in (1) to (4), the ratio of quantum yield to lifetime is constant and equal, within experimental uncertainty, to that of tryptophan. [No lifetime data are available for (5) and (6).] This indicates that the dominant radiationless deactivation processes are all first order with respect to the excited state.[79] For these classes of modification, the differences in quantum yield disappear at 91°K in a 50% ethylene glycol–water glass.[79]

5.2. Tyrosine Derivatives

The fluorescence of tyrosyl derivatives has also been studied in several laboratories. The results obtained indicate the following general trends:

(1) Fluorescence is strongly quenched upon the dissociation of the phenolic group.[69, 80] If the phenolic group is substituted, as in tyrosine O-phosphate, the fluorescence quenching typical of phenolate formation is not observed in the same pH range.[69]

(2) Protonation, amidation, or esterification of the α-carboxyl moiety reduces the fluorescence yield.[48, 62, 69, 80, 85]

(3) Acetylation of the protonated α-amino group has little if any effect on the fluorescence yield,[69, 85] in contrast to the case of tryptophan.

(4) Peptide bonds markedly quench the fluorescence of tyrosine.[67, 69, 80-82, 85, 86] They are somewhat more effective quenchers when substituted on the α-carboxyl rather than the α-amino end of tyro-

sine.[85] The fluorescence yields obtained are not very sensitive to the nature of the neighboring residues; in particular, the ionization of imidazole or glutamic γ-carboxylate did not significantly affect the fluorometric titration curve of histidyl tyrosine or glutamyl tyrosine.[86]

(5) In the series NH_3^+-$(Gly)_n$-Tyr-COO^-, the fluorescence efficiency increases with increasing values of n. While protonation of the α-amino group in NH_2-Gly-Tyr-COO^- almost halves the yield, there is no change upon protonation of NH_2-$(Gly)_2$-Tyr-COO^- or NH_2-$(Gly)_3$-Tyr-COO^-.[85]

(6) Iodination of tyrosine quenches fluorescence completely.[88]

(7) The quantum yield of tyrosine is greatly reduced by incorporation into L-cystinyl-bis-L-tyrosine or L-cysteinyl-L-tyrosine.[89]

5.3. Phenylalanine Derivatives

Because both the molar absorbancy and the fluorescence quantum yield of phenylalanine are low relative to those of tryptophan and tyrosine, significant phenylalanine emission is not observed in peptides or proteins containing tryptophan and/or tyrosine; accordingly, interest in the fluorescence of phenylalanine derivatives has lagged that in the other aromatic amino acids. Nevertheless, the following observations have been made:

(1) Protonation of the α-carboxyl group quenches fluorescence[42, 62]; α-carboxyl esterification also lowers the emission yield relative to the free zwitterion.[62] There is disagreement over the effect of α-amino ionization (cf. Bishai et al.[42] and Feitelson[62]).

(2) Phenylalanyl dipeptides yield the following results:[79] (a) There is no generalized quenching as a result of peptide bond formation; the quantum yield of a phenylalanine-containing dipeptide may be larger or smaller than that of phenylalanine, depending upon the nature of the second residue. (b) The presence of an N-terminal basic residue lessens the yield significantly. (c) The fluorescence of methionyl phenylalanine is appreciably quenched relative to phenylalanine.

5.4. Oligopeptides Containing Trytophan and/or Tyrosine

In addition to the correlations noted above, there is considerable evidence that radiationless transfer of excitation energy in oligopeptides occurs between un-ionized tyrosyl residues, between un-ionized and ionized tyrosyl residues, and between tyrosyl and tryptophanyl residues. There is great interest with regard to the occurrence of such energy transfer in proteins and in its relationship to the functional integrity of proteins and especially enzymes. Energy transfer in oligopeptides will be discussed in more detail in Section 9.

6. RADIATIONLESS DEACTIVATION OF THE EXCITED STATE

Equation [1] of Section 3 can be rewritten in the form

$$Q = k_f \tau = \frac{\tau}{\tau_0} \qquad [6]$$

or

$$\frac{Q}{\tau} = k_f \qquad [7]$$

Here, $\tau \, [= 1/(k_f + \Sigma k_i)]$ is the actual excited lifetime and $\tau_0 \, (= 1/k_f)$ is the value the lifetime would assume if radiationless deactivation of the excited state did not occur.

From Eq. [6], the constancy of Q/τ for a series of related compounds indicates that k_f is invariant and that any variation in quantum yield can be attributed to the differing magnitudes of the rate constants for radiationless deactivation. This has been shown to be the case for most of the tryptophan derivatives discussed in the preceding section[79] and, although data are lacking, is probably true for the corresponding derivatives of the other aromatic amino acids as well. An understanding of the nature of the radiationless deactivation reactions is thus basic to the interpretation of the relationship between molecular structure and fluorescence efficiency for these series of compounds. We shall focus attention here primarily upon derivatives of tryptophan, for which by far the most information is available.

It must be conceded in advance that a definitive picture of the relative importance of the various processes has not yet been attained. The obvious possibilities include:

(1) Intersystem crossing to a triplet state. While this certainly figures among the deactivation processes for tryptophan and its derivatives, there are no estimates available for the extent of intersystem crossing at ordinary temperatures. At 91°K the efficiency of intersystem crossing, as determined from fluorescence/phosphorescence ratios, has been found to be almost constant for a wide variety of tryptophan derivatives, whose quantum yields at room temperature differ considerably.[79] It is thus unlikely that variations in the rate of intersystem crossing are the primary cause of the differences in quantum yield of tryptophan derivatives at 25°C.

(2) "Tunneling" from excited to ground states. Radiationless deactivation by quantum mechanical tunneling from the first excited singlet state to the ground state is a definite possibility if the potential energy diagrams of the ground and excited states cross or approach each other. Eisinger and Navon[90] have suggested that this may be a significant source of radiationless deactivation for tryptophan in water, despite the large energy separation of the ground and excited states. However, in view of the similar spectral distributions and hence similar excited-state energies of tryptophan derivatives,[79] variations in this factor are unlikely to be major sources of differences in quantum yield.

Cowgill[87] has recently suggested that enhanced vibrational coupling between the excited and ground states may be a factor in the quenching of peptide derivatives of the aromatic amino acids.

(3) Electron ejection. The ejection of an electron from the excited indole to the solvent to form a free solvated electron has been proposed as an important quenching mechanism.[49, 92] The most direct evidence for the existence of this mechanism is derived from flash photolysis studies, which have shown that ultraviolet irradiation of aqueous solutions of indole and its derivatives, including tryptophan, results in formation of solvated electrons.[93, 94]

While accurate measurements of the quantum yield of solvated electrons will be required to assess the quantitative significance of this mechanism, there is considerable indirect evidence that it may be an important factor in the radiationless deactivation of the indole excited state. The quantum yield of indole in water is strongly temperature dependent. The thermal profile of quenching above 20°C may be analyzed in terms of a single dominant quenching process of high activation energy (~ 12 kcal).[55, 92] In nonaqueous solvents, such as dioxane and propylene glycol, where solvated electron production does not occur upon ultraviolet irradiation, the thermal dependence of quantum yield is very much reduced, corresponding to an activation energy of only 4 kcal or less.[92]

It is tempting therefore to identify the quenching process of high activation energy observed in water with electron ejection to solvent. The actual electron donor may be an "exciplex" formed by excited indole and water.[49] However, it must again be stressed that proof of this requires direct measurement of quantum yields of electrons.

Recently, Hopkins and Lumry[95] have made observations of the quantum yield of solvated electrons for indole as a function of temperature, using an N_2O scavenging technique. A definite correlation between electron yield and extent of thermal quenching was found.

Differences in the rate of electron ejection to solvent are not likely to be the explanation for the varying quantum yields of tryptophan derivatives, since the quenching substituents are in general electrophilic and should reduce, rather than raise, the yield of solvated electrons.

(4) *Intramolecular electron transfer.* Since electron scavengers are also quenchers of indole and tryptophan fluorescence and a rough parallel exists between quenching and scavenging abilities,[91] it appears feasible that electron transfer from the excited indole to a second molecule or group may be an efficient deactivation mechanism. Intramolecular electron transfer to an electron-accepting group may also be a factor in the quenching observed for many tryptophan derivatives. For example, cysteine, cystine, and protonated histidine are the only three amino acids which have very high efficiency as electron scavengers.[96, 97] Bis-indole-3-methylene disulfide, its sulfhydryl derivative, and tryptophanyl dipeptides containing protonated histidine have very low quantum yields.[84, 89]

The possibility also exists that the reduced quantum yield of tryptophan (indole-3-alanine) relative to indole arises from electron transfer to the substituent. Alanine, like the other amino acids, is an electron scavenger.[97] Dipeptides are more efficient scavengers than single amino acids.[97] This parallels the quenching accompanying the formation of a peptide bond by tryptophan.

(5) *Proton quenching.* An early investigation by White[69] of the effect of pH on indole, tryptophan, and related compounds revealed (a) protonation of free amino and carboxyl groups (where present), and (b) extremes of pH, caused quenching. Quenching at low pH (<2) was thought due to collisional quenching by protons, and at high pH (>11) to abstraction of a proton from the imino nitrogen of the indole moiety by hydroxyl ions. The correlation between electrometric and fluorometric titration curves of tryptophan was later interpreted by Weber[98] as indicating that intramolecular proton transfer from the protonated α-amino group to the excited indole quenches fluorescence. More recently, results from investigations on the pH dependence of fluorescence of substituted indoles, including tryptophan,[59] and from studies of the variation of fluorescence energy yield with molecular structure for a series of tryptophanyl derivatives and peptides[79] were tentatively interpreted on a similar basis (see also refs. 135 and 136).

If quenching by *intramolecular* proton transfer is to be an important factor, such transfer must occur rapidly enough to compete effectively with emission from the singlet excited state. Since ring protonation of tryptophan in the ground state requires highly concentrated sulfuric acid solvents,[99] effective quenching by proton transfer and attachment to indole would dictate that the indole moiety in the excited state be a much better proton acceptor than in the ground state. Theoretical studies of the ground- and excited-state electron density distribution of indole derivatives are available, and are helpful though not conclusive.[31, 100] Calculations indicate that the imino nitrogen atom should be an even poorer proton acceptor in the excited state than in the ground state. However, the electron density of the carbon-2 atom of indole and indole-3-acetate (a reasonable model for tryptophan) increases upon excitation to the highest value for any ring atom.[100] It would seem a possible, though not certain, candidate for a proton acceptor site.[134]

However, it is not essential to this model that an actual attachment of the proton to the indole ring occur. Quenching might also take place by transient collisional contact. In this context it is of interest that hydrogen ion is an exceptionally efficient electron scavenger.[96] Should fluorescence quenching occur *via* electron–proton reaction within the solvent cage for compounds possessing intramolecular proton donors, the formal quenching mechanisms of electron transfer *from* the ring and proton transfer *to* the ring become operationally equivalent.

Inspection of the literature reveals that there has not been great interest in general mechanisms of fluorescence quenching in tyrosine and phenylalanine derivatives. Feitelson[62] suggested on the basis of the data then available that quenching by electron donation from the excited ring to exogenous or endogenous quencher (for example, carboxyl groups) was likely. In the case of tyrosine fluorescence quenching by exogenous amides,[67] the data appeared to exclude the possibility of direct involvement of the phenolic hydroxyl, and Feitelson concluded that direct interaction between quencher and excited aromatic ring was responsible.

7. PHOSPHORESCENCE OF THE AROMATIC AMINO ACIDS

7.1. Temperature-Dependence and Solvent-Dependence

As is the case for virtually all organic fluorogens, the triplet states of aromatic amino acids are completely deactivated by nonradiative processes at room temperature. Phosphorescence is only observed at low temperatures in a rigid medium. Most studies upon the phosphorescence of amino acids and oligopeptides have been carried out near 77°K in an aqueous 0.5% glucose or 1 : 1 (v/v) ethylene glycol–water glass, although other media have occasionally been used, including a polyvinyl alcohol film.

Under these conditions, all three aromatic amino acids exhibit a phosphorescence spectrum. The temperature profile of phosphorescence intensity is similar in appearance for all three, exhibiting a sharp break over a critical temperature range, which for tryptophan in 0.5% glucose is

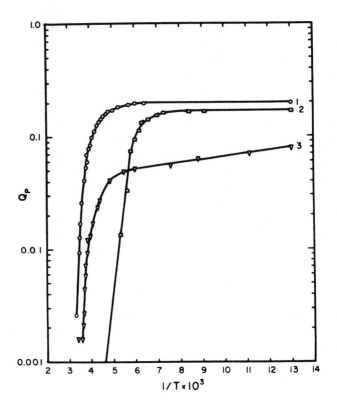

Figure 3. The dependence of the quantum yield of tryptophan phosphorescence upon temperature for (1) polyvinyl alcohol, (2) 0.5% glucose in H_2O, and (3) trypsin (from Kuntz[46]).

170–200°K.[108] Below the critical range the phosphorescence quantum yield is almost constant, while above it a rapid decrease with increasing temperature occurs (Figure 3). The critical zone presumably corresponds to the loosening of the rigid matrix, so that solvent reorientation becomes possible. At lower temperatures the molecule is effectively isolated electronically, while, at temperatures above the transition. interaction of the excited state with solvent can occur.

7.2. Tryptophan

The initial observations of the phosphorescence of tryptophan were made by Debye and Edwards,[7] and the phenomenon has subsequently been studied by many workers. In contrast to the tryptophan fluorescence band, the phosphorescence spectrum is highly structured (Figure 4). In aqueous media, distinct maxima are observed at about 406, 435, and 456 nm, with a shoulder at 480–490 nm.

The characteristics of the visible luminescence band emitted by tryptophan at low temperatures are uniformly those expected for a triplet → singlet transition. The decay time is long (~6.5 sec) in accordance with the forbidden nature of the transition. The decay is exponential, as was first observed by Debye and Edwards[7] and by Steele and Szent-Gyorgyi.[8] The decay of the triplet state may also be monitored by direct observation of the EPR signal.[109] The absorption spectrum of tryptophan in the triplet state has been determined by flash photolysis[110] and been found to be displaced to longer wavelengths than that of the ground state. The characteristic absorption spectrum of solvated electrons has been observed upon the irradiation of tryptophan in 10 M KOH at 80°K.[111] Kurtin and Song[56] have concluded from polarization studies of the luminescence of indole and several derivatives that the triplet → singlet emission originates from $^3(\pi,\pi^*)$ states of the ^3L_a type.

The α-phosphorescence (or delayed fluorescence) of tryptophan has also been reported.[112] This arises from the thermally activated reversion of the molecule from the triplet to the excited singlet state, followed by

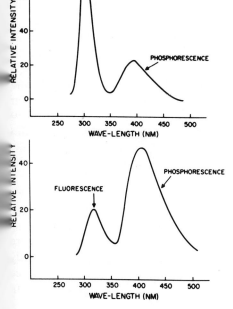
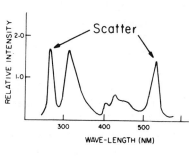

Figure 4. The total luminescence of tryptophan (upper right), neutral tyrosine (upper left), and alkaline tyrosine (lower left) in 50% ethylene glycol–water at 77°K.

Table 3. **Luminescence Properties of the Aromatic Amino Acids at 77°K**[a]

Compound	Fluorescence maximum (Å)		Phosphorescence maximum (Å)		F/P ratio		Average phosphorescence decay time (sec)	
	Powder	0.5% glucose	Powder	0.5% glucose	Powder	0.5% glucose	Powder	0.5% glucose
Phenylalanine	2950	2900	4150	3850	4.5	1.7	0.5[b]	5.5
Tyrosine	3050	3000	4500	3950	1.5	1.1	0.4[b]	2.6
Tryptophan	3300	3300	4900	4350	>15	3.9	1.5[b]	5.7

[a] From Nag-Chaudhuri and Augenstein.[43]
[b] Nonexponential decay.

a transition to the ground state with emission of radiation. The spectral distribution of the α-phosphorescence is equivalent to that of fluorescence, but the decay time is long ($\sim 10^{-3}$ sec) relative to that for fluorescence emission.

The phosphorescence of crystalline tryptophan powder has quite different characteristics from those observed in glasses.[43] The position of the maximum is shifted by over 50 nm to longer wavelengths, while the fluorescence/phosphorescence ratio increases to over 15 (Table 3). The decay of phosphorescence is nonexponential, the mean decay time being 1.5 sec. It seems clear that the properties of the triplet state are significantly perturbed in crystalline powders and that some heterogeneity of microenvironment exists.

Bishai et al.[42] have made a detailed study of the dependence of tryptophan phosphorescence upon the nature of the medium. In contrast to the behavior of the fluorescence at 25°C, the quantum yields of fluorescence and phosphorescence at 77°K in 0.5% glucose are constant and independent of pH (as measured at 25°C) between pH 12 and 2. At lower pH values a significant increase in the quantum yield of phosphorescence (but not of fluorescence) occurs. The phosphorescence quantum yield also increases significantly upon the addition of NaCl to 0.5 M. This was attributed by the authors to a change in the structure of the matrix.

7.2.1. Tryptophan Derivatives

The low-temperature luminescence of a series of tryptophan derivatives has been examined by Weinryb and Steiner.[79] In contrast to the fluorescence at 25°C, both the fluorescence and the phosphorescence at 91°K (in 1:1 ethylene glycol–water) are relatively insensitive to chemical substitution of the α-amino or α-carboxyl group of tryptophan. The spectral distribution and decay time of phosphorescence were essentially unchanged upon acetylation or carbobenzoxylation of the α-amino group, esterification of the α-carboxyl group, or formation of a peptide bond at either position. The fluorescence/phosphorescence ratio was likewise generally almost invariant, although a significant, though minor, increase may occur in the case of acetyl tryptophan. The decay of phosphorescence was strictly exponential for all the compounds studied. It thus appears that the indole triplet state is not greatly influenced by vicinal substitutions of this kind.

However, the phosphorescence of indole derivatives is sensitive to the position and chemical structure of substituents directly on the indole ring. Thus, in an ether–pentane–ethanol medium, the F/P ratio of indole-2-carboxylic acid is twice that of indole, while that of indole-5-carboxylic acid is decreased by a factor of 4.[56]

7.3. Tyrosine

The phosphorescence band of tyrosine, which was first described by Debye and Edwards in 1952, is devoid of resolvable structure.[7] Under conditions where the phenolic hydroxyl is un-ionized, the maximun in aqueous glasses is close to 387 nm (Figure 4). The decay time is about 2 sec.[113]

As in the case of tryptophan, the luminescence properties are different for crystalline powders (Table 3). The phosphorescence band is shifted to longer wavelengths, while the fluorescence/phosphorescence ratio in-increases. The decay curve is nonexponential, the mean decay time decreasing to 0.4 sec.[43]

In 0.5% glucose the quantum yields of fluorescence and phosphorescence have been reported to be independent of pH (as measured at room temperature) between pH 3 and 8 and thus appear to be insensitive to the state of ionization of the α-amino group.[42] Below pH 3 a significant decrease in fluorescence and increase in phosphorescence were observed. The summed quantum yields of fluorescence and phosphorescence were close to unity over the entire pH range.[42]

Ionization of the phenolic hydroxyl results in major changes in the luminescence pattern at 77°K. Phosphorescence is greatly enhanced relative to fluorescence, the fluorescence/phosphorescence ratio in aqueous glasses falling from about 0.9 to 0.16. The phosphorescence band is shifted to longer wavelengths, the maximum being close to 400 nm. The decay time of phosphorescence decreases to about 1.2 sec.

There are few published data on the phosphorescence of tyrosine oligopeptides. Longworth has reported that the phosphorescence of L-cystinyl-L-bis-tyrosine is almost entirely quenched.[114] At 77°K in 50%

ethylene glycol–water, the quantum yield of fluorescence is half the value found for N-acetyl-L-tyrosinamide, while the phosphorescence quantum yield is lowered to one-thirtieth of that of the latter compound. Apparently, a vicinal disulfide group blocks formation of the tyrosine triplet.[114]

7.4. Phenylalanine

The phosphorescence band of phenylalanine has a maximum at about 385 nm in 0.5% glucose[43] or in 50% ethylene glycol–water (Weinryb and Steiner, unpublished results). In both solvents the decay is exponential, with a decay time of about 5.5 sec. As in the cases of tyrosine and tryptophan, the crystalline powder has an increased fluorescence/phosphorescence ratio and a greatly reduced decay time of phosphorescence (Table 3).

The phosphorescence quantum yield is constant and independent of pH between pH 2 and 12, as is the decay time for phosphorescence. The summed quantum yields for fluorescence and phosphorescence are close to unity over the entire pH range at 77°K.[42]

Only minor differences in fluorescence/phosphorescence ratio were observed for a set of phenylalanine-containing dipeptides. The presence of a basic amino acid, or of methionine, appeared to decrease the ratio slightly.[79]

8. POLARIZATION OF LUMINESCENCE

8.1. Theory

If fluorescence is observed at an angle of 90° to the incident beam, the polarization of fluorescent radiation may, for practical purposes, be defined by

$$P = \frac{I_V - I_H}{I_V + I_H} \quad [8]$$

where I_V and I_H are the intensities of the vertically and horizontally polarized components, respectively. The exciting beam may be either unpolarized or vertically polarized.

Most aspects of the fluorescence polarization behavior of an organic molecule can be explained in terms of a simplified model which represents the fluorochrome as two linear oscillators corresponding to absorption and to emission which form a fixed angle, β.[28, 115, 116] Since the emission of fluorescence ordinarily occurs always from the same first excited single state, the orientation of the emission oscillator (or of the transition moment corresponding to fluorescence) is invariant and independent of the wavelength of exciting light. However, each absorption transition corresponds to a different absorption oscillator with its characteristic orientation. The value of β is thus dependent upon the absorption band used to excite fluorescence and hence upon the wavelength of exciting radiation.

If extraneous causes of depolarization of fluorescence, such as energy transfer and Brownian rotation, are absent, the observed polarization is determined by the angle between the directions of the effective transition moments for absorption and emission. In this case, the polarization is given by, for unpolarized exciting light,[117]

$$P_0 = \frac{3\cos^2\beta - 1}{7 - \cos^2\beta} \qquad [9]$$

and, for vertically polarized light,

$$P_0 = \frac{3\cos^2\beta - 1}{\cos^2\beta + 3} \qquad [10]$$

The complications mentioned above can be avoided if measurements are carried out in a rigid medium at sufficient dilution. Weber has employed propylene glycol at $-70°C$ as a solvent.[28, 98, 115] The viscosity of this medium is sufficient to immobilize the fluorescent species and eliminate depolarization arising from thermal motion.

If the emission is observed at a constant wavelength and the excitation wavelength is varied, a *polarization excitation spectrum* is obtained. Under these conditions, a variation in polarization with excitation wavelength may be attributed to the presence of two or more absorption bands, with different transition moments.

8.2. Phenol and Tyrosine

Phenol, cresol, and tyrosine all show very similar polarization excitation spectra, which are relatively uncomplicated.[28, 98, 115] In the case of phenol (Figure 5), a region of negative polarization occurs at low wavelengths, with the minimum at about 235 nm, and a region of positive polarization at long wavelengths. A definite plateau of positive polarization is approached at wavelengths well above the maximum of the primary absorption band ($>$ 280 nm).

This pattern can be readily interpreted in terms of two absorption transitions, the directions of whose moments form a large angle.[28, 115] If, as is probable, both transitions are of the π-π^* type, their moments are

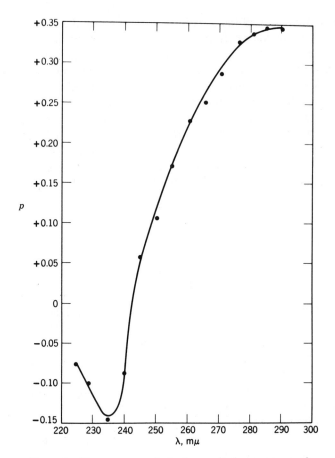

Figure 5. Fluorescence polarization excitation spectrum of phenol in propylene glycol at −70°C (from Weber[115]).

contained in the same plane and form an angle close to 90°. In this case, the absorption bands corresponding to the two transitions are well separated, so that no anomalies appear in the polarization excitation spectrum.

8.3. Indole and Tryptophan

In the case of indole and tryptophan, the polarization excitation spectra are much more complicated (Figure 6).[115] In the original measurements of Weber,[28] a minimum was observed at about 240 nm and a maximum close to 295 nm. However, there is in addition a sharp minimum near 290 nm. There must thus be at least three electronic transitions.

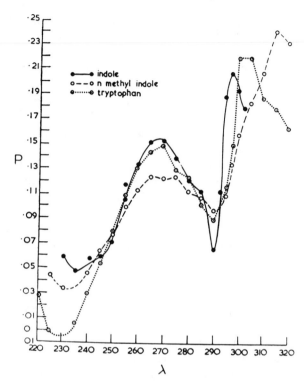

Figure 6. Fluorescence polarization excitation spectrum of indole and its derivatives in propylene glycol at $-70°$C (from Weber[98]).

To explain these results, Weber postulated that the long-wavelength absorption band of indole is composite and consists of at least two electronic transitions whose moments are oriented at a large angle.[28, 98, 115] For a given set of external conditions, only one of the two corresponding electronic levels is involved in emission; the other transfers absorption energy to it.

According to this model, the sharp minimum at 290 nm results from a narrow absorption band (^1L_b), which is superimposed upon a relatively broad band (^1L_a) corresponding to a different transition. The orientation of the moment corresponding to the 290 nm band is such that the polarization of the fluorescence arising from it is negative. Apart from a narrow region about 290 nm, the polarization excitation spectrum above 250 nm is dominated by the transition corresponding to the broad band, whose polarization is positive. At lower wavelengths (below 250 nm), a third transition, giving rise to negative polarizations, becomes important.

However, there are features of the polarization excitation spectrum which are not accounted for by this picture. In contrast to the behavior of the phenol derivatives, the polarization does not approach a definite plateau at long wavelengths but passes through a maximum near 295 nm. The subsequent decrease of polarization at higher wavelengths is not predicted by this model and may indicate the presence of an additional transition.

In addition, Weber[28] observed that, for glycyltryptophan, a minor maximum appeared near 288 nm. Lynn and Fasman,[118] using a device of higher spectral resolution, have subsequently reported that this small peak can also be detected in tryptophan itself. More recently, McKay[119] has confirmed this observation, both for tryptophan and indole, and has suggested that vibronic contributions from either or both of the ^1L_a and ^1L_b transitions may be responsible. The lack of a generally accepted explanation of the peak near 288 nm together with the anomalous behavior at long wavelengths indicates a need for a reappraisal of the electronic transitions of indole.

Kurtin and Song[56] have examined the polarization of indole fluorescence as a function of emission wavelength for a constant excitation wavelength (see section 3.3). A significant decrease in polarization with increasing emission wavelength was observed. As this result would not be predicted for emission from a single energy level, these workers were led to the conclusion that emission occurred from both the ^1L_a and ^1L_b levels in polar solvents.

9. ENERGY TRANSFER IN OLIGOPEPTIDES

9.1. Radiationless Exchange

Energy migration at the singlet level has been shown to occur for aromatic amino acids incorporated into oligo- and polypeptides. Radiationless transfer of excitation energy, which should be differentiated from the trivial process of direct reabsorption of emitted radiation, may occur between two groups as much as 100 Å or more apart, provided that a finite degree of overlap exists between the emission band of the donor and the absorption band of the acceptor.[120, 121] If the acceptor species is itself fluorescent, the transferred energy may appear as its characteristic fluorescence spectrum. If it is nonfluorescent, the energy transferred is ultimately degraded as heat and the process is manifested solely as a quenching of the fluorescence of the donor species.

$$D + h\nu \longrightarrow D^*$$
$$D^* + A \longrightarrow A^* + D$$
$$A^* \longrightarrow A + h\nu \quad \text{(A fluorescent)}$$

9.2. Intermolecular Transfer

Radiationless energy transfer by the Förster mechanism may occur between different molecules in a sufficiently concentrated solution. For example, the absorption and emission bands of phenol (and of tyrosine) show extensive overlap, making transfer by this mechanism a possibility. The efficiency of the process is of course dependent upon the mean separation of the molecules in solution and hence upon the concentration.

The effect of radiationless transfer between similar molecular species is to randomize the mutual orientations of the absorption and emission oscillators and hence to reduce the polarization of fluorescence. This has indeed been observed for concentrated solutions of phenol in propylene glycol.[28]

Weber[115] has derived the following quantitative relationship for concentration depolarization in a highly viscous medium, such as propylene glycol at $-70°C$, where depolarization by Brownian rotation is blocked:

$$\frac{1}{P} - \frac{1}{3} \cong \left(\frac{1}{P_0} - \frac{1}{3}\right)\left[1 + 1.7\left(\frac{R}{2a}\right)^6 C\right] \quad [11]$$

where P and P_0 are the polarizations (for vertically polarized exciting light) at concentration C (moles/liter) and at infinite dilution, respectively, R is the separation at which the probability of transfer is 50%, and $2a$ is the distance of closest approach of two molecules. The above equation was applied by Weber[115] to the observed concentration dependence of polarization for phenol in propylene glycol at $-70°C$. Data were obtained for 30 micron layers, for which radiative transfer by direct reabsorption of emitted fluorescence should be negligible. The value of R computed on this basis was about 16 Å. It is to be expected that a similar value will hold for tyrosine, whose absorption and emission spectra are similar to those of phenol, and hence that radiationless transfer will be important for adjacent tyrosine residues in an oligopeptide.

9.3. Intramolecular Transfer in Tyrosine Oligopeptides

Intramolecular energy transfer is an important factor in the luminescence properties of polymers and oligomers of tyrosine. Since the behavior of hexa-L-tyrosine is characteristic of all the tyrosine oligopeptides, the discussion here will be centered upon this oligomer.

Longworth et al.[122] have found that, at both 298°K and 77°K, isosbestic points exist in the absorption spectrum of $(Tyr)_6$ over the entire pH range in which ionization of the phenolic hydroxyl occurs. The oligomer can thus be treated as a two-component system, and the degree of ionization may be computed from the known absorption spectra which are characteristic of the ionized and un-ionized forms.[122]

At 298°K the fluorescence band arising from neutral tyrosine (λ_{max} = 303 mμ) is rapidly quenched as consequence of the fractional ionization of $(Tyr)_6$. The ionization of a single tyrosine is sufficient to quench the fluorescence of all the remaining un-ionized residues.[123]

The behavior at 77°K in a 1:1 ethylene glycol–water glass is analogous.[124] Again, the ionization of one of the six tyrosyl groups eliminated both the fluorescence and the phosphorescence characteristic of neutral tyrosine. As a consequence, the fluorescence/phosphorescence ratio and the decay time for phosphorescence attain values typical of ionized tyrosine at this degree of ionization.

Examination of the excitation spectra for phosphorescence indicates that, for fractional degrees of ionization at which the total emission pattern is characteristics of tyrosinate ion, much of the excitation energy still originates from neutral tyrosine. It thus is clear that extensive energy migration is occurring from neutral to ionized tyrosine.[124]

All of the above observations can be explained adequately in terms of transfer at the singlet level by the Förster mechanism. The fluorescence band of neutral tyrosine and the absorption band of ionized tyrosine overlap extensively, so that energy migration by this mechanism is feasible. From parallel observations upon $(Tyr)_2$ and $(Tyr)_3$, Knopp and Longworth[123] have concluded that transfer between tyrosine and tyrosinate groups is 100% efficient for these compounds. For the related case of polytyrosine, Longworth et al. have been able to account quantitatively for the observed transfer efficiency by calculations based upon the Förster equation.[122]

Since the fluorescence and phosphorescence characteristic of neutral tyrosine were quenched to an equivalent degree for partially ionized $(Tyr)_6$, triplet⟶triplet transfer does not appear to be an important contributing factor for this system.[124]

Edelhoch et al.[85] have studied Tyr⟶Tyr transfer between un-ionized tyrosines in the series Tyr-(Gly)$_n$-Tyr ($n = 0, 1, 2$). In a viscous medium the polarization of Tyr-Tyr is substantially less than that of tyrosine itself as a consequence of energy exchange between tyrosines. The efficiency of transfer decreases as n increases from 0 to 2, as predicted by Eq. [11]. An increase in separation of 2–2.5 Å per glycine unit was computed.

9.4. Intramolecular Transfer in Oligopeptides Containing Tryptophan and Tyrosine

Since the fluorescence band of neutral tyrosine shows extensive overlap with the absorption band of tryptophan, the formal requirements for radiationless energy exchange by the Förster mechanism are satisfied. For the dipeptide Trp-Tyr, the fluorescence of the neutral tyrosine group is completely suppressed both at 25°C and 77°K and only tryptophan fluorescence is observed.[114,125] Since the excitation spectrum for tryptophan fluorescence is identical to the absorption spectrum, which is the sum of those of tryptophan and tyrosine, radiant energy absorbed by tyrosine must appear as tryptophan fluorescence.[114] It has been reported that the degree of enhancement of tryptophan fluorescence is that expected for quantitative transfer of the excitation energy of tyrosine to tryptophan.[126]

The total luminescence pattern (fluorescence plus phosphorescence) of Trp-Tyr is also characteristic of tryptophan.[125,127] In this case, the suppression of tyrosine phosphorescence is presumably a consequence of radiationless energy transfer to tryptophan at the singlet level. The depopulation of the first excited singlet state of tyrosine by exchange with tryptophan blocks the subsequent formation of the triplet state.

The preceding refers to conditions where the tyrosine residue is unionized. Ionization of the tyrosine results in a pronounced quenching of the tryptophan fluorescence, both in water at 25°C and in a 1:1 ethylene glycol–water glass at liquid-nitrogen temperatures.[125,127] In the latter case, the suppression of tryptophan fluorescence is accompanied by a major enhancement of tryptophan phosphorescence. No important contribution of tyrosine phosphorescence is observed. The enhancement of tryptophan phosphorescence and the accompanying drop in fluorescence/phosphorescence ratio are correlated with the degree of ionization of tyrosine.

The ionization of tyrosine results in a shift of its absorption band to longer wavelengths, so that a finite degree of overlap with the emission band of tryptophan is present. The quenching of tryptophan fluorescence under these conditions probably arises from radiationless transfer to ionized tyrosine at the singlet level. The concomitant elevation of tryptophan phosphorescence indicates that a transfer in the reverse direction also occurs from the triplet state of ionized tyrosine to that of tryptophan. The efficiency of the transfer process is high, as very little phosphorescence characteristic of ionized tyrosine is observed.

Edelhoch et al.[83] have examined the quenching of tryptophan fluorescence by ionized tyrosine in water at 25°C for the series of compounds Trp-(Gly)$_n$-Tyr, where $n = 0$–4. The maximum degree of quenching falls from 85% for Trp-Tyr to 50% for Trp-(Gly)$_4$-Tyr. It is of interest that the backbone chain length of the latter compound, when fully extended, is 14.5 Å, as compared with a computed value for R_0 of 13.3.

The degree of quenching for this series of oligopeptides is unchanged in 35% ethanol and in 9 M urea. The latter observation is inconsistent with any alternative quenching mechanism involving direct interaction of the indole and tyrosinate groups and is in harmony with the idea that radiationless transfer by the Förster mechanism is the dominant process.

10. THERMOLUMINESCENCE OF THE AROMATIC AMINO ACIDS

If polycrystalline samples of the aromatic amino acids are irradiated while at temperatures near 77°K and, subsequent to cessation of irradiation, allowed to warm slowly, luminescence may be observed for characteristic temperature regions. This phenomenon is known as thermoluminescence, and has been induced by both X-ray and ultraviolet excitation. The mechanism by which the absorbed energy is trapped at low temperatures and later emitted as the temperature is raised is obscure, although a long-lived excited state which may lie below the lowest excited triplet state and which is formed after the initial absorption event has been postulated.[128] The reader is referred to other articles for details of research in this area.[128-133]

REFERENCES

1. R. L. Bowman, P. A. Caulfield, and S. Udenfriend, Spectrophotofluorometric assay in the visible and ultraviolet, *Science* 122, 32–33 (1955).
2. D. E. Duggan and S. Udenfriend, The spectrophotofluorometric determination of tryptophan in plasma and of tryptophan and tyrosine in protein hydrolysates, *J. Biol. Chem.* 223, 313–319 (1956).
3. D. E. Duggan, R. L. Bowman, B. B. Brodie, and S. Udenfriend, A spectrophotofluorometric study of compounds of biological interest, *Arch. Biochem. Biophys.* 68, 1–14 (1957).
4. H. Sprince, G. R. Rowley, and D. Jameson, Spectrophotofluorometric studies of 5-hydroxyindoles and related compounds, *Science* 125, 442–443 (1957).
5. V. G. Shore and A. B. Pardee, Fluorescence of some proteins, nucleic acids, and related compounds, *Arch. Biochem. Biophys.* 60, 100–107 (1956).
6. F. W. J. Teale and G. Weber, Ultraviolet fluorescence of the aromatic amino acids, *Biochem. J.* 65, 476–482 (1957).
7. P. Debye and J. O. Edwards, A note on the phosphorescence of proteins, *Science* 116, 143–144 (1952).
8. R. H. Steele and A. Szent-Gyorgyi, On excitation of biological substances, *Proc. Nat. Acad. Sci.* 43, 477–491 (1957).

9. H. M. Hershenson, "Ultraviolet and Visible Absorption Spectra. Index for 1930–1954," Academic Press, New York (1956).
10. H. M. Hershenson, "Ultraviolet and Visible Absorption Spectra. Index for 1955–1959," Academic Press, New York (1961).
11. D. B. Wetlaufer, Ultraviolet spectra of proteins and amino acids, *Advan. Protein. Chem. 17*, 303–390 (1962).
12. G. H. Beaven and E. R. Holiday, Ultraviolet absorption spectra of proteins and amino acids, *Advan. Protein. Chem. 7*, 319–386 (1952).
13. C. S. Hicks, The relationship of thyroxin to tryptophan, *J. Chem. Soc. 127*, 771–776 (1925).
14. F. C. Smith, The ultra-violet absorption spectra of certain aromatic amino-acids, and of the serum proteins, *Proc. Roy. Soc. (London) B104*, 198–205 (1929).
15. C. B. Coulter, F. M. Stone, and E. A. Kabat, The structure of the ultraviolet absorption spectra of certain proteins and amino acids, *J. Gen. Physiol. 19*, 739–752 (1936).
16. G. L. Brown and J. T. Randall, Low-temperature ultra-violet absorption spectra of biologically important compounds, *Nature 163*, 209–210 (1949).
17. B. G. Edwards, Ultraviolet spectra of some indole derivatives, including tryptophan and gramicidin, *Arch. Biochem. 21*, 103–108 (1949).
18. H. Grinspan, J. Birnbaum, and J. Feitelson, Environmental effects on the ultraviolet absorption spectrum of tyrosine, *Biochim. Biophys. Acta 126*, 13–18 (1966).
19. T. W. Campbell, S. Linden, S. Godshalk, and W. G. Young, The absorption spectra of some benzene derivatives with unsaturated side chains, *J. Am. Chem. Soc. 69*, 880–883 (1947).
20. L. Doub and J. M. Vendenbelt, The ultraviolet absorption spectra of simple unsaturated compounds. I. Mono- and p-disubstituted benzene derivatives, *J. Am. Chem. Soc. 69*, 2714–2723 (1947).
21. H. Sponer, G. Nordheim, A. L. Sklar, and E. Teller, Analysis of the near ultraviolet electronic transition of benzene, *J. Chem. Phys. 7*, 207–220 (1939).
22. A. L. Sklar, The near ultraviolet absorption of substituted benzenes, *J. Chem. Phys. 7*, 984–993 (1939).
23. J. R. Platt and H. B. Klevens, Absolute absorption intensities of alkylbenzenes in the 2250–1700 A. region, *Chem. Rev. 41*, 301–310 (1947).
24. A. L. Sklar, Theory of color of organic compounds, *J. Chem. Phys. 5*, 669–681 (1937).
25. G. N. Lewis and M. Kasha, Phosphorescence in fluid media and the reverse process of singlet–triplet absorption, *J. Am. Chem. Soc. 67*, 994–1003 (1945).
26. A. C. Pitts, Near ultraviolet absorption spectrum of liquid benzene from 2795 to 3560 A, *J. Chem. Phys. 18*, 1416–1417 (1950).
27. L. Pauling, "The Nature of the Chemical Bond," Cornell University Press, Ithaca, New York (1960).
28. G. Weber, Fluorescence-polarization spectrum and electronic-energy transfer in tyrosine, tryptophan, and related compounds, *Biochem. J. 75*, 335–345 (1960).
29. B. L. Van Duuren, Solvent effects in the fluorescence of indole and substituted indoles, *J. Org. Chem. 26*, 2954–2960 (1961).
30. J. C. D. Brand and A. I. Scott, in "Techniques of Organic Chemistry" (A. Weissberger, ed.), Vol. XI, Part I (D. W. Bentley, ed.), p. 98, Interscience, New York (1963).
31. E. Yeargers, A. self-consistent-field study of tryptophan, *Biophys. J. 8*, 1505–1510 (1968).

32. K. Rosenheck and P. Doty, The far ultraviolet absorption spectra of polypeptides and protein solutions and their dependence on conformation, *Proc. Nat. Acad. Sci. 47*, 1775–1785 (1961).
33. R. B. Setlow and W. R. Guild, The spectrum of the peptide bond and other substances below 230 mμ, *Arch. Biochem. Biophys. 34*, 223–225 (1951).
34. A. D. McLaren, Photochemistry of enzymes, proteins, and viruses, *Adv. Enzymol. 9*, 75–170 (1949).
35. K. Sommermeyer, V. Birkwald, and H. Pruetz, X-ray excitation of fluorescence of dilute aqueous solutions of aromatic compounds, *Naturwissenschaften 48*, 666–667 (1961).
36. L. Augenstein, E. Yeargers, J. Carter, and D. Nelson, Excitation, dissipative, and emissive mechanisms in biochemicals, *Radiation Res. Suppl. 7*, 128–138 (1967).
37. H. B. Steen, Excitation of tryptophan in solution during irradiation with x-rays and UV light between 77°K and 300°K, *Radiation Res. 41*, 268–287 (1970).
38. D. R. Nelson, J. G. Carter, R. D. Birkhoff, R. N. Hamm, and L. G. Augenstein, Yield of luminescence from X-irradiated biochemicals, *Radiation Res. 32*, 723–743 (1967).
39. J. G. Carter, D. R. Nelson, and L. G. Augenstein, Effects of temperature on X-ray-induced light emission from powders of amino acids and trypsin, *Arch. Biochem. Biophys. 111*, 270–282 (1965).
40. J. A. Gally and G. M. Edelman, The effect of temperature on the fluorescence of some aromatic amino acids and proteins, *Biochim. Biophys. Acta 60*, 499–509 (1962).
41. J. A. Gally and G. M. Edelman, Effects of conformation and environment on the fluorescence of proteins and polypeptides, *Biopolymers Symp. 1*, 367–381 (1964).
42. F. Bishai, E. Kuntz, and L. Augenstein, Intra- and intermolecular factors affecting the excited states of aromatic amino acids, *Biochim. Biophys. Acta 140*, 381–394 (1967).
43. J. Nag-Chaudhuri and L. Augenstein, Effect of the physical environment on excited states of amino acids and proteins, *Biopolymers Symp. 1*, 441–452 (1964).
44. E. Yeargers and L. Augenstein, UV spectral properties of phenylalanine powder, *Biophys. J. 5*, 687–696 (1965).
45. S. V. Konev, M. Y. Kostko, L. G. Pikulik, and I. D. Volotovskii, Possible sources of different activity of the excited state of fluorescence of protein solutions, *Dokl. Akad. Nauk Belorussk. SSR 10*, 500–502 (1966).
46. E. Kuntz, Tryptophan emission from trypsin and polymer films, *Nature 217*, 845–846 (1968).
47. E. Lippert, Spektroskopische Bestimmung des Dipolmoments aromatischer Verbindungen in ersten angeregten Singuletzustand, *Z. Elektrochem. 61*, 962–975 (1957).
48. R. W. Cowgill, Fluorescence and protein structure. X. Reappraisal of solvent and structural effects, *Biochim. Biophys. Acta 133*, 6–18 (1967).
49. M. S. Walker, T. W. Bednar, and R. Lumry, Exciplex studies. II. Indole and indole derivatives, *J. Chem. Phys. 47*, 1020–1028 (1967).
50. L. F. Gladchenko and L. G. Pikulik, Determination of dipole moments of indole and tryptophan in excited states, *Zh. Prikl. Spektrosk. 6*, 355–360 (1967).
51. N. Mataga, Y. Torihashi, and K. Ezumi, Electronic structures of carbazole and indole and the solvent effects on the electronic spectra, *Theoret. Chim. Acta (Berlin) 2*, 158–167 (1964).

52. B. L. Van Duuren, Effects of the environment on the fluorescence of aromatic compounds in solution, *Chem. Rev. 63*, 325–354 (1963).
53. M. S. Walker, T. W. Bednar, and R. Lumry, Exciplex formation in the excited state of indole, *J. Chem. Phys. 45*, 3455–3456 (1966).
54. V. P. Bobrovich, G. S. Kembrovskii, and N. I. Marenko, Indole luminescence peculiarities, *Dokl. Akad. Nauk Belorussk. SSR 10*, 936–940 (1966).
55. M. S. Walker, T. W. Bednar, and R. Lumry, Exciplex studies. III. Radiative and non-radiative relaxation of the fluorescence state of indole and methyl derivatives of indole, in "Molecular Luminescence" (E. C. Lim, ed.), p. 135, W. A. Benjamin, Inc., New York (1969).
56. W. E. Kurtin and P. S. Song, A spectroscopic study of the polarized luminescence of indoles, *J. Am. Chem. Soc. 91*, 4892–4906 (1969).
57. H. U. Schuett and H. Zimmerman, Polarization of electron bonds of aromatic compounds. VII. Indole, indazole, benzimidazole, benztriazole, and carbazole, *Bev. Bunsenges. Physik. Chem. 67*, 54–62 (1963).
58. G. S. Kembrovskii, V. P. Bobrovich, and S. V. Konev, Low-temperature luminescence spectra of indole, *Zh. Prikl. Spektrosk. 5*, 695–698 (1966).
59. J. W. Bridges and R. T. Williams, The fluorescence of indoles and aniline derivatives, *Biochem. J. 107*, 225–237 (1968).
60. G. M. Barenboim, Interaction of excited biomolecules with oxygen. I. Quenching of photoluminescence of biomolecules by oxygen and nitric oxide, *Biofizika 8*, 154–164 (1963).
61. G. M. Barenboim and A. N. Domanskii, Interaction of excited biomolecules with oxygen. II. Extinguishing of tryptophan and tyrosine X-ray fluorescence with oxygen and nitric oxide, *Biofizika 8*, 321–330 (1963).
62. J. Feitelson, On the mechanism of fluorescence quenching. Tyrosine and similar compounds, *J. Phys. Chem. 68*, 391–397 (1964).
63. W. M. Vaughan and G. Weber, Oxygen quenching of pyrenebutyric acid fluorescence in water. A dynamic probe of the microenvironment, *Biochemistry 9*, 464–473 (1970).
64. I. Weinryb, The effect of solvent viscosity on the fluorescence of tryptophan derivatives, *Biochem. Biophys. Res. Commun. 34*, 865–868 (1969).
65. G. Weber and K. Rosenheck, Proton-transfer effects in the quenching of fluorescence of tyrosine copolymers, *Biopolymers Symp. 1*, 333–341 (1964).
66. P. Cuatrecasas, H. Edelhoch, and C. B. Anfinsen, Fluorescence studies of the interaction of nucleotides with the active site of the nuclease of *Staphylococcus aureus*, *Proc. Nat. Acad. Sci. 58*, 2043–2050 (1967).
67. J. Feitelson, Environmental effects on the fluorescence of tyrosine and its homologues, *Photochem. Photobiol. 9*, 401–410 (1969).
68. J. W. Longworth, Conformation and interactions of excited states. II. Polystyrene, polypeptides, and proteins, *Biopolymers 4*, 1131–1148 (1966).
69. A. White, Effect of pH on fluorescence of tyrosine, tryptophan, and related compounds, *Biochem. J. 71*, 217–220 (1959).
70. G. Weber and F. W. J. Teale, Determination of the absolute quantum yield of fluorescence solutions, *Trans. Faraday Soc. 53*, 646–655 (1957).
71. R. F. Chen, Fluorescence quantum yields of tryptophan and tyrosine, *Anal. Lett. 1*, 35–42 (1967).
72. R. F. Chen, Some characteristics of the fluorescence of quinine, *Anal. Biochem. 19*, 374–387 (1967).

73. W. R. Dawson and M. W. Windsor, Fluorescence yields of aromatic compounds, *J. Phys. Chem.* 72, 3251–3260 (1968).
74. J. E. Gill, The fluorescence excitation spectrum of quinine bisulfate, *Photochem. Photobiol.* 9, 313–322 (1969).
75. A. N. Fletcher, Quinine sulfate as a fluorescence quantum yield standard, *Photochem. Photobiol.* 9, 439–444 (1969).
76. A. Weisstuch and A. C. Testa, A fluorescence study of aminopyridines, *J. Phys. Chem.* 72, 1982–1987 (1968).
77. R. F. Chen, G. G. Vurek, and N. Alexander, Fluorescence decay times: Proteins, coenzymes, and other compounds in water, *Science* 156, 949–951 (1967).
78. L. F. Gladchenko, M. Y. Kostko, L. G. Pikulik, and A. N. Sevchenko, Duration of the excited state of ultraviolet fluorescence of aromatic amino acids, *Dokl. Akad. Nauk Belorussk. SSR* 9, 647–650 (1965).
79. I. Weinryb and R. F. Steiner, The luminescence of tryptophan and phenylalanine derivatives, *Biochemistry* 7, 2488–2495 (1968).
80. R. W. Cowgill, Fluorescence and the structure of proteins. I. Effects of substituents on the fluorescence of indole and phenol compounds, *Arch. Biochem. Biophys.* 100, 36–44 (1963).
81. R. W. Cowgill, Fluorescence and the structure of proteins. II. Fluorescence of peptides containing tryptophan or tyrosine, *Biochim. Biophys. Acta* 75, 272–273 (1963).
82. H. Edelhoch, R. S. Bernstein, and M. Wilchek, The fluorescence of tyrosyl and tryptophanyl diketopiperazines, *J. Biol. Chem.* 243, 5985–5992 (1968).
83. H. Edelhoch, L. Brand, and M. Wilchek, Fluorescence studies with tryptophyl peptides, *Biochemistry* 6, 547–559 (1967).
84. M. Shinitsky and R. Goldman, Fluorometric detection of histidine–tryptophan complexes in peptides and proteins, *Europ. J. Biochem.* 3, 139–144 (1967).
85. H. Edelhoch, R. L. Perlman, and M. Wilchek, Fluorescence studies with tyrosyl peptides, *Biochemistry* 7, 3893–3900 (1968).
86. E. C. Russell and R. W. Cowgill, Fluorescence and protein structure. XIII. Further effects of side-chain groups, *Biochim. Biophys. Acta* 154, 231–233 (1968).
87. R. W. Cowgill, Fluorescence and protein structure. XVII. On the mechanism of peptide quenching, *Biochim. Biophys. Acta* 200, 18–25 (1970).
88. R. W. Cowgill, Fluorescence and protein structure. IV. Iodinated tyrosyl residues, *Biochim. Biophys. Acta* 94, 74–80 (1965).
89. R. W. Cowgill, Fluorescence and protein structure. XI. Fluorescence quenching by disulfide and sulfhydryl groups, *Biochim. Biophys. Acta* 100, 37–44 (1967).
90. J. Eisinger and G. Navon, Fluorescence quenching and isotope effects of tryptophan, *J. Chem. Phys.* 50, 2069–2077 (1969).
91. R. F. Steiner and E. Kirby, The interaction of the ground and excited states of indole derivatives with electron scavengers, *J. Phys. Chem.* 73, 4130–4135 (1969).
92. R. F. Steiner and E. Kirby, The influence of solvent and temperature upon the fluorescence of indole derivatives, *J. Phys. Chem.* 74, 4480–4490 (1970).
93. L. I. Grossweiner and H.-I. Joschek, Optical generation of hydrated electrons from aromatic compounds, *Advan. Chem. Ser.* 50, 279–288 (1965).
94. H.-I. Joschek and L. I. Grossweiner, Optical generation of hydrated electrons from aromatic compounds. II, *J. Am. Chem. Soc.* 88, 3261–3268 (1966).
95. T. R. Hopkins and R. Lumry, Energy transfer in proteins: Ejection of electrons from indole exciplexes, *Biophys. J.* 9, A-216 (1969).

96. M. Anbar, Reactions of the hydrated electron, *Advan. Chem. Ser. 50*, 55–81 (1965).
97. R. Braams, Rate constants of hydrated electron reactions with amino acids, *Radiation Res. 27*, 319–329 (1966).
98. G. Weber, *in* "Light and Life" (W. M. McElroy and B. Glass, eds.), pp. 82–105, The Johns Hopkins Press, Baltimore (1961).
99. R. C. Armstrong, A third dissociation constant for tryptophan, *Biochim. Biophys. Acta 158*, 174–175 (1968).
100. P.-S. Song and W. E. Kurtin, The charge distribution in the excited states of some indoles, *Photochem. Photobiol. 9*, 175–178 (1969).
101. H. V. Drushel, A. L. Sommers, and R. C. Cox, Correction of luminescence spectra and calculation of quantum efficiencies using computer techniques, *Anal. Chem. 35*, 2166–2172 (1963).
102. W. E. Blumberg, J. Eisinger, and G. Navon, The lifetimes of excited states of some biological molecules, *Biophys. J. 8*, A-106 (1968).
103. H. C. Borreson and C. A. Parker, Some precautions required in the calibration of fluorescence spectrometers in the ultraviolet region, *Anal. Chem. 38*, 1073–1074 (1966).
104. H. C. Borresen, The fluorescence of guanine and guanosine, *Acta Chem. Scand. 21*, 920–936 (1967).
105. H. Ley and K. V. Englehardt, Ultraviolet fluorescence and chemical constitution of cyclic compounds, *Z. Physik. Chem. 74*, 1–64 (1910).
106. R. A. Badley and R. W. J. Teale, Resonance energy transfer in pepsin conjugates, *J. Mol. Biol. 44*, 71–88 (1969).
107. I. B. Berlman, "Handbook of Fluorescence Spectra of Aromatic Molecules," Academic Press, New York (1965).
108. J. Koudelka and L. Augenstein, The importance of the microenvironment surrounding a chromophore in determining its spectroscopic behavior, *Photochem. Photobiol. 7*, 613–617 (1968).
109. Z. P. Gribova, *cited in* "The Fluorescence and Phosphorescence of Proteins and Nucleic Acids" (S. V. Konev, ed.), Plenum Press, New York (1967).
110. L. I. Grossweiner and W. A. Mulac, Primary processes in the flash photolysis of ovalbumin and constituents, *Radiation Res. 10*, 515–521 (1959).
111. S. L. Aksentsev, Y. A. Vladimirov, V. I. Olenev, and Y. Y. Fesenko, Impulse photolysis study of primary photoproducts of aromatic amino acids at 80°K, *Biofizika 12*, 63–68 (1967).
112. G. M. Barenboim, Short-lived phosphorescence of DL-tryptophan in solutions, *Biofizika 7*, 227–232 (1962).
113. J. W. Longworth, Tyrosine phosphorescence of proteins, *Biochem. J. 81*, 23p–24p (1961).
114. J. W. Longworth, Excited state interactions in macromolecules, *Photochem. Photobiol. 7*, 587–596 (1968).
115. G. Weber, Polarization of the fluorescence of solutions, *in* "Fluorescence and Phosphorescence Analysis" (D. M. Hercules, ed.), pp. 217–239, Interscience, New York (1966).
116. F. Perrin, Polarization of light in fluorescence, average life of molecules in the excited state, *J. Phys. Radium 7*, 390–401 (1926).
117. A. Jablonski, Theory of the polarization of photoluminescence of colored solutions, *Z. Physik. 96*, 236–246 (1935).

118. J. Lynn and G. D. Fasman, Conformational dependence of fluorescence polarization spectra of L-tryptophan-containing copolypeptides, *Biopolymers 6*, 159–163 (1968).
119. R. H. McKay, Effect of various environments on the intrinsic fluorescence polarization spectra of horse liver alchol dehydrogenase, *Arch. Biochem. Biophys. 135*, 218–230 (1969).
120. T. Forster, "Fluoreszens organisches Verbindungen," Vandenhoeck and Ruprecht, Gottingen (1951).
121. T. Forster, Transfer mechanisms of electronic excitation, *Dis. Faraday Soc. 27*, 7–17 (1959).
122. J. W. Longworth, J. J. Ten Bosch, J. A. Knopp, and R. O. Rahn, Electronic energy transfer in oligomers and polymers of L-tyrosine, in "Molecular Luminescence" (E. C. Lim, ed.), p. 529, W. A. Benjamin, Inc., New York (1969).
123. J. A. Knopp and J. W. Longworth, Energy transfer in oligotyrosyl compounds: Fluorescence quenching as a function of the ionization of the phenolic hydroxyl groups, *Biochim. Biophys. Acta 154*, 436–443 (1968).
124. R. F. Steiner, The phosphorescence of tyrosine oligopeptides, *Biochem. Biophys. Res. Commun. 30*, 502–507 (1968).
125. R. F. Steiner and R. Kolinski, The phosphorescence of oligopeptides containing tryptophan and tyrosine, *Biochemistry 7*, 1014–1018 (1968).
126. T. Cassen and D. Kearns, Investigation of energy transfer in peptides by excitation difference spectra techniques, *Biochem. Biophys. Res. Commun. 31*, 851–855 (1968).
127. C. Hélène, M. Ptak, and R. Santus, Optical and magnetic study of the excited states in the polynucleotides and polypeptides, *J. Chim. Phys. 65*, 160–166 (1968).
128. J. E. Gill and M. Weissbluth, Thermoluminescence of amino acids and proteins irradiated with ultraviolet light, *Biopolymers Symp. 1*, 433–439 (1964).
129. S. Prydz and T. Rogeberg, Spectral study of thermoluminescence from aromatic amino acids, *Phys. Norvegica 1*, 227–233 (1963).
130. M. Guermonprez, R. Santus, and M. Ptak, Delayed luminescence of aromatic amino acids in boric acid, *Comp. Rend. 261* (Group 6), 387–390 (1965).
131. D. I. Roshchupkin, Effect of gases on the thermoluminescence and prolonged afterglow of frozen proteins and aromatic amino acids irradiated by ultraviolet light, *Biofizika 11*, 167–168 (1966).
132. A. K. Kukushkin and A. N. Kuznetsov, Possible physical mechanisms of the thermal luminescence of some aromatic amino acids and proteins, *Biofizika 11*, 223–227 (1966).
133. E. E. Fesenko, E. A. Burshtein, and Y. A. Vladimirov, Biphotonic ionization of aromatic acids in alkaline medium at 80°K, *Biofizika 12*, 616–623 (1967).
134. R. W. Ricci, Deuterium-isotope effect on the fluorescence yields and lifetimes of indole derivatives—including tryptophan and tryptamine, *Photochem. Photobiol. 12*, 67–75 (1970).
135. S. S. Lehrer, Deuterium-isotope effects on the fluorescence of tryptophan in peptides and in lysozyme, *J. Am. Chem. Soc. 92*, 3459-3462 (1970).
136. J. Feitelson, Environmental effects on the fluorescence of tryptophan and other indole derivatives, *Israel J. Chem. 8*, 241–252 (1970).

Chapter 6

LUMINESCENCE OF POLYPEPTIDES AND PROTEINS

J. W. Longworth

*Biology Division,
Oak Ridge National Laboratory*
Oak Ridge, Tennessee*

1. HISTORICAL SURVEY

It is frequently advantageous to consider the evolution of ideas and results when surveying a particular area of reseach since the early results often contain the fundamental outlines of a problem which subsequently become obscured by the later accumulation of detailed knowledge. The history of the development of protein fluorescence followed a logical pattern and so is quite suitable to discuss with a linear time scale. Because there are many experimental complications presented to investigators of the luminescence of biological macromolecules, each advance in knowledge was and is intimately associated with technological breakthroughs in measurement rather than with advances in theory. Also, because of considerable inherent difficulties in luminescence measurements, many of the initial observations announced were later shown to be in error. So, with apologies to friends and former colleagues for remembering the errors, it is hoped that a brief account of the development of the field will prove to be of value to those who are tempted to enter this difficult area, which is one so susceptible to serious errors.

---

*Operated by the Union Carbide Corporation under contract with the United States Atomic Energy Commission.

1.1. Existence of Excited States

Fluorescence* and phosphorescence have been associated with proteins and proteinaceous material for more than a century, but it was only in the early 1950's that any quantitative studies were performed. Even with hindsight, it is not really possible to be sure whether much that was reported by these early workers was a result of the intrinsic fluorescence or phosphorescence from proteins or amino acids, or whether it was due to natural or adventitiously bound organic materials sensitized by the protein adsorption.

1.1.1. Photochemical Evidence

The existence of excited states that were potentially detectable by emission was implicit in the early photochemical studies of Warburg and Negelein,[3] though Warburg did not comment on this aspect at that time. Warburg[4] studied the action spectrum for the photodissociation of a carbon monoxide complex of cytochrome a and found that excitation with 280 nm light produced an efficient photodissociation of the complex (Figure 1). The predominant absorbers of light of this wavelength were the aromatic acids of the protein; heme a does not have significant absorption in this spectral region. Warburg and his colleagues extended their observations and showed that the photodissociation was a property of several other hemoprotein complexes, particularly the hemoglobin–CO complex, though here the heme absorption made a greater contribution at those wavelengths where the protein also absorbed light than it did with cytochrome oxidase–CO complex. Bucher and coworkers[6-8] reconfirmed and extended the observations of the Warburg group. In a detailed study of the photodissociation of myoglobin–CO complex, their work gave clear evidence for a sensitization by the protein absorption of the photodissociation. However, there were two complications: At 280 nm, the wavelength where protein absorption was maximal, the protein absorption only accounted for 40% of the total light absorbed, and there was an alteration in the hemoprotein absorption upon CO dissociation. Because of these difficulties, some workers considered the experiments to be not entirely convincing.[9]

We now know the atomic structure of myoglobin—only one of the two tyrosines is immediately adjacent to the heme, and the other tyrosine and the two tryptophans are considerably removed from the heme. So, in ret-

---

*G.G. Stokes (ca. 1852). Cited by Dhere.[1]

†J.B. Becconi,[2] a Bolognian physician and physicist, observed (in severe frost one January) that his freshly washed hand phosphoresced to his dark-adapted eyes after an exposure to sunlight.

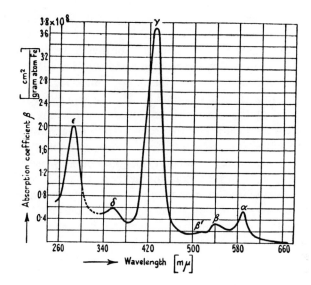

Figure 1. Photodissociation action spectrum of the carbon monoxide compound of the oxygen-transporting respiratory enzyme. (After Warburg.[4])

rospect, these experiments provided strong evidence for transfer of the electronic energy from the protein to the heme chromophore. To overcome the objections raised by Havemann and Wolff,[9] which were largely based on the small fractional absorption due to the aromatic amino acids of myoglobin, Broser and Lautsch and their coworkers[10-12] synthesized a porphyrin that was attached to polyphenylalanine-[mesohemin-IX-poly-DL-(Phe)$_{100}$Glu] in which essentially all the absorption at 260 nm was due to the phenylalanine. On excitation at 260 nm, the quantum efficiency was found to be unity for the dissociation of the CO complex of this hemin polypeptide. A related natural biological system was the photosynthetic pigment phycocyanin studied by Bannister.[13] Here the absorption of light by the protein was shown to cause a sensitization of the fluorescence of the phycocyanobilin prosthetic group. Since then, similar sensitizations have been demonstrated with a wide range of proteins either with luminescent prosthetic groups[14] or with synthetically bound chromophores.[15] The evidence of sensitization of chromophores bound to proteins by the protein absorption occurs with such a wide variety of proteins and chromophores that clearly there must be relatively long-lived excited states of these proteins. These ought to be emittent in the absence of the quenching chromophore.

Other experiments which pointed toward relatively long-lived excited states being formed after an exposure to ultraviolet light were those on ultraviolet-light inactivation of enzymes. Kubowitz and Haas,[16] and later Gates,[17] found that the photoinactivation spectra for the enzymic activity of urease[16] and pepsin[17] were almost identical with the protein-absorption spectrum. Light absorbed by the aromatic amino acid residues of the protein must have caused a sensitization of the photochemical reaction that occurred at the enzymic site, since it was unlikely that an aromatic amino acid was always involved at the site. With improved instrumentation, Setlow and Doyle[18] were subsequently able to reconfirm these observations, particularly with aldolase and RNase. However, Setlow[19] showed that the predominant contribution to photoinactivation of the majority of the proteins investigated was, rather, a result of direct absorption by the cystine residues; sensitization by the aromatic amino acids represented only a minor fraction of the total effect. Nevertheless, for certain proteins there was considerable sensitization by the absorption of the aromatic amino acids, particularly when there was no cystine in the protein (aldolase).

1.1.2. Phosphorescence by Proteins

With the advent of intense ultraviolet light sources and photomultipliers with good UV responses, a direct measurement of the emission from these excited states could be contemplated. In 1952, Debye and Edwards,[20] using a spectrograph with photographic detection, were the first to report the existence of a protein luminescence. They froze several globular proteins in aqueous matrices to 77°K, and, after exposing the proteins to short ultraviolet light radiated by a compact mercury arc, they were readily able to observe a blue phosphorescence upon cessation of the illumination. However, they could detect no fluorescence. A similar phosphorescence in the blue was found from only three of the amino acid constituents of proteins— the three aromatic amino acids phenylalanine, tyrosine, and tryptophan. The remainder of the amino acids did not exhibit any significant phosphorescence, an observation entirely consistent with photophysical studies of other organic chemicals. (Luminescence is invariably associated with aromatic ring structures, and is rarely observed from aliphatic compounds other than ketones or aldehydes.[21]) The blue phosphorescence decayed exponentially with a period of several seconds and, as suggested by Linschitz, was associated with triplet states.[20] Tyrosine and tryptophan had decay times comparable to those observed from the proteins, while phenylalanine had a considerably shorter decay constant. So it was considered that there was no phenylalanine contribution to the protein luminescence. This has since been fully justified, but now the lifetime of phenylalanine is known to be the longest of all the amino acid decay times; presumably, Debye and Edwards were studying microcrystals of phenylalanine. Debye and Edwards also found that there was a fine structure to the phosphorescence from both the amino acids and protein, and they noted that the intensity and spectrum depended on the pH value of the medium; both are features which have since been studied extensively.

1.1.3. Flash Photolysis—Radicals and Triplet States

A significant series of experiments substantiating the presence of triplet states in proteins after exposure to light was provided by Grossweiner[22] in 1956. Aqueous solutions of albumins, after deaeration, were exposed to an intense pulse of ultraviolet light at ambient temperatures. Species which absorbed in the visible formed. These were transient species, and they had lifetimes of several microseconds after the cessation of the actinic flash. For a time, the absorption was considered to have been caused by triplet states, but in 1959 Grossweiner and Mulac[23] showed that radical products were formed from the triplet states and it was these rather than the triplet which absorbed. Triplet states of tyrosine and tryptophan have only been observed by their absorption when flash-excited in rigid media at 77°K,[24,25] though even at this temperature considerable photoionization of the triplet occurred, forming radical species.

1.2. Protein Fluorescence

1.2.1. Prediction and Discovery

With the discovery of emitting triplet states in proteins in 1952, several workers suggested that there must be a singlet precursor to the triplets. Setlow and Doyle[26] discussed the temperature dependence of photoinactivation of enzymes by UV light in terms of both singlet and triplet states using a Jablonski diagram (see Figure 2), while Weber[27] and Bannister[13] proposed that proteins must have a fluorescence deep in the ultraviolet. Debye and Edwards had considerably simplified the problem: (a) only emissions from tyrosine and tryptophan occur from proteins, and (b) the properties of residues incorporated into the protein are not drastically different from those of isolated amino acids. With this in mind, several groups of workers in the Soviet Union, the United Kingdom, and the United States began the search for protein fluorescence and reported its existence almost simultaneously in 1955–1957, each without knowledge of the others' work. The reason for the success undoubtedly was the availability of suitable monochromators and photodetectors. Shore and Pardee[28] (1956) adopted the simplest approach; they merely used the Beckman photoelectric absorption spectrophotometer as a fluorometer. A complementary filter was placed between the fluorescence cuvette and the photomultiplier, and the fluorescence was excited by the monochromator in an inline configuration.

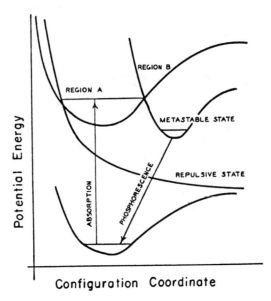

Figure 2. A Jablonski term scheme of the potential energies of excited states of proteins. (After Setlow and Doyle[26]; a fluorescence emission in the ultraviolet is implicit from this diagram, though not stated by the authors.)

Several significant conclusions were reached about the fluorescence of proteins by Shore and Pardee:

(a) Proteins fluoresced in the ultraviolet at ambient temperatures.
(b) The quantum yield of protein fluorescence was comparable to that of the free amino acids.
(c) Dissolved atmospheric oxygen had no major influence upon the value of the quantum yield.
(d) A fluorescence from all of the aromatic amino acids when they were incorporated into proteins was possible; proteins with only Phe, with Phe and Tyr, and with all three amino acids exhibited fluorescence.
(e) Different proteins, with different or similar aromatic amino acid compositions, showed widely differing yield values.
(f) The quantum yield of protein fluorescence was dependent on pH, decreasing at acid and alkaline values.
(g) The quantum yield of proteins was a function of the exciting wavelength, whereas that of amino acids was not.

The great difficulty in assessing the significance of the work of Shore and Pardee is that an inline geometry was used, a method notoriously susceptible to errors caused by scattered light. Nevertheless, much of the evidence presented has since proved substantially correct, and this paper provided the basis for much subsequent work. They had shown that proteins were fluorescent in aqueous solutions at ambient temperatures and that the solutions did not have to be degassed, a very convenient feature when working with proteins. Thus the stage was set for the second phase of development, a period which covers the years to 1960. The questions which now had to be answered were:

(a) Which amino acids are luminescent in proteins?
(b) How does the incorporation of the amino acid into a protein influence its emissive character?
(c) Is there any electronic transfer between the residues, both dissimilar and similar residues?
(d) Is there any relationship between the structure of proteins and their emissive character, and can fluorescence be used to study the functioning of enzymes?

1.2.2. Fluorescence Spectra of Amino Acids

It was to meet these more critical demands in studying protein fluorescence that the instrument described by Bowman et al.[29] proved so valuable. It added a major technical advance—the use of two monochromators, one to isolate the exciting light, the other to disperse the emitted light. To compensate for the small light intensities ultimately available to the light detector, an intense light source was used—the compact xenon arc. A spectrophotometer to measure luminescence directly having been designed, it was possible to view the sample cuvette perpendicularly to the direction of the exciting light beam and so minimize the problems from scattered light and stray light in the exciting light beam. This is basic design now universally adopted, and further improvements have consisted only in improving the stability of the light source, and using photomultiplier-detector systems which offer better stability, larger sensitivity, and improved signal-to-noise ratio.

With the Bowman spectrophotometer, the fluorescence emission spectra of tyrosine and tryptophan were determined (Bowman et al.[29]; Duggan and Udenfriend[30]) showing tryptophan has a significantly greater Stokes' shift than tyrosine; the fluorescence maxima of tyrosine and tryptophan were widely separated, tyrosine lying to the blue of tryptophan. This preliminary study was followed by a very detailed report from Teale and Weber,[31] which fully characterized the fluorescence of Phe, Tyr, and Trp

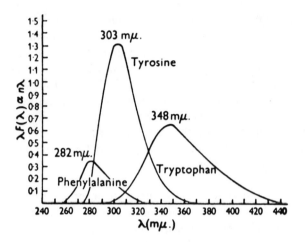

Figure 3. Corrected fluorescence spectra of the aromatic amino acids in water at 293°K. (After Teale and Weber.[32])

at neutral pH in aqueous media (presumably at ambient temperatures, though curiously not specifically stated, so the exact temperature is not known). The energy levels were shown to follow the sequence Phe>Tyr> Trp (Figure 3). Though Teale and Weber did not use two monochromators, their analysis of artifacts introduced by the instrument still constitutes the basis for fully calibrating spectrofluorometers today; it is recommended that the reader see this paper. Teale and Weber determined the fluorescence spectra in quanta versus wavelength (which can be converted to a wavenumber basis by multiplying the relative quantal intensity at a given wavelength by the square of that wavelength). Absolute quantum yields were accurately determined with the dipolar scattering method previously devised by Weber and Teale.[32] Phenylalanine was found to have a small quantum yield (0.04), while tyrosine and tryptophan had essentially equal values of 0.20. A significant feature of this paper was the determination of the excitation spectra and exact comparison of these with the corresponding absorption spectra (Figure 4). The coincidence of the two spectra provided two important facts which must be determined whenever a fluorescence of a biological compound is studied: (a) it proved that the emission was indeed a result of absorption by the compound under study, and (b) it defined the dependence of quantum yield on exciting wavelength—in these cases a constant. This is an essential requirement for any future discussion of energy transfer. The third feature emphasized by Teale and Weber was the invariance of the emission spectrum to the exciting wavelength. Much would be gained if these three features were customarily confirmed whenever a

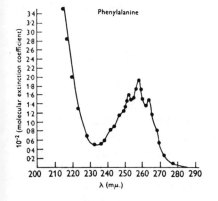
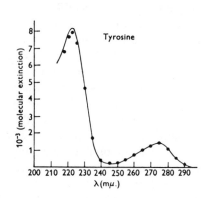
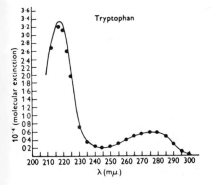

Figure 4. Corrected excitation and absorption spectra of aromatic amino acids. Solvent, water at pH 7; temperature, 293°K. Abscissa: wavelength (nm); ordinate: molar extinction coefficient ($10 \ m^2 \ mol^{-1}$). The continuous line is the absorbance spectrum; the dots represent fluorescence excitation values expressed as molar values derived from the briggsian logarithm of the fractional light absorption. (After Teale and Weber.[32])

luminescence study is reported. This paper sealed the fate of phenylalanine to be forever neglected, unless completeness was necessary; it has a small molar extinction coefficient ($19.7 \ m^2 \ mol^{-1}$) in comparison to tyrosine (142–$185 \ m^2 \ mol^{-1}$) or tryptophan (560–$617 \ m^2 \ mol^{-1}$) at their respective peaks of 258, 275, and 280 nm. When these values are taken together with the smaller quantum efficiency of phenylalanine, the small absorption of phenylalanine is never compensated for by a corresponding large Phe/(Tyr +Try) ratio in the amino acid composition of a protein and it makes a negligible contribution to the absorption of most proteins. Unless dramatic changes in yield occur on incorporating amino acids into proteins (which is not likely), the conclusion reached was that only an emission from tyrosine and tryptophan needed to be seriously considered in analyzing fluorescence from proteins.

At the same time as Teale and Weber were studying amino acid fluorescence in the United Kingdom, Vladimirov and Konev had begun a similar program in the Union of Soviet Socialist Republics, and a broad agreement is found between these groups of workers over the subsequent years. Konev[33] reported spectra of the aromatic amino acids obtained with solutions, and Vladimirov[34] gave them for the crystalline phase. There was also a somewhat earlier report of fluorescence from crystals by Brumberg,[35] who used an ultraviolet microspectrophotometer, and this was reported on in greater detail somewhat later by Barskii and Brumberg.[36] Since neither the United States workers nor the Soviet workers corrected for the transmission and detector responses at this time (though a good agreement for the wavelength of the maximum of tryptophan was reached among all workers), the maxima of uncorrected spectra from tyrosine and phenylalanine lay further to the red than the spectra given by Teale and Weber. This discrepancy is still often found and commented upon, though the differences are due to an instrumentally introduced technical artifact.

1.2.3. Electronic Energy Transfer Between Amino Acids

On the basis of the observed sequence of singlet energies, overlap of absorption and emission, and molar extinction, Teale and Weber[31] proposed that transfers

$$\text{Phe} \rightarrow \text{Tyr} \rightarrow \text{Trp}$$

were feasible as a result of dipolar resonance coupling. They suggested that these transfers would also occur in proteins since interresidue distances derived from the dimensions of proteins were commensurate with the known values for a dipolar resonance interaction. Vladimirov[34] was the first to report an experimental observation of such a transfer of electronic energy; he found that the crystals of phenylalanine slightly contaminated with tryptophan emitted only from tryptophan; the phenylalanine emission was quenched (Figure 5).

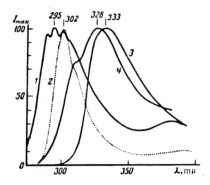

Figure 5. Fluorescence spectra of crystals of aromatic amino acids. (1) Phenylalanine; (2) tyrosine; (3) tryptophan; (4) tryptophan included into phenylalanine crystals in the molar ratio 1:100. Excitation at 240–280 nm. Abscissa: wavelength (nm); ordinate: fluorescence intensity. (After Vladimirov.[34])

1.2.4. Fluorescence Spectra of Proteins

Konev[33] was the first worker to study the fluorescence spectra of proteins. He showed that the fluorescence spectrum of a protein was not a summation of emissions from the constituent amino acids. Furthermore, the protein fluorescence spectrum did not alter when the exciting wavelength was changed, whereas the relative fractional absorptions by the amino acids did. This demonstrated that the protein fluorescence was an intrinsic property of that protein molecule and not a summation of its separate parts. Konev compared the fluorescence of the proteins with that obtained from their hydrolysate or from a synthetic mixture of aromatic amino acids of equal composition. He initially concluded that there was a broad similarity between these two spectra; all three amino acids made a contribution to the fluorescence of native proteins though that of the tryptophan predominated. He suggested that the tryptophan dominance resulted from a transfer from the tyrosine residues to tryptophan. Like Debye and Edwards, Konev found that proteins which lack any tryptophan, though they still possess tyrosine (class A), emit only with a tyrosine-like spectrum. He further showed that proteins which lack all three aromatic amino acids (the fish-egg protamines clupeine and sturnine) are nonfluorescent. There was an inherent discrepancy between the suggestion that both tyrosine and tryptophan emit and yet the spectrum was independent of the exciting wavelengths, but this could not be pursued further without an additional monochromator to provide monochromatic exciting light. The exciting wavelengths available were limited to those radiated by a mercury arc discharge and isolated by absorption filtering.[37] This seriously hindered exploration of this point. In 1959, both Vladimirov[38] and Konev[39] significantly improved their equipment by adopting single-photon counting techniques, and a much clearer picture of the nature of protein fluorescence emerged, but still a single monochromator in combination with a color filter was utilized—the monochromator used for isolating either exciting or emitted light. Vladimirov, both separately and with Konev (1959),[40] showed that the fluorescence of human serum albumin was only from the single tryptophan residue of the protein, even though there were 17 tyrosines in the protein, whereas the spectrum from the synthetic mixture of aromatic amino acids in equivalent proportions was dominated by the emission from tyrosine (Figure 6). A protein, zein (from *Zea mays*), which does not possess tryptophan, had a spectrum identical with that of tyrosine, and no phenylalanine contribution was detectable. This confirmed that tyrosine residues in proteins will fluoresce, and that a phenylalanine contribution is not detectable. But Konev (1959)[39] himself still maintained that both tyrosine and phenylalanine contributions could be detected in globular protein fluorescence, though the tryptophan contribution pre-

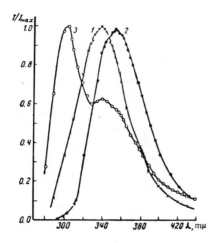

Figure 6. Aqueous fluorescence spectra of human serum albumin and a synthetic mixture of tyrosine and tryptophan of equal molar composition to the protein. (1) Human serum albumin; (2) tryptophan; (3) mixture of tyrosine and tryptophan in the molar ratio 18:1. Excitation at 240–250 nm. Abscissa: wavelength (nm); ordinate: fluorescence intensity. (After Vladimirov and Burstein.[34])

dominated. Moreover, he reported measurements of the excitation spectra of protein fluorescence which showed that the spectrum was identical to that obtained from the synthetic mixture for wavelengths above 250 nm. Below 250 nm, he found a peak at 240 nm present in a native protein but absent from the denatured protein, which he ascribed at the time to absorption by the amide groups. Later work has shown this band is not present, but was an artifact. These experiments provided the strongest support for the contention that the aromatic amino acids were the cause of the fluorescence of proteins; it could now no longer be in doubt. Another important observation made by Konev was a reconfirmation of Shore and Pardee's evidence that proteins alter their quantum efficiency of fluorescence when denatured or when the pH is altered; i.e., the intensity of protein fluorescence is sensitive to the conformational state of a protein.

1.2.5. Quantum Yields and Stokes' Shifts

The difference between these two Soviet workers was in part resolved in the preliminary report of Teale and Weber,[41] also made in 1959, which was subsequently published as a full account by Teale[42] in 1960. Teale and Weber measured the fluorescence spectra of 26 globular proteins; with every protein they investigated, no phenylalanine contribution could be detected, and tyrosine fluorescence was only found in the absence of tryptophan (i.e., from the class A proteins). When tryptophan was present, only an emission by tryptophan could be observed, supporting Vladimirov (Figure 7). Teale also measured the excitation spectra of these 26 proteins and found the class A proteins had the excitation spectra of tyrosine while the class B proteins had the spectrum of tryptophan (Figure 7). Hence the contribution from tyrosine to either excitation or emission spectra of proteins was small

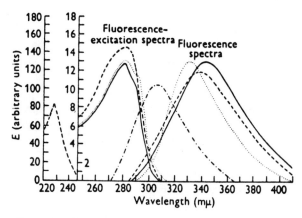

Figure 7. Fluorescence excitation spectra and fluorescence spectra of proteins in aqueous solution. (······) Chymotrypsin; (—·—) insulin; (– – – –) human serum albumin; (———) pepsin. (After Teale.[42])

or negligible. Teale found that the Stokes' shift of the tryptophan fluorescence from class B proteins was quite variable for different proteins, while the tyrosine fluorescence of several class A proteins always had the same Stokes' shift, with a peak wavelength equal to that of free tyrosine. The quantum yields of the fluorescence of class A proteins were five- to tenfold less than the value of the tyrosine, but for class B proteins yield values were found to range in an area approximately half that of tryptophan. There was significantly less quenching of the tryptophan residues than of the tyrosines in proteins. However, individual proteins from both classes did show appreciable variation. There was no discernible correlation between the yield and shift of tryptophanyl fluorescence. Denaturation of class B proteins with 8 M urea equalized the Stokes' shifts and quantum yields, and all the denatured proteins behaved similarly to free tryptophan in 8 M aqueous urea solvent. (Large variations in the apparent quantum yield of fluorescence were found. When they were corrected for filtering by the tyrosine absorption, however, equal values for the true yield were found.) Urea denaturation did not affect class A protein fluorescence. However, another denaturant, propane-1,2-diol, did not produce such simple results. Instead a large variability was found in the extent of the Stokes' shift and in the yield, even when they were corrected for tyrosine screening. (This we now know to be due to a large tyrosine fluorescence contribution in several of the proteins studied.) However, class A proteins showed a more consistent behavior on glycol denaturation. They doubled their quantum yield with no alteration in peak wavelength. Teale also determined the quantum yields of the globular proteins at different exciting wavelengths, and he was able to confirm the earlier observation of Shore and Pardee that the yield decreased to shorter wavelengths than 300 nm. The quantum yield measured with 300 nm excitation was entirely due to tryptophan absorption, and Teale was able to analyze the observed dependence of quantum yield on exciting wavelength on the basis of the fractional absorption of tryptophan at that wavelength derived from model absorption spectra. This was reasonable since the absorption of proteins is well approximated by the sum of the absorption of their monomers in proportion to the composition of the protein.[43,44] From the comparison with this synthetic spectrum, it was proposed that tyrosine absorption did not sensitize tryptophan emission but merely acted as an internal filter (Figure 8). A similar analysis of the excitation spectra of the class A proteins, insulin and RNAse, showed a small contribution to the tyrosine emission from an absorption by phenylalanine—largely detectable because of the fine structure in the phenylalanine absorption. The fluorescence of aqueous solutions of phenylalanine, unlike its absorption spectrum, has no distinct fine structural details, only shoulders.

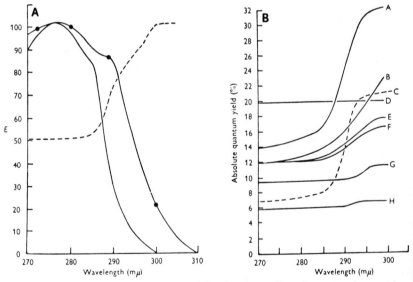

Figure 8. A. Wavelength dependence of fractional tyrosine absorption in proteins. Solid line is the relative absorbancy of tyrosine and the broken line is that of tryptophan between 270 and 310 nm. The fractional absorption of tyrosine in an equimolar mixture of tyrosine and tryptophan is given by the broken line. B. Wavelength-dependent variation in fluorescence quantum yield of tryptophan in proteins. (A) Bovine serum albumin; (B) pepsin; (C) human serum albumin; (D) tryptophan; (E) ovalbumin; (F) carboxypeptidase; (G) chymotrypsin; (H) lysozyme. (After Teale.[42])

However, when the antibiotic polypeptide polymyxin was investigated, Teale found that its fluorescence was fine-structured, so he suggested that the phenylalanine residues of this polypeptide were in a hydrophobic microenvironment and not available to water. This proposal was then extended to explain the variable Stokes' shift of tryptophanyl residues in proteins. Teale showed that the fluorescence of Stokes'-shift tryptophanyl compounds is strongly dependent on the dielectric constant of their environment. The change in dielectric from dioxane to water covered the range of Stokes' shifts exhibited by the globular proteins investigated (Figure 9). Thus, Teale proposed that it was a variability in the dielectric constant of the environments of the tryptophanyl residues which produced the different Stokes' shift exhibited by individual proteins.

From these studies the biochemist was offered a superficially simple measuring technique. Fluorescence measurements were more sensitive than absorption spectrophotometry and did not require or consume appreciable quantities of protein. In fact, as most proteins contain tryptophan residues, on average they have an absorption predominantly due only to their tryptophan content. Only a fluorescence from, and an absorption by, the tryptophan residue needed to be considered. Furthermore, this fluorescence

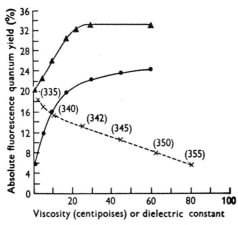

Figure 9. Influence of solvent viscosity and dielectric on quantum yield and Stokes' shift of fluorescence from tryptophan and glycyltryptophan. (▲) Tryptophan yield at different viscosities; (●) glycyltryptophan yield at different viscosities; (×) glycyltryptophan yield at different dielectric constants (the wavelength of the fluorescence maximum is given in parentheses). (After Teale.[42])

appeared to be specific to an individual protein, and reflected its properties as a whole rather than merely the summation of its parts: as the protein conformation was modified either by denaturation with organic compounds or heat, subjected to pH or ionic strength changes, or complexed with its substrate analogue, an inhibitor, or a metal cation activator, its Stokes' shift and quantum yield frequently altered. Plotting the change in these two fluorescence parameters against the magnitude of the perturber gave valuable insights into the structure and the mechanism of function of enzymes.

A large number of papers have been published since 1960 on these aspects, but it is not our intention to report on fluorescence spectra of proteins from the standpoint of analytic biochemistry, protein structure, or enzymology, but rather to discuss the photophysics of these excited states within proteins. For the photophysicist, much remains to be explained: there was a constant fascination with electronic energy transfer and with explanations for low quantum yields of tyrosine residues in proteins. Perhaps Teale's most disturbing discovery was that even though a protein possessed several tryptophan residues and these residues were known (from solvent perturbation and chemical modification studies) to exist in different environments, the emission spectrum did not exhibit any indication of this. The bandwidth of the fluorescence spectrum was constant for different Stokes' shifts and it was equal to that found for tryptophan. One might have expected dramatic evidence in emission for this heterogeneity because of the marked dependence on environment dielectric of the Stokes' shift of tryptophan. But this was definitely not found.

2. LUMINESCENCE OF SYNTHETIC POLYPEPTIDES

2.1. Chemistry and Stereochemistry of Polypeptides

One of the first questions to be considered in the investigation of the photophysics of aromatic amino acids is what influence does the incorporation into a polypeptide have on their luminescent characteristics. It is largely for this reason that luminescence of polypeptides has been studied. For our purposes, three chemical groups of polypeptides are available:

(a) Aromatic homopolypeptides, where all the residues are identical.
(b) Aromatic heteropolypeptides, where two or three of the aromatic amino acids have been copolymerized—randomly, in a defined sequence, or as blocks.
(c) Heteropolypeptides, where a single aromatic amino acid is copoly-

merized with a nonaromatic amino acid (random, sequenced, or blocks).

The other aspect of polypeptides which is specified is the stereochemistry. The polymers can be (a) stereoregular (composed of a single stereoisomer), (b) stereosequential (e.g., syndiotactic, where there are stereoisomers alternating in sequence), or (c) random. Needless to say, not all possible combinations of stereochemical and chemical specifications of polypeptides have been made as yet, particularly the sequential systems. The majority of the work has been performed with random chemical sequences with either regular or random stereochemical sequences. A recent review of polypeptides is that by Fasman,[45] and for aromatic amino acids a survey of the optical properties has been made by Goodman et al.[46]

Polypeptides are not usually soluble in aqueous media, unless there is a charged polar side-chain present. They can, however, be dissolved into several organic solvents. The most common solvents are aprotic (dimethylsulfoxides, dimethylformamide), acidic fluoroalcohols (hexafluorisopropanol, hexafluoroacetone sesquihydrate, and trifluoroethanol), and polar ethers (tetrahydroufuran and diglyme). Organic acids such as trifluoroacetic acid or methanesulfonic acid have been used as solvents, but they do not maintain ordered structures. Organic bases are also good solvents and support random structures (ethane-1,2-diamine, hydrazine). There are extensive studies on the conformations possessed by polypeptides in both aqueous and organic solvents. The conformational state is assessed mainly by circular dichroic studies by a comparison with reference spectra.[47,48] The experimental data have more recently been supported by theoretical treatments of the configurational energies[49] and circular dichroism[50] of various conformers.

The amide-repeating backbone chain of a polypeptide polymer is known to be able to adopt several conformations: ordered structures— (a) α-helix, where right- or left-handed conformations are possible, though the right-handed form is predominant and (b) β-structure pleated sheet, an extended conformation which exists in solution as an antiparallel sheet, and unordered structures—(a) the "random" coil, where there is no long-range order, (b) extended helix (polyproline II, transamide), three-fold left-handed helix.

It is the last structure which is found when charged side-chains induce an order-to-disorder transition of the polypeptide.

Order-to-disorder transitions can be induced by altering the solvent composition, by adding either organic compounds or salts, by heating, by titrating side-chains to give them a charge, by altering amide solvation by titrating with strong acids in aprotic solvents, or by dissolving in basic organic solvents such as ethanediamine or hydrazine. The transition is a cooperative process, occurring over a small range of values of the perturbing agency magnitude.

2.2. Homopolypeptide Luminescence

2.2.1. Conformation of Polytyrosine

Here we are concerned only with three polymers—polyphenylalanine, polytyrosine, and polytryptophan. There is obviously the additional complication with these three polymers of a high effective local concentration, since a single aromatic amino acid will have other aromatic amino acids in its local environment. The particular advantage to studying these three polymers is that the conformation of the stereoregular polymers formed from the L-isomers is well established.[51,52] The majority of the work has been performed with poly-L-tyrosine, but this polymer is believed to be representative. The principal conclusion is that the amide backbone chain is in a right-handed α-helix when the side-chains are in the neutral state.[53] Theoretical calculations of either the circular dichroism[54] or the configurational energy[55] confirmed that the amide backbone formed a right-handed α-helical conformation and furthermore that the lowest energy conformation calculated placed the aromatic rings of the side-chain in a helical array that possessed the repeating features found in the α-helix. The aromatic rings were so arranged as to form another helix around the amide backbone helix, and their planes were approximately perpendicular to the helix axis. Experimental support for these theoretically predicted structures was provided by X-ray radial diffusion studies on a dibromo derivative of poly-L-tyrosine.[56] The presence of α-helix-repeating reflections in the bromine-scattering intensities indicated a regular helical array for the side-chain residues. Additional support came from linear absorption dichroism studies by Troxell and Scheraga,[57] who oriented poly-L-tyrosine in an electric field and then observed that the linear dichroism was consistent with the proposed helical side-chain conformation. Analyses of the circular dichroism of poly-L-tryptophan and poly-L-phenylalanine also led to the conclusion that these polymers have a right-handed α-helical amide backbone conformation, presumably with a similar conformation at the β-carbon to that found in poly-L-tyrosine. The aromatic amino acid polymers undergo a transition to an unordered extended helix when they are titrated with trifluoroacetic acid or dichloroacetic acids.[85] This is of little value for fluorescence studies as strong acidic conditions result in quenching of fluorescence at ambient temperatures,[58] while most quantum yields are equal at 77°K, where the quenching is suppressed. Equally well, thermal melting would be of little value in a fluorescence study as quantum yields of fluorescence are very dependent on temperature. But this latter point is of little significance since the aromatic amino acid homopolymers do not undergo melting below 400°K; their helical conformations are very stable structures.

Poly-L-tyrosine is soluble in aqueous media when it is partially ionized. Fasman et al.[53] showed that, at a fractional ionization of 0.45, a cooperative transition to an unordered state occurred. This disorder transition did not dramatically influence the absorption properties of the system. The ionized polymer was also significantly hypochromic, like the neutral form. Reasonable isosbestos or isodichroic points can be observed in the respective absorption and circular dichroic spectra throughout the whole ionization process. Spectroscopically, only two components need be considered— neutral and ionized phenoxyl residues. This disorder transition also cannot be used to interpret the influence of conformation on luminescence properties. The luminescence is greatly complicated by an extensive transfer from tyrosine to ionized tyrosine, and it will be discussed later in detail.

Therefore, it was not possible to investigate the relationship between fluorescence and conformation in aromatic homopolypeptides. Only one conformation was available for study, the α-helix. To obtain another conformation, stereorandom polymers must be used, and these do not have well-defined conformational states.

2.2.2. Luminescence Spectra

The molar extinction coefficients of tyrosine[59] and tryptophan[60] polymers are considerably smaller than the values obtained from suitable monomers; these polymers exhibited hypochromism; poly-L-phenylalanine is hyperchromic.[61] There was a small shift to longer wavelengths in the absorption peak wavelength value (1–3 nm), the spectral distribution remained identical to that of the monomer spectra, and there were no new absorption bands. The situation that prevailed with the emission of the polymers with the exception of ambient fluorescence from poly-L-phenylalanine,[61] was consistent with the absorption spectral changes. The ambient fluorescence and low-temperature fluorescence and phosphorescence spectra were broadly similar to these of the monomers[62] (Figure 10 and Table 1). There was only a small red shift to the fluorescence at 77°K, and no new bands were found. At ambient temperatures, fluorescence from poly-L-tryptophan had no solvent dependency to its Stokes' shift, though a large dependence was found with monomers. Polytryptophan fluorescence was consistently peaked to shorter wavelengths than monomer fluorescence, while fluorescence from poly-L-tyrosine showed a small red shift in all the solvents used. The Stokes' shift of the polymer fluorescence was increased slightly in value from that of the monomer at 77°K, but there was a significant increase for

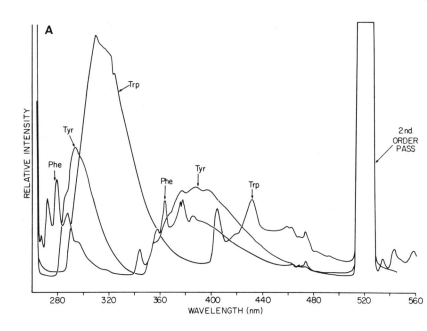

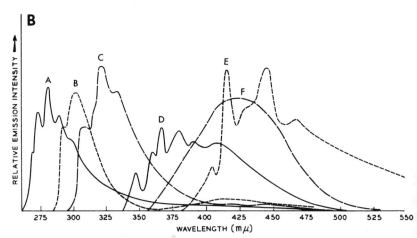

Figure 10. A. Total emission spectra of aromatic amino acids at 77°K. Solvent, ethylene glycol: water glass (EGW) at *p*H 7; excited at 260 nm. The fractional absorption was equal at 260 for all three solutions. B. Total emission and phosphorescence spectra of aromatic polyamino acids. Solvent; diglyme; temperature, 77°K; concentration, 5×10^{-4}. Total emission spectra: (A) poly-L-phenylalanine (10 mv); (B) poly-L-tyrosine (100 mv); (C) poly-L-tryptophan (200 mv). Phosphorescence spectra: (D) poly-L-phenylalanine (1 mv); (E) poly-L-tryptophan (1 mv); (F) poly-L-tyrosine (10 mv). The amplifier gain setting is given in parenthese. (After Longworth.[62])

Table 1. Luminescence Properties of Aromatic Polypeptides

Polymer	Extinction Absorption coefficient maximum ($m^2 mol^{-1}$) (nm)		Fluorescence				Phosphorescence			
			298°K		77°K		77°K			
			Maximum (nm)	Yield	Maxima (nm)	Yield	Onset (nm)	Maxima (nm)	Yield	Lifetime (sec)
Poly-L-phenylalanine[a]	22.4	259	258 (195)	0.006	275 282 292	0.05	340	347 360 367 380 390 408	0.004	—
Poly-L-tyrosine[b]	145.0	279	308	0.070	289	0.35	356	356 387 415 440	0.150	2.80
Poly-L-tyrosinate[c]	190.0	298	346	0.015	321 315 320 330	0.07	371	413 417 445	0.700	0.90
Poly-L-tryptophan[d]	568.0	282	333	0.300		0.70	395	404 460	0.010	2.85

[a] Trifluoroethanol.
[b] Dimethylsulfoxide : ethane-1,2-diol : water (1 N NaCl), 4:5:1.
[c] Dimethylsulfoxide : ethane-1,2-diol : water (1 N NaOH), 4:5:1.
[d] Dimethylsulfoxide : ethane-1,2-diol, 1:1.

the corresponding phosphorescence shift. At low temperatures, fluorescence quantum yields of the polymers were all lower than those found with the respective monomers. The polymer phosphorescence yields were significantly lower—one to two orders of magnitude less. This was an indication of the greater perturbation on the triplet than on the singlet state by the incorporation of a residue into a homopolymer system. More evidence of this effect is provided by the altered intensity distribution of phosphorescence. The emission maximum of poly-L-tyrosine phosphorescence occurs to much longer wavelengths than does the onset wavelength compared to a monomer, while with poly-L-tryptophan phosphorescence emission, there is an additional fine-structure component on the blue side, and its vibrational structuring is entirely different from that of tryptophan (Figure 11). The decay times of the monomers were affected when they were incorporated in a polymer, but no changes in the D^1 value for the triplet state resonance have been found. The lifetime of poly-L-tyrosine is 2.80 sec for the neutral state and 0.9 sec for the ionized; poly-L-tryptophan has a decay constant of 2.85 sec.

Another demonstration of the greater perturbation of the triplet state compared to the singlet state is provided by the polymer poly-DL-tryptophan. Here, since probably there is largely a random distribution of stereoisomers, little structure is expected for the polymer. Inspection of space-filling molecular models does show that a stereoregular structure is feasible for a syndiotactic polymer, but the tacticity of the random polymers is not es-

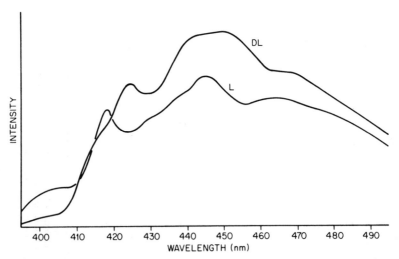

Figure 11. Phosphorescence of poly-L-tryptophan and poly-DL-tryptophan. Solvent, dimethylsulfoxide : ethylene glycol; temperature, 77°K; concentration, $E_{250\,nm}^{10\,mm} = 0.3$; excited at 280 nm.

tablished. Certainly, a DL-polymer does not have the same conformation as an L-polymer. The fluorescence yield of poly-DL-tryptophan, both at 77 and 300°K, is similar to that of poly-L-tryptophan, the DL-polymer having a slightly smaller yield and a slightly greater Stokes' shift at 300°K, but identical shift and yield at 77°K. The phosphorescence spectra of the two polymers, though not their yields, differ greatly. The emission from the DL-structure has different fine structure and a greater Stokes' shift to the onset and the peak intensity. The lifetime of the phosphorescence is also greater, a value of 4.6 sec.

2.2.3. Excimer Formation

Poly-L-phenylalanine differed from the other two aromatic amino acid polymers in having an additional fluorescence band at ambient temperatures.[61,62] This band lay to the red of a monomer spectrum and had no fine structuring (Figure 12). The second band was absent from poly-L-phenylalanine at 77°K[62] and is not found in random L-Glu, L-Phe copolymer fluorescence at 298°K.[61] It was suspected to arise by an excimer formation between adjacent phenyl rings. To test this suggestion, the two stereoisomers of a cyclic dipeptide of diphenylalanine were investigated. One was L-*cis*-diphenylalanine diketopiperazine, which has two benzyl substituents on the same side of the diketopiperazine ring; it had an ad-

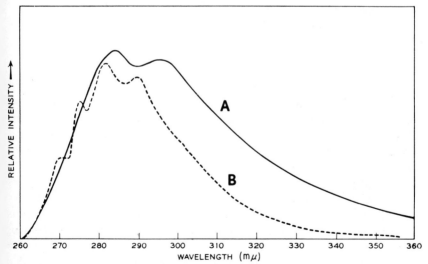

Figure 12. Fluorescence spectra of poly-L-phenylalanine at 298°K. Solvent, diglyme; concentration, 10^{-4} M; excited at 255 nm. (A) 298°K, amplifier setting 10 mv; (B) 77°K, amplifier setting 50 nm. (After Longworth.[62])

ditional fluorescence band at ambient temperatures. The second compound was the corresponding trans-substituted diketopiperazine. This trans-compound did not have an additional fluorescence band, but did have fluorescence identical to the monobenzyl-substituted diketopiperazine. Since the phenyl rings in the trans-substituted compound were sterically prevented from forming an excimer, it was possible to associate the presence of the additional band in poly-L-phenylalanine fluorescence with the occurrence of an excimer formed between two adjacent residues in the polymer during the excited state.

The lack of any excimer fluorescence from poly-L-tyrosine and poly-L-tryptophan and their related L-cis-diketopiperazines is not satisfactorily explained. It may be related to a special local conformation in poly-L-phenylalanine, which is absent in the former two polymers.

2.2.4. Quantum Yields of Luminescence

The quantum yields of fluorescence at ambient temperatures of both the polymers and the monomers were found to be very solvent dependent.[62] There was no adequate explanation for this. Activation energies were not reported, nor was the yield of triplet formation or solvated electron production measured. There are indications that a major effect of incorporating an amino acid into a homopolymer is drastically to influence singlet internal conversion, since, though the phosphorescence yields were significantly reduced, the lifetimes were not seriously affected. This remains to be investigated experimentally, as does the role of the solvent in perturbing the internal conversion, since the conformation is not considered to be altered by these changes in solvent.

2.2.5. Energy Transfer

2.2.5(a). Singlet. Another phenomenon which is readily investigated with the synthetic homopolymers is that of transfer of energy. Polarization of the fluorescence has been measured at several exciting wavelengths (FAPS), and since no two adjacent residues have identical coordinates, a transfer of energy between them leads to a depolarization. At wavelengths shorter than 285 nm, a considerable depolarization of the fluorescence was found from both poly-L-tyrosine[63] and poly-L-tryptophan.[64] A series of oligomers of tyrosine also formed an intermediate family of curves between those of the polymer and of the monomer.[65] At wavelengths longer than 285 nm, the magnitude of the depolarization decreased, until at 300 nm polarization values equal to those of the respective monomers were found (Figures 13A and 13B). The explanation for this effect (Weber's repolarization effect) is the presence of a lower-energy trap which is sensitized by all but long-wavelength excitations where the trap

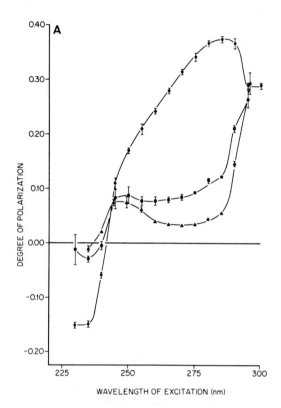

Figure 13A. Fluorescence absorption polarization of tyrosine, hexatyrosine, and polytyrosine. Fluorescence was observed through a cutoff filter with pass 310 nm. Solvent, propylene glycol : dimethylsulfoxide 10:1; temperature, 220°K; concentration, $E^{5\,mm}_{280\,nm} = 0.3$. (●) N-acetyl-L-tyrosine amide; (■) L-hexatyrosine; (▲) poly-L-tyrosine. (After Knopp et al.[65])

absorbs light directly. Since the conformation of the polymer poly-L-tyrosine is reasonably well established, Knopp et al.[65] calculated, using Förster's formula, the expected depolarization. The molecular structure[54] defined the interresidue distances and their coordinates. The transition dipole direction was known from independent crystal linear dichroic studies. It only remained to take into account a diffusive motion of the transfer in the course of the fluorescence decay. Remarkable agreement was found by Knopp et al. between the values calculated for polarization and values measured for the depolarization at the absorption maximum. By one lifetime, energy had migrated over the whole polymer, of length 100 residues.

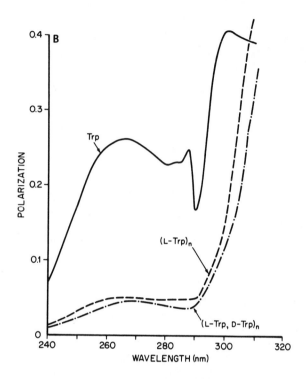

Figure 13B. Fluorescence absorption polarization of tryptophan and poly-L- and DL-tryptophan. Experimental conditions as in *A*. (———) *N*-acetyl-L-tryptophan amide; (– – – –) poly-L-tryptophan; (—·—) poly-DL-tryptophan.

2.2.5(b). Triplet. Another method for investigating transfer, this time at the triplet level, is based on a triplet–triplet annihilation. When two triplets diffuse together they can annihilate each other, and the resultant product is a singlet in the excited state and another in the ground state. Therefore, a fluorescence is produced, but with a lifetime determined by the triplet state through the annihilation process, and so it is a delayed fluorescence, easily separated from prompt fluorescence by light choppers. Such a delayed fluorescence has been observed from poly-L-tyrosine.[72] It has several of the expected features: (a) it depends on the second power of the exciting light; (b) it decays with a long lifetime. The decay kinetics are complex; the initial decay is considerably shorter (60 msec) than that of the phosphorescence, and accounts for 70–90% of the intensity (Figure 14). The remaining intensity has a decay similar to that of the phosphorescence. A similar delayed fluorescence is not detectable from a hexamer, so this suggests that the triplet states have to migrate over considerably more

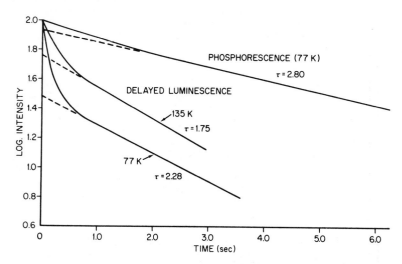

Figure 14. Decay of phosphorescence and delayed fluorescence of poly-L-tyrosine. Solvent, propanediol : dimethylsulfoxide 1:1; concentration, 1 mg/ml; excited at 280 nm. Intensity decay after shutter closure (25 msec) followed by a conversion to logarithm (base 10) and storage in an averaging multichannel scaler (512 channels). Typically, 10 sweeps of the channels stored. Phosphorescence measured at 420 nm (500 mv); delayed fluorescence at 308 nm at 77°K (50 mv) and 135°K (1 mv). Abscissa time, 8-sec sweep.

than six residues to account for delayed fluorescence from poly-L-tyrosine. A delayed fluorescence can also be found from poly-DL-tryptophan but not poly-L-tryptophan, which shows that there is an intertryptophan triplet transfer in the DL-polymer too. The decay from poly-DL-tryptophan is nonexponential, with an initial slope of 87 msec, and reaches a limiting slope of 2 sec. The nonexponential decay is a result of the presence of traps which quench the triplet excitons, and traps are the predominant source of phosphorescence and delayed fluorescence.

2.3. Heteropolymer Luminescence: Aromatic Amino Acid Systems

2.3.1. Energy Transfer in Ionized Polytyrosine

For our purposes we will consider only those polymers with two different aromatic amino acids. Though there would be many advantages to studying polymers of defined chemical and stereochemical sequences, the only polypeptides studied to date have had random chemical sequences with either a single stereoisomer or a random mixture of the two stereoisomers. The most convenient polymer to study was derived from poly-L-tyrosine.

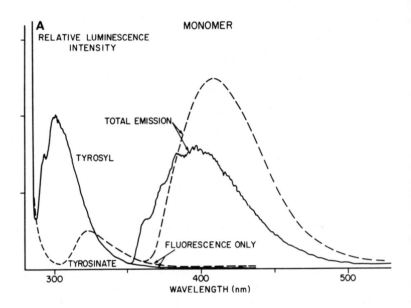

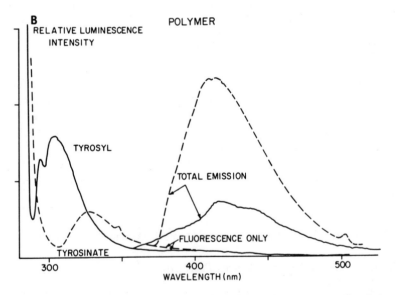

Figure 15. A. Luminescence of *N*-acetyltyrosine amide. Solvent, dimethyl-sulfoxide : ethylene glycol : water 4:5:1 (DEW); temperature, 77°K; concentration, 0.3 mg/ml; excitation at 281 nm (isosbestos). (———) Neutral, pH 7; (– – – –) ionized, pH 13. *B.* Luminescence of poly-L-tyrosine. Similar conditions to above (isosbestos is now at 283 nm). (———) Neutral; (– – – –) ionized. (After ten Bosch *et al.*[69])

By ionizing the phenoxyl group of the tyrosine residues, another species, the tyrosinate residue, could be continuously incorporated in a random manner into the polymer. Therefore, by simply altering the pH of the solvent, polymers with all compositions of tyrosine/tyrosinate were obtainable. A consequence of introducing ionized tyrosine residues in poly-L-tyrosine is that significant spectroscopic changes are produced. The tyrosinate residue has lower excited singlet and triplet energy levels than tyrosine (Figure 15). Hence separate tyrosine or tyrosinate contributions to both absorption and luminescence can be determined in a mixture.[67] The quantum yield of tyrosinate fluorescence is significantly smaller than that of tyrosine, both at 77 and 298°K, and the phosphorescence yield is much larger.[68] The partially ionized polymer provided a suitable system in which it was possible to study a transfer of electronic energy from tyrosine to tyrosinate at both singlet and triplet levels. Though there is a potential complication from a cooperative transition from an α-helix to an unordered extended helix occurring in the polymer at 0.45 fractional ionization, this is not accompanied by significant alterations in the absorption properties of the polymer, and the isosbestos points are maintained through the ionization. Nevertheless, the fractional ionizations from 0 to 0.45 do not alter the conformation, and this proved to be the range of interest for energy transfer studies.[63] The presence of an isosbestos in the absorption was critical to the study of the dependence of tyrosine or tyrosinate quantum yield of luminescence on the tyrosine/tyrosinate ratio. Excitation at an isosbestos ensured that a constant fraction of light was absorbed as the fractional ionization was altered. As measurements were also made at 77°K,[69] the isosbestos served as a normalization point. Cooling glassing solutions from 300 to 77°K produced two complications, (a) an alteration in the fractional ionization, and (b) variable fractional absorption due to fracturing or path length. However, the absorption measurements were normalized to the isosbestos and corrected for the variable effective path length, and from these corrected spectra a fractional ionization could be obtained. There is in emission an equivalent intersection point to the absorptive isosbestos, termed an isoemissive point. It is only observed when the excitation wavelength is the isosbestos. The isoemissive point then served as the normalization point to obtain the emission spectra corrected for the varying fractional absorption, and so the true intensity could be measured.

The demonstration of energy transfer in oligomers and polytyrosine depended on finding that (a) the tyrosine fluorescence was quenched by a small amount of tyrosinate (approximately 3% caused a 50% reduction in tyrosine quantum yield in the polymer) (Figure 16), (b) the tyrosinate yield was equivalently enhanced (a direct proof was the observation of an isoemissive point on isosbestic excitation), and (c) excitation spectra showed the

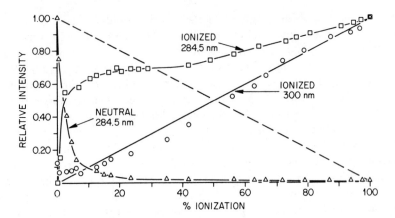

Figure 16. Fluorescence titration of poly-L-tyrosine. Solvent, DEW; temperature, 298°K; concentration, 5 μg/ml. (△) Excitation at 284.5 nm (isosbestos), observation, 294 nm (neutral species); (○) excitation at 300 nm, observation 353 nm (ionized species); (□) excitation, 284.5 nm, observation 353 nm (sensitized fluorescence). The sensitized fluorescence intensity is corrected for the direct contribution from the neutral species. The intensity plotted is the difference between an observed value and a contribution estimated from the 294 to 353 nm ratio for the neutral polymer fluorescence. (After Longworth and Rahn.[67])

sensitization of tyrosinate luminescence by tyrosine absorption (a polymer which was 10% ionized luminesced essentially only from the ionized form, but had an excitation spectrum predominantly due to tyrosine absorption). The more general and direct proof again was based on the isosbestos point. When the excitation spectra were determined by measuring luminescent intensity at the isoemissive points of fluorescence or phosphorescence for several fractional ionizations and then superimposed, all excitation spectra intersected at a single wavelength which was the wavelength previously found as the isosbestos in absorption.

Oligomers with degrees of polymerization 2, 3, and 6 were investigated at both room temperature[70,71] and 77°K.[63,72] The presence of a single ionized molecule was sufficient to quench all the un-ionized tyrosine residues, so the fluorescence and phosphorescence came entirely from the tyrosinate (Figure 17). Because of a perturbation of the phosphorescence spectrum of tyrosine when it is incorporated into poly-L-tyrosine, it was not possible to separate neutral and ionized phosphorescence effectively. However, the electron spin resonance of the two species occurred at different field values,[73] and by quantitative ESR measurements it was found that the triplet state of the neutral residues in poly-L-tyrosine was quenched by tyrosinate identically to the fluorescence quenching.[74]

Therefore, though there was direct evidence for transfer from tyrosine

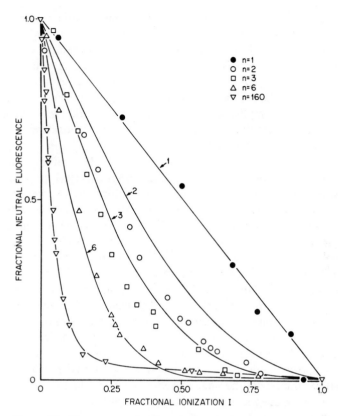

Figure 17. Fluorescence titration of oligotyrosines. Solvent, DEW; temperature, 298°K; excitation at 284 nm (isosbestos), observation 305 nm; concentration, $E_{280\,nm}^{10\,mm} = 0.3$. Fractional ionization determined from 305 nm absorbancy. (●) N-acetyl-L-tyrosine amide; (○) L-tyrosyl-L-tyrosine; (□) L-tyrosyl-L-tyrosyl-L-tryosine; (△) L-hexatyrosine; (▽) poly-L-tyrosine. Solid lines labeled $n = 1$–6 are plots of $(1-I)^n$, where I is fractional ionization. (After Longworth et al.[63])

to tyrosinate at the singlet level, it was not possible to detect any transfer at the triplet level between these two residues.

As we mentioned previously, the conformation of neutral poly-L-tyrosine was known, and furthermore the presence of ionized tyrosine residues below fractional ionizations of 0.45 did not affect the conformation. Hence, it was feasible to calculate directly the magnitude of transfer in the same manner as with the neutral polymer intertyrosine transfer.[75] The interresidue distances and orientation obtained from the molecular model, together with spectral overlap data and quantum yields, were inserted into

Förster's equation. An additional property which could be investigated was the influence of temperature, since fluorescence quenching studies had been performed at 77 and 298°K. The overlap integrals for both tyrosine transfers and tyrosine-to-tyrosinate transfer alter when the temperature is reduced to 77°K as does the tyrosine quantum yield. Temperature changes, therefore, significantly alter the individual interresidue rates of transfer. The details of the transfer are complicated by the presence of both intertyrosine and tyrosine-to-tyrosinate transfers taking place in the helical array simultaneously. The reader is referred to the paper of ten Bosch and Knopp[75] for details of how this was treated. The agreement between the theoretical magnitude of transfer and that which had been observed at both temperatures was remarkably good. A surprising feature of the theoretical analysis was the conclusion that transfers occur directly from tyrosine to tyrosinate from relatively distant residues, rather than diffusing by intertyrosine transfers to the nearest tyrosine–tyrosinate pair to make the final transfer to the trap. Ten Bosch and Knopp found that the proportion of intertyrosine transfers to tyrosine-to-tyrosinate transfers in the total transfer altered between 77 and 298°K; there was less effective transfer to the tyrosinate residues predicted at room temperatures. Nevertheless, the observation was that the tyrosinate residue was as effective a quencher at 77 as at 298°K, even though the values used for the overlap integrals and the donor quantum yield differed greatly. This in itself provided support for the application of Förster's approximation to an electronic energy transfer process, and we will consider transfer at the singlet level to occur only by this transfer mechanism in all the subsequent related transfers in proteins.

2.3.2. Energy Transfer in Tyrosine–Tryptophan Copolymers

Another transfer investigated in heteropolymers was tyrosine-to-tryptophan transfer. The energies for both the singlet and triplet levels of tryptophan are known to be lower than those of tyrosine. Therefore, transfers between the two residues are expected to occur whenever favorable distances and orientations coexist. Unlike the previous polymer system, it is not possible to alter continuously the tyrosine-to-tryptophan ratio. In our work[76] only two polymers were studied, one with the ratio 4:1 and the second with a 1:1 ratio. A large number of tyrosine-tryptophan ratios and stereorandom polymers were investigated by Wada and Ueno.[77] But, since these workers did not report the emission spectra, the significance of their work is uncertain. The luminescence of both polymers we studied came only from their tryptophan residues. This conclusion was based on the characteristic energy level, spectral distribution, and phosphorescence fine structure and phosphorescence lifetime. Tryptophan was easily distinguished from tyrosine or ionized tyrosine by all these criteria. The strongest proof that the emission was solely from tryptophan was the coincidence in the spectrum found when the polymer was excited at 295 nm with that obtained when it was excited at 278 nm. At 278 nm both tyrosine and tryptophan absorb, but at 295 nm only tryptophan absorbs. To show that the absence of tyrosine emission is a result of transfer to tryptophan, it was essential to demonstrate a sensitization of tryptophan emission by tyrosine absorption. To do this, the excitation spectra of the two polymers were compared with the absorption spectra, and by normalizing the ratios of these two spectra at wavelength 295 nm, the values obtained at other wavelengths defined the dependence of the relative quantum yield on the exciting wavelength. With the 1:1 polymer, this ratio was constant. No dependence to the quantum yield could be found. This indicated that there was a complete transfer from tyrosine to tryptophan in the 1:1 polymer. The 4:1 polymer, with its greater tyrosine fractional absorption, still emitted only from tryptophan, but the tryptophan quantum yield depended on the exciting wavelength. An analysis of this dependency (see later) by a comparison with suitable model system spectra indicated that the fractional transfer from tyrosine to tryptophan was 0.5. There was no evidence for emission from tyrosine at either the singlet or the triplet at 77°K. No difference could be observed between the fluorescence and phosphorescence excitation spectra, so the remainder of the tyrosine-absorbed energy must be undergoing an internal conversion in tyrosine.

2.3.3. Energy Transfer in Ionized Tyrosine–Trytophan Copolymers

The tyrosine–tryptophan polymers also permitted an investigation of another transfer, that of tryptophan to tyrosinate. Again, model system energy level studies, particularly with dipeptides,[78-80] have shown that the singlet of tyrosinate is lower than that of tryptophan. The reverse situation applies to the triplet level. Now tryptophan has the lowest level. That the expected transfers occurred was confirmed in a study of the luminescence and of the electron spin resonance of the triplet states of dipeptides. At the singlet level, tryptophan transferred to tyrosinate; at the triplet level, tyrosinate sensitized the tryptophan triplet. The ionized polymers are not believed to be in an ordered conformation, but rather to form a helix that is unordered, related to the extended left-handed helix of polyproline II. The interresidue distance for closest approach is estimated as 0.8 nm. This distance is still not unreasonably large for a triplet-to-triplet transfer by an electron exchange process.

The fluorescence of tryptophan is quenched by the ionization of tyrosine in tyrosine–tryptophan polymers. A quantitative treatment is possible when the polymers are excited at the isosbestos. The fluorescence intensity is measured at the tyrosine–tyrosinate isoemissive to eliminate a tyrosine

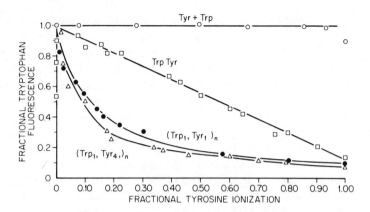

Figure 18. Fluorescence titration of tryptophyltyrosine and copoly-tyrosinetryptophan. Solvent, DEW; temperature, 298°K; concentration, 1×10^{-4} in residues; excitation at 282 nm (isosbestos), observation 355 nm. (–○–○–) Equimolar mixture of N-acetyl-L-tyrosine amide and N-acetyl-L-tryptophan amide; (–□–□–) L-tryptophanyl-L-tyrosine; (–△–△–) poly-L-tyrosine (4), L-tryptophan (1); (–●–●–) poly-L-tyrosine (1), L-tryptophan (1). Fractional ionization determined from 305 nm absorbancy. Abscissa: fractional phenoxyl ionization; ordinate, fractional tryptophanyl fluorescence intensity.

contribution, if any is present. It was found that the quenching occurred and was greater than that expected for a single tyrosinate quenching a single tryptophan residue at both 298 and 77°K (Figure 18). The tyrosinate was a less effective quencher at 77°K because of the reduced overlap between tryptophan fluorescence and tyrosinate absorption.

The weak fluorescence from the fully ionized polymer (7.5% of the neutral polymer) can be shown to be due to emission from tyrosinate. Identical spectra are obtained when ionized polymers are excited at 305 nm, where only tyrosinate absorbs, and at 280 nm, where tryptophan also absorbs. The phosphorescence, however, is entirely due to tryptophan for all fractional ionizations. The yield is enhanced by the ionization, and a small red shift occurs. The excitation spectra of both of the tyrosine–tryptophan polymers' fluorescence and phosphorescence were measured and found to be identical to the respective absorption spectra at 77°K. The transfer from tryptophan to tyrosinate was found to be 100% effective at 298 and 77°K at the singlet level. The most dramatic experiment is to excite the completely ionized polymers at 305 nm. Here only tyrosinate absorbs. Then the fluorescence can only be from tyrosinate. The phosphorescence is entirely from tryptophan (Figure 19). The only process that could lead to a tryptophan

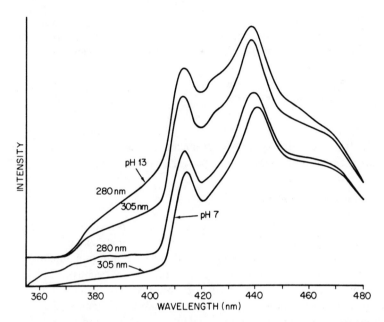

Figure 19. Phosphorescence of copolytyrosinetryptophan. Solvent, DEW; temperature, 77°K; concentration, $E^{10\ mm}_{280\ nm} = 1.5$; excited at 280 and 305 nm. Phosphorescence intensity equalized 480 nm for excitation at 280 and 305 nm.

emission is a transfer at the triplet level from a tyrosinate triplet to a tryptophan triplet. In the 4:1 polymer, not all tyrosinate residues are adjacent to a tryptophan, so when the intersystem crossing to a triplet level occurs in such a tyrosinate residue, for this triplet energy to reach a tryptophan a transfer to another tyrosinate takes place. This is because triplet–triplet transfers proceed by an exchange rather than a resonance and therefore require contact with an adjacent residue.

A depolarization of the fluorescence absorption polarization spectra of the neutral and ionized tyrosine–tryptophan polymer has been measured.[64] Such a depolarization is consistent with transfers between residues including tyrosine-to-tryptophan transfers within the copolymers. The magnitude of the depolarization is less in the ionized polymers than the neutral polymers, perhaps because of greater individual interresidue distances in these polymers.

2.3.4. Summary of Energy Transfers Found in Aromatic Polypeptides

To summarize, transfers which have been found to occur in homo- and heteropolymers of aromatic amino acids are the following:

(a) Singlet level

Tyrosine→tyrosine; tryptophan→tryptophan; tyrosinate→tyrosinate.
Tyrosine→tryptophan; tyrosine→tyrosinate; tryptophan→tyrosinate.

(b) Triplet level

Tyrosine→tyrosine; tryptophan→tryptophan; tyrosinate→tyrosinate.
Tyrosinate→tryptophan.

The missing transfer is the triplet tyrosine–tyrosinate.

In every case, the transfer is predicted on the basis of the energy levels obtained from monomers and so presents no great novelty.

2.4. Quenching Studies

The third group of polymers which have been investigated are copolymers of an aromatic amino acid and either glutamic acid or lysine, where the aromatic amino acids are only 5% of the total amino acid composition. These particular charged amino acids were chosen because model system studies involved their side-chains in quenching both tyrosine and tryptophan. The general conclusion reached from the model system studies was that COO^- and NH_2 quench tyrosine, while —COOH and —NH_3^+ quench tryptophan residues. Fasman et al.[81] eliminated the complication of conformational changes affecting the conclusions. They found that copolymers made with DL-Glu have the same quantum yield values as those made from L-Glu, but the DL-Glu polymers cannot form any helical structure. The quantum yield of the carboxylate polymers with tyrosine was 0.02, extremely small. Fasman et al. showed that the discharging of the carboxylate group led to an enhancement of tyrosine yield in the polymer (L-Glu$_{95'}$, L-Tyr$_5$)$_n$ (Figure 20). The low quantum yield of the carboxylate form was consistent with the model studies that had shown that carboxylate was a potent quencher for tyrosine. One detailed mechanism suggested for carboxylate quenching was excited-state protolysis of tyrosine phenoxyl to the carboxylate acceptor. The anion formed by the protolysis, tyrosinate, is essentially nonfluorescent in these aqueous solutions at the ambient temperatures used. However, in dimethylsulfoxide, an appreciable ambient fluorescence can be found. When the fluorescence of sodium salt of a 95% Glu-Tyr copolymer was investigated in DMSO, no tyrosinate fluorescence was observed. Even in this solvent, though, the quantum yield of the tyrosine residues was considerably reduced from the monomer value. A tyrosine monomer, N-acetyl tyrosinamide, is quenched by sodium formate in dimethylsulfoxide, and a tyrosinate fluorescence component can be observed. I concluded from these results that the quenching by carboxylate in polymers was not through an excited-state protolysis. Both Feitelson[84] and Cowgill[85] have reached the same con-

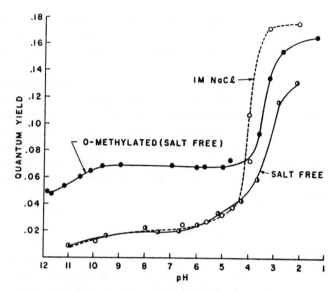

Figure 20. pH dependence of absolute quantum yield of copolymers of tyrosine and glutamic acid (Glu_{96}, Tyr_4)$_n$. Abscissa: solution pH; ordinate: quantum yield of tyrosine fluorescence. (After Weber and Rosenheck.[86])

clusion with model system studies. O-methyl-substituted phenols are quenched by the carboxylate group as effectively as phenols. Weber and Rosenheck[86] found the O-methyltyrosine copolymers were quenched too.

The equivalent lysine–tyrosine copolymers have not proved to be as simple to study since the discharge of the lysine ammonium side-chain group overlapped the ionization of the phenoxyl hydroxy of tyrosine. The quantum yield of the lysine copolymer below pH 8 was 0.07.[83] Weber and Rosenheck[86] concluded from the salt dependency of the alkaline quenching of (L-Lys_{95}, L-Tyr_5)$_n$ that the discharge of the ammonium group produced a quenching of the tyrosine residues. In the amino acid, discharge of the α-amino group does not affect the quantum yield. When the proton of the phenoxyl group was replaced by a methyl group, no alkaline quenching occurred in either polymer or monomer. However, the discharge of the α-amino group of dipeptides does enhance the tyrosine dipeptide quantum yield (NH_3^+ is a quencher at the α-terminus).

A similar series of tryptophan copolymers was investigated by Fasman et al.[87] The quantum yield of the tryptophan copolymers was 0.12–0.15, and protonation of the carboxylate group or discharge of the ammonium group caused a 50% quenching of the tryptophan residues. The observation of quenching by the uncharged amino group is in contradiction to titration studies of tryptophan. Discharge of the ammonium group of the zwitterion causes a large increase in the quantum yield of tryptophan at ambient temperature, together with a shift to longer wavelengths in the fluorescence peak.

Studies of copolymers of aromatic amino acid and potential quenching groups have not led to simple analyses of the mechanism for quenching in either the more complicated polypeptides and proteins or in monomers of dipeptides. However, one clear conclusion was that incorporation of tyrosine into a random polypeptide chain led to quenching by an order of magnitude, while the yield of tryptophan was reduced only by half. The mechanism of the quenching appeared to be an enhanced internal conversion process rather than a chemical reaction.

2.5. Photochemistry of Polytyrosine

The photochemical reactions of tyrosine and tryptophan either free or in proteins are not well understood, and many products have been found.[88] However, some preliminary studies on the photochemistry of poly-L-tyrosine appear to be simply interpretable. Samples of polypeptides containing tyrosine, when dissolved in aprotic solvents, were found to exhibit an additional fluorescence in the visible and to peak at 420 nm.[82,89] Though originally thought to be an excimer, when excitation spectral studies were performed, the sample was found to have absorption maxima at 320 and 350 nm, with the 320 nm predominating.[62] Inspection of the absorption of poly-L-tyrosine samples shows the presence of a weak absorption beyond that due to the tyrosine.[60] Lehrer and Fasman[90] have shown that a photoproduct of tyrosine formed during the preparation of the polymer accounted for the fluorescence and the absorption. The product was shown to be dityrosine, an O,O'-dihydroxydiphenyl compound. The free dityrosine has been fully studied by Andersen,[91] who showed that the first phenoxyl ionization occurs at pH 7 and that both the neutral molecule and the half-ionized molecule fluoresce from the next ionized excited state and not their ground state as a result of an excited-state protolysis. The excited states are more acid than their ground states and are able to eject the proton within their lifetime. This, then, accounts for the large Stokes' shift of the photoproduct. In a study of the photochemistry of phenols, Dobson and Grossweiner[92] and Joschek and Miller[93] were able to demonstrate that a radical precursor, resulting from a photoionization of triplet state, would account for the production of diphenyl compounds. In independent studies, Fessenko et al.[94] and Santus et al.[95] have shown the ready photoionization of the triplet of tyrosine compounds, particularly when they are ionized. The molecular model of poly-L-tyrosine indicates that the dityrosine compound can be readily formed in the helical conformation of poly-L-tyrosine. A plausible mechanism for the photoproduct formation is the formation of a triplet state of neutral tyrosine and subsequent photoejection of an electron, leaving a cation radical tyrosine. This could then undergo protolysis to form the neutral phenoxyl radical that attacks a neighboring tyrosine residue to form the dityrosine photoproduct.

3. LUMINESCENCE OF NATURAL POLYPEPTIDES: HORMONES AND ANTIBIOTICS

After the synthetic polypeptides, the next stage of complexity is provided by simple protein structures which have a small number of amino acids and only one or a small number of aromatic amino acids. Several polypeptide hormones and antibiotics fit into this class and are characterized by having little-organized tertiary structure.

3.1. Phenylalanine Systems

3.1.1. Polypeptides with Phenylalanine as the Sole Aromatic Amino Acid

Several cyclic polypeptides are known which have phenylalanine residues as the only aromatic amino acid. Examples are the bacterial antibiotics polymyxin B_1, gramicidin S-A, and bacitracin A. The structures are the following[96]:

$$\text{1pel—DAB—Thr—DAB—DAB} \genfrac{}{}{0pt}{}{\diagup \text{DAB—D-Phe} \diagdown}{\diagdown \text{Thr——DAB} \diagup} \genfrac{}{}{0pt}{}{\text{Leu}}{\text{DAB}}$$

Polymyxin B_1*

$$\text{Val} \genfrac{}{}{0pt}{}{\diagup \text{D-Phe—Leu—Orn} \diagdown}{\diagdown \text{Orn—Leu—D-Phe} \diagup} \genfrac{}{}{0pt}{}{\text{Pro}}{\text{Val}}$$
Pro

Gramicidin S-A

---

*1pel = isopelargonic acid −(+)−6-methyloctan-1-oic acid; DAB = diaminobutyric acid.

$$\begin{bmatrix} & & \text{H} & \text{H} & \text{S}\!\!-\!\!-\!\!-\!\!-\!\!\text{CH}_2 & \\ & & | & | & | & | \\ \text{CH}_3\text{CH}_2\!-\!\!\text{C}\!\!-\!\!-\!\!-\!\!\text{C}\!\!-\!\!\text{C} & & \text{CH}\!-\!\text{CO}\!- \\ & & | & | & \diagdown\!\!\diagup & \\ & & \text{CH}_3 & \text{NH}_2 & \text{N} & \end{bmatrix} \begin{bmatrix} \text{Ileu}\!-\!\text{Cys-} \end{bmatrix} \!-\!\text{Leu}\!-\!\text{D-Glu}\!-\!\text{Ileu}\!-\!\text{Lys}\!-\!\text{D-Orn}\!-\!\text{Ileu} \\ \text{NH}_2\!-\!\text{D-Asp}\!-\!\text{Asp}\!-\!\text{His}\!-\!\text{D-Phe}$$

Bacitracin A

The Ileu-Cys is present in bacitracin A as a thiazoline which is complexed with the histidine residue.

A suitable linear phenylalanine polypeptide is secretin [27 amino acids, 1 phe (6)], a polypeptide hormone. Though some circular dichroic studies have been performed on secretin[97] which indicate a partial structure for the hormone, no fluorescence studies have been reported. Other possible systems are the S-peptide obtained by the subtilisin proteolysis of RNase A [20 residues, 1 phe (8)] and bradykinin ArgProProGlyPheSerProPheArg. Again, fluorometric studies have not been reported.

Teale[98] studied the fluorescence of polymyxin B_1 and found that the quantum yield of the antibiotic was identical with that of free phenylalanine. The fluorescence spectrum obtained at ambient temperatures was more finely structured than the spectrum of free phenylalanine. Teale proposed that this resulted from the residue being in a hydrophobic environment. The luminescence spectra at 77°K of both gramicidin S-A and polymyxin B are also similar to the spectra of phenylalanine[99] (Figure 21). The luminescence from polymyxin B is identical to the emission from phenylalanine, aside from a small red shift of both the fluorescence and the phosphorescence. The fluorescence at 77°K from gramicidin S-A is unlike that of either polymyxin or phenylalanine. It peaks 8 nm to the red of phenylalanine and has no fine structure, and there is a larger fluorescence-to-phosphorescence ratio. Weinryb and Steiner[100] found that the fluorescence-to-phosphorescence ratio of phenylalanine dipeptides is dependent on sequence. More detailed studies on these cyclic structures are in order, since the systems have a defined conformation. A structure has been proposed for gramicidin S-A by Stern et al.[101] (from high-resolution ^{1}H- and ^{13}C-NMR studies); polymyxin B_1 by Ohnishi and Urry[101]; and for bacitracin A by Correll and Guirey.[101]

3.1.2. Polymers Where Phenylalanine Occurs with Tyrosine

Collagen represents the only other example of a polypeptide where an emission from phenylalanine has been reported. Vertebrate collagen possesses no tryptophan residues, and only small amounts of tyrosine and phenylalanine. The protein consists of three chains of 1000 residues, which form a composite structure with the three chains parallel to each other, twisted around each other to form an extended left-handed helix (polyproline

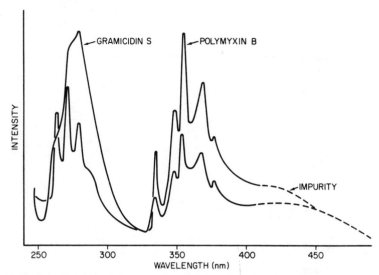

Figure 21. Total luminescence of gramicidin S and polymyxin B. Solvent, EGW at pH 7; temperature, 77°K; concentration, $E^{10\,mm}_{260\,nm} = 1.0$; excited at 260 nm. Note the presence of an impurity in the long wavelength region of phosphorescence which is also excited by wavelengths greater than 280 nm.

II). The phosphorescence from collagen at 77°K has two components which have been identified with phenylalanine and tyrosine, respectively, on the basis of fine structure, lifetime, and electron spin resonance behavior[102] (Figure 22). The phenylalanine contribution was red-shifted, and the lifetime was less than that reported for free phenylalanine (there appears to be no reliably established value for the lifetime of phenylalanine; reported values range from 4 to 9 sec).

The luminescence of collagen is of some historic interest since biological materials largely composed of collagen were among the first proteins to be investigated for any luminescence. A luminescence can be found from collagon at ambient temperatures since the material forms a glasslike structure. Grimes *et al.*[102] have found that the ambient phosphorescence is also composed of two decay components, with a lifetime of approximately one third the values found at 77°K. This would suggest that the earlier workers who studied purified collagen material had in all likelihood observed an emission from phenylalanine and tyrosine. However, much of the unpurified material also contains the tendon protein elastin. Elastin has as a constituent the compound desmosine, which absorbs up to 380 nm and fluoresces in blue with a peak at 450 nm.[103] The presence of elastin in some of the early studies would also have accounted for the blue fluorescence reported.

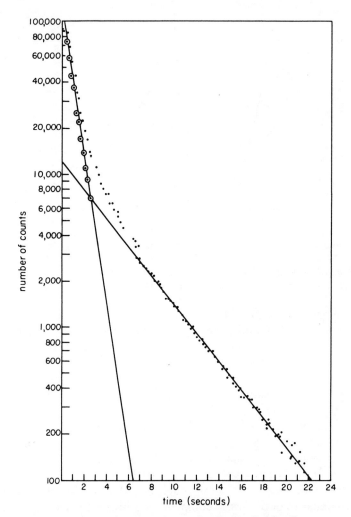

Figure 22. Phosphorescence decay of collagen. Solid collagen at 77°K was excited at 254 nm, and the phosphorescence was isolated by choppers and a cutoff filter with pass 280 nm. Decay was digitized and counts stored in a multichannel scaler, and several sweeps are summed. Ordinate: phosphorescence intensity on a $\log_{10}$ scale. Circled points were obtained by subtracting the long component (4.5 sec) and are fitted by a line with slope of 2.3 sec. (After Grimes.[102])

Kallman et al.[104,105] have proposed that there is an emission from the amide group on the basis of studies on the luminescence of collagen. On exciting collagen at 365 nm, they observed an emission, and since there was no absorption at 365 nm attributable to aromatic amino acids, they suggested that it was a result of a residual absorption by the amide group. Fujimori[106] has shown that collagen is photochemically labile and forms photoproducts which absorb at 365 nm. Vladimirov[107] and Lehrer and Fasman[108] have found that many of these photoproducts are luminescent; hence it appears that Kallman and coworkers were studying the photoproducts introduced into collagen, and not a luminescence of the amide group.

3.2. Tyrosine-Containing Polypeptides

Again two groups of polypeptides, cyclic and linear, are available which have only a single tyrosine group. The cyclic polypeptides which fit in this category are the pituitary hormones oxytocin and vasopressin and the bacterial antibiotic tyrocidine A.

$$
\begin{array}{c}
\text{Ile}\text{---}\text{Gln} \\
\text{Tyr} \diagup \qquad \diagdown \text{Asn} \\
\diagdown \text{Cys S---S Cys} \diagup \\
| \\
\text{Pro} \\
| \\
\text{Leu} \\
| \\
\text{Gly NH}_2
\end{array}
$$

Oxytocin

$$
\begin{array}{c}
\text{Phe}\text{---}\text{Gln} \\
\text{Tyr} \diagup \qquad \diagdown \text{Asn} \\
\diagdown \text{Cys S---S Cys} \diagup \\
| \\
\text{Pro} \\
| \\
\text{Arg} \\
| \\
\text{Gly NH}_2
\end{array}
$$

Arg vasopressin

```
         Val—Orn—Leu
        /            \
      Tyr             D-Phe
       |               |
      Gln             Pro
        \            /
         Asn—D-Phe—Phe
```
Tyrocidine A

A linear polypeptide is the heptapeptide hormone, angiotensin II (AsnArg ValTyrIleuHisProPhe).

3.2.1. Polypeptides Where Tyrosine is the Sole Aromatic Amino Acid

Of the available cyclic polypeptides, oxytocin is the only one that has been studied. As in all tyrosine-containing polypeptide systems, its fluorescence at ambient temperature is identical to that of free tyrosine, but the quantum yield is considerably reduced. Cowgill[109] reports a yield of 0.055 at pH 7 and 298°K. The low-temperature fluorescence spectrum is also identical to that of tyrosine, although the quantum yield is half the value found for free tyrosine.[110] The unusual feature of the luminescence of oxytocin is the phosphorescence. The peak wavelength and the onset wavelength are both red-shifted, and the quantum yield is one quarter that of free tyrosine. Hence the ratio of fluorescence to phosphorescence is significantly different than that found for free tyrosine, but the phosphorescence lifetime is unaffected. The special feature of oxytocin is that the tyrosine is adjacent to a disulfide bridge. To investigate whether there is any effect of the adjacent disulfide linkage on the tyrosine luminescence, Cowgill[111] and later Longworth[110] investigated model tetrapeptide systems. The model used was a dimer of the dipeptide CysTyr, L-cystinyl-bis-L-tyrosine. This tetrapeptide has a small ambient quantum yield, and this is considerably enhanced on cooling to 77°K, reaching a yield of 0.2, though this is still reduced by half from the yield of N-acetyltyrosylamide (0.4). The phosphorescence has an extremely small quantum yield (0.01) some thirtyfold less than that of monomer tyrosine, but the lifetime is not seriously altered (1.7 sec) (Figure 23). On the basis of the long lifetime, it is attractive to suggest that the effect of a vicinal disulfide linkage is to alter the internal conversion processes in tyrosine rather than intersystem crossing. Circular dichroic studies of oxytocin have shown that the tyrosine contribution to the spectrum is altered when the free α-ammonium N-terminus group of cystine is discharged with pK=6.5.[112] However, little is known about the conformation of oxytocin, though detailed proton NMR studies have recently been reported.[113] Cowgill has found that the reduced form of the model compound, the dipeptide CysTyr, had a greater quantum yield at ambient

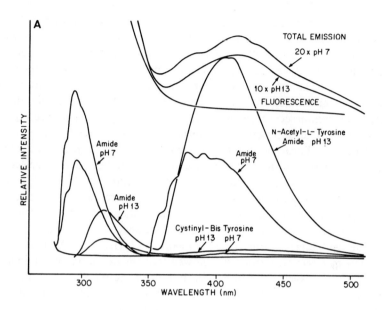

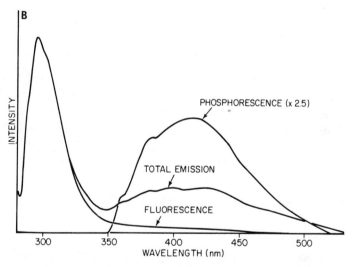

Figure 23. A. Total luminescence of tyrosine and cystinyl-bis-tyrosine. Solvent, EGW; temperature, 77°K; concentration, $E^{10\ mm}_{280\ nm} = 1.0$; excited at 279 nm (isosbestic). Insert is at an increased sensitivity for the phosphorescence, with a fluorescence intensity baseline. Solution at pH 7 given at twentyfold increase in sensitivity, the pH 13 spectrum at a tenfold increase. B. Total luminescence phosphorescence of oxytocin. Solvent, EGW at pH 7; temperature, 77°K; excitation, 278 nm. Phosphorescence is given at 2.5-fold greater sensitivity.

temperatures than the disulfide oxidized forms. Reduction of the disulfide linkage of oxytocin also led to an increase in the quantum yield of oxytocin. However, it is not possible to explain the low quantum yield observed at ambient temperatures as due simply to an enhanced internal conversion, until other effects such as enhanced intersystem crossings or photoionizations have been eliminated. Wampler and Churchich[114] have found that, at 77°K, the luminescence yields of reduced cysteinyl tyrosine are equal to those of tyrosine, supporting an internal conversion mechanism.

The only linear polypeptide to have been investigated is angiotensin II.[115] Circular dichroic and thin-layer dialysis studies have suggested that this hormone does not have any significant structure.[116] Cowgill[115] found that the fluorescence yield at ambient temperatures was 0.07, i.e., small compared to free tyrosine (0.21), and the Stokes shift and peak location were identical to that of free tyrosine. The sequence of angiotensin II shows that the next adjacent amino acid is histidine (TyrIleuHis). A model system titration of HisTyr dipeptide[117] showed that the discharge of the imidazolium cation between pH 5 and 7 did not affect the tyrosine quantum yield, but the yield changes that were observed were associated with titrations of the amino terminus. The pH titration of angiotensin II showed no indication of dependence on histidine titration, or a titration of the N-terminus; the yield was independent of pH until phenoxyl titration occurred.

3.2.2. Polypeptides with Tyrosine and Phenylalanine

Vasopressin and tyrocidine A have not been investigated. Another system of interest is the heptapeptide LysProThrTyrPhePheGly obtained by tryptic digestion of insulin,[118,119] which has been studied fluorometrically by Wampler and Churchich.[114] The polypeptide has no structure, according to an analysis of its circular dichroic spectrum,[120] and its luminescence at 298 and 77°K is indistinguishable from tyrosine zwitterion.[114]

3.3. Tryptophan-Containing Polypeptides

3.3.1. Chemical Features

The majority of simple polypeptides, either cyclic or linear, which possess tryptophan also have either tyrosine or phenylalanine or both residues present. However, this does not lead to an overly complicated situation. Two linear polypeptides are known which possess only tryptophan: a venin, mellitin, from the honey bee, which has 26 amino acids with a single tryptophan at position 19 [mellitin (26), 1 trp-(19)], and the bacterial antibiotic gramicidin A (HCO Val Gly Ala D-Leu Ala D-Val Val D-Val Trp D-Leu Trp D-Leu Trp $NHCH_2CH_2OH$). Neither polypeptide has had its fluorescence spectrum investigated, but mellitin is known to have a tryptophan fluorescence. Gramicidin A readily aggregates in aqueous media, but not in organic solvents, and it is thought to have little structure from an analysis of its CD spectrum.[121] A similar conclusion has been reached with mellitin from analysis of its anisotropy temperature dependency. Two cyclic polypeptides which are potentially available for study are tyrocidine B and C; both are bacterial antibiotics from *Bacillus brevis*:

```
           Val——Orn——Leu
          /              \
        Tyr              D-Phe
         |                |
        Gln              Pro
          \              /
           Asn—D-Phe(X)Trp
```
Tyrocidine B

Tyrocidine C is similar in structure to tyrocidine B, but a D-Trp replaces a D-Phe(X) group. The conformation of tyrocidine B is known to be similar to that of gramicidin S-A from CD,[122] thin-layer dialysis,[123] and NMR studies.[124]

Other structures involving tryptophan peptides, where the sulfhydryl group of cysteine substitutes for tryptophan at position 2, are found in the toxins of the death-cap fungi such as phalloidin and amanitin:

```
    Ala——Trp—e-DihyLeu        HyPro—HyTrp—β-MeDihyLeu
     |    |                     |         |
     |    S    Ala               |         S
     |    |    |                 |         |
 a-HyPro—L-Cys—D-Thr            Asp——Cys—Gly—Ileu
        Phalloidin                    α-Amanitin
```

Of these suitable systems, only mellitin has been investigated fluorometrically,[125] though the absorption and circular dichroic spectra of all the other compounds have been reported.

Slightly more complicated polymers have been investigated fluorometrically. These are all linear polypeptides and are hormones, the majority of which have been chemically synthesized. Again, all have known circular dichroic spectra, and thin-layer dialysis studies have been performed. The general conclusion reached from structural studies is that these polypeptides possess little, if any, tertiary structure, though small segments of α-helical secondary structural elements are not ruled out. These simple hormones are attractive fluorometrically since another of their suitable features is that they often possess only one or two tyrosine residues, separated by either a small number (≤ 5) or a large number (≥ 15) of amino acid residues. By total chemical synthesis, fragments can be obtained which possess the respective pairs that occur in the intact hormone structure. A study of the individual transfer steps between a single tyrosine and tryptophan residue can be investigated in the fragments. The results, when compared to transfer studies with the intact hormones, will provide further insight into the tertiary structure of the polypeptides, particularly since the fragments will almost certainly lack any tertiary structure, and any differences could be attributed to secondary structure, i.e., helical regions.

The following is a brief summary of the hormones which have been studied. The sequences are for bovine or porcine hormones, though any exchanges between species in general do not involve the aromatic amino acids.

	Amino acids	
Corticotrophin (ACTH)	39	Tyr-2; Trp-9; Tyr-23; Phe-7; Phe-35; Phe-39
α-Melanotrophin (α-MSH)	13	Tyr-2; Trp-9; Phe-7
β-Melanotrophin (β-MSH)	18	Tyr-5; Try-12; Phe-10
Glucagon	29	Tyr-10; Tyr-13; Trp-25; Phe-6; Phe-22
Gastrin	17	Trp-4; Tyr-12; Trp-14; Phe-17
Parathormone	84	Trp-23; Tyr-43; Phe-7; Phe-34

The ultraviolet absorption of these polypeptides is largely due to the tryptophan and tyrosine content; the phenylalanine contribution is small even at its maximum. At the main absorption peak, the preponderant absorption is due to tryptophan residues. To the first approximation, these polymers can be treated as tryptophan-emitting systems, the tyrosine contribution to the emission being negligibly small.

3.3.2. Polypeptides Where Tryptophan Is the Sole Aromatic Amino Acid

Bernstein et al.[125] measured a fluorescence from mellitin fragments, and, though a spectrum was not given, the fluorescence obviously originated in the tryptophan residues. Three fragments were studied: a decapeptide (10-mer), residues 18–27, where the single Trp is adjacent to the Ser N-terminus; a tridecapeptide (13-mer), residues 15–17; and a unoduodecamer (21-mer), residues 7–27. Though these workers do not state the quantum yield of fluorescence, it appears from the data given to be comparable to that for free tryptophan, unlike tyrosine polypeptides where the ambient yield is significantly less than the free amino acid. Bernstein et al. did, however, report the peak wavelength. It decreased as the polymer chain elongated, the values being 353, 350, and 345 nm, respectively. Therefore they concluded, from the chain-length dependency of the Stokes' shift, that there were indications of some structure in the longer hormone fragment. Support for this contention was obtained by addition of guanidinium chloride, a

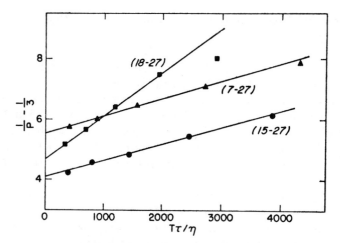

Figure 24. Einstein–Perrin analysis of the fluorescence anisotropy from three fragments of mellitin. Solvent, glycerol : water mixtures, where the glycerol concentration is varied between 75 and 95%; temperature, 298°K; excitation, 280 nm; observation, 345 nm. Abscissa: viscous friction, 10 Km s kg^{-1}; ordinate: fluorescence anisotropy, the Perrin function $(1/P - 1/3)$ with polarized excitance. (After Bernstein et al.[125])

reagent which disrupts a helical structure. The 21-mer fluorescence peak is shifted by 5 M guanidinium to 350 nm, which is the monomer location in that solvent, and it reduces the quantum yield by 10%. Limiting polarization values for the fluorescence of the fragments were also measured by Einstein— Perrin plots (Figure 24). The values differed from one another, and from that of a monomer. This unusual observation was explained by suggesting there was a conformational perturbation to the polarization spectrum. Until a complete polarization spectrum (FAPS) is determined, it is difficult to interpret these results.

The fluorescence spectra of polypeptides which contain only tryptophan and no tyrosine residues have not been extensively studied, but the preliminary conclusions drawn from the one example are that (1) the quantum yields are comparable to those from monomers, (2) the Stokes' shift of the tryptophan is dependent on the polypeptide conformation, and (3) from the magnitude of the shift the tryptophan residues are suggested to be freely exposed to the equeous medium.

3.3.3. Polypeptides Which Contain Both Tyrosine and Tryptophan

The fluorescence spectra of polypeptide hormones which contain both tyrosine and tryptophan residues, with one or two residues of each amino acid per chain, have been studied in greater detail than those polypeptides that lack tyrosine.

3.3.3(a). Fluorescence Spectra. The fluorescence is predominantly from tryptophan residues when 280 nm light is used to excite the hormones, though a tyrosine contribution can readily be observed.[126] This tyrosine emission band is absent from the fluorescence spectrum excited at 295 nm, and by suitable normalization the tyrosine spectrum can be obtained by the difference between the two fluorescence spectra excited at 280 and 295 nm. An example of the fluorescence spectra is shown for ACTH in Figure 25.

3.3.3(b). Quantum Yield of Fluorescence. The quantum yield of tryptophan does not have an established value. Before discussing the quantum yield values for these hormones, it is worth digressing to comment on the value for the tryptophan zwitterion yield. Teale and Weber[127] obtained

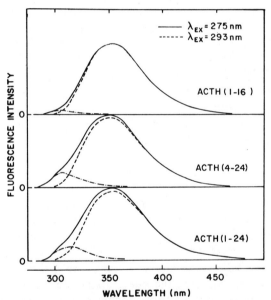

Figure 25. Fluorescence spectra of adrenocorticotrophin and two fragments. Solvent, water at pH 7; temperature, 298°K. (———) ACTH excited at 275 nm; (– – – –) excited at 293 nm; (–·–·–) a difference spectrum between 275 and 293 nm spectra after their normalization at 370 nm. This difference spectrum has a shape characteristic of tyrosine fluorescence. Ordinate: relative number of quanta and constant wavelength resolution. (After Eisinger.[126])

a value of 0.20±0.01 for the yield of the zwitterion of tryptophan in water. This was at an ambient temperature presumed to be 20°C, and at pH 7 in aqueous solutions. The value has since become widely accepted, been confirmed on several occasions, and until recently been accepted as a reliable value. Badley and Teale[128] have quite recently reconfirmed the value. Furthermore, in their study of the lifetime and quantum yield of many indole compounds and in several solvents, they found there was a linear relationship between quantum yield and fluorescence lifetime. Tryptophan zwitterion fitted this function with a quantum yield of 0.2. Other workers have reached significantly different conclusions. Chen (0.13),[129] Borresen (0.11),[130] and Bridges and Williams (0.15)[131] all suggest much lower (though different) quantum yields. Their values are based on the quantum yield of quinine (0.55), which itself has been shown to be a somewhat unreliable standard compound unless care is taken in its application.[132] The comparison to quinine was made by integrating the area of the fluorescence spectrum in units of quanta at different wavelengths at a constant wavelength resolution. This must be corrected to a resolution that is constant in wavenumber to obtain the quantum yield. Such a correction is accomplished by the inclusion of a wavelength-squared multiplier in the integration. All of these workers did not include this step in determining their values for the tryptophan quantum yield, so their values must be discounted until they have been suitably corrected. The correction is of the order of 35%. It is to be remembered that Teale and Weber utilized quantum-counting screens, and so they did not require this type of correction to their results. Eisinger and Navon[133] have used another standard, p-terphenyl in air-equilibrated cyclohexane. The quantum yield currently accepted as reliable is 0.72, and differs from the value used by Eisinger and Navon. Moreover, as p-terphenyl emits in a similar spectral region to tryptophan, it is convenient to use in this application. Eisinger and Navon did apply the wavelength-squared correction in the comparison of the standard and unknown, so their yield was determined on the correct constant wavenumber basis. They obtained a value for tryptophan zwitterion of 0.115 when corrected to 298 K for the newer standard value. However, they did not include a correction for differences in the refractive indices of the two solvents, which would alter their value to 0.135, distinct from Weber and Teale's value.[134-136]

Suffice it to say that with this particularly crucial reference point being in doubt there are serious handicaps to any quantitative assessment of quantum yield data for tryptophan compounds, including proteins. The trend is for values for the zwitterion to approach 0.16–0.18. For convenience in this review, where possible, all values are referred to the Teale and Weber value, without prejudice to the final outcome. In many ways it is the value for the yield at low temperature, where a limiting value is attained that is more diagnostic for a specific compound. Low-temperature quantum yields, too, cannot be regarded as well established. There are additional corrections to consider: (a) refraction changes in cyclindrical tubes,[135] (b) concentration changes, and (c) polarized emission.[136] But there is a better agreement; tryptophan compounds are reported to possess similar yields at 77°K,[137-140] and proteins have comparable yield values.[140-142] To use low-temperature quantum yield values adequately, it is necessary to analyze the temperature dependent quenching of the luminescence and to specify activation energies of the thermally induced quenching reactions. Since these energies too are not necessarily constant over the range of temperatures from a temperature where the limiting value prevails to ambient temperature, this in itself represents a somewhat arbitrary attitude. Hence the quantum yield at 298°K, though empirical in nature, is a convenient number in practice. To quantify the situation further, the quantum yields of triplet formation and photoionization are needed, and work along these lines is expected to be forthcoming.

To return to the point previously raised, the quantum yield of the tryptophan residues of polypeptide hormones at ambient temperature is estimated to be 0.14 ± 0.01, i.e., approximately 70% of the tryptophan zwitterion value.[126,143] The magnitude of a tyrosine contribution to the emission depends largely on the fractional absorption at the exciting wavelength; the values for the quantum yield of the fluorescence from tyrosyl residues range from 0.02 to 0.05, which are comparable to the yields of synthetic tyrosine–glutamate or lysine copolymers, or to those of the small tyrosine-only polypeptides. This illustrates the pattern found regularly with the "average" protein and polypeptide hormones—there is a tryptophan-dominated fluorescence whose yield lies close to 50% of tryptophan zwitterion. In addition, there is a small tyrosine contribution with a yield that is 10% of the tyrosine zwitterion. The fractional contribution of tyrosine to the emission spectrum is determined by the fractional absorption of tyrosine at the exciting wavelength chosen.

3.3.3(c). pH Titrations. Since these polypeptides are regarded as possessing little tertiary structure in aqueous media, the linear polypeptides offer a suitable system in which near-neighbor interactions can be investigated. The interresidue distance is assumed to be related directly to the number of intervening residues. Adopting this attitude, Shinitsky and Goldman[143] measured the pH dependence of the tryptophan quantum yield of α-MSH and ACTH. α-MSH represents the 13-amino-acid N-terminal sequence of ACTH. An identical titration behavior was found for the two polypeptides (Figure 26). Hence the conclusion was that the remainder of the chain of ACTH did not influence the luminescence prop-

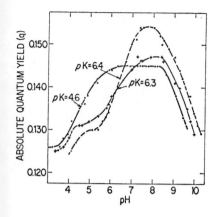

Figure 26. Fluorometric titration of glucagon, α-melanotrophin, and adrenocorticotrophin. Solvent, water at pH 7, 0.1 M NaCl; temperature, 298°K; concentration, $E_{280\,nm}^{10\,mm}=0.5$; excitation, 280 nm; observation, 340 nm. (–·–·–) Glucagon; (--------) α-MSH; (+++) ACTH. (After Shinitzky and Goldman.[143])

erties of the single tryptophan residue of ACTH, which is position 9. The tryptophan yield was quenched on acid titration from neutrality by 20%, with an apparent pK of 6.5. A similar pH-dependent quenching had been found with both histidine–tryptophan dipeptides and their corresponding cyclic diketopiperazines. Therefore, the polypeptide titration effect was interpreted to result from the titration of the single histidine residue that was separated from the typtophan residue by two other amino acids. At alkaline pH values, where the tyrosyl–phenoxyl groups were titrated, the tryptophan yield was significantly reduced. A pK of 9.5 was found for the alkaline quenching of α-MSH, ACTH, and glucagon.

Edelhoch and Lippoldt[144] performed similar alkaline titrations on both glucagon and ACTH, but they reported an entirely different behavior (Figure 27). They also found that the tryptophanyl and the tyrosyl quantum yields were uninfluenced by pH from pH 2 to 9, with no indication of a histidine-induced quenching. Only small changes were found in the tryptophan yield when the phenoxyl groups were titrated. With glucagon, no change whatsoever was reported. pH values had to reach >12 before tryptophan quenching as a result of the enamine proton ejection occurring from the excited singlet state. However, Edelhoch and Lippoldt did find that the tryptophanyl residues of the more complicated hormone parathormone were quenched by the phenoxyl ionization. This quenching was suppressed by the presence of 5 M guanidinium hydrochloride. In model system studies, Edelhoch et al.[145] investigated the quenching by tyrosinate of tryptophan of the series of oligopeptides Tyr (Gly)$_n$ Trp, where $n=0$, 1–6. The magnitude of the phenoxyl quenching decreased with increasing value for n, and became independent of n at $n>4$. From these results, Edelhoch and Lippholdt conclude that in the sequence of parathormone there is a tyrosine separated from the single tryptophan by four residues,

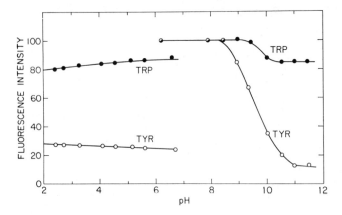

Figure 27. Fluorometric titration of tyrosine and tryptophan residues of an adrenocorticotrophin fragment ACTH (1–25). Solvent, water, 0.09 M KCl; 0.02 M lysine in alkaline section; 0.01 M acetate in the acid section of the titrations; temperature, 298°K; concentration, $E^{10\ mm}_{280\ nm} = 0.05$; excitation, 270 nm (isosbestos). (–•–•–) Intensity at 350 nm; (–∘–∘–) intensity at 300 nm. The 350 nm fluorescence intensity is associated with tryptophan and the 300 nm intensity with tyrosine (see Figure 25). (After Edelhoch and Lipphold.[144])

whereas the interresidue separation determined by the primary sequence is considerably greater than with α-MSH, ACTH, or glucagon, it is separated by twenty residues.

3.3.3(d). Stokes' Shift and Solvation. Stokes' shifts of tryptophan model compounds have been used to indicate the extent of solvation of the tryptophanyl residues.[98] Tryptophanyl fluorescence maxima located at a wavelength greater than 340 nm are taken to reflect an exposure to the aqueous solvent. Peak wavelength values less than 335 nm indicate a tryptophanyl residue situated in a hydrophobic environment. On this basis, the Stokes' shift of tryptophanyl residues of the polypeptide hormones suggests that the tryptophan is freely available to the solvent.

Another means of investigating the extent of exposure is to study the temperature dependency of the Stokes' shift. In their temperature study of model systems in polyalcohol : water glasses, Eisinger and Navon[133] found that above the second glass transition the Stokes' shift of tryptophan compounds progressively increased with temperature at a constant quantum yield, until a limiting value was reached. Subsequent increases in temperature then led to a decrease in the quantum yield without changes in the Stokes' shift. When similar studies were performed on ACTH or glucagon, a different

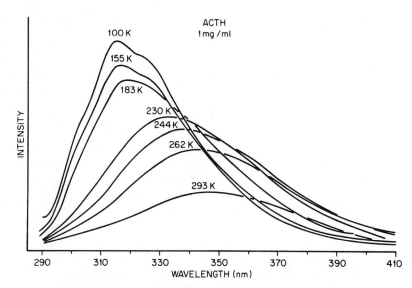

Figure 28. Temperature dependency of the fluorescence spectrum of adrenocorticotrophin. Solvent, propylene glycol : ethylene glycol : water 1:1:2 (PEW) at pH 7; concentration, 1 mg/ml; excitation, 280 nm. Spectra obtained at various temperatures, with a fluorometer functioning in a ratio mode with a reference derived from the excitance.

picture prevailed[146] (Figure 28). Both polypeptide hormones progressively shifted their peak wavelengths and decreased their quantum yields of fluorescence simultaneously at temperatures above the second glass transition (180°K). The activation energy in the high-temperature region (above ambient temperatures) is 1.7 kcal mol^{-1} for ACTH, 2.7 kcal mol^{-1} for glucagon. These activation energies are to be compared to that of tryptophan in the same glasses, which is 7.0 kcal mol^{-1}. Edelhoch and Lippoldt reported a similar behavior; the activation energies of the hormones are less than those found with monomers (Figure 29).

Another manner of investigating ths extent of solvation explored by Eisinger and Navon was study of a solvent isotope effect. Stryer[147] had observed that replacing H_2O by D_2O enhanced the fluorescence of tryptophan, and Eisinger and Navon confirmed this effect and further showed that there was the same thermal activation energy for thermally activated quenching of the fluorescence in both solvents around ambient temperatures (280–380°K). The enhancement ratio $\Phi D_2O/\Phi H_2O$ for tryptophan was 2.14. When the solvent isotope effect was measured for the hormone ACTH, a value of 1.45 was found. This value was also found for the other proteins studied, chymotrypsin and staphylococcal endonuclease. These proteins,

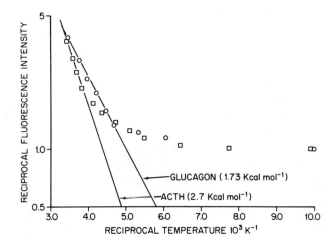

Figure 29. Arrhenius activation analysis of the intensity of fluorescence of glucagon and adrenocorticotrophin. Abscissa: reciprocal temperature, $10^3 K°$; ordinate: reciprocal fluorescence intensity, measured at the respective emission maximum.

however, do not have large Stokes' shifts, and the tryptophan residues are believed to be in a hydrophobic region. The conclusion from the temperature dependence of Stokes' shift and the solvent isotope effect on the quantum yield is that, though the tryptophan groups of the polypeptide hormones are exposed to water, the local environment is not identical to that found for free tryptophan; there must be some restrictions placed on the solvation shells compared to a free residue. This is not at all unreasonable considering that the residue is now incorporated into a polymer chain. But the lack of a solvent isotope effect is still inadequately explained.

3.3.3(e). Energy Transfers. (1) Sensitization: The hormone ACTH has proved to be a particularly valuable system in which to study electronic energy transfer between tyrosine and tryptophan at neutral pH values. Fragments of corticotrophin are available by a total synthesis, and these then allow one to study separately the transfer from each of the two tyrosine residues to the single tryptophan. In ACTH (4–24) there is the transfer from Tyr-23 to Trp-9; in ACTH (1–16) the transfer is from Tyr-2 to Trp-9. Eisinger[126] investigated the magnitude of transfer in both fragments and in ACTH. They measured the sensitization of the tryptophan fluorescence by tyrosine absorption. The method of analysis was one first used by Teale.[98] It requires the measurement of a quantum yield for the tryptophan fluorescence for all exciting wavelengths, and either relative or absolute yield can be determined. Eisinger used the high concentration limit to measure the quantum yields.

The fractional absorption of tyrosine is defined in the following manner:

$$f_{\text{Tyr}} = \frac{N_{\text{Tyr}} \langle \varepsilon_{\text{Tyr}} \rangle}{\varepsilon_{\text{Pr}}}$$

where N is the number of tyrosines in the polymer, ε_{Pr} is the polymer molar extinction coefficient, and $\langle \varepsilon_{\text{Tyr}} \rangle$ is the tyrosine residue average molar extinction coefficient in this polymer. To obtain f_{Tyr}, a value for ε_{Tyr} is taken from model systems and assumed to be equal to that prevailing in the polymer. Beaven and Holiday[148] had shown several years ago that the absorption spectrum of a protein can be well approximated by the sum of the absorptions of the constituent aromatic amino acids as the total digest of a protein absorbed within 10% of the intact protein. There are small shifts to be taken into account, and in constructing a mixture of amino acids to mimic the polymer, the absorption of histidine and cysteine and cystine must not be neglected. Some idea of the magnitude of the error can be found in the recent work of Bailey et al.,[149] Edelhoch,[150] Herskovitz and Sorensen,[151] and the review by Donovan.[152] For ACTH, Eisinger shows the excellence of the fit between the absorption of a mixture of the N-acetyl amide derivatives of phenylalanine, tyrosine, and tryptophan in the correct proportion with the absorption spectra of ACTH (Figure 30). Inspection of the plot of the fractional absorption of tryptophan determined from these model systems at wavelengths greater than 295 nm shows all the light is absorbed by tryptophan. Hence, the intensity of fluorescence and the quantum yield determined at these wavelengths greater than 295 nm can be ascribed entirely to an emission by the tryptophan residues of ACTH and its fragments. To obtain the extent of transfer (e) from a tyrosine to the tryptophan, it is possible to write the observed tryptophan quantum yield as

$$\Phi_{\text{Trp}}^{\text{obs}} = \Phi_{\text{Trp}} (f_{\text{Trp}}^{\text{Pr}} + e f_{\text{Tyr}}^{\text{Pr}})$$

Φ_{Trp} can be obtained by exciting with 295 nm light. Then plots of $\Phi_{\text{Trp}}^{\text{obs}} / \Phi_{\text{Trp}}^{295\text{nm}}$ at different exciting wavelengths can be fitted by $(f_{\text{Trp}}^{\text{mon}} + e f_{\text{Tyr}}^{\text{mon}})$, which is obtained from standard absorption curves of monomers and the known composition of the polymer. Thus e is the variable parameter and is chosen to give the best fit to the experimental data. The results Eisinger obtained with ACTH fragments are shown in Figure 31 to illustrate the nature of the results which can be obtained. With a value for e, the fractional tranfer, it is then possible to obtain the true quantum yield of the tyrosine emission contribution to the emission. As mentioned earlier, excitation at 295 nm

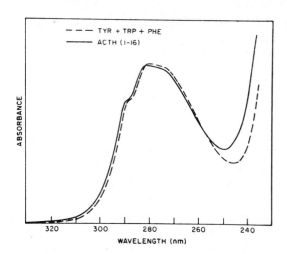

Figure 30. Comparison of an aqueous absorption spectrum of a polypeptide with a mixture of aromatic amino acids of equal molar proportions. (———) ACTH (1–16); (– – – –) equimolar mixture of phenylalanine, tyrosine, and tryptophan at equal residue molarity to ACTH solution. (After Eisinger.[126])

excites only the tryptophan residues, excitation at shorter wavelengths leads to both tyrosine and tryptophan components. The pure tryptophan spectrum can now be defined, and by a suitable normalization at a wavelength in the emission where the tyrosine makes no appreciable contribution ($\geqslant 370$ nm), the tyrosine contribution can be determined by the difference between two spectra at wavelengths less than that used for the normalization. The apparent quantum yield is determined from the area by the integration

$$\Phi_{\text{Tyr}}^{\text{obs}} = \int \lambda^2\, q_{\text{Tyr}}(\lambda)\, d\lambda$$

where $q_{\text{Tyr}}(\lambda)$ is the tyrosine difference spectrum in spectral quantal yield values at constant wavelength. The equivalent quantity $q_{\text{Trp}}(\lambda)$ is defined for the tryptophan band. Then

$$\frac{\Phi_{\text{Tyr}}}{\Phi_{\text{Trp}}^{\text{obs}}} = \frac{f_{\text{Trp}}(\lambda)(1 - e_{\text{Tyr}})\Phi_{\text{Tyr}}}{[f_{\text{Tyr}}(\lambda) + e_{\text{Tyr}} f_{\text{Tyr}}(\lambda)]\Phi_{\text{Trp}}}$$

Since we now have determined both Φ_{Trp} and $e_{\text{Tyr}}, \Phi_{\text{Tyr}}$ is directly found on substitution into the previous equation. Eisinger investigated the two ACTH fragments and ACTH, and found e (Tyr-2) $= 0.5 \pm 1.15$; e (Tyr-23)

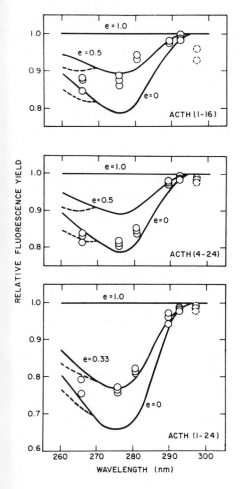

Figure 31. Wavelength dependency of tryptophanyl fluorescence quantum yield. The excitation spectral values are used to determine the relative quantum yield dependency by comparison with the absorption by ACTH fragments. The curves are theoretical values obtained from an analysis of synthetic absorption spectra of amino acids and correspond to different efficiencies of transfer by tyrosine to tryptophan (E). The lower solid line corresponds to a 100% transfer between Phe-7 and Trp-9, while the dashed segment represents no phenylalanine transfer contribution. The best-fit line to the experimental points is given for each ACTH fragment. (After Eisinger.[126])

$= 0.15 \pm 0.1$; e (Tyr) $= 0.33 \pm 0.18$. The e obtained from ACTH was equal to the average transfer determined from the fragments.

This study clearly substantiated that there was a transfer between the nearest tyrosine–tryptophan pairs. In an analysis of the critical distance obtained from Förster's equation assuming a random orientation, Eisinger was able to conclude that there was an α-helical conformation for the N-terminal region of ACTH, but not for the remaining C-terminal region beyond the tryptophan residue at position 9.

(ii) Polarization: Polarization of fluorescence studies would independently confirm the existence of transfer in these polymer hormones. The only values currently reported are at a single wavelength, and the

FAPS has not been measured. Bernstein et al.[125] performed polarization measurements at the single wavelength 270 nm, using values determined by an Einstein–Perrin plot of anisotropy extrapolated to an infinite viscosity so as to obtain the fundamental polarization value. They found that there was depolarization and concluded that this was consistent with the transfer from tyrosine to tryptophan.

(iii) *p*H titration: The discrepancy between Shinitsky–Goldman[143] and Edelhoch–Lippoldt[144] as to whether there is or is not a transfer from tryptophan to tyrosinate deserves further investigation. On the basis of Eisinger's study of tyrosine-to-tryptophan transfer in ACTH fragments, and the known overlap integral, extensive quenching by tyrosinate formation would be expected in all the polymers, as Shinitsky and Goldman reported.

3.4. Summary

The overall conclusion of the studies on both synthetic polypeptides and natural polypeptides is remarkably consistent, and there is also an excellent agreement with conclusions obtained previously from dipeptides:

(a) Incorporation of tryptophan into a polypeptide does not drastically alter its luminescent properties.
(b) Because there is large variation in the Stokes' shift of tryptophan with solvent, the Stokes' shift of residues in polypeptides is conformationally dependent.
(c) Incorporation of tyrosine into polypeptides drastically reduces its quantum yield. The reduction is an order of magnitude at 300°K.
(d) Electronic energy transfers between tyrosine and tryptophan can be demonstrated.
(e) The magnitude of transfer does not account for the majority of the quenching of the tyrosine residues at ambient temperatures unless the tyrosine residues are immediately adjacent in primary sequence to a tryptophan residue.

4. LUMINESCENCE OF PROTEINS—CLASS A PROTEINS

Teale[153] grouped proteins into two fluorometric classes:

(a) Class A—proteins which lack tryptophan but possess tyrosine.
(b) Class B—proteins which possess tryptophan and also tyrosine and phenylalanine.

Luminescence of Polypeptides and Proteins

There is no fundamental difference in either structure or function of these two groups of proteins; they are simply a convenience for interpreting the luminescent properties. However, the selection for no tryptophan does favor proteins with a small total number of amino acids, and of the well-characterized proteins available few fit into class A. Another point worth making at this stage is that the intrinsic properties of tyrosine residues in class B proteins are not altered significantly by the presence of the tryptophanyl residues. It is just that the tryptophanyl absorption and emission dominates, and any tyrosine contribution is small.

Two pancreatic proteins have been the most extensively investigated of the class A proteins—these are an enzyme, ribonuclease A, and a hormone, insulin. With the availability of molecular models derived from X-ray diffraction analysis of their single crystals, it is now feasible to offer a detailed analysis of the emissions from these two proteins.

4.1. Fluorescence Spectra

The fluorescence at all temperatures, including both ambient and low temperatures of all the class A proteins so far studied, is essentially identical to that of the zwitterionic tyrosine or other suitable monomers. There is a red shift to the fluorescence monomers at 77°K, but none is present at

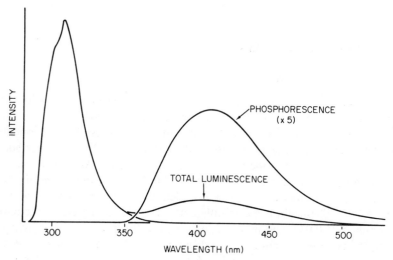

Figure 32. Luminescence of ribonuclease A. Solvent, EGW at pH 7; temperature, 77°K; concentration, 4.0 mg/ml; excited at 280 nm. Fluorescence (high peak); total luminescence, and phosphorescence (sensitivity increased five-fold) are shown.

298°K (Figure 32). The absorption maxima of the monomer are 277 nm at 298°K, 275 nm at 77°K. Emission maxima are 304 and 294 nm, respectively. The Stokes' shift is 0.32 μm^{-1} at 298°K, 0.25 μm^{-1} at 77°K. For proteins the absorption maxima are similar, 0.5 nm red shift, and Stokes' shifts of 0.31–0.32 μm^{-1} are found at 298°K and 0.29–0.30 μm^{-1} at 77°K. The quantum yield of tyrosine zwitterion is not in serious dispute; the value obtained by Teale and Weber for the zwitterion at 298°K and pH 7 is 0.21 and is generally taken.[154] The quantum yields of the class A proteins at ambient temperatures fall into the range of 1–5%. Therefore, the most striking feature of class A protein fluorescence is its small quantum yield. For RNase A, Teale gave a value of 0.017, Cowgill 0.014, and Wampler[155] 0.023; for insulin, Teale's value was 0.037, Cowgill[156] confirmed this value, Vladimirov[157] quoted 0.03, and Wampler gave 0.06; ovomucoid was given the value 0.012 by Teale, 0.007 by Cowgill,[156] and 0.006 by Donovan.[158] There is excellent agreement on the yield values from several different workers. Other proteins which have been studied are zein (Teale,[153] Cowgill,[156] Vladimirov and Burshtein[159]); malic dehydrogenase (Thorne and Kaplan[160]); ovomucoid (Teale,[153] Cowgill,[156] and Donovan[158]); Δ^5-3-ketosteroid isomerase (Wang et al.[161]); pancreatic trypsin inhibitor (Cowgill[156]); paramyosin and tropomyosin (Cowgill[162]). Two well-characterized class A proteins which have not so far been investigated are tryptophan synthetase α and aspartate transcarbamylase R of E. coli.

4.2. Fluorescence Quenching

4.2.1. Quantum Yield

There is an unusually small value for the ambient quantum yield of the fluorescence of RNase A [153, 156] compared to the values found for other tyrosine dipeptides[156] and synthetic polymers.[163] (These range from 0.04 to 0.07.) Denaturation of RNase with 8 M urea,[153, 156] sodium dodecyl sulfate,[156] or acid[156] increased the quantum yield. Acid increased the value to 0.021, the other denaturants to 0.037. Reduction of the disulfide linkages in the presence of 8 M urea led to a further increase of yield (0.055).[155, 156] Insulin behaved differently; the quantum yield of the native hormone was 0.037, and this was unaffected by 8 M urea[153, 156] or by detergent,[156] but the reduction of the disulfide links led to an enhancement in yield.[155, 156] The quantum yield of the native RNase increased on acid titration with pK 2, but in 8 M urea there was no effect of acid over this range of pH values and the yield remained constant up to alkaline pH values when phenoxyl ionization occurred.

4.2.2. Abnormal Tyrosine Residues

Ribonuclease A has six tyrosyl residues. Aqueous alkaline spectrophotometric titration studies have demonstrated that only three of the phenoxyl hydroxyls titrate normally, and the protein has to be denatured for the other three residues to undergo titration with base.[164] Another means of investigating the exposure to solvent of the tyrosyl groups is to add sucrose to the solution. This subjects the absorption spectrum of the exposed groups to a change, shifting them slightly to longer wavelengths. This is observed by difference absorption spectrophotometric methods devised by Herskovits and Laskowski.[165] They concluded that 3.4 tyrosyl groups were exposed in RNase. Bello adopted another approach to the same problem. Rather than alter the composition of the medium, he measured the melting curve for the protein.[166] Tyrosine has an absorption spectrum which is susceptible to large changes in molar extinction coefficient when the nature of the solvent environment is altered, though the spectral shifts are small.[167, 168] Thus, when the absorption spectra of native and heat-denatured proteins are compared, differences are found because all tyrosine residues are fully exposed to solvent in the denatured protein above its melting temperature but not below this temperature. The fraction of buried residues in the native protein below the melting temperature is determined by comparison with the heat-denatured absorption and suitable model systems. Bello concluded from his melting curves that 3.7 residues were exposed in native RNase at 298°K. From an analysis of the molecular model for RNase available to him, Bello reported that the three buried tyrosine residues were Tyr-92, Tyr-25, and Tyr-97, and he estimated the exposure for each individual residue as 0.7–0.8, 0.3–0.4, and 0.0, respectively. The three exposed tyrosines were identified as Tyr-73, Tyr-76, and Tyr-115. However, these residues are not completely exposed either; the equivalent exposure was put at 2.5. There is excellent agreement with the values determined from the molecular model and from spectroscopic studies, an exposure of 3.6 average with limits from 3.2 to 3.8 obtained from the molecular model. Previous group-specific chemical modifications are quite consistent with these results. Thus iodination of the tyrosines, or O-acetylation by N-acetylimidazole, modified Tyr-73, Tyr-76, and Tyr-115. When more extreme conditions were used, Tyr-92 became modified. Tyr-25 and Tyr-97 never reacted in the native protein.[169]

4.2.3. Molecular Structure

The molecular model shows that Tyr-25 is bonded to the carboxylate side-chain of Asp-14, while Tyr-92 is bonded to the amide chain at Asp-38, and Tyr-97 is completely buried and is bonded to the amide chain at Lys-41.

In the primary sequence, Tyr-25 and Tyr-73 are adjacent to a disulfide linkage and Tyr-73 is placed directly above the Cys-58–Cys-110 disulfide link. The molecular structure can be used to interpret the fluorescence by drawing upon previous model system studies. Carboxylate anions, either free[153, 170, 171] or present as side-chains of Glu[163, 172] or Asp,[174] have been shown to be effective as collisional quenchers for tyrosine. Carboxylates[174] and amides[175] can also form weak complexes with the ground state of tyrosine through the phenoxyl group, and these are nonfluorescent. The ground-state complex is favored by a low dielectric constant. Amides do not function as quenchers in the presence of water[176] but do form nonfluorescence complexes in its absence, though the complexes are not as stable as a carboxylate H-bonded ground-state complex. Amide and carboxylate complexes are present largely in the hydrophobic environments in the interior of proteins.

4.2.4. Disulfide Quenching

A third quenching mechanism which affects tyrosyl residues in protein is associated with vicinal disulfide linkage.[177] Little is known of the mechanism of the disulfide quenching reaction. Perhaps it is related to the photoionization quenching process studied by Steiner and Kirby with tryptophan.[178] The basic features of vicinal disulfide quenching were discussed previously, when quantum yield of oxytocin was considered. Therefore, Tyr-25 and Tyr-73 are probably quenched by being vicinal to disulfides. Tyr-25 is additionally quenched by forming a carboxylate complex, and Tyr-92 and Tyr-97 are quenched as amide complexes. Tyr-76 appears to have no potential quenching group in its environment, and Tyr-105 has Asp-113 in its environment, which may act as a collisional quenching agent. This suggests that Tyr-76 is the source of the fluorescence.

4.2.5. Electronic Energy Transfer

This interpretation neglects any influence that electronic energy transfer may have on the quenching process by transferring energy to or from a given residue. Weber[179] measured the fluorescence polarization spectra of insulin and RNase at different exciting wavelengths (FAPS), and the polarization values found were depolarized compared to a monomeric tyrosine spectrum (Figure 33). Since the only cause of a depolarization can be a transfer of electronic energy from one residue to another,[180] these processes must be occurring in RNase. When the buried tyrosine residues are complexed to either a carboxylate or an amide group, there is a small shift to longer wavelengths in their absorption peaks. Therefore, it is possible for these quenched residues to act as energy traps for other uncom-

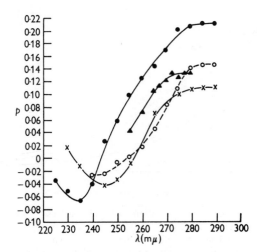

Figure 33. Fluorescence absorption polarization of insulin, zein, and ribonuclease A. Solvent, propylene glycol; temperature, 205°K; fluorescence isolated with a cutoff filter with 310 nm pass. (◯) Tyrosine; (△) zein; (×) insulin; (◉) ribonuclease. Abscissa: exciting wavelength (nm); ordinate: degree of polarization, unpolarized excitance used. (After Weber.[179])

plexed tyrosines in RNase, which thereby explains the observation of a small yield from the enzyme.

4.2.6. Buried Residues

To investigate the role of buried residues on the fluorescence quenching of RNase, Cowgill[181] analyzed a fluorescence quenching that accompanies a titration of phenoxyl groups. Ionization of the phenoxyl groups at 298°K replaces the 304 nm fluorescence with a very weak fluorescence from tyrosinate residues with a peak intensity to longer wavelengths. Thus, the major effect of phenoxyl ionization is to quench the tyrosine fluorescence. When 304 nm intensity is chosen, the quenching is total. When three tyrosine residues were ionized, there was 90% quenching of the tyrosine fluorescence, and from the dependence of the fluorescence intensity on the fractional ionization, Cowgill showed that there was no transfer between residues, and only a small amount of transfer to the exposed groups from buried tyrosine residues (Figure 34).

Iodination of tyrosine also leads to a drastic reduction in its fluorescence yield.[182] The absorption spectra of both the mono- and di-iodoty-

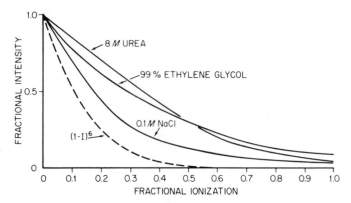

Figure 34. Fluorescence titration of ribonuclease. Temperature, 298°K; concentration, $E_{280\ nm}^{10\ mm} = 0.3$; excitation, 282 nm (isosbestos); observation, 305 nm. Water, 8 M urea, 0.1 M NaCl; 99% ethylene glycol : water, 0.001 M NaCl; water, 0.1 M NaCl. The broken line is a theoretical line for 100% transfer between six residues within a molecule.

rosines are altered. Both are red-shifted from that of tyrosine.[183] Perlmann et al.[184] have shown theoretically that iodotyrosine could act as a good trap for tyrosine fluorescence. Adopting a similar approach to that used for ionization, Cowgill[182] iodinated RNase to varying degrees. From the fluorescence intensity dependence on extent of iodination, he was able to demonstrate that there was only a small degree of transfer from the exposed groups to iodotyrosine. When three tyrosine residues in RNase had been iodinated, there was no residual fluorescence from the buried residues. Therefore, Cowgill suggested that the buried groups did not contribute to the fluorescence.

A more distinctive proof that the fluorescence from RNase was largely due to emission by exposed residues was provided by solvent perturbation studies. Organic solvents enhance the quantum yield of tyrosine compounds in aqueous mixtures.[185] When 40% dioxane was added to RNase, the fluorescence was enhanced by an amount equal to that exhibited by a monomer[186] (Figure 35). Absorption and circular dichroic spectral measurements by Bigelow and Krinitsky[187] had previously shown that 40% dioxane did not change the conformation of RNase, but affected only the absorption proerties of the three exposed groups. Hence, from the fluorescent enhancement, Cowgill concluded that 90% of the fluorescence from RNase originated within the exposed tyrosyl reidues. Further proof came from acetylation studies. Cowgill[188] found that acetylation of the three exposed tyrosines led to a total quenching of the fluorescence from RNase. Riordan and Vallee[189] had previously shown that the O-acetyl ester of

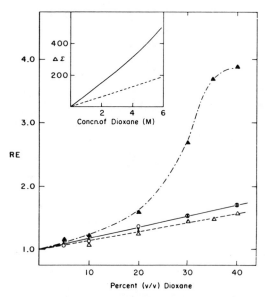

Figure 35. Pertubation of protein fluorescence by neutral organic solvents. The effects of *p*-dioxane on the fluorescence of ribonuclease and on model compounds are compared. Solvent, water : dioxane mixtures, 0.01 M tris at pH 8.0, and 0.1 M NaCl. (×) 4 μM *p*-cresol; (◯) 6 μM anisole; (△) 0.2 mg/ml ribonuclease A; (▲) 0.2 mg/ml ribonuclease S. The insert, taken from Bigelow and Krinitsky, is of the molar difference in extinction at 278 nm for (———) tyrosine and (– – – –) ribonuclese A at various water : dioxane compositions. (After Cowgill.[186])

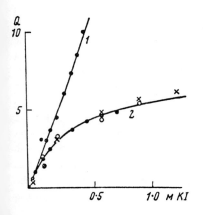

Figure 36. Stern–Volmer analysis of the iodide quenching of ribonuclease fluorescence. (1) Tyrosine; (2) ribonuclease. The solid line in (2) is a theoretical line, where a fraction 0.88 of protein tyrosines is quenched with a Stern–Volmer constant of 23.2 liter mol^{-1} and a fraction 0.12 is unaffected by iodide anions. Solvent, water, pH 7; temperature, 298°K; concentration, $E^{10\,mm}_{280\,nm} = 0.2$; excited at 280 nm. ◯ and ● measured at 305 nm; × measured at 325 nm. Abscissa: molarity of KI; ordinate: reciprocal of relative fluorescence intensity. (After Burstein.[190])

tyrosine absorbed to shorter wavelengths than tyrosine, and so the O-acetyl groups cannot act as energy traps, but only as energy donors.

Burstein[190] has provided even more convincing proof for exposed groups as the source of the fluorescence of RNase. Potassium iodide quenched 88% of the fluorescence of native RNase, leaving a residual 12% of the initial intensity which could not be quenched (Figure 36). Burstein used three as the number of exposed residues, and he concluded from his quenching data that the fluorescence quantum yield of an exposed residue had an average value of 0.025–0.03. The average for the buried tyrosines was 0.003–0.004. Any alteration in the number of exposed residues to fractional values such as those determined by Bello does not seriously affect the principal conclusion.

All the evidence points to the fluorescence of RNase preponderantly coming from its three exposed tyrosyl residues, with the buried residues contributing no more than 10% to the total fluorescence intensity.

To explain the small fluorescence yield of RNase, it is suggested that exposed tyrosines on the surface of the protein are quenched by collisons with adjacent amide side-chain residues, and by being vicinal to a disulfide linkage. Some transfer of energy occurs between the exposed residues, and this possibly enhances the quenching reactions. The interior tyrosines are extremely weak emitters of fluorescence, and are quenched by forming a carboxylate–phenoxyl complex or an amide carbonyl hydrogen bond to the phenoxyl, and this complex is quenched. Whether the details of the photophysical mechanism causing the fluorescence quenching are the same for each type of complex is not known. Proton transfers or electron transfers could account for all the observed quenching reactions, and this can be tested in principle by measuring the products of the quenching reaction. Alternatively, a photophysical perturbation could occur; the singlet internal conversion rate could be enhanced in the complex. Currently this latter mechanism is favored.

4.3. Phosphorescence

A further refinement to this model for the quenching of tyrosyl fluorescence in RNase is gained from low-temperature fluorescence and phosphorescence measurements. The fluorescence of RNase at 77°K is red-shifted from that of tyrosine by 5 nm, and has a peak at 298 nm, with a quantum yield of 0.20, one half the value of free tyrosine. The phosphorescence spectrum is significantly different. It is of the oxytocin type and not that of free tyrosine residues (Figure 37). There is a different spectral distribution from the protein compared to the tyrosine zwitterion spectrum: both the peak and onset wavelength were red-shifted. There is also a much

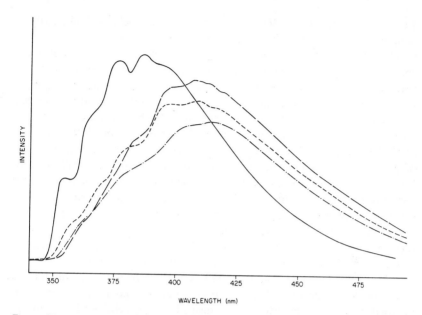

Figure 37. Phosphorescence of ribonuclease A. Solvent, EGW at pH 7; temperature, 77°K; excitation, 278 nm. (———) N-acetyl-L-tyrosine amide; (---------) oxytocin; (- - - -) ribonuclease A; (—·—) L-cystinyl-bis-L-tyrosine. (After Longworth.[200])

smaller phosphorescence quantum yield than tyrosine—approximately 0.05. Though the yield is reduced sixfold, the phosphorescence lifetime is halved to 1.40. The emission from RNase has all the characteristics previously found for phosphorescence from a tyrosyl residue that is adjacent to a disulfide linkage. The only tyrosyl residue at the surface of the protein and next to a disulfide link is Tyr-73; it is, in fact, next to two disulfide links. The other tyrosine residue (Tyr-25) next to a disulfide is involved in a carboxylate complex, and since it will not fluoresce it probably does not phosphoresce. Perhaps circumstantial, but attractive, is the observation that the quantum yield of RNase at 77°K is approximately one half that of tyrosine. The surface quenching reactions do not involve permanent complex formation, but are believed to be collisional in nature, and they would, therefore, be predicted to be ineffective at 77°K. Hence, only the buried residues are still quenched at 77°K. We suggest that the majority of the luminescence of RNase is from residue Tyr-73. This residue is sensitized by transfer from Tyr-76 and Tyr-115, which lie in the same region of RNase and are within 1.0 nm of each other.

Wampler and Churchich[155] have informed me in a personal communication that they found that the phosphorescence of insulin was also of an oxytocin type. As for RNase, the fluorescence yield is significantly enhanced at 77°K. Since only one out of the four tyrosine residues in insulin is adjacent to a disulfide link, the result points to the phosphorescence coming entirely from this particular residue. The large fluorescence yield variance with temperature suggests that only exposed groups fluoresce. Energy transfer was also shown to tqke place within the hormone both by polarized fluorescence measurements and by base titration studies involving an intact hormone and its oxidized fragments A and B.[191] Wampler and Churchich also investigated phosphorescence from several fragments of insulin. A desalanine heptapeptide sequence from the terminus of the B chain of the hormone is obtained by a tryptic digestion. The heptapeptide contains a single tyrosine, Tyr-B26, and the phosphorescence from this fragment is identical with that of free tyrosine. The larger fragment produced by the digestion, termed the desoctapeptide, contains the tyrosine residue, Tyr-A19, that is adjacent to the disulfide link. The phosphorescence of the desoctapeptide had an oxytocin-like spectrum.

Another way to fragment insulin is to oxidize the disulfide links to form S-sulfonic groups. This reaction separates the hormone into two polypeptide chains which have been termed A and B chains. They can then be separated from each othr chromatographically or electrophoretically. Chain A has two tyrosines, one adjacent to a former disulfide link (Tyr-A19) and the other removed from this link (Tyr-A14). Chain B also has two tyrosines (B16 and B26), both remote from S-sulfocysteines. Chain A phosphoresces with an oxytocin-like spectrum, and chain B with a normal tyrosine phosphorescence. Hence one is tempted to conclude that the phosphorescence of insulin originates in Tyr-A19.

Solvent perturbation studies on insulin have shown that there are two exposed residues in the hormone,[192] though more recent work from the same group indicates that one of the exposed groups is partially buried.[193] Ayoma et al.[194] found that tyrosines A19 and B16 are reactive with cyanuric fluoride. Menendez et al.,[195] from their solvent. perturbation studies with the desoctapeptide and urea-denatured insulin, suggested that residue A19 is the partially blocked group. This is confirmed by iodination studies which show that A19 is blocked by the disulfide link.[196]

An atomic structure of insulin has been described by Adams et al.,[197] and the molecular model shows that A19 and A14 are internally H-bonded to Gly-A1 and Gln-A5; B16 tyrosine residue is exposed and is involved in the metal complexation together with His-B5 and Phe-B1; B26 is a buried tyrosine and forms an aromatic cage with two phenylalanines, B24 and B25.

A conclusion which can be drawn from these studies that fits with the luminescence data is that B16, though not involved in any direct chemical reaction which induces quenching, may be transferring its energy to A19, the partially exposed tyrosine, because of a lower singlet level energy. Tyrosine-A19 would then become the major source of fluorescence.

Of special interest is the finding that Tyr-B26 is adjacent to two phenylalanines in the sequence, Phe-B24 and Phe-B25. In excitation spectral studies on insulin, Teale found that there were fine-structure components in the excitation of tyrosine fluorescence. These were clearly identifiable with those in the phenylalanine absorption spectrum. Teale concluded that phenylalanine resdues in insulin were transferring electronic energy to tyrosine. Perhaps it was this pair of Phe residues which were involved in the transfer to B26 tyrosine. However, this would require Tyr-B26 to transfer its energy to A16 also.

Another experiment which can be used to support the suggestion that emission originates from tyrosines which are next to disulfide links in both native insulin and RNase is the observation of the luminescence from the denatured proteins. In strong alkali, when all the tyrosine residues have become ionized, there is a relatively weak fluorescence by the tyrosinate residue, but there is a very strong phosphorescence. The emission spectrum and quantum yields are approximately the same as those obtained with free tyrosine, and there is no indication of any unusual features. At these pH values, the polymers are undoubtedly in an unordered state, and there is little likelihood of an interaction of the excited singlet state with the disulfide link. Urea or guanidinium denaturation restores the low-temperature luminescence to within 80% of the monomer values at 77°K. Reduction of the disulfide links leads to luminescence yield, lifetime, and Stokes' shift identical to those of tyrosine.

4.4. Fluorescence Lifetime

Blumberg et al.[198] determined the fluorescent lifetime of insulin (1.4 nsec) and RNase (1.9 nsec). Since the quantum yields are known, the natural lifetimes can be calculated to be 35 nsec for insulin and 95 nsec for RNase. Tyrosine is known to have a natural lifetime of 16 nsec, which is considerably less than that observed from these two proteins. To account for the long natural lifetime observed, it is necessary to suggest that not all the absorption leads to an emission by a given residue. Because of the uncertainty with respect to yield and lifetime, the accuracy of the natural lifetime is of the order of 30%. For insulin, this requires that two of the residues not contribute to the emission; the two others contribute with an average yield of 0.08. For RNase, if one tyrosine emitted and the remaining five were nonemitting, then the yield of the emitting tyrosine would be 0.61. The observed long natural lifetimes for the proteins are therefore consistent with the previous discussion and confirm that not all the tyrosine residues of proteins are involved in contributing to the easured fluorescence. Extensive internal conversion must occur in many of the residues.

4.5. Phosphorescence Lifetime

The lifetime of tyrosine zwitterion is not accurately established; reliable values range from 2.5 to 2.8 sec for the neutral form and from 1.3 to 1.5 sec for the ionized species. We find values of 2.90 ± 0.02 for the zwitterion, and 1.42 ± 0.02 for the anion.[199] Incorporation into a polypeptide does not significantly alter these values. For example, the desalanine heptapcptide from insulin was found to have a decay constant of 2.1, while reduced carboxymethylated RNase A has a lifetime of 2.6 sec.

The lifetime found for oxytocin is 1.7 sec,[200] while insulin and RNase A have lifetimes of 1.4 sec.[155] Reduction of the disulfide linkages of insulin and RNase A, or persulfate oxidation to separate the insulin chains, increases the lifetime to 2.0 sec, and carboxymethylation of RNase A restores the lifetime to the value level previously found for tyrosine zwitterion.[155]

A comparable situation is found for a model system which mimics tyrosine phosphorescence from RNase of insulin. The disulfide dimer, cystinyl-bis-tyrosine, has a lifetime of 1.7 sec,[200] the reduced dipeptide cysteinyl tyrosine has one of 2.6 sec.[155] Thus, the shortened lifetime is associated with tyrosine residues adjacent to disulfide links.

Phosphorescence lifetime studies have not been performed on the high quantum yield class A proteins. However, the lifetime of phosphorescence from tyrosine residues in class B proteins, which are clearly of a tyrosine zwitterion type and not of the oxytocin type, have values from 2.2 to 2.5 sec.[199] Though there are tyrosine components with a low phosphorescence yield in class B protein phosphorescence spectra which have oxytocin-like lifetime values (1.4–1.7 sec), their phosphorescence spectra are very difficult to study because of the small yield. Incorporation of tyrosine into a protein appears to shorten the lifetime 10–15%; being adjacent to a disulfide shortens the value by at least 50%.

4.6. Temperature-Induced Quenching

A thermally activated quenching of class A protein fluorescence has been found, and RNase has been the protein most thoroughly investigated.[201, 102] As the temperature of glycol : water glass solution is increased above the second glass trasition (180°K)—where the glass softens—the fluorescence maximum shifts to longer wavelengths, and the quantum yield decreases in a progressive manner. At 250°K, the fluorescence has reached a limiting value for the Stokes' shift, and further increases lead only to reductions in the quantum yield. It is not possible to fit the temperature dependency of the quantum yield with a single value for an activation energy. Above ambient temperature, Galley and Edelman[201] found the yield to decrease with increasing temperature until the melting temperature is reached, when an enhancement in the yield occurs, though further heating produces a progressive decrease in yield. Later, Galley and Edelman[202] studied the temperature dependence of RNase fluorescence as a function of pH and found that the acid denaturation pH was shifted to more neutral pH values as the temperature was decreased (Figure 38).

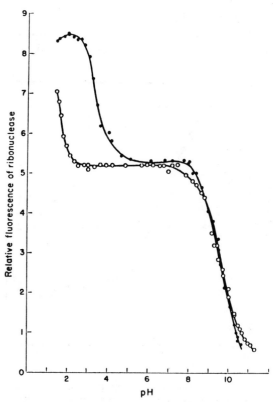

Figure 38. The influence of pH on the temperature dependency of the fluorescence of ribonuclease A. Solvent, water; concentration, 0.2 mg/ml; excitation, 268 nm; observation, 303 nm. (○) 299°K; (●) 327°K.

4.7. Acid Denaturation

Few fluorometric studies have been performed with class A proteins. However, insulin,[156] steroid isomerase,[161] and ovomucoid,[158] like RNase, have all been shown to increase their quantum yield when acid denaturation occurs (Figure 39). Esterification of the carboxyl groups of ovomucoid eliminated the acid-induced enhancement in fluorescence, and the protein quantum yield maintained the acid-denatured yield at all pH values until the alkaline titration of the phenoxyls occurred. These results are taken to show that, in the native protein, the buried tyrosines are associated with carboxylate residues in a complex which also quenches the fluorescence. The steroid isomerase is very unusual in having an exceptionally large value for the quantum yield of fluorescence (0.14), and there is no satisfactory explanation available for this.

Absorption spectral studies on RNase A and insulin[203] have shown that the buried tyrosine residues absorb at longer wavelengths than the exposed residues. The excitation spectra of RNase A and insulin were

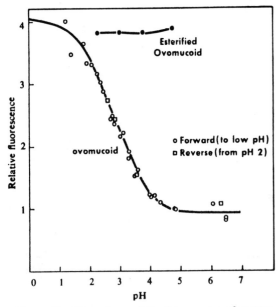

Figure 39. Dependence of the fluorescence of ovomucoid and esterified ovomucoid on solution pH. Solvent, water, 0.25 M KCl; temperature, 298°K; concentration, 3.3×10^{-5} M; excitation, 278 nm; observation, 305 nm. The fluorescence intensity is reported as a fraction of the value at pH 7. (After Donovan.[158])

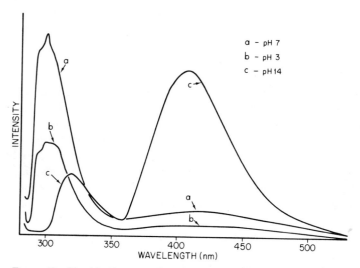

Figure 40. Total luminescence from denatured ribonuclease A. Solvent, EGW; temperature, 77°K; concentration, 1.0 mg/ml; excitation, 280 280 nm. (a) *p*H 7; (b) *p*H 3, 0.1 M NaOAc; (c) *p*H 14, 0.1 M NaOH.

found to be identical with the absorption spectrum of the monomer.[153] Though Teale did not interpret the excitation spectra of class A protein in the same fashion as class B, it can be concluded that there is a large decrease in fluorescence quantum yield on the long wavelength edge of the excitation spectrum of native proteins. This is not found with the denatured proteins; the quantum yield shows no wavelength dependency.

Small changes in the wavelength of the fine-structure peak at approximately 285 nm can be noted in excitation spectra at 77°K when natural denatured forms are compared. The denatured form lies to the blue by up to 1 nm. The fluorescence spectra at 298°K also reflect small shifts in the peak wavelength of 0.5 nm. The effect becomes more pronounced at 77°K. The denatured RNase has an emission maximum at 295 nm like the monomer, the native protein has a larger Stokes' shift, and the peak wavelength is at 298 nm at 77°K[204] (Figure 40).

4.8. Muscle Proteins

Two proteins of rabbit muscle, paramyosin and tropomyosin, are quite distinctive class A proteins. They have moderately large values for the quantum yield of fluorescence; Cowgill[162] reports a yield of 0.08 for paramyosin and 0.055 for tropomyosin. When both proteins are denatured by acid, their yields rise to 0.14 at *p*H 2. Unlike insulin or RNase, these two muscle proteins are believed to have a large helical content; a value of 90% has been suggested, even at acid pH values.[206] Because of the large yield increase on acidification of the proteins, which was not accompanied by any conformation change, Cowgill suggested that the acid-side-chains of Asp and Glu residues are responsible for quenching tyrosines at neutral pH. This was supported by the observation that heat-denatured proteins had lower yields, comparable to synthetic polymers of tyrosine, where amide groups are believed to be a major contributor to the quenching (Figure 41). The more convincing evidence for quenching by carboxylate side-chains of the surface tyrosine residues was provided by blocking the reaction by substituting the carboxylate group. Esterification of the carboxylate groups increased the yield of fluorescence. The same value was reached when the carboxylate groups were titrated to discharge all the anions. Another feature of the tyrosine residues in these helical proteins was the lack of extensive energy transfer. Cowgill found no evidence for transfer from iodination studies, though a small degree of transfer was indicated from the alkaline fluorescence titration studies. No polarization studies have been performed. Cowgill suggests that these two muscle proteins represent the behavior of isolated surface tyrosine residues, where the only quenching mechanism is a collisional interaction with a neighboring carboxylate side-chain. No backbone amide group quenching occurs until the proteins are denatured.

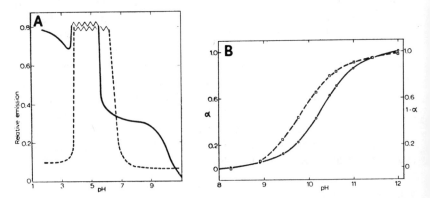

Figure 41. A. Fluorometric titration of tropomyosin. Solvent, water, 0.2 M KCl, 0.001 M imidazole; temperature, 298°K; concentration, 0.4 mg/ml; excitation, 275 nm; observation, 305 nm. (———) Fluorescence intensity; (- - - -) scatter intensity at 275 nm. B. Spectrophotometric and fluorometric titration of tropomyosin. Solvent, water, 0.2 M KCl, 0.005 tris; temperature, 298°K; concentration, 1.0 mg/ml. Solid line is absorbancy at 305 nm; dotted line is the fluorescence intensity at 305 nm on excitation at 281 nm (isosbestos). (After Cowgill.[162])

4.9. Summary

Mechanisms for quenching of tyrosine residues which have been discovered to play a role in protein fluorescence are:

(a) Carboxylate side-chain collisional quenching.
(b) Backbone amide group quenching (in random coils).
(c) Vicinal disulfide links.
(d) Carboxylate complex formation with buried residues.
(e) Amide complex formation with buried residues.

The weight of evidence points toward the fluorescence of tyrosine residues in proteins being associated with surface tyrosyl residues, and phosphorescence with residues that are adjacent to disulfide linkages. Though situations have been discovered where quenching can be explained by reference to model systems, it must be emphasized that the detailed mechanism of the quenching is still not established. Two conflicting photo-

Table 2. **Luminescence Properties of Class A Proteins**

	Protein	
	Bovine insulin	Bovine ribonuclease A
No. of amino acids		
Tyrosine	4	6
Phenylalanine	3	3
Cysteine	2	4
Molar mass (kg mol^{-1})	5.733	13.683
Molar extinction (m^2 mol^{-1})	600	960
No. of exposed tyrosines	1	3
Absorption maximum (nm)	277	277.5
Emission maximum (nm)	304	304
Fluorescence (298°K)		
Stokes' shift (μm^{-1})	0.321	0.311
Quantum yield	0.04	0.02
Lifetime (nsec)	1.4	1.9
Natural lifetime (nsec)	35.0	95.0
Fluorescence (77°K)		
Emission maximum	296	298
Yield	0.16	0.20
Phosphorescence (77°K)		
Onset (nm)	353	355
Maximum (nm)	405	408
Yield	0.05	0.05
Lifetime (sec)	1.4	1.4

chemical proposals have been made: (a) a proton ejection and (b) electron ejection. The observation of a suitable H-bonded complex would appear to support a proton ejection mechanism but does not in itself preclude the possibility that electron ejection causes the quenching. Until the products of a quenching reaction have been determined, it is not possible to distinguish between these two mechanisms and the third possible source of quenching, a photophysical process—enhanced singlet internal emission. In this process, the large Arrhenius activation energy to the quenching process is consistent with a photochemical process rather than a photophysical phenomenon. However, with the observation of heterogeneity in contribution to the emission, buried groups are almost certainly quenched because of an enhanced internal conversion. Exposed groups may be subject to both enhanced internal conversions and photochemical reactions. Table 2 summarizes the luminescence characteristics of ribonuclease A and insulin.

5. LUMINESCENCE OF PROTEINS—CLASS B PROTEINS

5.1. Introduction

Proteins have been given a term, class B, for spectroscopic reasons when they possess tryptophan.[206] Proteins which possess tryptophan also possess both phenylalanine and tyrosine residues. The overwhelming contributors to the emission and often the major cause of ultraviolet absorption of class B proteins are the tryptophanyl residues[206] (Figure 42). A tyrosyl contribution to the emission can be found with many globular proteins. It rarely comprises more than 10% of the emission intensity.[207] An exception is subtilisin carlsberg from *Bacillus subtilis*[208]; here the predominant emitter is tyrosine (Figure 43). The other extreme is found with azurin[209] and ribonuclease T_1[210]; no tyrosine contribution could be discovered in their emission spectra. It has not been possible to detect even a small tyrosine contribution to the fluorescence of hen egg avidin or hen egg lysozyme[208] though a weak tyrosine phosphorescence can be detected below 390 nm. Since these proteins have a large tryptophan-to-tyrosine ratio (20:4 and 6:3, respectively), this is predominantly a result of the small tyrosine fractional absorption, rather than a true lack of emission from tyrosine.

Each protein must really be treated individually since currently there is no effective means of systematizing the emission properties. However, for convenience, an attempt will be made to classify the tryptophanyl residues and their fluorescence into three groups which are based on the local en-

Luminescence of Polypeptides and Proteins 397

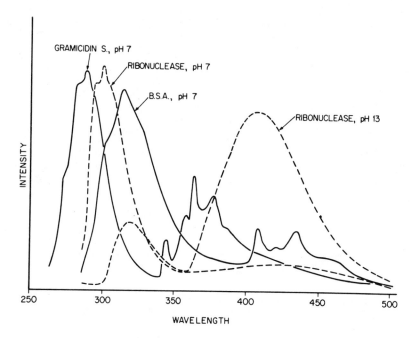

Figure 42. Protein luminescence. Solvent, EGW at pH 7; temperature, 77°K; excitation, 260 nm.

vironment—exposed, buried, and partially exposed. This classification has previously been developed from other excited-state chemical studies which relied on the solvent dependency of the protein's absorption.[211] Group-specific chemical modification has substantiated the presence of differently reacting residues in many proteins, and broad agreement is invariably found between chemical reactivity studies and excited-state investigations.[212]

It is necessary to preface any discussion of the experimental studies of class B protein fluorescence with a caveat. The quantitative data which are available are not usually published in a form in which it is easy to compare different studies. Since most workers use similar instruments, there is of necessity a degree of agreement on many occasions, but this is often superficial.

Few of the fluorescence spectra reported have been corrected for the technical artifacts introduced by the measuring instrument. There have been three studies which have provided corrected fluorescence spectra on the same protein.[206, 213, 214] Though there is a broad agreement between the studies, considerable difference still remains. For example, with human serum albumin Teale gives the peak wavelength as 339 nm, Chen as 342 nm, and Vladimirov and Ziminia report 338 nm. Closer inspection shows

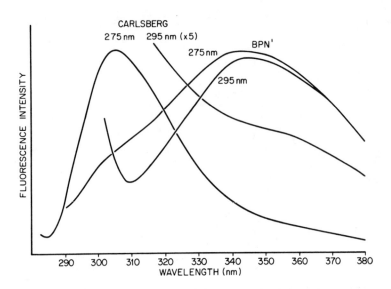

Figure 43. Fluorescence of subtilisin carlsberg and BPN'. Solvent, water, pH 7; temperature, 298°K; concentration, $E_{280\,nm}^{10\,mm} = 0.5$. Both samples excited at 275 and 295 nm. Fluorescence from BPN' (right) normalized at 370 nm; fluorescence from carlsberg (left); the sensitivity for 295 nm excitation is increased fivefold.

that the tyrosine contribution to the emission is clearly present in the spectra but the amount is subject to large relative variation. Vladimirov and Ziminia show an approximately equal contribution at the respective peak wavelengths. Chen's spectra indicate tyrosine contribution is approximately 30% of the tryptophan intensity at the respective peak intensities, whereas Teale finds negligible tyrosine contribution. Perhaps the differences are due in part to the sample preparations used. Bovine serum albumin, though the tyrosine contribution is not as easily measured, also shows similar variations in relative amounts of tyrosine to tryptophan when the spectra given by the three groups are compared; the variation in peak wavelength between workers is comparable to that reported for the human protein. It is advisable at present not to regard the reported spectra as absolute spectra but rather to treat them as only semiquantitative data. Chen has emphasized that since most workers choose to work with 300 nm blazed gratings when measuring protein fluorescence, the monochromator transmission and photocathode response combine to give an instrument response which is constant within 20% from 280 to 400 nm[215]—this is the spectral region of immediate concern.

A collation of the fluorescence peak wavelength reported by several workers for the tyrosine zwitterion and other tyrosyl derivatives shows that the peak wavelength is agreed to within 2 nm, with a median of 304 nm. A comparable catalogue for tryptophan is given in Table 3. By tradition, aqueous solutions of the zwitterion of tryptophan at 25°C (298°K) are used as the standard fluorescence reference whenever this is feasible. There does not appear to be a major effect on the peak wavelength value of substituting either the amino or carboxylate groups, though the quantum yield is dependent on the chemical nature of substituents. Values for the three ionic forms of tryptophan in aqueous media at 25°C, taken from Vladimirov and Li Chin-kuo,[216] are representative, and I will regard these as provisional reference values for the purposes of this chapter:

NH_3^+ CHR COOH 347 nm, $\Phi = 0.09$

NH_3^+ CHR COO$^-$ 353 nm, $\Phi = 0.20$

NH_2 CHR COO$^-$ 360 nm, $\Phi = 0.51$

Until spectra are reported in spectral quantal yields at different wavelengths or, better, different wavenumber values, not too much emphasis should be placed on particular wavelength values. For convenience of other investigators, an internal reference standard should always be given, and

Table 3. Emission Maxima

Wavelength of maximum (nm)	References
	Zwitterionic tryptophan
(a) 348	F. W. J. Teale and G. Weber, *Biochem. J.* **65**, 476 (1957).
(b) 350	J. A. Galley and G. M. Edelman, *Biochim. Biophys. Acta* **60**, 499 (1962).
	M. Shinitzky and R. Goldman, *Europ. J. Biochem.* **3**, 139 (1967).
	M. J. Kronman, *Biochim. Biophys. Acta* **133**, 19 (1967).
(c) 352	J. W. Bridges and R. T. Williams, *Biochem. J.* **7**, 2488 (1968).
(d) 353	Y. A. Vladimirov and E. A. Burshtein, *Biofizika* **5**, 385 (1960).
(e) 355	I. Weinryb and R. F. Steiner, *Biochemistry* **9**, 135 (1970).
	R. W. Cowgill, *Arch. Biochem. Biophys.* **100**, 36 (1963).
	P. Rosen and G. M. Edelman, *Rev. Sci. Instr.* **36**, 809 (1965).
(f) 356	J. Eisinger and G. Navon, *J. Chem. Phys.* **50**, 2069 (1969).
	N-acetyl-L-tryptophan amide
(g) 354	H. Edelhoch, R. S. Bernstein, and M. Wilcheck, *J. Biol. Chem.* **243**, 5985 (1968).
(h) 355	J. W. Longworth, *Photochem. Photobiol.* **7**, 587 (1968).
	Glycyl-L-tryptophan
(i) 355	F. W. J. Teale, *Biochem. J.* **76**, 14 (1961).

many workers have recently chosen to report yields relative to tryptophan zwitterion directly.

As was mentioned earlier, the quantum yield of the tryptophan zwitterion is in dispute, so it is premature to be dogmatic about any absolute values for protein quantum yields and their accuracy. There have been few attempts to measure absolute quantum yields of proteins aside from the survey included in Teale's paper in 1960. Hence it is not possible to state values for the absolute yield of protein luminescence of individual proteins with any great confidence. When values are collected, it is clear that literature values are not in great conflict either, and the overall pattern is well established.

The low-temperature emission spectra and luminescence yields and their ratio have often been reported for tyrosine and tryptophan and their derivatives.[216-222] There appears to be broad agreement between workers on values, and these have been tabulated in Table 4; here ethanediol : water glasses have been used as the solvent of choice. The peak wavelengths and yields do not appear to be extremely dependent on the nature of the medium or on substitution at amino and carboxyl groups. So the values for the zwitterions can be taken as convenient reference values, and are equal to those found for the N-acetyl carboxamides. Another quantitative measurement frequently reported for proteins is the phosphorescence lifetime. The values obtained for the three amino acids by several workers are collected in Table 5. Though the differences may in part be due to temperature, solvent, and pH differences, the variation seems to be largely due to other causes. The variation in reported values for a given amino acid zwitterion

Table 4. Luminescence Properties of Tyrosine and Tryptophan[a]

	Tyrosine		Tryptophan
	pH 7	pH 12	pH 7
Fluorescence			
Maximum (nm)	294	320	312
Shoulders (nm)	291, 297	—	296, 320
Quantum yield	0.38, 0.47	0.14	0.64, 0.72
Phosphorescence			
Onset (nm)	348	360	398
Maxima (nm)	387	405	405, 432
Shoulders (nm)	358, 365, 375, 398	—	417, 456
Quantum yield	0.31, 0.53	0.86	0.16, 0.17
Luminescence			
Energy loss	0.31, 0.00	0.00	0.20, 0.11
Ratio of yields	0.82, 1.12	0.61	0.31, 0.15

[a] Solvent—ethane-1,2-diol : water at 77°K.

Table 5. Phosphorescence Lifetimes of the Aromatic Amino Acids

Reference	Phenylalanine	Tyrosine	Tyrosinate	Tryptophan
a	0.1	3.0	0.9	3.0
b	—	2.2	1.3	6.3
c	5.6	—	—	6.3
d	—	2.7	—	6.6
e	—	3.0	1.3	6.4
f	—	2.0	1.3	6.8
g	7.3	2.1	2.0	5.3
h	—	2.8	1.2	5.8
i	5.6	2.5	—	6.3
j	8.4	—	—	—
k	—	2.5	—	5.9
l	—	2.8	1.2	5.4
m	—	2.7	1.5	7.3
n	—	2.4	—	5.6
o	—	—	—	6.8
p	—	2.6	—	5.0
q	9.20	2.8	—	6.0
r	7.70	2.90	1.42	6.65

[a] P. Debye and J. O. Edwards, *Science 116*, 143 (1952) (we now know that the decays were from crystals).
[b] R. F. Steiner and R. Kolinski, *Biochemistry 7*, 1014 (1968).
[c] I. Weinryb and R. F. Steiner, *Biochemistry 7*, 2488.
[d] H. B. Steen, *Radiation Res. 38*, 260 (1969).
[e] C. Helene, M. Ptak, and R. Santus, *J. Chim. Phys. 65*, 160 (1968).
[f] S. L. Aksentsev, Y. A. Vladimirov, V. I. Olenev, and Y. Y. Fesenko, *Biofiziko 12*, 63 (1967).
[g] J. W. Longworth, *Biochem. J. 81*, 23 (1961).
[h] F. Bishai, E. Kuntz, and L. Augenstein, *Biochim. Biophys. Acta 140*, 381 (1967).
[i] L. Augenstein and J. Nag-Chaudhuri, *Nature 203*, 1145 (1964).
[j] M. W. Grimes, D. R. Graber, and A. Haug, *Biochem. Biophys. Res. Commun. 37*, 853 (1969).
[k] R. H. Steele and A. Szent-Gyorgyi, *Proc. Nat. Acad. Sci. 44*, 540 (1958).
[l] T. F. Truong, R. Bersohn, P. Brumer, C. K. Luk, and T. Tao, *J. Biol. Chem. 242*, 2979 (1967).
[m] Y. A. Chernitskii and I. D. Volotovski, *Biofiziko 12*, 624 (1967).
[n] B. Rabinovitch, *Arch. Biochem. Biophys. 124*, 258 (1968).
[o] T. Shiga, H. S. Mason, and C. Simo, *Biochemistry 5*, 1877 (1966).
[p] J. E. Churchich, *Biochim. Biophys. Acta 92*, 194 (1964).
[q] T. Shiga and L. H. Piette, *Photochem. Photobiol. 3*, 223 (1964).
[r] J. W. Longworth and S. S. Stevens, unpublished values obtained using a multichannel scaler, mean averaging 10 successive decays.

in the same solvent is as large as when different pH values or solvents are compared. Again, a convenient reference is the zwitterion in equivolume polyalcohol : water glasses. Lifetimes have been measured from both the decay of the phosphorescence and the decay of electron spin resonance amplitudes after cessation of illumination. Values for the three amino acid zwitterion lifetimes suggested as the true values recently measured in my laboratory over a 40 db intensity range using a multichannel analyzer are phenylalanine 7.70 sec, tyrosine 2.90 sec, tryptophan 6.65 sec, and ionized tyrosine (tyrosinate) 1.42 sec. The precision we quote is ± 0.05 sec; the accuracy is believed to be comparable.[223] The nature of the solvent may also change the lifetime value, though the evidence for this is not conclusive; e.g., Chernitskii and Volotovskii[224] used ethanol as a solvent and obtained much greater values than any other workers have reported from polyalcohol : water glass studies. Steiner and coworkers have found that there are only small changes in lifetime of the amino acids when the α-amino and carboxyl groups are substituted with different groups.[221]

The fluorescence lifetime is another quantitative parameter, and several proteins have had their decay of fluorescence measured.[225-229] The lifetime values of the zwitterions of tyrosine and tryptophan have been measured by several methods. The values found by different workers and by different methods are in excellent agreement, in marked contrast to the values of the corresponding quantum yields. The published data are collected together in Table 6; theoretical values derived from the oscillator strength and quantum yield are also given. Weinryb and Steiner,[221] Ricci,[221] and Badley and Teale[128] have measured lifetimes and quantum yields of many tryptophanyl and indole compounds in aqueous solution at 298°K. The ratio of these two values was found to be constant within 10%. The quantum yields ranged from 0.05 to 0.70. Since the fluorescence maxima, absorption maxima, and molar extinction coefficients were also constant within 10%, the values given for the absolute yields by Badley and Teale are probably accurate.

By tradition, the zwitterions have been taken to be the reference com-

Table 6. Fluorescence Lifetimes: Zwitterions of Aromatic Amino Acids[a]

	Natural lifetime quantum yield product	Radiative lifetime	
		Pulse method	Phase-shift method
Phenylalanine	8.0[b,c]	6.8[c]	6.4[i]
Tyrosine	2.5,[b] 7.5[d]	2.6,[e] 3.6[f]	3.2[i]
Tryptophan	8.0[d]	2.6,[e] 2.8,[f] 2.9,[g] 3.0,[h] 3.3,[k] 2.8[l]	2.5,[j] 3.0[i]

[a] Solvent—water at pH 7 and 298°K.
[b] G. Weber, *Light and Life*.
[c] E. Leroy, H. Lami, and G. Laustriat, *Photochem. Photobiol.* **13**, 411 (1971).
[d] J. Feitelson, *J. Phys. Chem.* **68**, 391 (1964).
[e] R. F. Chen, G. G. Vurek, and N. Alexander, *Science* **156**, 949 (1967).
[f] W. E. Blumberg, J. Eisinger, and G. Navon, *Biophys. J.* **8**, A106 (1968); M. Fayet and Ph. Wahl, *Biochem. Biophys. Acta* **229**, 102 (1971).
[g] I. Weinryb and R. F. Steiner, *Biochemistry* **9**, 135 (1970).
[h] W. B. DeLauder and P. Wahl, *Biochemistry* **9**, 2750 (1970).
[i] L. F. Gladchenko, M. Y. Kostko, L. G. Pikulik, and A. N. Sevchenko, *A.N. Dokl. Akad. Nauk Belorrusk. SSR* **9**, 647 (1965).
[j] R. A. Badley and F. W. J. Teale, *J. Mol. Biol.* **44**, 71 (1969).
[k] R. W. Ricci, *Photochem. Photobiol.* **12**, 67 (1970).
[l] E. P. Kirby and R. F. Steiner, *J. Phys. Chem.* **74**, 4480 (1970).

pound, though N-acetyl carboxamide derivatives are better choices. The latter compounds are not subject to a marked pH dependence in their fluorescence properties. The following quantities require better definition before accurate quantitative analysis of protein luminescence can be attempted: (a) The ambient fluorescence spectra are best reported as spectral quantal yield at a wavenumber. The band shape can be analyzed to provide information. (b) The fluorescence quantum yields require accurate definition at ambient temperatures. (c) Phosphorescence lifetimes have been subject to a large uncertainty among individual investigators.

5.2. Tyrosine Fluorescence

Though Teale[206] found that tyrosyl residues are capable of fluorescing when incorporated into a protein that lacks any tryptophan residues, the prescence of tryptophan in a protein was thought to prevent tyrosyl fluorescence. At least in the most favorable case studied by Teale, human serum albumin, in which there are 17 tyrosines and one tryptophan, no tyrosine component was noted in his paper in 1960. The lack of any transfer of electronic energy to tryptophan as measured by excitation spectra and polarization spectra showed that this could not cause the negligible tyrosyl fluorescence intensity. Clearly there was an inherent discrepancy between the studies with class A and those with class B proteins. More recently, subtilisin carlsberg has been found to emit strongly from tyrosine and weakly from tryptophan, but the related enzyme substilisin BPN' has a predominant tryptophan emission.[208] Phosphorescence measurements offered the most attractive means to search for any tyrosine emission from a class B protein. The phosphorescence maximum of tyrosine lies to shorter wavelengths than the onset wavelength of tryptophan emission. Also, tyrosine and tryptophan have different triplet lifetimes at 77°K.

In 1957, Steele and Szent-Gyorgyi[230] measured the decay times of tyrosine and tryptophan phosphorescence together with the lifetime of an emission from a protein obtained from ox lenses. All the phosphorescences studied were found to decay with a single exponential constant, and the lens-protein lifetime was equal to that found for tryptophan but different from that of tyrosine. Therefore, there was no evidence for a tyrosine component in the phosphorescence of the proteins. In their next paper, Steele and Szent-Gyorgyi[231] reported the spectrum for the phosphorescence of the three amino acids. They confirmed that the sequence for the triplet energy levels followed the same sequence found previously with singlets; the energy sequence was phenylalanine > tyrosine > tryptophan. A phosphorescence spectrum was determined from bovine serum albumin, obtained at a low resolution, and this spectrum had no evidence for a tyrosine emission component. Freed et al.[232] subsequently obtained high-resolution spectra of the emission from both class A and class B proteins. They too concluded that class A proteins do emit from their tyrosine residues, while class B emits only from their tryptophan residues; no discernible tyrosine emission was reported. These workers only summarized their observations and did not offer any detailed results. In 1960 both Fujimori[233] and Vladimirov and Litvin[234] presented phosphorescence spectra they too had obtained from bovine serum albumin. Again it was apparent that there was no tyrosine component to the spectrum. Vladimirov and Burstein[235] reached a similar conclusion when they measured the total emission spectra of both class A and class B proteins. In a study of the phosphorescence of indole and tryptophan, Freed and Salmre[236] had shown that there was a very sharp onset to the phosphorescence of indole compounds at 395 nm and no emission occurred at shorter wavelengths. Today, the tyrosine contribution to the phosphorescence spectrum at wavelengths shorter than 395 nm can be readily found in proteins so far investigated,[210, 217, 220] with rare exceptions such as RNase T_1,[210] (Figure 44). Why it was not observed by these early workers is not clearly known;

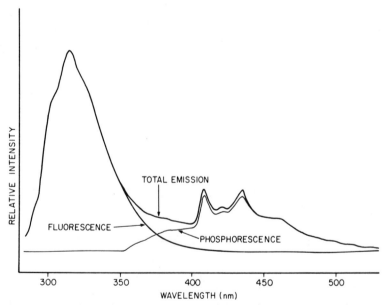

Figure 44. Fluorescence, luminescence, and phosphorescence of bovine serum albumin. Solvent, EGW at pH 3; temperature, 77°K; concentration, $E_{280\ nm}^{10\ mm}$ = 0.3; excited at 280 nm.

perhaps the background emission from the solvent obscured the tyrosine band.

The anomalous situation between class A and class B was cleared up by several workers almost simultaneously in 1961.[237-239] A tyrosine contribution to the fluorescence could be observed with many proteins, particularly from the serum albumins with their large Tyr/Trp ratio. Weber is usually credited with the first report, which was in the spring of 1961.[237] He developed and then applied a matrix analytic procedure to the emission spectra of proteins excited at different wavelengths. The procedure disclosed that there were two components in the fluorescence emission of serum albumin, including a major component clearly associated with emission from the tryptophan residues, but this was accompanied by a second small component with an intensity that peaked on the short wavelength side of the tryptophan emission. This was ovbiously due to emission from the tyrosine residues. Acid pH values, heat, organic solvent, or detergent denaturations led to a large enhancement of the tyrosyl contribution. There could be little doubt concerning the origin of the second component with the denatured proteins, which in certain cases dominated the protein spectrum when excited at 280 nm.[214] The short wavelength component was always absent from spectra excited at 295 nm. To obtain the spectrum of the tyrosine component, a difference spectrum between spectra excited at 295 nm and those excited at shorter wavelengths was used by all workers, and is the principal method of study adopted; however, an increasing application of matrix analytic procedures can be expected in the future.

Longworth[240] found in 1961 that tyrosine residues contributed to the phosphorescence spectrum obtained from class B proteins. There was an easily discerned tyrosine phosphorescence band with a maximum at shorter wavelengths than the onset wavelength of the fine-structured tryptophan band. The tyrosine component could be identified on the basis of its spectral position and distribution, its lifetime, and its excitation spectrum: all were similar to the equivalent parameters found with tyrosine. The tyrosine phosphorescence yield was enhanced on denaturation of most globular proteins (Figure 45). A tyrosine component is entirely absent from spectra excited at <295 nm, so the pure spectrum due to emission by the tryptophan residues can always be obtained. An analysis of the decay of intensity confirms this conclusion.[241] Excitation at >295 nm always gives a single exponential decay process.

A thorough discussion of the nature of the tyrosyl fluorescence contribution to serum albumin fluorescence has been given by Vladimirov and Zimina.[214] Two methods that have been used to demonstrate a tyrosine component are (1) the matrix analytic procedure or (2) a difference fluores-

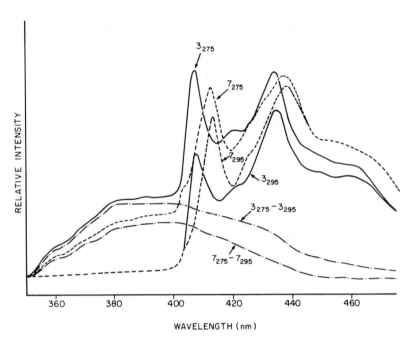

Figure 45. Phosphorescence of bovine serum albumin. Solvent, EGW; temperature, 77°K; concentration, $E_{280\,nm}^{10\,mm} = 0.3$; excited at 275 and 295 nm. (——) pH 3; (- - - -) pH 7; (—·—) difference between 295 and 275 nm spectra, normalized at 480 nm.

cence spectrum determination.[237] An example of the second method of measurement is illustrated in Figure 46. Two factors contribute to permitting a separate determination of tyrosine and tryptophan spectra from the spectra of a mixture. Tryptophan has appreciable absorption at longer wavelengths than does tyrosine, and it also has a greater Stokes' shift, though the wavelength of the absorption maxima are frequently equal. To indicate the quantitative parameters involved, an equimolar mixture of tyrosine and tryptophan is considered at a temperature of 298°K. The two wavelengths commonly used to excite fluorescence or phosphorescence of mixtures are 278 and 295 nm. At 278 nm, the molar extinction coefficient of tyrosine, in water, is 138 m² mol⁻¹; for tryptophan the value is 555 m² mol⁻¹; therefore, the fractional absorption ratio is 4:1 Trp : Tyr. At 295 nm, the tyrosine extinction is 3 m²mol⁻¹, tryptophan 220 m²mol⁻¹; hence the corresponding fractional absorption ratio at 295 nm is 73:1. The equivalent ratio of spectral quantal yields at 370 nm is 500:1; i.e., the fluorescence is effectively due only to tryptophan emission; at 304 nm, the ratio favors tyrosine and is 1:25. Nevertheless, at 304 nm, on excitation

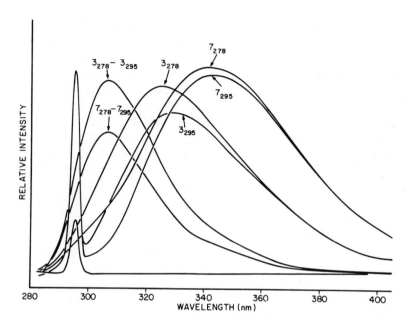

Figure 46. Tyrosine contribution to bovine serum albumin fluorescence. Solvent, water, pH 7 and 3; temperature, 298°K; concentration, $E_{280\,nm}^{10\,mm} = 0.3$. Spectra are excited at 275 and 295 nm for each pH. The difference spectrum, at tenfold increased expansion, is shown on the left.

at 295 nm, only 30% of the fluorescence intensity in the equimolar mixture would be caused by tyrosine emission. To a first approximation, there is negligible tyrosine fluorescence at all emission wavelengths on excitation at 295 nm or greater. Clearly then, there is a straightforward method to obtain the tyrosine contribution in a mixture of tyrosine and tryptophan. Two fluorescence spectra are first measured, spectrum I excited at 278 nm and spectrum II at 295 nm. Spectrum II is taken to constitute emission purely from tryptophan. This is then normalized to spectrum I at 370 nm, as tyrosine makes no significant contribution to spectrum I at this particular wavelength. The difference spectrum, spectrum I–(370 nm normalized) spectrum II, can therefore be associated entirely with tyrosine. An example is shown with the fluorescence of bovine serum albumin, analyzed in the above fashion, and clearly a tyrosine spectrum is obtained. To extend the treatment to proteins, it must be remembered that the extinction coefficients that prevail with the model compounds are only a moderate approximation to the values for residues with a protein structure—though the changes are not large on incorporation into a protein. The magnitude of the change in extinction ($\Delta\varepsilon$) at 278 nm is 9 for tyrosine and tryptophan at the maximum, while at 285 nm it is 1.3 for tyrosine and 34 for tryptophan. These values were obtained from the model systems studies of Herskovits and Sorenson.[242] The fractional absorption at 278 nm due to tryptophan is of course largely determined by the composition of the particular protein; values range from 0.2 to 0.95, but a representative value is 0.70 ± 0.05. The above model system analysis was based on the amino acids having equal quantum yields. This rarely applies to residues in proteins. A ratio for quantum yield of Trp to Tyr of 10:1 is quite commonly found. Therefore, the fractional tyrosine contribution at 304 nm, the wavelength where it reaches a maximum contribution, is still small. Measurement of the tyrosine contribution to protein fluorescence requires large signal-to-noise ratios and low scattered light, and usually can only be studied with laboratory-constructed instruments.

A tyrosine fluorescence has been observed from class B proteins when all the tryptophan residues have been removed by oxidation with either N-bromosuccinimide[238] or by a dye-sensitized (singlet oxygen) photo-oxidation. Both Chen[243] and Churchich[244] have found that, when the tryptophanyl fluorescence of a protein is quenched by a transfer to a dye bound to the protein, either a prosthetic group or a synthetic dye, a residual tyrosine fluorescence band remained—tyrosine does not transfer effectively to bound chromophores.

A tyrosine contribution can be easily observed in phosphorescence spectra.[240] Several features contribute to this: (a) There is no scattered light from the exciting beam to contend with; (b) there is a sharp onset to the phosphorescence of the tryptophan, so the tyrosine contribution can be measured completely uncontaminated by any tryptophan emission intensity. To obtain the contribution at all wavelengths of emission due to the tyrosine resdues in a protein phosphorescence spectrum, it is necessary to determine difference spectra. An example of the results which can be obtained with phosphorescence is shown in Figure 45.

5.3. Tyrosine Fluorescence and Phosphorescence Spectra

The fluorescence of the tyrosine residues in class B proteins obtained by the difference measurement methods has a spectral distribution identical to that found for tyrosine compounds and class A proteins.[214] The Stokes' shift does not appear to vary appreciably from protein to protein, though the peak wavelength appears to range from 304 to 306 nm.[210, 237-239] The spectrum is unaffected by the conformation of the proteins; when native or denatured conformers are compared, only the yield is influenced by the difference in conformation. A tyrosine phosphorescence spectrum has not been as frequently measured, though of the spectra reported, all are similar

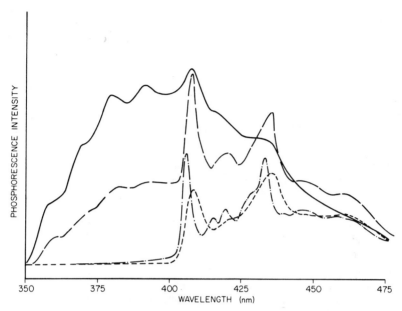

Figure 47. Phosphorescence of *Staphylococcus endonuclease* and ribonuclease T_1. Solvent, EGW at pH 7; temperature, 77°K; concentration, $E_{280\ nm}^{10\ mm} = 0.3$. Excitation, 278 nm: (———) Mixture of amino acids, 7 Tyr, 1 Trp; (- - - -) staphylococcus endonuclease; (—·—) ribonuclease T_1. Excitation, 295 nm: (--------) staphylococcus endonuclease.

to the monomer emission but there are some small indications of shifts and differences between proteins[210, 245] (Figure 47). A particularly clear tyrosine phosphorescence is observed from the endonuclease of *Staphylococcus aureus* and subtilisin carlsberg (Figure 48). A special feature of these enzymes is lack of any disulfide links in the protein. Bovine serum serum albumin has a tyrosyl phosphorescence which appears to be identical in distribution to the spectrum of tyrosine, though the maxima and onset are shifted by 8 nm.[245] This protein has 20 tyrosine residues and 18 disulfide links; there is no indication that an oxytocin-like tyrosine phosphorescence component is present, only that the maximum is at longer wavelengths (Figure 45).

The fluorescence of many proteins at alkaline pH values is largely from the ionized tyrosine residues.[246-248] There is a close similarity between tyrosinate[249, 250] and tryptophan fluorescence spectra, so a convincing distinction cannot be made on the basis of the spectrum; lifetime measurements are required. Phosphorescence offers a more effective procedure to distinguish tyrosinate from tryptophan. The tyrosinate is highly phos-

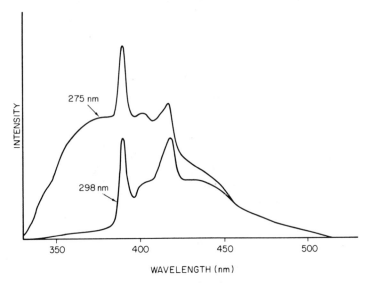

Figure 48. Tryptophan phosphorescence of subtilisin carlsberg. Solvent, EGW at pH 7; temperature, 77°K; concentration, $E_{280\ nm}^{10\ mm} = 0.5$. Excited at 275 and 298 nm.

phorescent, emits to shorter wavelengths than tryptophan, has no fine structure, and has a much reduced lifetime.[241, 151] The phosphorescence of class B proteins in alkaline solutions always has as its major contribution a tyrosinate emission.[241] This component is easily identified on the basis of lifetime, spectrum, and D' value for the ESR resonance. Nevertheless, the phosphorescence spectrum also contains a tryptophan component which is identified by the fine-structure peaks and the presence of a long-lifetime decaying component. An example of a protein phosphorescence at alkaline pH is bovine serum albumin at pH 12.

The phosphorescence and fluorescence spectra obtained when 305 nm light is used to excite emission of proteins at alkaline pH values are the same as when 280 nm light is used (Figure 49). A conclusion from this result is that the fluorescence is entirely due to emission from tyrosinate.[252] Such a conclusion had previously been derived from the alkaline pH fluorescence quenching at 298°K. The fluorometric and spectrophotometric titrations of tyrosine were correlated.[246–248] A second conclusion which can be stated is that the tryptophan triplet states are populated by a triplet sensitization from tyrosinate triplets. As triplet transfers proceed by electron exchange processes requiring orbital overlap,[253] the implication is that in bovine serum albumin there is a tyrosinate adjacent to a tryptophan. A partial N-terminal sequence is available for this protein.[254] The first 54

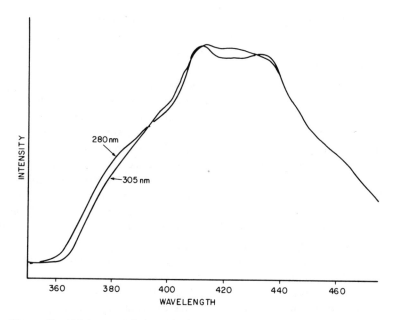

Figure 49. Triplet-to-triplet transfer in bovine serum albumin. Solvent, EGW at pH 13; temperature, 77°K; concentration, $E^{10\,mm}_{305\,nm} = 0.5$; excitation, 280 and 305 nm.

residues sequenced include the two tryptophanyl residues of the protein, but there are no tyrosine residues. Therefore, no tyrosine is adjacent to tryptophan in the primary sequence. There are a large number of disulfide linkages in bovine serum albumin (18), and the polypeptide chain must have tertiary structure even at pH 12 and cannot be treated as an extended polypeptide chain. Therefore, a tyrosine must be spatially adjacent to tryptophan as a result of a disulfide link, even though there is no disulfide link in this part of the primary sequence either.

5.4. Tyrosine Quantum Yield

The quantum yield of phenol in aqueous solutions at 77°K is 0.21,[255-257] and the zwitterion of tyrosine has a yield of 0.21.[255] Formation of an amide link, tyrosine amide (0.10),[256] or a peptide link, N-acetyl tyrosine (0.012),[256] or both, N-acetyltyrosinamide, (0.009)[258] leads to a major alteration in yield. Incorporation into dipeptides or polypeptides also causes significant quenching; yields of 0.08 ± 0.01 are found with dipeptides[256] and 0.02–0.05 with polymers.[259] Other quenching groups are the α-amino group,[257] lysine, and ammonium[259, 260] and glutamate,[259, 261] aspartate,[262] carboxylate, and their corresponding carboxamides (Gln, Asn).[263]

The quantum yield values for tyrosine residues in proteins at 298°K range from 0.02 to 0.05 with native proteins.[237-239] The yield is increased by up to threefold when the proteins are denatured, either by heat, pH, detergent, or organic solvents or by organic and inorganic denaturants. Denaturations which lead to the formation of highly helical structures produce even greater yield enhancements. Values up to 0.08–0.15 are observed on detergent[238] or 2-chloroethanol denaturations.

Binding of a substrate inhibitor can produce an additional quenching of tyrosine fluorescence in a native conformer. Staphylococcal endonuclease is an example of this behavior.[264] Solvent perturbation,[265] group-specific modification[266] and difference-fourier X-ray diffraction analysis[267] of single crystals of the endonuclease all indicated that the binding of the substrate inhibitor, pTp, involved tyrosine residues at the binding site. The inhibitor binding, though it quenches tyrosine, does not affect the yield of the tryptophan fluorescence.[264]

The peptide or amide quenching which affects surface tyrosine residues is suppressed in organic solvents. The extreme is, however, dimethylformamide or acetamide; they do not quench tyrosine in the absence of water.[268] The quantum yield of N-acetyl tyrosinamide is increased by the addition of organic cosolvents to water.[269] Cuatrecasas et al. found that 30% ethanol did not affect the conformation of the staphylococcal endonuclease as judged by the invariance of the CD, though this concentration was sufficient to perturb the absorption of the exposed tyrosine residues (7).[265] There was a significant enhancement of the tyrosine quantum yield of the enzyme in this solvent mixture. The quantum yield of the free enzyme and its complex with the inhibitor were doubled by 30% ethanol. Analogous to Cowgill's previous conclusion on RNase A, Cuatrecasas et al. suggest that the emission of the exposed tyrosyl residues must dominate the tyrosyl contribution to the endonuclease fluorescence at 298°K. The tryptophanyl absorption, emission yield, and Stokes' shift are unaffected by the organic solvent addition, indicating that the residue is located in the interior of the enzyme. Solvent perturbation studies of the absorption together with group-specific chemical modification, showed that three exposed tyrosine residues 85, 27, and 115 are buried by complexation of the endonuclease with pTp. These observations, together with a quenching of fluorescence, strongly suggest that specific complexes are formed involving the phenoxyl hydroxyl resdues in the conformation changes induced by inhibitor binding. Residues 91 and 93 are permanently buried in both complexed and uncomplexed enzymes, and residues 113 and 54 are permanently exposed. The single tryptophan at 140 is always buried. The molecular structure has been determined, and is completely consistent with the chemical and photophysical studies.

Though the reduced Stokes' shift and smaller difference in shift between tyrosine and tryptophan at 77°K frequently prevents any quantitative separation of the tyrosine from the tryptophan contribution to the total fluorescence at the low temperatures, on a semiquantitative basis it appears that, like class A proteins, the tyrosine yield is enhanced on cooling to 77°K; it is in the range of 0.05–0.20. Bovine serum albumin fluorescence at 77°K can be readily resolved into its tyrosine and tryptophan components. As previously noted for the class A proteins, the tyrosine Stokes' shift at 77°K is greater in the protein than in the native monomer but not in a denatured conformer. Several phosphorescence yields also have been reported, though not in great detail nor with absolute values.[217, 270] The yields reported are not comparable to the yield of free tyrosine phosphorescence, but they are significantly reduced. Denaturation of the protein leads to an enhancement in tyrosine phosphorescence yield of up to threefold. It appears that the relative quantum yield of tyrosine fluorescence and phosphorescence at 77°K in native enzymes is 0.5–0.2 of that of the free amino acid, and denaturation increases the yield, but the values do not equal that of isolated amino acid. Tyrosine residues are still quenched at 77°K,[241] unless the structure of the protein is totally removed by reducing all disulfide links, and by placing the polypeptide chain in denaturing solvents such as 6 M guanidinium chloride.

The mechanism of tyrosine quenching in proteins has previously been discussed when class A proteins were considered. The major reactions which have to be entertained for quenching a tyrosine residue in proteins are for exposed residues (a) a collisional quenching by a carboxylate or carboxamide side-chain; (b) a collisional quenching by an aqueous peptide link; (c) the presence of a vicinal disulfide link. The buried tyrosine residues are usually involved in complexes through their phenoxyl group with (d) carboxylate groups or (e) amide carbonyl of the peptide chain. These complexes are believed to be nonemissive at 298°K and perhaps this applies even at 77°K.

In the class B proteins there is also the additional quenching by a transfer to tryptophan residues, though this rarely exceeds more than 30% of the total quenching. The evidence from model system studies points toward the quenching of tyrosine proceeding through an enhanced internal conversion rather than a chemical mechanism such as the excited-state protolysis. The latter mechanism cannot be ignored, particularly for the buried tyrosine–carboxylate complexes of proteins. There is some evidence for the presence of an excited-state protolysis from the tyrosine emission spectrum of human and bovine serum albumin reported by Vladimirov and

Zimina.[214] To prove existence of the chemical quenching mechanism it is required to identify products formed. In the case of the protolysis, this would be to find an emission from tyrosinate. Tyrosinate is emissive in nonaqueous solvents with a large Stokes' shift, and the magnitude of the shift is highly solvent dependent, because there is an exciplex interaction with a polar solvent. Vladimirov and Zimina found that there was an additional peak to the tyrosine contribution located at 320 nm in the tyrosine fluorescence of native BSA. This peak was absent from the spectrum obtained from an acid-denatured protein. In the denatured conformer there are no buried residues, while in the native conformer BSA has only six of its 20 tyrosine residues exposed to the solvent. From an analysis of the absorption and emission spectra of tyrosine and tyrosinate, Vladimirov and Li Chin-kuo[216] showed that the excited state of tyrosine is more acidic than the ground state. It has a $pKa = 5.4$ in the excited singlet state and 10.1 in the ground state. It is possible for the 320 nm component in the native tyrosine spectrum to result from an emission by buried tyrosinate residues involved in a carboxylate complex. Weber's matrix analysis,[237] however, only disclosed two components in the fluorescence of native BSA. But this is still quite consistent with an excited-state protolysis mechanism, as Weber showed in the same paper for the β-naphthol model system. The method cannot distinguish between parent and daughter product formed by an excited state. However, the preponderant evidence is still in favor of an enhanced internal conversion accounting for the quenching rather than a chemical mechanism. The tyrosine fluorescence spectra of other proteins have never been found to have a second long-wavelength component.

5.5. Excitation Spectra of Tyrosine

An excitation spectrum of the fluorescence from protein tyrosine residues is not readily obtained because of the difficulty in differentiating between the tyrosine and tryptophan contributions for each of the exciting wavelengths used. This complication is absent in phosphorescence. There is a greater Stokes' shift to the onset wavelength of tryptophan phosphorescence compared to tyrosine. Longworth[240] relied on this feature, and used UV-transmitting glass filters to separate tyrosine from tryptophan phosphorescence. The UV filter passed light only between 340 and 385 nm; to isolate the tryptophan emission, a filter which passed light longer than 450 nm was used and the tyrosine contribution was negligible (Figure 50). For proteins with a significant tyrosine absorption, the emission wave-length can be selected by a monochromator, but for many proteins the greater light transmissions provided by color filters compared to a monochromator is an asset.

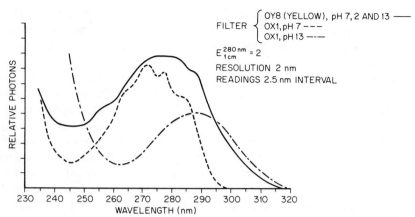

Figure 50. Phosphorescence excitation spectrum of tyrosine in bovine serum albumin. Phosphorescence is isolated by two filters: OX1, pass 240–380 nm; OY8, pass 350 nm. Solvent is an ice matrix at 77°K; concentration, $E_{1cn}^{280\,nm} = 2$; resolution 2 nm, readings 2.5 nm interval.

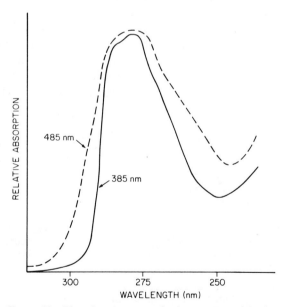

Figure 51. Phosphorescence excitation spectra of bovine serum albumin in polyalcohol glass. Solvent, EGW at pH 7; temperature, 77°K; concentration, $E_{280\,nm}^{10\,mm} = 0.3$. Observed at 385 ± 1.0 and 485 ± 1.0 nm.

The excitation spectrum of the tyrosine phosphorescence observed at 385 nm is quite distinct from a corresponding tryptophan excitation determined at 485 nm with the protein bovine serum albumin (Figure 51). Longworth also measured the tyrosine excitation spectrum in aqueous matrices at 77°K of both native and acid-denatured bovine serum albumin. The spectrum of the native form, after an arbitrary normalization at the peak amplitudes, absorbs to longer wavelengths than the same spectrum obtained from an acid-denatured protein. There is a shoulder to the long-wavelength edge of the native protein excitation, but the peak wavelength is the same in both conformer (Figure 52). This result was expected since acid denaturation was known to result in a blue-shifting of the tyrosyl absorption of the buried tyrosyl residues, and these residues amount to 70% of the total tyrosine absorption in the protein. A solvent perturbation of the excitation spectrum from the native protein tyrosyl phosphorescence occurs with 20% sucrose or with 50% ethanediol. There is a shift in the peak wavelength caused by the sucrose or glycol to longer wavelengths (Figure 53). This causes the long-wavelength shoulder seen in the aqueous spectrum to disappear.

The spectra of native and acid denatured BSA are not readily distinguished in the presence of either perturber, particular in ethanediol : water glasses[245] (see Figure 51).

Cassen and Kearns[271] have determined a tyrosine excitation spectrum with difference spectral measurements. When proteins have an appreciable absorption due to their tyrosine residues, there is an addition of a contribution from tyrosine absorption resulting from energy transfer to the excitation spectrum of tryptophan emission. Absorption by tyrosine

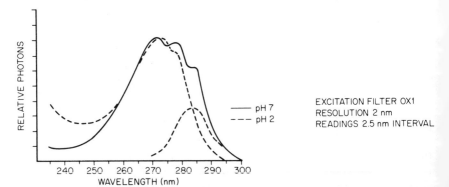

Figure 52. Acid denaturation difference spectrum of tyrosine in bovine serum albumin. Tyrosine phosphorescence excitation spectra are determined in ice matrices at 77°K at pH 7 and pH 2. The difference between the two spectra is shown on the right of the figure.

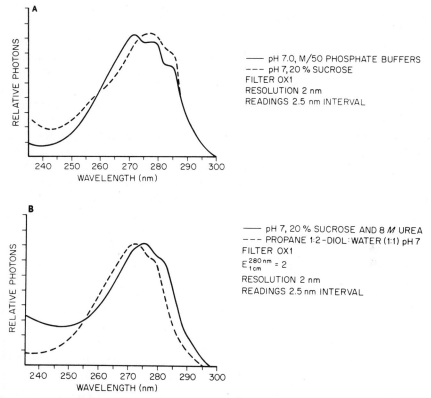

Figure 53. Solvent perturbation of tyrosine absorption in bovine serum albumin. Phosphorescence excitation spectra are determined in ice matrices for (A) native and (B) 8 M urea denatured BSA. The perturbation is produced by adding 20% sucrose to the ice matrix.

residues sensitizes the phosphorescence of the tryptophanyl residues. The contribution from tyrosine sensitization in the excitation spectrum of phosphorescence was determined by subtracting from the spectrum an excitation spectrum for pure tryptophan. The two spectra were normalized at 295 nm; tyrosine does not have appreciable absorption at this wavelength. Cassen and Kearns used the excitation spectrum of the tryptophan phosphorescence of chymotrypsin as the pure tryptophan spectrum. This spectrum is overwhelmingly due to tryptophan absorption; they neglected the contribution of the single Tyr/Trp pair in the sequence. They reported the excitation spectra of tyrosine in several proteins, and were able to conclude that the tyrosine absorption was red-shifted from the aqueous spectrum location, suggesting that there was a nonaqueous environment for protein tyrosine residues.

5.6. Tyrosine Phosphorescence Yield and Decay Time

Longworth[217, 223, 240] and Augenstein and Nag Chaudhuri[272] have measured the lifetime of tyrosine phosphorescence from several proteins. They quote a value of 2.1 sec for the decay constant. This is less than the monomer value, which is 2.85 sec. Because of an uncertainty in the absolute values for the reported lifetime values, it is not clear whether there are any individual variations between proteins, or whether there is any relationship of lifetime to either yield or to Stokes' shift. The general conclusion is that tyrosine phosphorescence decays with a single exponential value which is less than the monomer value.

Recent studies in our laboratory indicate the presence of individual variation among proteins, and among proteins of differing conformation. Proteins which lack disulfide links or denatured proteins have decay times in the range 2.1–2.7 sec. Other proteins with disulfide links have decays ranging from 1.4 to 1.7 sec, which are elongated by denaturation. We have not yet discovered evidence for a heterogeneity of the tyrosine decay process in any protein or particular ion forms of an individual protein.

The fractional intensity of tyrosine to tryptophan in several proteins has been reported and estimates have been made of the tyrosine phosphorescence yield. These are summarized in Table 7. Truong et al.[241] report similar values. They obtained their values from an analysis of the decay of the phosphorescence at a wavelength where there were both tyrosine and

Table 7. Tyrosine Phosphorescence Yields of Class B Proteins[a]

	pH	Fractional intensity	Phosphorescence yield (%)
Human serum	2	0.05	7.50
Albumin	7	0.25	3.80
	13	1.40	21.00
Bovine serum	2	0.35	7.90
Albumin	7	0.17	3.80
	13	1.00	22.00
Ovalbumin	2	0.60	3.00
	7	0.10	0.50
	13	1.20	6.00
Chymotrypsin	2	0.10	0.40
	7	0.05	0.20
	13	0.30	1.20
Lysozyme	1	0.15	0.60
	7	0.02	0.02
	13	0.40	3.50

[a] Aqueous medium, 77°K; excitation at 280 nm (3 nm bandwidth).

Table 8. Tyrosine Fractional Phosphorescence Intensity of Modified Bovine Serum Albumin

Neutral, pH 7.0	0.18	Sodium dodecylsulfate,	
Acidified, pH 2.0	0.35	100 moles, bound	0.35
50% propylene glycol, pH 7.0	0.42		
80% 2-chloroethanol, pH 7	0.81	Reduced BSA with sulfite + mercuric ion	0.25
8 M urea, pH 7	0.38	(Reduction of 17-sulfides)	
Methylated (complete)	0.36	do + 8 M urea	0.38

tryptophan components. Extrapolation back to initiation of the decay process gave the relative intensities. Denaturation of the proteins leads to an enhancement of the phosphorescence yield, and results obtained with several BSA systems are given in Table 8.

It is apparent that the phosphorescence yield of protein tyrosines is reduced from the individual amino acid yield value. This conclusion has also been reached by several workers who compared phosphorescence spectra of a protein with those of an equivalent mixture of amino acids.[228, 271] There does not appear to be any dependence of the yield on the Trp : Tyr ratio for the individual proteins.

McCarville and McGlynn[272] have studied the phosphorescence emission spectrum from proteins as the temperature is increased. There is a quenching as the temperature increases above 77°K. This is a result of an O_2 interaction with the triplet levels, rather than an intrinsic temperature coefficient of the phosphorescence yield; it sets in above the glass transition and is characterized by a nonexponential decay behavior in the quenching and by an Arrhenius activation characteristic of the solvent viscosity. Phosphorescence can be observed at room temperatures from proteins in sucrose glasses since, in the sucrose glass, the viscosity is sufficiently large so as to prevent extensive O_2 quenching. McCarville and McGlynn found that the tyrosine component of bovine serum albumin was totally quenched at a lower temperature than was necessary to totally quench the tryptophan component. The tyrosine contribution could be selectively removed. These results suggest an alternate quenching process for these tyrosine residues in their triplet state, possibly a chemical reaction.

The relationship of the fluorescence yield to the phosphorescence yield is not reliably known for tyrosine residues located in proteins. This is because of the difficulties in obtaining a value for the tyrosine fluorescence yield at 77°K. Since there is considerable variation from protein to protein in the phosphorescence yield, more reliable information is required on the low-temperature luminescence yields of tyrosine in class B proteins before the question of origin of the yield variation can be settled. In particular, the role energy transfer plays at 77°K remains to be determined for each protein. Internal conversion at the singlet level appears to be the predominant process, even in class B proteins. Nevertheless, transfer at the singlet and triplet levels must also be carefully investigated for each protein studied.

5.7. Electronic Energy Transfer

Three transfer processes take place within class B proteins. There are intertyrosine and intertryptophan transfers. However, the one which has captured most attention from photophysicists is the tyrosine-to-tryptophan transfer. No phenylalanine emission is found from proteins, either fluorescence or phosphorescence. Hence, there is a possibility of phenylalanine-to-tyrosine and phenylalanine-to-tryptophan transfers. The small fractional absorption of phenylalanine precludes any assessment of the process, and, to all intents, it is possible to totally disregard phenylalanine when considering emission and transfer of class B proteins, with rare exceptions.

Two methods for studying transfer have evolved: One is restricted to the tyrosine-to-tryptophan transfers; a second method is of general application to any transfer step, including the heterotransfer. Method 1 is a demonstration of a sensitization of the acceptor (tryptophan) by the donor absorption (tyrosine).[206] This has been previously discussed in detail for simple polypeptide hormones. The basis of the method is to determine the quantum yield of tryptophan, either absolute or relative, at different exciting wavelengths. Comparison with a standard absorption spectrum obtained from model systems, either synthetic or proteins with no transfer, then allows direct estimation of the extent of tyrosine sensitization (Figure 54). A general conclusion applicable to the majority of proteins is that the process occurs, though it is rarely a significant contributor to the total yield of tryptophan from proteins; an average contribution of 10–20% of the total intensity is found. Method 2 is measurement of polarization of emission to show that there is an anisotropy depolarization.[273] Since no two residues will in general have identical coordinates in the molecular structure, absorption by one residue and emission by another will cause a depolarization of the emission compared to absorption and emission by the same residue. Observation of depolarization of emission from a polymer compared to a monomer immediately implies the presence of electronic energy transfers within the polymer. The method applies equally to transfer between like and dissimilar residues, though it tends to be used most frequently to show the presence of homotransfers.

Kinetic analysis is also feasible, but so far has not been successfully applied. Method I would require detection of an initial nonexponential

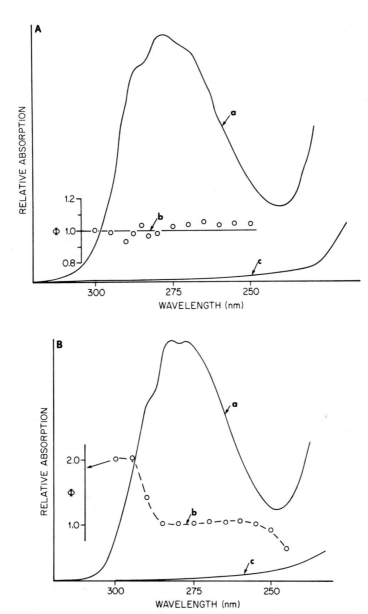

Figure 54. Excitation spectra and wavelength dependence of quantum yield of tryptophan and ribonuclease T_1: tyrosine-to-tryptophan energy transfer in a protein. Solvent, water, pH 7; temperature, 298°K; concentration, $E_{280\ nm}^{10\ mm} = 0.05$. *A*. *N*-acetyl-L-tryptophan amide. *B*. Ribonuclease T_1. (a) Excitation spectra; (b) relative quantum yield; (c) baseline.

decay at initiation of a decay. Method 2 would measure the time dependency of anisotropy. Both could be applied to singlet and triplet decays.

5.8. Fluorescence Polarization Spectra

Two types of polarization spectra are available for study, absorption (FAPS) and emission (FEPS). Most measurements reported to date have been FAPS studies: the fluorescence was typically isolated by a color filter and so averaged the emission polarization. The influence of emission polarization on the absorption polarization spectra is not well studied with proteins, and may account in part for discrepancies among workers.

Critical transfer distances had been calculated using Förster's theoretical equation from absorption and emission data. The distances were comparable to an average interresidue distance calculated from the composition and volume of a protein.[106, 274] Hence it was expected that transfer would readily occur within proteins. Weber[275] measured the critical distances directly from polarization measurements with indole and with phenol for self-transfers, and used sensitization studies to determine the distance for phenol-to-indole transfer in mixtures. Thin films of concentrated solutions of these compounds were used in the experiments, and the critical distances found were in excellent agreement with theoretically calculated critical distances previously obtained using only spectroscopic data. Vladimirov[276] studied the fluorescence from concentrated solutions of the amino acids, and he found there was a self-quenching with increased concentration with both tyrosine and tryptophan solutions. Since the quenching constant was independent of the exciting wavelength, Vladimirov concluded that transfer was occurring to a nonluminescent aggregate. Konev et al.[277] studied the concentration depolarization in tryptophan solutions and clearly demonstrated that transfer occurred between the tryptophans in these concentrated solutions. However, tyrosine-to-tryptophan and phenylalanine-to-tryptophan transfers could be demonstrated only with cocrystals.[239, 278] All these studies pointed toward transfer occurring within a protein, and these processes would significantly influence any luminescent behavior from a protein.

To get ahead slightly, the excitation spectral measurements of tryptophan emission performed by Teale and by Vladimirov and his colleagues did not disclose significant transfer from tyrosine to tryptophan in the proteins they studied, and cast doubts on the role energy transfer plays in affecting protein luminescence.

Weber[273] measured polarization absorption spectra (FAPS) of several class A proteins. He found there was appreciable anisotropy depolarization of the fluorescence. Hence intertyrosine transfers occurred in the proteins he studied (insulin, zein, and ribonuclease). Since the tyrosine quantum yield is small in these proteins, and is equally small from tyrosine residues in class B proteins, it appears likely that tyrosine-to-tryptophan transfer is not a predominant cause of the low yield of tyrosine in class B proteins, but at least transfers did occur between tyrosines at these low quantum yield values.

The fluorescence absorption polarization spectra of several class B proteins, both native and denatured, were also reported by Weber.[273] Weber was not able to adequately interpret the observed spectra by considering both transfers occurring from tyrosine to tryptophan and among tryptophan residues. The FAP spectrum of tryptophan itself exhibited fine structure, and the range of polarization values covered between 250 and 300 nm was large. The spectrum for the monomer has been interpreted on the basis of two perpendicular transitions absorbing in this region, each possessing a fine-structured absorption. In polar media, the fluorescence is believed to occur from an emission from a single electronic origin. However, in nonpolar media and perhaps in glass solutions at 77°K, there are suggestions that two electronic states are involved in the emission process. Thus the intrinsic behavior of tryptophan is in itself complex and not well understood. The additional complication of tryptophan(s) within a protein does not simplify the situation.

The fluorescence absorption polarization spectra of proteins were in broad outline similar to that of tryptophan but individual details differed—absolute values, wavelength of minima and maxima, etc. Three related proteins (human, bovine, and hen albumins) were investigated. They form a convenient series with one, two, and three tryptophans. It was hoped that perhaps this series would show progressively increasing intertryptophan transfer. In particular, we now know that the two tryptophan residues of the bovine protein are eight residues apart in the primary sequence, two turns of an α-helical structure. To the first approximation all three proteins were identical to each other and to tryptophan. The denatured proteins showed a greater difference between each other and the monomer. There is expected to be less transfer in the denatured conformers than in ths native structures. In fact, no tyrosine-to-tryptophan transfer has been demonstrated in denatured proteins. The polarization at 305 nm, which is the wavelength where there is a maximum polarization in tryptophan, was the same for all of the three albumins, and also equal in values to tryptophan polarization. Weber felt that it was premature to conclude that there was or was not transfer simply from FAP spectra, since two complicating factor had to be considered:

(a) Was there an environmental perturbation of the spectrum? This

is particularly relevant to the comparison of the FAP spectra of different conformers.
(b) Model system studies had shown that there was a wavelength dependency on concentration depolarization of anisotropy. At the longest wavelengths, that is, when excitation was at the edge of the lowest absorption band, there was less depolarization. At the limiting wavelength, 305 nm, no depolarization occurred. This effect is now termed edge repolarization, and it has been found in many model systems, particularly with the homopolymers poly-L-tyrosine[279] and poly-L-tryptophan.[280] The effect is a result of the presence of a lower-energy trap which is selectively excited.

Konev and Katibnikov[281] were able to confirm several of the FAP spectra published by Weber. However, unlike Weber, they interpreted their data on the basis of intertryptophan transfers. Model system studies performed by Konev and his colleagues indicated that there was no marked environmental dependence of the FAP spectrum of tryptophan.[283] Furthermore, Konev and Katibnikov found that a KEMF protein, keratin, had a marked anisotropy depolarization at all exciting wavelengths, including 305 nm. Konev and Katibnikov[284] subsequently used the anisotropy at 270 nm as a measure of the extent of transfer between tryptophans.

The true situation may lie somewhere between these two points of view. Some degree of resolution of the points of disagreement is possible. Firstly, the wavelength 270 nm is a poor choice. At this wavelength, tyrosine-to-tryptophan transfers have also to be included with intertryptophan transfer. Many proteins have appreciable fractional absorption due to tyrosine at 270 nm, though for all the examples given by Konev the fractional absorption by tyrosine was small. Secondly, Weber did report examples where there was an appreciable depolarization at 305 nm, in particular lactic dehydrogenose and several proteases. Since then, these results have been repeated by other workers. A final point which may be relevant is that the edge repolarization is not complete with poly-L-tryptophan; there is a depolarization at 305 nm. The wavelength 305 nm must be used to assess for intertryptophan transfer, and the influence of an edge repolarization on protein FAP spectra from a lower-energy tryptophan is not known reliably.

Konev et al.[284] offered more definitive evidence for intertryptophan transfer in 1965. The apoprotein of hemoglobin, globin, was compared with the holoprotein having the bound heme prosthetic group, hemoglobin Binding of heme to globin quenches the globin fluorescence (quantum yield $= 0.14$) approximately a hundredfold. Konev et al. report a yield of 0.00

for tryptophan fluorescence from hemoglobin. The effect of the reduced yield must be to drastically shorten the fluorescence lifetime of the tryptophan residues from the 3.0 nsec value for the globin. Now intertryptophan transfers will not be able to compete with deactivation by a transfer to the heme. The effect on the FAP spectrum is marked; the anisotropy values for the globin are increased and reach the values observed from tryptophan. Hence the lower polarization of globin fluorescence results from intertryptophan transfers.

Another means of demonstrating transfer was also provided by Konev et al.[285] They measured the emission polarization spectrum (FEPS) for a given exciting wavelength. These workers found that the polarization values were dependent on the wavelength of emission. In dilute solutions, the short wavelength edge was less polarized than the remainder of the band. In concentrated tryptophan solutions, all wavelengths were less polarized because of the transfer depolarization, but the short wavelength region was more polarized than the main band. This was explained by the presence of two components to the emission, one of which has a much shorter lifetime than the other. Chen[286] has measured the FEPS of tryptophan, and though he did not find the same magnitude of changes as those reported by Konev et al., there was qualitative agreement. Konev et al. found that the FEP spectra of chymotrypsin and keratin were similar to that obtained from concentrated tryptophan solutions. Hence they concluded that in these proteins there was transfer between tryptophan residues.

Bobrovich and Konev[287] found that there was a constant polarization for the BSA FEP spectrum. Since there was an appreciable tyrosine emission contribution at the shorter wavelengths, the Soviet workers drew the conclusion that the tyrosine component compensated for the change in the FEP spectrum of tryptophan. In the absence of a tyrosine component, the change would consist of a decreased polarization at the wavelengths where tyrosine emits. The tyrosine contribution in the absence of transfers would be polarized, hence the net result would be an increased polarization. Since a constant and small value was found, Konev et al. concluded that the tyrosine fluorescence was depolarized by intertyrosine transfer. Longworth[217] has performed similar measurements with BSA; a value $P = 0.15$ was found at 350 nm, and 0.27 at 308 nm for excitation at 280 nm. Chen has also studied the FEP spectrum of bovine serum albumin and had similar results. So, unlike Konev, these workers found the tyrosine region to be more polarized than the tryptophan region. Nevertheless, the value is still less than expected for tyrosine monomer (0.38). Hence all workers have suggested that intertyrosine transfers occur in bovine serum albumin. Chen also investigated both ovalbumin and lysozyme, and, like Konev, he reports wavelength dependent values, but again the magnitude of the change is noticeably less than the ones given by the Soviet workers. Clearly, there is need for more detailed studies on this point.

Weber emphasized the significance of the ratio of anisotropies measured at 305 and 270 nm. (In his paper, the ratio is given as polarization values for unpolarized exciting light rather than anisotropy values and so must be corrected.) The wavelength 270 nm was chosen because it is a broad maximum in the FAP spectrum, and 305 nm represented the maximum value attained. The difference in this ratio between tryptophan and proteins is taken to indicate tyrosine-to-tryptophan transfers. Weber did not consider this at the time because no sensitization data were then available to suggest the presence of this heterotransfer. Most globular proteins showed a different $A(305)/A(270)$ ratio from tryptophan in their native conformation. For example, the ratio with tryptophan was 1.4, and in bovine serum albumin it was 2.3. When denatured proteins were studied, the ratio became equal to that of the monomer measured in the same denaturing solvent. Hence there was negligible tyrosine-to-tryptophan transfer in denatured conformers of several proteins.

Badley and Teale[128] have considered the role of intertryptophan transfer in the enzyme pepsin. Several chromophores were attached to a substrate to give chromophoric substrate analogues. These bind specifically and stoichiometrically to the catalytic site of pepsin. The spectral overlap with tryptophan differed from dye to dye, and so different degrees of quenching were observed for the different substrate analogues. The fluorescence lifetime was quenched to a different extent for a given dye than was the quantum yield. Hence they conclude there must be several paths of transfer to the single dye molecule located at the active site from the array of protein tryptophans. This is reasonable since, of the several tryptophan residues within the protein, not all would be equidistant from the binding site. The fluorescence decay was always fitted by a single exponential value for each of the individual pepsin complexes and for pepsin. This showed that each of the tryptophan residues in the protein has an equal quantum yield and lifetime.

The fluorescence anisotropy of the pepsin–dye conjugates was measured for several exciting wavelengths, and the anisotropy values were extrapolated to a zero value for the tryptophan lifetime. This eliminated any rotational introduced depolarization. The residual depolarization of the protein compared to a monomer value could then only originate from interresidue transfers within the protein. The anisotropy values at wavelengths greater than 290 nm, where tyrosine transfer depolarization is negligible, were noticeably smaller than the monomer values. Badley and Teale, however, still maintained that it was not possible to conclude that intertryptophan transfers occur within pepsin. This is undoubtedly based on the observation of the exponential nature of the fluorescence decay from each conjugate. The cause of the reduced polarization of protein tryptophanyl residues was not explained by them, but we may presume that, like Weber, they were considering environmental effects which are very poorly understood.

It is perhaps premature to attempt to reconcile both the conflicting data and the opinions drawn from the data at this stage of analysis of protein FAP and FEB spectroscopy. Like the quantum yield of tryptophan, until a satisfactory explanation for the electronic nature of tryptophan and how it is modified by transfer is available, extension to the situation prevailing in proteins must be approached with caution. However, it does appear likely that intertyrosine, intertryptophan and tyrosine to tryptophan transfer can be disclosed by polarized emission measurements.

5.9. Phosphorescence Polarization Studies

Both absorption (PAP) and emission (PEP) spectra have been determined for tryptophan by Chernitskii et al.[288] The only reported polarization study with a protein is one by Konev and Bobrovich[289] with chymotrypsin, where the PEP spectrum was obtained at 77°K.

The PAP and PEP spectra of tryptophan are both negative, with a value of -0.12, except when either the excitation or the emission involves the fine-structure band that lies at the edge of the spectrum (red in absorption, blue in emission). Excitation and observation at the o–0' transition region reaches -0.20. The PEP spectrum for chymotrypsin showed no fine structure, which was like the monomer behavior, but had a much smaller polarization value, -0.09. Konev et al. also determined the time dependence of the emission anisotropy at both 0.2 and 10 sec after cessation of excitation. There was no alternation with time of the polarization. The constant value implies that triplet–triplet transfers are negligible in chymotrypsin, and since no tryptophan has tryptophan as a neighbor this is expected.

The small polarization values found for phosphorescence suggested that there was transfer between tryptophan residues in chymotrypsin at the the singlet level. Tyrosine-to-tryptophan transfers have been shown to be insignificant with this enzyme by both singlet and triplet excitation studies, even though a tyrosine is adjacent to tryptophan in the primary sequence.

Chernitskii et al. analyzed PAP and PEP spectra from indole according to the simplified treatment given previously. For indole, both FEP and PEP spectra show fine structure. The equivalent fine structure from tryptophan is less distinct, and differences between solvents exist. Single crystals of α-amylase at 77°K, however, have a FEP spectrum comparable to that of indole; the PEP spectrum was not reported.[290] The phosphorescence polarization of indole could be explained by assuming that there was a transition perpendicular to the molecular plane predominating, and only a small long-axis component in the molecular plane. The equivalent analysis for the fluorescence was complicated by the suspected ability of indole to emit from two electronic levels simultaneously. This did not apply to tryptophan when dissolved in a polar glass, and Konev et al.[291] interpret the FAP spectrum for tryptophan on the basis of two transitions, one long axis and the second short axis, with an angle between them of 85°. The more complete treatment developed by Czekella[292] has not been applied to the FAP and FAP spectra.

5.10. Tryptophan Excitation Spectra

Excitation spectra for tryptophan fluorescence from many class B proteins have been determined at 298°K[206, 214, 293] and at 77°K for fluorescence[208, 235] and phosphorescence.[217, 234, 271] The luminescence has been isolated by either a color filter or a monochromator. No major difference between the spectra determined from the singlet or triplet, with or without dispersion of the emission, has been recorded. The color filters used are chosen to reject the tyrosine contribution.

All workers have been in agreement, and have emphasized the semiquantitative agreement between the excitation spectra and the absorption spectra. A feature of the absorption spectrum of tryptophan is the presence of a fine-structure shoulder on the long-wavelength edge of the absorption. This band is particularly well defined at 77°K, in absorption spectra. Although their study was not systematic, Beaven and Holiday[294] report that the wavelength of the maximum was not significantly affected by cooling to 77°K. This point should, however, always be considered for each protein, particularly since absorption spectra at 77°K can be readily determined today. A long phosphorescence lifetime complicates measurements of the phosphorescence excitation spectrum. But fluorescence and phosphorescence spectra are always considered to be identical to each other.

Two electronic transitions, with differing vibronic fine structure, comprise the 280-nm absorption of indole compounds, including tryptophan. The ^1L_a transition interacts more strongly with polar solvents than does the ^1L_b transition, and examination of the absorption spectra in several solvents has permitted a dissection of the two overlapping transitions. The prominent shoulder observed at the long-wavelength edge of the absorption bands of all indoles, including tryptophanyl residues in proteins, results from the 0–0' vibronic component of the ^1L_b transitions, and its wavelength has been found not to depend on the polarity of the solvent. For indoles in hydrocarbon solvents, the ^1L_b, transition 0–0' component lies at longer wavelengths than that of the ^1L_a and the fluorescence originates primarily from the latter level. When indoles are hydrogen bonded with polar solvents, or dissolved entirely in polar solvents, additional absorption occurs at longer wavelengths than the 0–0' component of the ^1L_b transition. These bands are particularly prominent for several proteins which possess only a single tryptophan and can be resolved when the absorption is measured at 77°K. The shoulder due to the 0–0' component of the ^1L_b transition is found between 287 and 294 nm, the additional bands from the ^1L_a transition lie at 290–295 nm and 298–302 nm. The presence of these weak transitions has been noted frequently in spectral studies[295, 335, 372]. At room temperature and in polar solvents, the bands are particularly diffuse, and cannot be resolved even at 77°K. For proteins with single tryptophan residues, and which have this tryptophan in a nonpolar environment, e.g., RNase T_1 and endonuclease, the bands are clearly resolved, and can be found in the 77°K spectrum; Strickland *et al.* report in great detail on the spectra of protein model. The influence of the weak 0–0' band of the ^1L_a transition on the excitation spectrum is not clear; its presence is not disclosed in the excitation spectrum of RNase T_1, but can be observed in the excitation spectrum of horse liver alcohol dehydrogenase and aplysia apomyoglobin.

Two points need to be made concerning the excitation spectra: (1) Aside from the region where there is rapid change in extinction with wavelength, which is where tyrosine makes an insignificant contribution to the absorption and excitation spectra, both the singlet and the triplet emissions are superficially similar in excitation spectra to the absorption spectrum of tryptophan. Tryptophan obeys Vavilov's law—there is no excitation wavelength dependency to either its emission spectrum or its quantum yield of luminescence. (2) The spectra are obtained at low resolution compared to absorption spectral studies. This limits the confidence which can be placed on the first statement in detail. This is particularly applicable in the region of the fine-structure shoulder 287–294 nm. The shoulder is at 288 nm for tryptophan in absorption at 77°K, and the same value is obtained when the phosphorescence excitation spectra are measured.[217, 271] For bovine serum albumin, the shoulder lies at 292 nm in absorption,[294] and this is the wavelength obtained from the phosphorescence excitation spectra.[217] For other albumins and the proteolytic enzymes chymotrypsin, trypsin, and pepsin, agreement is found between absorption and phosphorescence excitation.[217, 271]

Denaturation of three albumins, human, ox, and hen, with acid did not alter the wavelength of the shoulder in phosphorescence excitation spectra. Acid denaturation of chymotrypsin shifted the peak to the blue by 2 nm, and with trypsin and chymotrypsinogen the shift was 3–4 nm to the blue.

Alkaline denaturation cannot be readily studied because of the presence of a strong tyrosinate phosphorescence which overlaps the tryptophan emissions. Trypsin is a favorable protein, however, with only a small tyrosinate contribution to its phosphorescence at pH 12. The shoulder is unaffected by alkaline denaturation.

Purkey and Galley[295] report that they have been able to separate the tryptophan phosphorescence and its excitation spectrum into two components with horse liver alcohol dehydrogenase. In these experiments, a particular component in the emission is isolated by choice of emission wavelength, and the excitation spectrum shows at this wavelength a different shoulder wavelength than when other wavelengths in the emission spectrum are used. A greater application of this approach can be expected, now that use of increased resolution in both emission and excitation is feasible. The work performed by the author was obtained using color filters; Purkey and Galley used a second monochromator to isolate the emitting light.

Anderson and coworkers report results similar to those of Purkey and Galley for aplysia apomyoglobin. The fluorescence from this protein was from two tryptophans, one located in a polar environment, the other in a nonpolar environment. The single nonpolar residue emitted in the blue region of the fluorescence spectrum, and the polar residue contributed predominantly to the red region. Excitation spectra obtained from the fluorescence intensity derived from the blue region of the spectrum were different from the analogous spectra obtained with light derived from the red region of the total protein fluorescence. Different maxima and wavelengths for the long-wavelength shoulder were observed.

Rather than analyze the excitation spectrum of tryptophan of proteins to discover the heterogeneity of tryptophan environments, most workers have emphasized whether or not there is any tyrosine sensitization. At wavelengths longer than 370 nm in fluorescence and 470 nm in phosphorescence, a tyrosine contribution to the emission intensity can be neglected. Hence a wavelength where the intensity is entirely due to the energy acceptor can be readily isolated from emission spectra of proteins. To demonstrate a sensitization, all that is required is to determine the quantum yield at several exciting wavelengths. The analysis of the procedures used has been given previously, and is generally applicable to all the various approaches that have been adopted to obtain the yield dependence on wavelength.

The most direct method is to determine an absolute quantum yield value using either a quantum counting screen to integrate the fluorescence (Weber and Teale[206]) or else sufficiently concentrated solutions of the protein so that it too acts as a quantum counting solution. The protein quantum counter can be compared with a true dye quantum counter, and thereby the yield wavelength dependency is directly determined (Vladimirov[296] and Eisinger[297]). Relative methods fall into two classes and use low-concentration solutions where the fractional absorption does not exceed 0.1. Since the emission spectrum is independent of the exciting wavelength, neglecting the tyrosine contribution, the quantum yield for excitation at wavelengths greater than 295 nm can be used as the internal standard for the pure tryptophan yield. Yields at other wavelengths can be obtained by (a) using solutions which have approximately equal, but known fractional absorption at all exciting wavelengths,[206] (b) determining the excitation spectrum with a monomer reference to determine the excitance (its yield acts as a secondary standard) or else using a quantum counting screen to obtain the excitance.[245] Some workers choose to determine the excitation spectrum of a tryptophan standard and subtract this from the protein, after normalization of the two spectra at 295 nm.[208, 271] Others fit the dependency of the relative quantum yield to the 295 nm yield with fractional absorption data determined from monomer spectra.[206, 296] Whichever approach is adopted, identical information is obtained. The excitation spectrum is fitted by a summation of standard tryptophan and tyrosine absorptions.

Teale[206] was the first worker to treat excitation spectra of tryptophan in this manner, and he concluded that there was negligible transfer from tyrosine to tryptophan occurring in the albumins, proteases, and globulins he studied. He contended that tyrosine absorption merely acted as an inner filter to tryptophan absorption. This was largely confirmed by Vladimirov working with Litvin[234] and with Roshchupkin.[293] Konev and Bobrovich[298] were the first to point out that Teale had in fact underestimated the situation when he considered bovine serum albumin. The relative yield at 280–295 nm found by Teale was 0.07. That predicted from model system absorption was 0.04. Hence there is transfer of 20% of the absorption by tyrosine to tryptophan. Since then, several other examples have emerged; ribonuclease T_1 has from 30–50% transfer,[245, 299] pepsin 25% (Badley and Teale[229]), papain 56% (Weinryb and Steiner[229]), *B. subtilis* α-amylase 15%,[271] and yeast alcohol dehydrogenase 32%.[271] All were analyzed by the same procedure. Recalculations of the results of Teale, and those of other workers, are collected in Table 9 and 10; e.g., Teale's data for carboxypeptidase A when recalculated using the known composition of the enzyme gives a transfer yield of 0.55.

Tryptophan excitation spectra from proteins were fitted well at all wavelengths above 260 nm. Below 260 nm, the excitation spectrum is consistently less than the absorption as though the yield decreased. This is particularly well illustrated when relative yield is plotted against wavelength. No minimum occurs in these plot[206, 208, 297]; the yields remain constant from 275 to 240 nm, the shortest wavelength reached. Since Teale[206] and Vladimirov and Litvin[234] obtained excellent fits down to 200 nm with monomers, the failure below 260 nm in proteins lacks an adequate explanation. Teale found that human serum albumin had an excitation spectrum identical with its absorption down to 220 nm, and Vladimirov

Table 9. **Luminescence Properties**

Protein	Source species	Number of chromophore residues per molecule			
		Trp	Tyr	Phe	Cys_2
Serum albumin	Man	1	17	33	18
Serum albumin	Cattle	2	20	24	18
Egg albumin	Hen	3	9	21	1
Egg lysozyme	Hen	6	3	3	4
α-Lactalbumin	Cow	4	4	4	4
Corticotrophin	Cattle	1	2	3	0
Glucagon	Pig	1	2	2	0
Ribonuclease T_1	*Aspergillus oryzae*	1	9	4	2
Azurin	*Pseudomonas fluorescens*	1	2	6	0
Endonuclease	*Staphylococcus aureus*	1	7	3	0
Chymotrypsinogen A	Cattle	8	4	6	5
α-Chymotrypsin A	Cattle	8	4	6	5
Trypsin	Cattle	4	10	3	6
Pepsin	Cattle	5	17	12	3
Carboxypeptidase A	Cattle	7	19	15	1
Elastase	Pig	7	11	3	4
Papain	Papaya	5	19	4	3
Subtilisin carlsberg	*Bacillus subtilis*	1	13	4	0
Subtilisin BPN	*Bacillus amyloliquefaciens*	3	10	3	0
Metmyoglobin	Sperm whale	2	3	6	0
Apomyoglobin	Sperm whale	2	3	6	0
Hemoglobin	Man	6	12	30	0
Apoglobin	Man	3	6	15	0
Avidin	Hen	16	4	28	4
Aldolase	Rabbit	12	47	—	0
β-Bactoglobulin A	Cow	4	8	8	4
γ-Globulin	Cattle	22	56	44	8
Glyceraldehyde 3-phosphate dehydrogenase	Rabbit	12	36	56	0
Glutamate dehydrogenase	Cattle	18	108	138	—
Lactate dehydrogenase	Cattle	20	32	—	—
Carbonic anhdrase	Man	7	9	11	0
Pyruvate kinase	Rabbit	14	—	—	—
Growth hormone	Cattle	2	16	26	4

and Litvin observed a corresponding behavior with phosphorescence spectra. Several other proteins, particularly those discussed previously that show tyrosine-to-tryptophan transfer, were found by Vladimirov[296] to have a fluorescence excitation spectrum with a lesser contribution at 240 nm than found in the absorption spectra.

It is apparent, perhaps with exceptions such as carboxypeptidase A,

of Class B Proteins

Molar mass (kg mol^{-1})	Number of exposed residues		Molar extinction at 280 nm (10^3 m^2 mol^{-1})	Fraction absorbed by tyrosine at 280 nm	Wavelength of shoulder in absorption (nm)
	Trp	Tyr			
68.460	1	5	3.970	0.850	290
69.000	2	8	4.600	0.750	290
43.500	2	5	2.830	0.340	291
14.300	3	2	3.800	0.110	291
14.400	1	4	3.050	0.190	290
4.390	1	2	0.836	0.330	289
3.482	1	2	0.831	0.330	290
11.100	0	2	2.110	0.680	295
15.000	0	0	1.500	0.330	292
16.800	0	5	1.560	0.690	293
25.700	2	2	5.100	0.120	292
25.200	3	2	5.000	0.120	292
24.000	3	2	3.480	0.340	292
35.000	2	6	5.100	0.480	290
34.400	3	16	6.420	0.450	290
25.900	3	10	4.620	0.290	—
20.700	2	12	5.180	0.510	291
27.500	1	9	2.360	0.790	—
27.500	2	9	3.160	0.470	—
17.800	2	1	3.100	0.310	—
17.800	2	1	1.590	0.310	292
67.000 (T)	4	8	4.950	0.350	291
33.500 (D)	—	—	2.600	0.350	—
62.400 (T)	12	1	8.480	0.087	—
158.000 (T)	2	18	14.820	0.530	—
36.800 (D)	2	6	3.660	0.380	288
160.000	17	22	22.000	0.420	292
135.000 (T)	4	12	13.700	0.480	—
312.000 (H)	13	78	32.000	0.200	—
156.000 (T)	8	20	22.600	0.170	—
31.000	0	4	6.200	0.280	—
237.000 (T)	—	—	12.800	—	—
45.650 (D)	0	14	3.250	0.680	—

Table 10. **Luminescence Properties**

Protein	Maximum (nm)	Stokes' shift (mm^{-1})	Ambient Fluorescence Quantum yield for 280 nm	Fractional tyrosine transfer	Average tryptophan yield
Serum albumin	339	622	0.110	0.23	0.35
Serum albumin	342	647	0.210	0.20	0.51
Egg albumin	332	538	0.190	0.38	0.24
Egg lysozyme	342	634	0.070	0.00	0.08
α-Lactalbumin	342(328)	647(522)	0.050	0.12	0.06
Corticotrophin	350	727	0.080	0.33	0.10
Glucagon	345	673	0.140	0.30	0.18
Ribonuclease T$_1$	325	432	0.260	0.40	0.44
Azurin	308	300	0.100	0.00	0.13
Endonuclease	334	540	0.220	0.30	0.42
Chymotrypsinogen A	331	520	0.075	0.00	0.10
α-Chymotrypsin A	334	550	0.120	0.00	0.13
Trypsin	335	561	0.125	0.78	0.13
Pepsin	342	650	0.220	0.33	0.31
Carboxypeptidase A	327	513	0.100	0.55	0.125
Elastase	335	—	—	—	—
Papain	340(328)	620(488)	0.400(0.200)	0.10(0.50)	0.58(0.25)
Subtilisin carlsberg	305(375)	—	—	—	—
Subtilisin BPN	345	—	—	—	—
Metmyoglobin	328	—	0.002	—	—
Apomyoglobin	328	476	0.150	0.80	0.16
Hemoglobin	335	574	0.001	—	—
Apoglobin	335	—	0.100	0.00	0.14
Avidin	338	—	—	—	—
Aldolase	328	522	0.100	0.34	0.12
β-Lactoglobulin A	330	565	0.080	0.71	0.09
γ-Globulin	332	535	0.038	Yes	—
Glyceraldehyde 3-phosphate dehydrogenase	345	—	0.200	—	—
Glutamate dehydrogenase	332	—	0.300	—	—
Lactate dehydrogenase	345	—	0.380	Yes	—
Carbonic anhydrase	336	—	0.170	—	—
Pyruvate kinase	334	—	0.200	—	—
Growth hormone	325	—	0.150	—	—

of Class B Proteins

temperature			Low temperature (77°K)		
Fluorescence			Phosphorescence		
Lifetime (nsec)	Natural lifetime	Fluorescence yield (nsec)	Yield	Lifetime (sec)	Peak wavelengths (nm)
3.3(7.8)	14.5(14.0)	0.11	0.030	6.00	411, 437
4.6	11.0	0.29	0.070	6.40	412, 438
4.5	33.0	0.12	0.030	5.60	412, 438
2.6	32.5	0.14	0.005	1.15(0.5) 3.95	416, 445
—	—	—	—	—	—
—	—	0.50	0.100	—	407, 435
—	—	0.60	0.120	—	407, 435
—	—	0.70	0.150	6.00	406, 434
—	—	—	—	—	—
—	—	0.65	0.130	6.00	408, 437
1.6	16.0	—	—	5.90	413, 440
3.0	22.0	0.18	0.030	6.00	412, 438
2.0	7.0	—	—	6.30	413, 440
4.5	14.5	—	—	—	412, 438
—	—	—	—	(1.6)4.90	412, 438
—	—	—	—	(1.7)5.65	414, 441
6.2(28)	11(11)	0.50	0.070	(2.4)5.20	408, 436
—	—	—	—	(2.49)6.10	(395)409, 438
—	—	—	—	(2.44)6.20	409, 439
—	—	—	—	—	—
3.1	19.0	—	—	—	—
—	—	—	—	—	—
3.0	21.0	—	—	—	—
—	—	—	—	0.70(0.22), 4.0	413, 439
—	—	—	—	—	413, 439
—	—	—	—	—	—
3.2	—	—	—	—	—
—	—	—	—	—	—
4.6	—	—	—	—	—
—	—	—	—	—	—
2.6	—	—	—	4.80	412, 438
—	—	—	—	—	—
—	—	—	—	—	—

papain, and RNase T_1, that transfer to tryptophan does not account for the small fluorescence yield of tyrosine in class B proteins.

The role of temperature on energy transfer has not been investigated in detail. Longworth reports that the fluorescence excitation spectrum of RNase T_1, which does not possess any tyrosine emission, is unaffected by temperature other than the usual enhancement of fine structure on cooling to 77°K. The phosphorescence excitation spectrum is also identical to the fluorescence excitation. This implies that internal conversion from the singlet occurs in competition with transfer from the singlet of tyrosine. For other proteins, no great differences between fluorescence and phosphorescence excitation spectra at 298 nm and 77°K have been noted. Though the yield of tyrosine fluorescence is not known reliably at 77°K, the phosphorescence yield has been measured. Assuming an unaltered fluorescence-to-phosphorescence ratio, the yield values expected suggest that in many native proteins, even at 77°K, internal conversion from the tyrosine singlet still proceeds and thus largely accounts for the low tyrosine quantum yield. Transfer to tryptophan does not account for a significant fraction of deactivation of the energy absorbed by tyrosine.

5.11. Quantum Yields of Tryptophan Residues

The quantum yields of the tryptophan residues of proteins are quite variable at both 298 and 77°K. At 298°K, the values range from 0.04 to 0.50, and literature values have been collected for many proteins,[206, 302] together with their composition and spectroscopic properties, in Table 9. To obtain the quantum yield value for the tryptophan residues, the tyrosine contribution must be either subtracted or assumed to be negligible. This is usually accomplished by determining the excitation spectrum and an absolute yield for 295 nm excitation. In the yield values tabulated, in the majority of cases, the tyrosine transfer contribution to the total intensity under these conditions is small and can be neglected in a first approximation.

The observed yield is quoted for excitation at 280 nm. Corrections have to be applied for the inner filtering by tyrosine absorption and for transfer from tyrosine. The correction is obtained by applying the methods of Teale and Eisinger *et al.* described earlier and discussed in detail when transfer was considered in class B proteins. The inner filtering at a given wavelength, usually 280 nm, is determined from model system molar extinction coefficients, where *N*-acetylcarboxamide derivatives of the tyrosine and tryptophan are used, in the same proportion as they occur in the protein. To adequately account for the absorption spectrum of a protein, the solvent dependency of the molar extinction coefficients must be taken into account. We have chosen to use the extinctions in water and ethylene glycol for the requisite exposed and buried residues: exposed Tyr 140, Trp 550, CySSCy 13 m^2 mol^{-1}; buried Tyr 180, Trp 600 m^2 mol^{-1}. The fractional tyrosine absorption at 280 nm is given for 33 proteins in Table 10. When the ratio of the quantum yield at 280 nm to that at 295–300 nm is known, it is possible to obtain the fractional transfer from tyrosine to tryptophan for that protein. This too is tabulated in Table 10. Since wavelength dependency of quantum yield is not known for many proteins, Teale chose, when he surveyed protein fluorescence data, to assume that transfer is negligible, and to correct the observed yield at 280 nm for tyrosine absorption filtering only. This is probably unjustified in many proteins, particularly those where there is appreciable tyrosine absorption which can go up to 0.70 of the light absorbed at 280 nm. The average fractional tyrosine absorption appears to be 0.30–0.40 and the average fractional transfer from tyrosine to tryptophan 0.20–0.30. This suggests that an inner filtering of 0.2–0.3 of the absorbed light occurs on average. As the tyrosine contribution to the emission is also neglected, and can amount to 0.1 to 0.2 of the total photons emitted, the observed fluorescence yields at 280 nm are to first approximation representative values for the yield for the average tryptophan residue in the protein.

The quantum yield values are usually obtained by using tryptophan as a standard value. For the purposes of the table, the value of 0.20 for the zwitterion at pH 7 in water at 25°C (198°K) was used. To compare protein with standard, the method of Parker and Rees[300] was used by most workers, which integrates the spectrum with respect to wavelength. Therefore, a small error enters into the calculated values since the integration ought to be performed in wavenumbers. Since the difference in peak wavelength values is never large, this error probably is comparable to the accuracy of the measurements (workers quote a precision of 3%; accuracy appears to be 20–30% when values from different laboratories are compared at a common reference value for tryptophan standard). Use of the quantum counting screen method developed by Weber and Teale[301] eliminates the necessity to perform corrections for the technical artifacts introduced by the spectrofluorometer. However, eliminating the contribution from tyrosine requires the use of spectrofluorometers, since scattering and the loss of anti-Stokes' fluorescence intensity complicates the use of 295 nm excitation with a quantum counting screen.

Denaturation with 8 M urea or 5 M guanidinium hydrochloride causes a change in yield for many proteins, either an increase or a decrease. When denatured proteins are investigated, the quantum yields typically have a common value; with urea, yields range from 0.20 to 0.25; with guanidinium, from 0.12 to 0.15.[302] As both denaturants enhance tyrosine fluorescence, its contribution becomes appreciable. This must frequently be removed from the spectrum to obtain the true tryptophan yield. Electronic energy transfer is also drastically reduced with most proteins, unless tyrosine and tryptophan are vicinal in the primary sequence or as a result of the presence of secondary structures of the disulfide links. The tryptophan yield can often be safely obtained by assuming there is no tyrosine-to-tryptophan transfer, and only an inner-filter correction need be applied to the observed yield. The quantum yield of the tryptophan residues from the random coiled conformation in the presence of urea is roughly equal to the value for the monomer in the same solvent.

Reduction of a disulfide link can cause additional increase in the fluorescence yield, but for many proteins reduction does not affect the yield of denatured proteins. Sulfhydryl groups are known from Cowgill's model studies to quench tryptophanyl residues, and their involvement in quenching of protein tryptophans must be considered. Substitution of sulfhydryl hydrogen, e.g., in methionine, lead to no quenching in model compounds. After reduction to form sulfhydryl groups, a reaction forming either carboxymethyl or cyanoethyl derivatives of proteins is commonly encountered. Examples are known where the carboxymethylated reduced protein is less quenched than either the native or the denatured conformers. Since carboxymethylated reduced proteins are comprised of single polypeptide strands, by choice of suitable solvent it is feasible to investigate the polymer in either random-coiled or helical conformations. The influence on electronic transfers between tyrosine and tryptophan can then be followed. The general pattern which has emerged is that reduction of disulfide links and subsequent blocking of the sulfhydryl group leads to yield values comparable to those of simple oligopeptides. The tyrosine yield is greatly enhanced, and tyrosine-to-tryptophan transfer drastically reduced, but not entirely eliminated in water, but probably negligible in random-coil supporting solvents.

Detergents such as sodium dodecyl sulfate bind to proteins, and when there are 10–100 detergent molecules, the protein becomes denatured[303] and forms polypeptide chains which are largely helical, unless a large number of disulfide links prevent this occurring. Again the quantum yields of the tryptophans are normalized and fall in the range 0.08–0.12.[304]

Organic solvents, at large mole fractions, will denature proteins, and in so doing drastically alter the yield. There has not been any systematic collection of the yields in organic solvents, other than in propane-1,2-diol. Teale observed with the globular proteins he investigated that the tryptophan yield was quite variable in the glycol solvent, but the Stokes' shift was also variable, and he did not determine the tyrosine contribution. In several cases, it appears as though there was appreciable tyrosine fluorescence, and this may in part account for the variability in peak wavelength and yield.

The conformation of polypeptides in the organic solvents is not well established; perhaps glycol and methanol, ethanol and dioxane support unordered conformation. 2-Chloroethanol and trifluoroethanol support a helical conformer, while ethane-1,2-diamine and dimethylsulfoxide favor the random coiled conformers.

Alterations in pH, as is known from circular dichroic and absorption spectral studies as well as hydrodynamic and hydrogen exchange studies, alter the conformation of proteins. Many of these conformational changes also affect the quantum yield, and fluorescence intensity measurements have frequently been used to study the influence of pH on protein structure.[303, 304] Alkaline pH values invariably lead to a reduction in quantum yield (quenching), and this pH region has received particular attention. The fractional ionization of the phenolic groups of a protein can be determined by absorption spectrophotometry. A comparison of the fluorometric quenching with the absorption spectrophotometric titration of the phenoxyl shows that fluorescence quenching goes hand in hand with tyrosine ionization.[257] In many proteins the phenoxyl ionization proceeds in two waves, the first with $pK = 10$, and the second with pH $>$ 11.[246, 248, 305] The protein fluorescence quenching usually accompanies the titration of the exposed tyrosine residues.[305] Titration of the majority of the exposed residues is not accompanied by major conformational changes and is fully reversible. Protein stability can be enhanced by the presence of 0.1 M NaCl and 20% sucrose, if necessary. Titration of the buried residues is not always reversible. A reaction with disulfide links and hydroxyl at the high pH values can occur, leading to desulfurization and group-specific chemical modification of the protein.[306, 307] The reaction between hydroxyl anions and disulfide links also alters the absorption spectrum since a cysteinyl anion is produced.[308] This absorbs appreciably even at 295 nm and complicates quantitation of the fractional tyrosyl ionization by absorption measurements.[309]

A detailed analysis of the alkaline quenching is somewhat complicated since phenoxyl anions also fluoresce,[249] though with a yield that is an order of magnitude less than tryptophan but in the same spectral region as tryptophan.[250] The tyrosinate fluorescence is sensitized by tryptophan and this enhances the yield.[310] Therefore, it is only possible to semiquantitatively analyze the transfer of tryptophan residues to tyrosinate. A usual assumption taken is to state that all the fluorescence originates in tryptophan.[246-248] Plots of the fluorescence intensity, preferably measured at the isoemissive wavelength of tyrosine–tyrosinate on excitation at the isosbestos of tyrosine–tyrosinate, when plotted against the fractional tyrosine ionization show that a single tyrosinate quenches more than a single tryptophan when several tryptophans are present in the protein.[311] Additional quenching can occur when the buried residues are titrated at more alkaline pH values. But it is not possible from measurements such as these to ascribe the quenching explicitly to tyrosinate; both conformational induced quenching or enhancement may occur, and a direct quenching of tryptophan by hydroxyl anions can take place.

Model systems have shown that the enamine proton of the tryptophan indole ring is ejected from the excited singlet state to a hydroxyl anion.[257, 311] The enamine proton undergoes a protolysis in the excited state with hydroxyl ion, with a $pK_a = 12$, though indolium anion is not ionized in its ground state in water (pK 16). The indolium anion fluoresces at 410 nm with a quantum yield of 0.10 when excited directly.[312] However, the fluorescence from tryptophan is quenched in water, with half-quenching occurring at pH 12.[257] The quenching is abolished by substitution at the enamine proton with methyl radical; the fluorescence of N-acetyl-1(N)-methyl tryptophan amide is unaffected by hydroxyl ions up to 1 M (pH 14).[311] The conclusion from the substitution of methyl for proton at the enamine nitrogen was that an excited-state protolysis occurred, and the spectra indicated a pK of 12 for the excited singlet. However, the indolium anion is fluorescent itself, so the absence of the fluorescence is not adequately explained. No evidence is available for the presence of indolium anion fluorescence from proteins.

In summary, the alkaline quenching of proteins is largely accounted for by a transfer from tryptophan to the weakly fluorescent tyrosinate. Since tyrosinate anions absorb to longer wavelengths than tryptophan and also fluoresce to slightly longer wavelengths, it is possible in principle to determine the fractional transfer to tyrosinate, and also to investigate the role that excited-state protolysis plays at pH values greater than 12. No reports of such studies are available.

Other pH-dependent quenching interactions which occur with tryptophan compounds involve the following functional groups:

(a) The neutral amine moiety of lysine[313, 314] and arginine,[313] pH 9.5–10.5.

(b) The cationic form of the α-amino group, pH 7–8.[257, 262, 306, 312, 315]

(c) The cationic form of histidine (imidazolium cation), pH 6–7.5.[316, 317]

(d) The neutral carboxylic acid group of glutamate[314] and aspartate,[313] pH 4–5, and the α-carboxyl C-terminus, pH 3–4.[257, 262, 311, 312, 315]

(e) Neutral sulfhydryl group, pH 8.[321]

All these interactions have been studied in model systems involving amino acids or substituted derivatives, dipeptides, diketopiperazine (cyclic dipeptides), and copolymers. Other means of studying the interactions have involved the addition of suitable model compounds such: hydroxylamine to tryptophan solution.[305] In the majority of cases, the quenching observed is of the order of threefold, i.e., not as large as with tyrosine compounds.

pH-dependent quantum yield changes not accompanied by conformational changes have been found in several proteins, and both carboxylate and imidazolium quenching interactions have been implicated. One example of a discharge of an α-amino terminus affecting the yield is known with the bovine growth hormone.[318] The primary sequence indicates that the residues are well separated in sequence (Trp-10), so the two residues must be associated a restult of the tertiary structure. An established example involves the tyrosine of chymotrypsin in the dimer state in the crystal.[319] Titration of the carboxyl group between pH 5 and 2 alters the quantum yield of lysozyme;[246–248] Asp-101 is adjacent to Trp-63 and could account for the quantum yield change in this pH region. A particularly clear example of imidazolium quenching is provided by papain, where His-159 is in contact with Trp-177 and is adjacent to Trp-181. The titration of the histidine residue probably accounts for the large pH dependency of papain between pH 4.5 and 7.5.[229, 320, 321] Other examples of imidazolium quenching have been described by Shinitzky and Goldman,[316] and by Hélene et al.[322–324] with an aminoacyl tRNA synthetase.

Acid titration of many proteins leads to only a modest reduction in quantum yield; there are few clear examples of a carboxyl-induced quenching occurring. More dramatic changes which occur on acidification are associated with conformation changes. A particularly well-studied phenomenon occurs with human and bovine serum albumin, but not with ovalbumin. At pH 3.5 a transition to an expanded conformation occurs which exposes several tyrosyl groups.[325] The quantum yield of tryptophan decreases and that of tyrosine increases.[207, 213, 214] Another acid-induced phenomenon is an association of β-lactoglobulin A dimers to form octamers below pH 5. This association is accompanied by an enhancement in tryptophanyl quantum yield.[326] Both serum albumin and β-lactoglobulin show a modest quantum yield change between pH 7–9, where with albumin another conformational change occurs and with β-lactoglobulin A the dimer dissociates into monomers.[246, 247]

Steiner[321] has found that for activated papain (free sulfhydryl group at the active site) there is, in addition to an imidazolium quenching of a tryptophan, another quenching process characterized by a pK of 8.0. The second acid-quenching reaction disclosed is associated with the sulfhydryl of the active-site cysteinyl residues. When the sulfhydryl function is blocked, as in unactivated papain (where it exists as a disulfide complex), no quenching with a pK of 8.0 occurs, but the imidazolium cation formation on acid titration leads to quenching.

The magnitude of tryptophan quantum yields in proteins has been considered in detail by Cowgill,[313] and his treatment will be followed. Cowgill has emphasized the need to determine tryptophan yields relative to tryptophan zwitterion rather than quote absolute values, and many workers have followed his advice. Van Duuren[327] studied the fluorescence of indole in several solvents and concluded that water is a quencher for indole. In a hydrocarbon, indole has a yield of 0.48, but on forming exciplexes in water the yield drops to 0.35. A similar pattern of yields can be observed with N-acetyl carboxamide derivatives of tryptophan. Cowgill terms the hydrocarbon environment class I and the aqueous environment class II, i.e., where there are no additional quenching processes from vicinal groups. The high tryptophan yields of several proteins are associated with tryptophan in either class I (RNase T_1) or class II (bovine serum albumin).

Complexation of the enamine proton with amides does not lead to tryptophan quenching as it does with tyrosine.[311, 328] However, like tyrosine, incorporation into a polyamide structure in an aqueous medium reduces the yield, values ranging from 0.10 to 0.15. Tryptophan residues in this group are termed class III, and several proteins may well have tryptophans which fit into this group.

Particular quenching reactions which occur in proteins and have been identified by means of suitable model system studies are:

Class IV, quenching by vicinal disulphide links, yield 0.05.[329]
Class V, quenching by imidazolium cation, yield 0.04.[316, 317]
Class VI, quenching by an α-ammonium group of the N-terminus, yield 0.15–0.12.
Class VII, quenching by vicinal ε-amino groups, yield 0.20.
Class VIII, quenching by vicinal sulfhydryl groups, yield 0.08.[329]

Examples of quenching by charged groups are known from pH titration studies; examples of quenching by disulfide groups can be identified from the primary sequence or from the molecular structure. No clear example of quenching by disulfide in proteins is established; perhaps the low yield of lysozyme results from the quenching of Trp-62 and Trp-63 by the vicinal disulfide link adjacent to the active site, involving Cys-64 and Cys-80, and Trp-123 is vicinal to disulfide link Cys-127–Cys-6. Trp-28 is near Cys-30–

115, and has neighbors Trp-108 and -111, which could transfer readily since they form a boxlike structure. Churchich[330] has found that full reduction of lysozyme enhances the fluorescence yield, and Steiner and Edelhoch[247] have shown that urea denaturation does likewise.

Denaturation by altering the conformation, and thereby affecting interresidue distances and disrupting complexes involving the enamine nitrogen, invariably affects the quantum yield and the Stokes' shift. Fluorometric studies have been used to study the dependence of denaturation equilibria on concentration of added denaturant and to follow the kinetics. When the studies are performed at different temperatures, allowing for temperature effects on the quantum yield of tryptophan itself, it is possible to investigate the thermochemistry and stability of proteins.[331] Edelhoch and Steiner[269, 332] have investigated the properties of several proteins using fluorescence to monitor the conformational populations present.

5.12. Solvent Perturbation

Neutral organic solvents influence the quantum yield of fluorescence of several proteins.[206] Steiner et al.[269] found that two regions prevail: (a) a region at low organic cosolvent concentrations where the conformation is unaffected but the environment of the external residues is changed and (b) a region where conformation changes occur and the proteins denature. The fluorescence studies complement the absorption studies frequently employed to investigate the availability of tyrosine and tryptophan residues to their solvent environment. Steiner and Edelhoch found that the yield of monomer[333] and protein is enhanced by organic solvent addition, typically ethanol or p-dioxane or dimethylsulfoxide. The extent of the enhancement compared to monomer (N-acetyl carboxamides) is used to determine the fraction of the fluorescence due to exposed groups. In the region of denaturation, yield enhancement or quenching can occur, and the tyrosine contribution is also enhanced significantly. Again comparison is made with the equivalent absorption studies and with circular dichroic measurements of the conformational states and populations.

5.13. Solvent Isotopic Effect

A further refinement of the solvent perturbation method has been to replace protonium oxide (water) by deuterium oxide (heavy water) as the solvent. Deuterium exchanges with the enamine proton of tryptophan together with any alteration in solvent property. Stryer[334] and, more recently, Eisinger and Navon,[229] Steiner and Kirby,[334] and Ricci[221] have all found a greater yield for tryptophan zwitterion when dissolved in heavy water rather than in water at pH (pD) 7.0. Enhancement ratios for the yield found by the respective workers were 1.65, 2.15, 1.9, and 2.3. Eisinger and Navon investigated the solvent isotopic yield-enhancement ratio with corticotrophin, chymotrypsin, ribonuclease T_1, and staphylococcal nuclease; Weinryb and Steiner studied papain (an unfractionated sample). In all the examples of proteins investigated, an enhancement ratio of 1.2 was observed, though their yields, Stokes' shifts, tryptophan environments, and content all differ. In particular, no distinction was found among proteins with exposed tryptophan (corticotrophin), those with buried tryptophan (ribonuclease T_1), or those with partially buried tryptophan (staphylococcal endonuclease).

Lehrer[334] has studied the pH dependency of tryptophan's yield in both water and heavy water. From the measurements obtained both he and Ricci, who investigated only specific pH values, conclude that there is an isotope effect upon the α-ammonium group's quenching of tryptophan fluorescence. Substitution of deuterium results in a reduced rate of deuteron reaction with the indole ring excited state compared to that of the corresponding proton reaction. Such a mass isotopic effect is expected for a protolytic reaction where either protons or deuterons add to the excited indole ring. Substitution of the enamine proton by a deuteron does not significantly alter the fluorescence yield; Eisinger and Navon found an enhancement ratio with N-methyl tryptophan at pH 7 comparable to that found for tryptophan zwitterion. Both Lehrer and Ricci suggest that the predominant cause of the solvent isotopic effect on tryptophan zwitterion is a result of a mass isotopic effect on the protolytic reaction between the N-terminus ammonium and the excited indole ring; Lehrer further suggests that similar reactions may occur in proteins, and that their perturbation by heavy water accounts for the observed solvent isotopic effect. Ricci noted that there still remains a solvent isotopic effect upon the yield of tryptophan with an uncharged amine (at pH 10). Here the enhancement ratio of 1.5 was ascribed to deuteron substitution of the enamine proton; the replacement of a proton by a deuteron perturbs internal conversion processes, thus altering the fluorescence yield. Because the enhancement ratio found for the proteins investigated is more comparable to a deuterium substitution than to a mass isotopic perturbation of ammonium quenching, it is the latter mechanism which is favored to account for the influence of deuterium oxide on protein fluorescence yields. Undoubtedly, proteins will be found where the protolytic quenching of tryptophan predominates, and its perturbation by replacement of water for heavy water will cause a large isotope effect.

5.14. Temperature Dependence of Quantum Yields

A study of the temperature dependency of the yield can be used to investigate the causes of quenching and the conformation transitions which occur as the temperature is altered. Two temperature regions of interest have been found, a low-temperature region[301,335] and one centered around ambient temperatures.[336,337] The causes of quenching are investigated by studying the yield and Stokes' shift from 100 to 300°K. Structural transitions occur between 280 and 380°K, and fluorescence quantum yields often alter, indicating the change in conformational state. Difference absorption spectral studies and circular dichroic studies which compare low-temperature and high-temperature spectra in the range 280–380°K show that proteins typically undergo a sharp cooperative transition to an unordered state—this is termed melting.[338,339] This transition to the disordered conformer, which may be a random coil or a β-structure, also typically occurs above 330°K and takes place over only a few degrees Kelvin. Other transitions can occur between different ordered conformations at lower temperatures; association of or dissociation into subunits is also influenced by temperature. As temperature is increased, the tryptophan residues have their fluorescence yield decreased. When a disorder transition occurs, there is a change in the rate of change in yield value with temperature; the yield can be subject to a significant increase or decrease in value, and a new temperature dependence sets in.

The quantum yield of fluorescence is defined by considering three rate processes which compete with the fluorescence rate. The rate constant which describes the emission of fluorescent photons from the excited singlet, k_1, is independent of temperature. A second rate constant, k_0, which is the sum of the separate rates of internal conversion and intersystem crossings is also temperature independent. A third constant, termed k_2, which accounts for another nonradiative deactivation has an activation energy E_A and a frequency factor A. This rate can also be regarded as the single temperature-dependent term; more complex situations can be developed if the accuracy of the data requires them.

The quantum yield of fluorescence Φ is then defined as

$$\Phi = \frac{k_1}{k_1 + k_0 + k_2}$$

$$\Phi^{-1} - 1 = \frac{k_0}{k_1} + \frac{1}{k_1} A e^{-E_A/RT}$$

For temperature close to ambient temperatures, k_0/k_1 is small compared with $\Phi^{-1} - 1$ and can be neglected; then the convenient expression is

$$\ln\left(-\frac{d\varPhi^{-1}}{dT^{-1}}\right) = \ln\frac{AE_A}{k_1 R} - \frac{E_A}{\ln 10 RT}$$

Frequently, a single value for E_A covers a wide range of temperatures, and it is only necessary to plot log $(\varPhi^{-1})$ against T^{-1} or simply $1/F$ versus $1/T$ to obtain the Arrhenius critical energy from the slope of these plots taken in the high-temperature region. More generally,

$$\log\left[(\varPhi^{-1} - 1) - \frac{k_0}{k_1}\right] = \log\frac{A}{K_1} - \frac{E_A}{\ln 10 RT}$$

is solved for linear plots of $\log[(\varPhi^{-1} - 1) - k_0/k_1]$ against T^{-1} at given values of k_0/k_1 utilizing appropriate curve fitting codes. Measurements of the fluorescent lifetime simplify the analysis since the separate rate constants can be directly measured when yield and lifetimes are also known. However, the only yield–temperature plot so far reported for proteins which fits the Arrhenius equation is one for lysozyme.[337]

Several workers have studied the Arrhenius critical energy for temperature-induced quenching of the monomers in water. Values reported for E_A of tryptophan are 7.5,[315] 8.1,[229] 8.45,[340] 8.1,[340] and 6.6[334] kcal mol^{-1}. Galley and Edelman[315] and Barenboim et al.[340] determined A/k_1 from the intercept at the reciprocal yield axis and k_0/k_1 by difference. Since values are known for k_1 at 298°K (6.7×10^7 sec^{-1}), $A = 1.3 \times 10^{14}$ and by difference in the first equation $k_0 = 9.3 \times 10^7$ sec^{-1}. For tyrosine the corresponding terms are $E_A = 7.1$ kcal mol^{-1}, $k_1 = 6.2 \times 10^7$ sec^{-1}, $A = 2.7 \times 10^{13}$ sec^{-1}, and $k_0 = 5.6 \times 10^7$ sec^{-1}. Barenboim et al. also studied the temperature behavior of the fluorescence of several indole compounds and found there was a linear relationship between E_A and $\varPhi$; the greater the yield, the greater the Arrhenius energy. The corresponding figures for phenylalanine are $k_1 = 4 \times 10^6$ sec^{-1}, $A = 2.4 \times 10^{12}$ sec^{-1}, $E_A = 5.9$ kcal mol^{-1}, and $k_0 = 4.1 \times 10^7$ sec^{-1}.[341]

Since lifetimes and quantum yields are fitted by the same curve, i.e., $\varPhi_F/\tau_F$ is constant, then k_1 is temperature independent. The current precision of the measurements permits a fully adequate analysis by considering only a single activation rate constant. Since the frequency factor is 10^{14}–10^{12} for the three amino acids, it is considered to be associated with an internal conversion. The second temperature-independent radiationless rate constant is of the order of 5×10^7 sec^{-1} at 300°K; it is consistent with other intersystem crossing rates for aromatic hydrocarbons, suggesting that the energy loss at 77°K is due to a triplet internal conversion.

Activation energies have not been as thoroughly studied with protein fluorescence. The fluorescence from a small number of proteins has been studied between 250 and 360°K; the temperature dependence fitted an

Arrhenius plot. There was a single activation energy with a considerably smaller value than was found with the monomers. Values ranged from 2 to 3 kcal mol^{-1}. Though the value for k_1 is poorly established with proteins, and the frequency factor has not been reported, a reasonable estimate for k_1 is an equal value to the monomer. It appears that the major change between proteins and monomer is accounted for by an increased k_0 value, since the critical increment is reduced, with little effect on the frequency factor.

Conformational transitions, which occur above 300°K in many proteins, have complicated the analysis of thermal dependency of the fluorescence quantum yield of proteins. Galley and Edelman reported that lysozyme was the only protein they studied that fitted the Arrhenius analysis for fluorescence quenching. No analysis in terms of a critical energy and frequency factor was possible with other proteins. The trend is always a reduced dependency of the yield on temperature below the melting temperature.

The other region of interest in the quantum yield–temperature plot is the low-temperature limit. Here k_0/k_1 becomes comparable to $\Phi^{-1} - 1$, though k_2 becomes negligible. Eisinger and Navon[299] and Longworth and Battista[335] have studied this region with both monomers and proteins, using glassing solvents. With glassing solvents, two temperatures of particular importance influence the structure of glass. The first is the temperature where all translational motion ceases; above this temperature, the glass softens, and this is termed the devitrification temperature. Below this temperature, rotational motions of the solvent persist until the second-order transition to the glass state occurs; below this latter temperature *all* motion by the solvent ceases. The lower temperature is the true glass transition temperature, and it is at this temperature that severe strains set in the glasses, and fracturing takes place. With equivolume mixtures of ethane- and propane-1,2-diol, the two critical temperatures are approximately 150 and 180°K. Below 150°K, the quantum yield is not dependent on temperature. Above 180°K, variable Stokes' shifts can be found when the excited state becomes solvated because of its increased dipolar moment. Eisinger and Navon found that, in the temperature interval 180–250°K, the quantum yield of tryptophan and its derivatives remains constant, within 20%, but the Stokes' shift increases from the limiting value which prevails at low temperatures to the limiting value that prevails at high temperatures. At temperatures above 250°K, no changes in Stokes' shift take place, but the quantum yield is subject to a thermal deactivation.

Proteins follow this general trend; they increase their yield on cooling until a limiting value is reached, which is approximately at a temperature of 180–250°K, and at lower temperatures no further significant change in fluorescence yield or spectrum can be detected. The value for this yield, like the room-temperature values, is variable from protein to protein, and does not follow amino acid composition, Stokes' shift, or room-temperature yield. To date, no careful analysis of the temperature dependency of the yield has been reported. Preliminary studies in the author's laboratory suggest that a single value for the activation energy adequately explains the temperature dependency up to 300°K. Aside from lysozyme, the proteins investigated demonstrated similar frequency factors and reduced activation energies from the monomer values. The predominant result of incorporation of tryptophanyl residues into a protein appears to be the addition of a new internal conversion process which reduces the fluorescent yield from the monomer value. Though there is little direct experimental support, the constancy of the phosphorescence lifetime for most proteins compared to the monomer and the single scavenging study with chymotrypsin[342] strongly suggest that the intersystem crossing rates are not affected when tryptophan residues become included into the structure of a polypeptide. To perform an accurate analysis of the temperature dependence of quantum yield, it is necessary to take into account the varying Stokes' shifts which occur accompanying the yield changes, and to remove the tyrosine contribution. It is largely for these reasons that few attempts have been made to adequately investigate the thermal behavior. In the high-temperature region where many of the problems are reduced, conformational changes occur which complicate the analysis.

Edelhoch et al.[332] have used the temperature-induced conformational perturbation of the Stokes' shift and the yield to study the thermochemistry of the conformational transition. In this fashion, they were able to determine the transition state free energy, enthalpy, and entropy for the heat denaturation of pepsinogen. Previously, Galley and Edelman[337] had performed similar analyses of the heat denaturation of Bence-Jones protein from their thermal fluorescence studies.

The cause of the temperature-dependent quenching of tryptophan fluorescence is not established. Most workers assume that it is a thermally activated internal conversion, though there have been no quantitative scavenging studies of the temperature dependency of the triplet yield. Barenboim has found for several tryptophan compounds that there is a linear relationship between the Arrhenius critical energy of activation and the absolute quantum yield of the derivative. There are not sufficient data to analyze whether a similar relationship prevails with protein activation energies or with the frequency factors. Two independent studies indicate that the mechanism of the internal conversion process may be of a chemical rather than a physical nature. Hopkins and Lumry[343] have studied the yield of solvated electrons formed by irradiating indole and tryptophan compounds. Solvated electrons were measured with scavengers, either N_2O or $CHCL_3$. They found that there was a reciprocal relationship between fluorescence yield and solvated electron production, when either the chemical or ionic nature of the tryptophan compound was altered. Furthermore, they measured the Arrhenius activation energies of fluorescence quenching and solvated electron formation and found that they were equal. To date, comparable studies with proteins have not been reported.

Steiner and Kirby[344] noted another correlation which implicated the formation of solvated electrons. They measured and collated the Stern–Volmer quenching constants of a wide variety of compounds which quenched tryptophan fluorescence. The Stern–Volmer constant was related to the rate of reaction of that compound with the solvated electron. The reaction with the solvated electron had been separately measured in pulse radiolytic measurements. Compounds which did not react with the solvated electron usually did not quench fluorescence. Hence, Steiner and Kirby suggest that the quenching reaction for the compounds studied may involve a transient ejection of an electron from the excited state to form the radical cation of indole and the corresponding radical anion of the quencher. The radical cation is then suggested to be nonfluorescent under the conditions used, and internal conversion to a ground state proceeds, or else the total energy available is consumed in the formation of the two radicals. Subsequent recombination of the radicals does not lead to any recombination luminescence, but restores the two initial partners.

The electron ejection mechanism proposed to account for quenching of indole fluorescence is to be contrasted with the facile quenching of indoles by α-ammonium groups or by protons. Both reactions are believed to involve the capture of a proton by the excited indole ring, and this leads to fluorescence quenching in some unexplained way. Other reversible nucleophilic ring substitutions by appropriate vicinal groups may also account for quenching reactions. The exact role played by exciplexes, solvation shells, ejected electrons, captured protons, ejected protons, and substituting nucleophiles in tryptophan quenching mechanisms cannot be stated with any confidence—how indoles are quenched by relevent mechanism to protein native conformers is certainly very far from clear. Only glimpses of plausible quenching reactions are found from model studies, and these can be extended to proteins only by a great deal of extrapolation, even when atomic models are investigated. There is the possibility, and a disturbing one too, that all feasible quenching mechanisms may occur in proteins, and even in a single protein, though their relative proportion will undoubtedly be dependent on conformation, solvent, and temperature, permitting attempts to successfully unravel several competing processes.

5.15. Energy Loss at 77°K

The quantum yields of luminescence of indole at 77°K add to unity; there is no energy loss $\Phi_{Fl} = 0.80$, $\Phi_{Ph} = 0.20$. The decay time for indole is 5–6 sec. Tryptophan has an energy loss at 77°K. Longworth[217] reports an energy loss of 20%: $\Phi_{Fl} = 0.64$; $\Phi_{Ph} = 0.16$; $\Phi_{Fl}/\Phi_{Ph} = 0.40$. Augenstein et al.[219] report an energy loss of 11%: $\Phi_{Fl} = 0.72$; $\Phi_{Ph} = 0.17$; $\Phi_{Fl}/\Phi_{Ph} = 0.42$. Both workers find a phosphorescence lifetime for tryptophan between 5 and 6 sec. Quantitative ESR studies performed by Shiga and Piette[345] suggest that approximately one third of the excited-state population can be found as triplet states. This result indicates that the energy loss is accounted for by an internal conversion process at the triplet level of tryptophan rather than at the singlet level. This would place the natural lifetime of the tryptophan triplet in the range of 10–12 sec.

The luminescence yields of several globular proteins did not add to unity; energy losses of greater than 50% were frequently found. The simple proteins with a single tryptophan such as corticotrophin and RNase T have energy losses comparable to tryptophan.

The only protein in which the triplet yield has been investigated is chymotrypsin.[342] The protease binds proflavine at its active site.[346] The protein triplet states are quenched by a triplet-to-singlet transfer from tryptophan to proflavin. A quantitative analysis of the quenching of tryptophan phosphorescence and the sensitized dye delayed fluorescence permitted Galley and Stryer to determine the intersystem crossing yields and obtain the natural lifetime of the triplet. They found that there was a $\Phi_{Fl} = 0.18$, $\Phi_T = 0.09$, $\Phi_{Ph} = 0.03$. They determined the measured lifetime as 0.23 of the natural lifetime, which would place the natural lifetime at 20–25 sec. The preponderant contribution to the energy loss at 77°K in chymotrypsin was a singlet internal conversion which accounted for 73% of the absorbed energy. Though there are no studies with other proteins, the Φ_{Fl}/Φ_{Ph} ratio of native proteins and the Φ_{Ph} do not differ widely from each other or from the monomer value. Hence, it appears likely that internal conversion at the singlet level accounts for a large fraction of the absorbed energy in many globular proteins at 77°K. At higher temperatures, the formation of solvated electrons could occur, and this reaction may explain the temperature-dependent thermal quenching of fluorescence.

Measurements of quantum yields at 77°K pose several problems, none of which have been satisfactorily solved. The photophysical effects which enter into measurment of absolute light intensities at 77°K are:

(a) The emitted light is polarized. Corrections have to be applied to account for this, since in any observing direction only two of the three polarized components will be measured. This problem has been treated by Singleterry and Weinberger[347] and by Teale and Weber.[301] More recent analyses are due to Aurich and Lippert[348] and Almgren.[344] When detector systems are used, such as monochromators, then account has to be taken of the role the instrumental anisotropy plays in modifying the polarized light intensities. This problem has been treated by Weill and Calvin[350] (with a minor correction noted by Galley[351]) and by Albrecht[352] and by Kalantar.[353]

(b) The refractive index is greater in the solutions at 77°K than at 298°K because of the greater density of the solvents at 77°K. The role refractive index corrections play in absolute light intensities has been considered by Shepp.[354]

(c) The integrated area of the spectrum changes, since both band shape and Stokes' shift alter.

(d) Molar extinction coefficients at the exciting wavelengths alter.

Some technical problems, which overwhelm the accuracy of the measurements after correcting for the photophysical phenomena are:

(e) Concentration increases due to solvent volume contraction.
(f) Fracturing of the glass occurs below the glass transition temperature.

The fracturing causes (1) increased effective optical pathlengths and (2) enhanced scattering of the exciting light.

There are no simple and direct solutions to many of these problems, and it is for these reasons that few quantum yield determinations at 77°K have been attempted, and the reported values can only be regarded as approximate values.

When proteins are involved, there is the additional complication of an emission by two different species, and the tyrosine and tryptophan contributions to the spectrum have to be effectively separated.

The quantum yield at 77°K of monomer and protein is not sensitive to the environment. The monomer tryptophan does not significantly alter its quantum yields of luminescence when different derivatives and ionic states are observed.[219] Equally, the tryptophan quantum yield of proteins does not appear to be particularly sensitive to the nature of the protein, or its composition or the conformation. An exception occurs when the protein has a small quantum yield. The example is lysozyme. The low yield of lysozyme is believed to be largely a result of a vicinal disulfide link. Reduction of the linkage chemically restores the quantum yield at both 298 and 77°K to values comparable to that of the monomer.[330]

The role of electronic energy transfer in the luminescence at 77°K is not known. Low temperatures do not seriously affect the absorption spectrum of tryptophan nor do they alter the peak wavelength significantly when these residues are incorporated within a protein structure. There is, however, a pronounced effect of low temperatures on the absorption of tyrosine and on the fluorescence of tryptophan. So intertyrosine and intertryptophan transfers will be strongly influenced by temperature, tyrosine-to-tryptophan transfers much less so. Increases in quantum yield of donor states occur on cooling; this too will affect the magnitude of transfer, and in part compensates for reduced overlapping of acceptor absorption with the donor emission. The FAP spectrum of several proteins has been reported at 298 and 220°K. There was no dramatic indication of an altered extent of electronic energy transfer to be noted. No wavelength-dependent quantum yield studies of fluorescence or phosphorescence have been reported at 77°K. However, excitation spectra have been obtained at 298 and 77°K, and no significant differences between the room-temperature fluorescence excitation spectra and the luminescence excitation spectra were reported by the workers who measured the low-temperature spectra. Cassen and Kearns,[271] who have subtracted a normalized protein excitation tryptophan spectrum from the excitation spectrum of several proteins, in some cases found evidence for tyrosine sensitization, in others not. To a first approximation, it has been considered that energy transfer occurs to the same extent at 77°K as it does at 298°K. However, more experimental evidence to support this opinion is needed.

The quantum yields of phosphorescence of tryptophan residues have been reported for several globular proteins,[217,218] and these are collected in Table 10. The tyrosine contribution to the phosphorescence can be readily determined, and the yield of both tyrosine and tryptophan can be separately measured. The phosphorescence yield of tryptophan, like the fluorescence yield, at 77°K is not strongly dependent on protein conformation. Lysozyme is an exception; denaturation of lysozyme restores the low yield which the native enzyme possesses to a normal value found with other proteins. Total reduction of the disulfide links further increases the yield to a value close to that of the monomer.[330]

The ratio of phosphorescence to fluorescence for the monomer tryptophan appears to be 0.4 ± 0.05. For proteins, the values found range from 0.4 to 0.2 in the native proteins, but more nearly approximate the monomer in the denatured proteins. Again lysozyme deviates. The ratio is approximately 0.02.[220] Denaturation restores the value to 0.4. The contribution of tyrosine to the fluorescence and its large dependence on protein conformation complicate the measurement of fluorescence yield at 77°K, and also as a result influence the measured yield ratio.

From an analogy with tyrosine residues, a particular quenching reaction which must be considered and one which may proceed at 77°K is that produced by the disulfide link. Lysozyme is a particularly fine example; there are six tryptophan residues, five of which are vicinal in the native conformer to disulfide links, but only two in primary sequence. Perhaps the large energy loss in luminescence at 77°K is accounted for by the presence of tryptophan residues which are adjacent to disulfide links, either in the primary sequence or in the tertiary structure. These residues may not require a large Arrhenius energy of activation to undergo an electron ejection into the acceptor—the disulfide link—and so the reaction proceeds at 77°K. Lysozyme may represent the extreme example of this behavior; other proteins having residues in this situation also have others remote from disulfides. For example, RNase T_1 with its single tryptophan shows behavior similar to the monomer; there is no major energy loss in its luminescence. The primary sequence indicates that there are no disulfide links in the vicinity of this tryptophan. Other quenching reactions, in particular the imidazolium and cysteinyl quenching, are suppressed by cooling to 77°K. Hence, it is feasible that only a disulfide link quenching proceeds at this low temperature and entirely accounts for the singlet internal conversion at 77°K.

The temperature dependence of the phosphorescence yield has been investigated by several workers.[272,355,356] The quenching was dominated by O_2 quenching. Since the O_2 concentration cannot be adequately reduced to insignificant levels, it was not possible to study the intrinsic thermal dependence of the triplet phosphorescence yield.[357] Therefore, the data obtained to date are empirical in nature. Differences are found between the thermal behavior of tyrosine and tryptophan. Tyrosine phosphorescence is quenched more effectively than tryptophan phosphorescence. Differences in O_2 quenching effectiveness are found between different proteins and different conformers of the same proteins. Until the combination of O_2 quenching, thermal deactivation of the triplet, and thermal dependence of intersystem crossing can be separated it is not possible to understand the processes that take place as the temperature is increased above the 100°K temperature where the limiting values prevail.

5.16. Luminescence Lifetimes

5.16.1. Fluorescence

Several proteins have had their fluorescence lifetimes determined at

298°K.[228,229,358-361] No lifetimes have been measured at low temperatures. The values are collected together in Table 10. The natural lifetime has also been calculated using a tryptophan quantum yield value corrected for both transfer and inner-filter-effect absorption screening.

The workers who have used pulsed fluorometers report that the decay of fluorescence intensity can be fitted by a single exponential value. This is confirmed by the good agreement between values, where the same protein has been studied, when modulation phase shift and degree of modulation methods are compared to values obtained with pulsed light sources. The natural lifetimes range from 11 to 32 nsec, but many proteins do not differ significantly from 15 nsec, which is the value of the indole natural lifetime. A value of the natural lifetime that is greater than 15 nsec can be explained by one of two phenomena:

(a) There can be tryptophan residues which do not emit any fluorescence, and so simply act as inner absorption filters to those residues that do emit.
(b) An exciplex can form, and since these new species constitute an emission with an entirely different electronic origin, the natural lifetime can alter.

Mataga et al.[362] have studied model systems in which he found that there was a greater natural lifetime as the Stokes' shift of an exciplex fluorescence increased. We lack sufficient data to properly distinguish between these two possible causes of an enhanced natural lifetime in protein tryptophan fluorescence. Since there does not appear to be an obvious correlation between Stokes' shift and the magnitude of the natural lifetime, the presence of an enhanced natural lifetime for protein fluorescence is best interpreted by suggesting that there are tryptophan residues which do not contribute to fluorescence.

DeLauder and Wahl[362] have found that the intensity decay of the fluorescence of human serum albumin can be fitted with two distinct exponential decay times, even though the protein has only a single tryptophan residue. There is a shorter decay component with a lifetime of 3.3 nsec, which comprises 65% of the emission intensity, and a second component with a lifetime of 7.8 nsec. Lehrer[362] has reported that iodide quenches human serum albumin by two separate processes, and confirms the presence of an inherent heterogeneity to the tryptophan fluorescence of this protein. Iodide anions bind to HSA, and on doing so quench 40% of the fluorescence. The remainder of the fluorescence is quenched collisionally. Different conformations alone can account for such a heterogeneity since only a single tryptophan residue is possessed by this protein; the different conformations place tryptophan in different microenvironments.

5.16.2. Phosphorescence

In the earlier discussion of the properties of the monomers it was emphasized that the lifetimes of the three monomers are not known to better than ±16%. The value for the decay constant of the zwitterion of tryptophan which appears to be the most reliable is 6.65. Lifetime values reported for proteins also are subject to a large variation among workers. [217,218,229,241,363,364] Values ranging from 5.5 to 6.0 sec have been reported most frequently. It is probable that the lifetime of tryptophan in a protein is smaller than the monomer lifetime, and individual variations among proteins are also small. All workers who have measured the decay of tryptophan phosphorescence from proteins have found that the decay is fitted by a single exponential value. The decay constant has not been found to depend on the wavelength of either emission or excitation. Only when tyrosine is also excited, and the wavelength selected from the emission is chosen so that both tyrosine and tryptophan have appreciable emission, can two decay constants be found.[217,241]

Lysozyme is a unique protein. It decays with a significantly reduced decay constant from the monomer when it is in its native conformation. A 1.4 sec decay time has been found by both Longworth[220] and Churchich.[330] Both workers also found that this shortened decay value accounts for 5/6 of the intensity, but that there is a component with a longer value that is more typical of other protein tryptophan residues; 4.4 sec was estimated to be the decay constant of the long-lived component. Denaturation of the lysozyme with urea,[220] or base,[220] or reduction of the disulfide links[330] restored the decay constant of tryptophan to a more usual value of 5.6 sec. There is no adequate explanation for the reduced phosphorescence lifetime. Perhaps it is related to the presence of several tryptophan residues that occur with close proximity, although the decay constant of poly-L-tryptophan does not differ greatly from that of the monomer. Another explanation may be the proximity of disulfide links to five out of the six tryptophan residues in the native conformation of lysozyme, though the mechanism of shortening the lifetime is not specified.

The lifetime of phosphorescence from proteins does not appear to be greatly influenced by changes in environment of the tryptophan; the lifetimes of RNase T_1 and corticotrophin are equal. Truong *et al.* report some dependency on conformational state in the lifetime.[241] Until the accuracy of phosphorescence lifetime measurements is increased, it is not clear whether significant differences have been found. The precision quoted by several workers for their lifetime measurements is sufficient for them to state that small changes in lifetime accompany differences in conformations. Until accurate values are reported, it is not feasible to discuss the significance of these observations.

A triplet yield for tryptophan residues in a protein has been obtained only for chymotrypsin. Nevertheless, the relative constancy of lifetime from protein to protein, even though the environments of the tryptophan residues differ greatly, suggests that the intersystem crossing rate to the triplet is not dependent on the conformation or environment.

5.17. Fluorescence Spectra of Protein Tryptophan Residues

5.17.1. Room Temperature

The fluorescence spectra of class B proteins are indistinguishable from that of tryptophan, when the spectra are plotted on a wavenumber scale and superimposed. A significant tyrosine contribution is found only from a few proteins. The tyrosine component can be minimized by exciting the protein fluorescence with light longer than 290 nm.

Matrix analysis of the fluorescence spectrum of several class B proteins has disclosed the presence of only two components, tyrosine and tryptophan. The tryptophan emission is therefore independent of the exciting wavelength.

Exceptions are known; the best-understood example is papain. Weinryb and Steiner[229] found that though tyrosine made a small contribution to the emission, it has a large fractional absorption at 280 nm, and electronic energy transfer from tyrosine to tryptophan largely accounted for the small quantum yield of tyrosine fluorescence from the protein. Varying the exciting wavelength altered the tryptophan emission spectrum when tyrosines were excited. The effect is found with activated papain, pH 5.0–7.5, or with unactivated papain. The phenomenon is not found when the intensely fluorescent tryptophan is no longer quenched by the cysteinyl residue of the active site—this situation prevails at pH 9. Now the fluorescence is dominated by the emission from a single polar tryptophan located in the active-site region; this tryptophan is not significantly sensitized by tyrosine absorption. The fluorescence peak shifted to the blue compared to the emission found when 295 nm light was used to excite tryptophan directly. Weinryb and Steiner suggested that in papain, tyrosine absorption selectively transfers to tryptophan residues which are not on the surface of the protein, and so these are residues which fluoresce with a smaller Stokes' shift than the exposed tryptophan. Direct excitation leads to emission mainly from the exposed tryptophan residues. Hence, as the fractional tyrosine contribution increased, the fluorescence peak blue-shifted. Another example is found with aplysia myoglobin but not, in contrast, with sperm whale myoglobin. The aplysia apoprotein has a fluorescence with two peaks, one centered at 328 nm, the other at 345 nm; a single peak is found for sperm whale apomyoglobin.

It is well known that the absorption spectrum of tryptophan is susceptible to alteration in energy as the environment is altered. From solvent perturbation studies, several proteins have been shown to possess tryptophans, some of which are exposed to the environment, and others are not. The fluorescence peak and the Stokes' shift of tryptophan monomers are very sensitive to the nature of the dielectric constant of their environment as a result of exciplex formation when polar complexes can form. Therefore, in those proteins which possess both buried and exposed tryptophan residues, it would be expected that the bandwidth of the emission spectrum would be greater because of the presence of overlapping emission bands with different Stokes' shifts. The effect will largely be shown by the wavelength of the peak intensity, since the variation in this value is considerably greater (an order of magnitude) than the variation in absorption peak. Proteins are known which have fluorescence peaks between 325 and 350 nm; their Stokes' shift varies from 0.43 to 0.73 μm^{-1}. Though treatment of the bandwidth by moment analyses such as those described by Stepanov and Gribkovski[365] has not been performed, on preliminary inspection it appears that the emission spectrum shape is independent of the magnitude of the Stokes' shift. A well-defined function has been found which fits absorption and emission spectra; it is the log-normal distribution.[366] Accurate analyses and fit to the measured absorption and emission spectra of complex biological compounds can be performed in principle, and an application of these methods to proteins appears profitable when fully corrected protein fluorescence spectra are available. Exceptions are papain, aplysia apomyoglobin, lysozyme, pyruvate kinase, and horse liver alcohol dehydrogenase.

5.17.2. Low Temperature

At low temperatures a comparison with monomer is simplified; now Stokes' shifts are little different among various proteins and the respective monomer. The fluorescence from the majority of globular proteins has a tyrosine contribution which can be minimized by use of long-wavelength excitation (>290 nm). Such a fluorescence is indistinguishable from fluorescence obtained from several tryptophan compounds. There is a peak at 325 nm, and there are two pronounced shoulders at 315 and 335 nm. There are, however, qualitative differences between native and denatured protein spectra. The native protein spectrum has clearly defined shoulders, like the monomer spectrum, and the two spectra are almost identical with each other. Emissions from denatured proteins are less well resolved; small inflections are the only indications of the presence of any structuring to the spectrum. There are indications of small differences in peak wavelengths, but the presence of tyrosine and scattered light complicates accurate measurements on this point.

Several proteins have been found which do not fit into the above pattern: ribonuclease T_1, staphylococcal endonuclease, avidin–biotin complex, elastase,[367] TMV coat protein (on the virus and separated), and *Drosophila* alcohol dehydrogenase.[368] Their fluorescence is distinct from the monomer spectrum. The first example was provided by Konev,[369] who found a fine-structured fluorescence from single crystals of α-amylase. All the above proteins have a fluorescence, and a phosphorescence, which shows well-defined fine structure; defined vibrational bands rather than shoulders exist (Figure 55). Such a fine-structured emission cannot be observed with monomers unless resort is made to the Shpol'ski effect, where indole is included as a guest in a crystalline host structure, typically a cyclohexane crystalline matrix. When included into a crystalline lattice most organic compounds, including indole, exhibit a fine-structured fluorescence with extremely small bandwidths. The individual components have widths approaching fundamental limitations. Several satellite bands giving a multiplet structure to each individual component of vibrational elements are typically found. This hyperfine structure reflects the presence of indoles in different crystalline sites. As there are several proteins which show fine-structured fluorescence, including proteins with more than a single tryptophan, this suggests that the broader vibrational pattern found with a majority of globular proteins investigated results from multiple environments for the individual tryptophan residues. The measured spectrum is a result of an overlapping of different fluorescent spectra from each tryptophan site. Since there are proteins with single tryptophan residues which also have unresolved fluorescences at 77°K, conformational populations must also be considered. A distinction between the individual residue sites may be feasible by choice of exciting wavelength, because small differences would also be expected to apply to the absorption spectrum. The presence of any electronic energy transfer would seriously complicate an analysis, as would differing quantum yields. These are topics which have only begun to be investigated and it is premature to be dogmatic on any aspect. Denaturation of all of those proteins which give a highly finestructured emission in their native state restores the emission to the pattern found from other denatured proteins. Now the emission is not distinguishable from that of the monomer in the same glassing solvent; it is not fine-structured, since no unique site is likely for an individual residue. The effect of denaturation eliminating fine structure is regularly observed with many proteins.

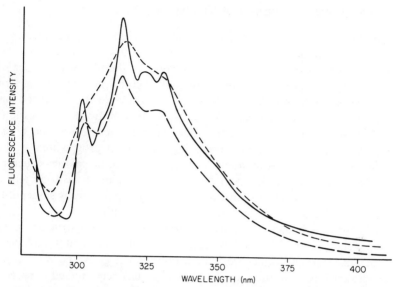

Figure 55. Low-temperature fluorescence of ribonuclease T_1 and staphylococcus endonuclease. Solvent, EGW at pH 7; temperature, $77°K$; concentration, $E^{10\,mm}_{280\,nm}$ = 0.5; excitation, 278 nm. (———) Ribonuclease T_1; (– – – –) staphylococcus endonuclease; (---------) ribonuclease T_1 in 8 M urea.

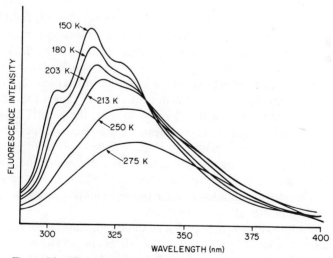

Figure 56. Fluorescence of endonuclease at different temperatures. Solvent, PEW at pH 7; concentration, 1 mg/ml; excitation, 280 nm.

5.18. Phosphorescence Spectra of Tryptophan

Unlike the fluorescence from a tryptophan derivative in polar glasses at 77°K, the phosphorescence is fine-structured, with several resolved vibrational bands. A detailed vibrational analysis of the phosphorescence of indole has been reported,[370] and though the corresponding analysis of the absorption and Shpol'ski fluorescence is not available in detail it too has been determined.[371] A conclusion was that the vibrational progressions involved and their electronic origin differ between fluorescence and phosphorescence. There are several overlapping vibrational progressions in the phosphorescence. With the monomer tryptophan, the principal fine-structure peaks are located at 405, 432, and 456 nm. These values have been reproduced by many workers to within 1 nm; they are not particularly susceptible to instrumental artifacts because of their narrow width.

Whether or not there is a solvent variation to the fine-structure peak location (phosphorescence Stokes' shift) has not been well established. The previously quoted values were for a polyalcohol glass, the most commonly chosen solvent in use today. Indole itself has been studied in several glasses and matrices. It has a peak at 404 nm in a hydrocarbon glass, and a peak at 406 nm in formamide matrix or dimethylformamide–glycol glass mixture. Therefore, the solvent dielectric can only cause small changes in the wavelength (Stokes' shift) of phosphorescence maxima at 77°K. Changes of a comparable magnitude are found with the absorption and fluorescence spectra at 77°K too.

The shape of the phosphorescence from native proteins is found to be indistinguishable from that of tryptophan, when the tyrosine contribution is absent. The fine-structure peak wavelengths are red-shifted and lie at 5–7 nm longer wavelengths (410–412 mm). All peaks are equally influenced. For convenience, the well-defined peak on the blue edge of the emission is chosen as a reference index. Denaturation with acid or alkali moves the peaks toward the monomer location with a 5 nm blue shift; the peak is at 407 nm ± 0.5, 1–1.5 nm to the red of the monomer. The tryptophan peaks can be clearly resolved at 407 nm in both acid and alkaline pH value emissions from proteins. This is true for the alkaline conditions even though the phosphorescence is dominated by the tyrosinate emission, which has a very broad structureless band. The vibrational width of the tryptophan component is increased 50–70 mm^{-1} at alkaline pH values.

Denaturation with 8 M urea or 5 M guanidinium chloride, or else neutral organic solvents such as propane-1,2-diol, shifts the emission to the blue to a lesser extent; there was a 3 nm shift found, so that the first peak lies at 409 nm ± 1 nm. Organic solvents which also disrupt the tertiary structure but support highly helical conformers of polypeptides, such as 2-chloroethanol, shift the phosphorescence peaks by 5–6 nm to the blue. Organic solvents such as ethylene diamine that support instead random coiled conformers also produce a similar perturbation of the tryptophan phosphorescence. Anionic detergents such as sodium dodecyl sulfate can cause both 3 nm and 5 nm shifts toward the blue in phosphorescence. Bovine serum albumin has binding sites with large affinity constants for five detergent molecules. The binding of five detergent anions to BSA causes a 3 nm shift in phosphorescence to the blue. There are also other binding sites with weaker affinity constants with the albumins and all globular proteins. These sites are capable of binding upward of 80 detergent anions per molecule. When many of these sites have been occupied, the protein phosphorescence is shifted to the monomer location; there is a 5 nm blue shift. Proteins are believed to be denatured by binding the detergent and to be in highly helical conformations in the presence of a large number of bound detergent anions.

The linewidth of the fine-structure peak of several globular proteins in their native conformations is 30 mm^{-1} ± 2 but narrows to 20 mm^{-1} ± 2 in acid-denatured conformers, though the width is unaffected by urea or guanidinium salts. The monomer has fine-structure widths of 18.5 mm^{-1} ± 0.2 at all pH and solvent conditions. There are proteins which differ from this pattern. Examples are RNase T_1, lysozyme, elastase, avidin, and subtilisin carlsberg. Both their fluorescence and phosphorescence have fine-structure bandwidths of 10 mm^{-1}; RNase T_1 has a peak wavelength of 406 nm, while lysozyme has a peak at 416 nm. These represent two extremes; the other proteins have peaks between 409–412 nm. Denaturation of RNase T_1 causes the phosphorescence spectrum to move to the red and broadens its bandwidth to 30 mm^{-1}. The emission from the RNase T_1 results from an emission by a tryptophan residue buried within the hydrophobic environment of the protein. With lysozyme, possible interactions are that of a tryptophan having another tryptophan as a neighbor (Trp-62, -63, and -108) or else having a disulfide link as a neighbor. These may cause the large red shift for the phosphorescence band of this enzyme.

It is not known whether complexing of the enamine nitrogen of tryptophan with an amide or another residue can affect the phosphorescence Stokes' shift, vibrational width, or yield. The origin of the red shift to native protein phosphorescence is not known.

Purkey and Galley[295] have suggested that phosphorescence from horse liver alcohol dehydrogenase can be interpreted as a superimposition of two overlapping spectra separated by 35 mm^{-1}. They were able to resolve the phosphorescence spectrum because of selective quenching of one component when NADH was bound, or by a selective triplet-to-triplet sensitization of tryptophan emission from sensitizer bound at the active site. Differences were also found between excitation spectra determined at selected emission wavelengths. This situation cannot necessarily be true for all proteins. The bandwidths of human serum albumin with only its single tryptophan, or else corticotrophin, are 30 and 20 mm^{-1}, respectively. Comparable bandwidths are found from most proteins, but these proteins possess multiple tryptophan residues, and these are in distinct environmental classes in several cases. The other extreme is provided by the phosphorescence of tobacco mosaic virus protein or α-amylase. Here there is a narrow bandwidth yet more than one tryptophan residue to the proteins. The role that electronic transfer plays in modifying vibrational linewidths and the relationship of the location of tryptophanyl residues in the different sites (exposed or complexed or buried) remains imperfectly understood.

5.19. Stokes' Shift of Fluorescence

A fundamental characteristic of fluorescence from a molecule is its Stokes' shift. The Stokes' shift has been found to depend on the nature of the surrounding solvent dielectric constant by a relationship first elaborated by Lippert[372] and by Mataga et al.[373] The expression given by Kawski is the one most faithfully obeyed by polar compounds.[374] This treatment involves a complicated description of the solute; an ellipsoid is used.

For many molecules, the use of a spherical cavity is an adequate approximation, and is the one commonly adopted. Teale[206] found that the fluorescence peak of tryptophan compounds was strongly dependent on the solvent dielectric; the absorption was an order of magnitude less dependent. This result has since been reconfirmed by other workers. The Stokes' shifts have been analyzed according to equations similar to these of Kawski. For tyrosine, the dipolar moment increase is found to be 0.7×10^{-28} Cm, and for indole 3.5×10^{-28} Cm (where 1 Debye=3.3×10^{-29} Cm). The greater dipole moment of the excited state of indole causes a reorientation of the solvation shell of a polar solvent. The process competes with the duration of the excited state. The relaxation causes a lowering of the energy level of the excited state, with an increase in the energy level of the ground state that is formed by emission from the solvated excited state. The two effects combine to give an increased Stokes' shift.

Liptay[375] has considered the free energy of solvation and derived expressions for the corresponding enthalpy and entropy. It is possible to analyze the temperature dependence of the Stokes' shift and obtain the magnitude of the dipole moment enhancement.[376] Another means of studying the effect is to determine the rate of shifting of the Stokes' shift in the course of fluorescence decay by time-window pulse fluorometry. This too has not so far been applied to the indole fluorescence in polar solvents.

Walker et al.[377] applied the equations derived by Mataga[373] and by

Lippert,[372] which are similar equations to the expression of Kawski.[374] They gave a poor fit to the indole data. Weber[378] found good agreement with phenol. To alter the dielectric constant of the solvent, Walker et al.[377] added alcohols to hydrocarbon solvent, both of which had been carefully dried of water to allow miscibility. At low alcohol concentrations, there was a rapid change in Stokes' shift. The magnitude of the change was greater than predicted by the solvation models. The value obtained from the alcohol–hydrocarbon mixtures by extrapolation to the pure hydrocarbon situation gave a Stokes' shift which was twice the measured value. The equations also predict an asymptotic limit at high dielectric constant values. Walker et al. found a linear dependence on fractional alcohol concentration up to pure alcohol.

Walker et al. felt that there must be some other process occurring besides a solvation of the excited state. When the complete spectra were examined, rather than just the Stokes' shift, it was apparent that a drastic alteration in the electronic nature of the fluorescence transition occurs with small concentrations of alcohol added to the hydrocarbon. Indole compounds in a hydrocarbon exhibit a fluorescence with a fine structure. There is a Stokes' shift comparable to that observed with other aromatic hydrocarbons. In alcohol, the fluorescence lacks any fine structure; only poorly defined shoulders can be found even at 77°K. There is a large Stokes' shift at ambient temperatures, and this is significantly less at 77°K. When spectra were obtained at different alcohol concentrations, and then were superimposed, it was apparent that the hydrocarbon-type fluorescence was being progressively replaced by the alcohol spectrum. No intermediate spectra occurred. This is best illustrated by using the Job variational procedure. An isoemissive point is found in the fluorescence, while the absorption spectrum does not undergo any significant alteration when there is complete conversion to the alcohol-type fluorescence. The existence of an isoemissive point implies that there is a stoichiometric chemical reaction occurring. A complex forms between the excited indole and the alcohol molecule, and it is this species that is responsible for the emission in the polar solvent. Complexes have to be formed by the excited state since no alteration in the absorption was detected. Such complexes have come to be known as exciplexes. The pseudoequilibrium constant for indole-n-butyl alcohol complex is 110 liter mol^{-1}. Chignell and Gratzer[379] have studied the absorption spectrum of indole in hydrocarbon–alcohol mixtures and find that at higher alcohol concentrations a complex will form between the alcohol and the indole ground state, the complex presumably involving H-bonding between the alcohol group and the enamine nitrogen proton. The equilibrium constant for indole-n-butyl alcohol H-bonded complex reported by Chignell and Gratzer is of the order of 1 liter mol^{-1}. With other polar cosolvents, such as dioxane, it is not possible to separately study exciplex and ground-state complexes since both have equal equilibrium constants, so in these cases the term exciplex is inapplicable. Rather, from the extent of the absorption and emission shifts, it is possible to obtain information on the difference in H-bonding energy between ground and excited state. The excited state must have a far stronger H-bonded complex than the ground to account for the large Stokes' shift.

Weller et al.[380] have studied the magnitude of the dipole moment of charge transfer exciplexes by altering the solvent dielectric. They found that, for the exciplexes they investigated, the change in dipole moment on excitation was comparable to that reported for indole.

One of the most obvious features of protein fluorescence, when several proteins are compared, is the large variability in their Stokes' shift. This is largely reflected in the wavelength of the peak of the fluorescence at 298°K, since the absorption maximum does not change significantly. Values for representative proteins are collected in Table 10. The range of Stokes' shift found at 298°K is from 0.52 to 0.73 μm^{-1}. Ribonuclease T_1 has an unusually small Stokes' shift, 0.44, and azurin is smaller still, 0.30 μm^{-1}. The Stokes' shifts of these last two proteins are similar in value to those for all proteins at 77°K. Denaturation of the protein with urea or with guanidinium chloride eliminates the variability in the Stokes' shift. All globular proteins investigated have had a Stokes' shift of 0.72 μm^{-1}. Therefore, the large variability in shift from protein to protein is associated with the native tertiary structure of the protein and is not a result of either the primary or the secondary structure in any major degree.

There are two mechanisms known which could account for both the magnitude and the variability of Stokes' shifts. One is based on a solvation model and suggests that there is variability in the extent of solvation due to differences in local environment of tryptophan in proteins. The second suggests that exciplexes form in proteins, and, as they are lower in energy, they become the principal source of emission. Since the exciplex has a large dipole moment and will also lie in different environments for individual proteins, different Stokes' shifts are expected. The crucial difference between these two models is that the solvation model does not require any stoichiometry, whereas the exciplex model does. To distinguish between the two mechanisms only requires that evidence for or against stoichiometric reactions be found. The exciplex model can be regarded as a photochemical explanation, while the solvation is principally a photophysical explanation.

As a preamble, it is necessary to show that there are indeed proteins to which neither model applies. These proteins would possess a tryptophan residue entirely surrounded by nonpolar groups. X-ray crystallographic analysis has shown that such residues do occur within protein structures, though invariably there are other tryptophan residues in these proteins which are exposed to the environment, or else they are involved in polar complexes at their enamine nitrogen's proton atom. The first indication of a hydrocarbon-like environment for protein tryptophans was found in the fluorescence of single crystals of α-amylase at 77°K. Bobrovich and Konev[369] discovered that the fluorescence from crystals of this enzyme was highly fine-structured, and that there was also a small Stokes' shift to the intensity maximum. The fluorescence was typical of the Shpol'ski fluorescence of indole in a cyclohexane crystal matrix. A second protein which exhibits a fine-structure fluorescence in solution, even at 298°K, is ribonuclease T_1 of *Aspergillus oryzae*. This protein has a single tryptophan; there is no detectable tyrosine fluorescence, or phosphorescence from the protein. A temperature study of the fluorescence shows that there is a progressive decrease in yield and in the degree of fine structuring as the temperature is raised from 100 to 300°K, but this is not accompanied by any significant change in the Stokes' shift (Yamamoto and Tanaka[210]; Longworth and Battista[381]). When the protein melts at 335°K, there is a large increase in the Stokes' shift (0.72 μm^{-1}), and tyrosine fluorescence can be readily observed. Denaturation by acid, base, or urea or guanidium chloride also produces the same increase in Stokes' shift, and obliterates the fine structure in the fluorescence at 77°K. All these studies suggest that the single tryptophan residue of native RNase T_1 is in a pure hydrophobic environment, and that the fluorescence represents emission from a tryptophan that has not interacted with any polar group. Another protein which has an even smaller Stokes' shift is the copper protein azurin of *Pseudomonas*; both the holoenzyme and apoenzyme exhibit similar spectra, though the presence of bound copper reduces the quantum yield. The fluorescence maximum of azurin is at 308 nm, and there is no indication of any tyrosine contribution to the fluorescence at 298°K. Finazzi Agro et al.[209] conclude that the tryptophan is in a hydrophobic environment. The smaller Stokes' shift suggests that in RNase T_1 there are still interactions which lower the singlet energy level. There have been no 77°K fluorescence and phosphorescence studies performed with azurin, and whether there is a fine-structure fluorescence is not known, but the spectra of Finazzi Agro et al. suggest that there is residual fine structure at 298°K.

Vladimirov and Burstein[30] had shown that the Stokes' shift of both tryptophan and tryptophan residues in proteins is dependent on temperature. At 298°K tryptophan has a Stokes' shift of 0.76 μm^{-1}; at 77°K this is reduced to 0.43 μm^{-1}. Eisinger and Navon[94] investigated both the yield and Stokes' shift of tryptophan above 180°K, which was the devitrification temperature of their glass. As the temperature increased, the Stokes' shift increased to 0.76 μm^{-1}, and the yield was, effectively, a constant. No unique intersection points are observed when the spectra obtained at the various temperatures between 180°K and 250°K are superimposed. Following the analysis of Hamilton and Raz-Naqvi[177] this suggests that there is no stoichiometric complex formed which has a greater Stokes' shift. Rather, the continuous change in peak-intensity wavelength with temperature suggests compliance with the solvation model, indicating that tryptophan has a greater dipole moment in its excited state than its ground state. Below 180°K, Longworth and Battista[130] observed a unique intersection point. They suggested that the fluorescence studied by Eisinger and Navon was that of a tryptophan–glycol complex, which was then progressively solvated, rather than the fluorescence from uncomplexed tryptophan residues.

Longworth and Battista[130] have found that the endonuclease of *Staphylococcus aureus* exhibits the directly opposite behavior when the fluorescence spectra are studied over the same temperature range (Figs. 5–15). Superimposition of spectra obtained at different temperatures between 180°K and 240°K showed that there was a unique intersection point. The temperature-dependent behavior of the endonuclease can be interpreted by considering only two emitting spectral types, a low- and a high-temperature type. The low-temperature species has a fine-structured fluorescence similar to that of RNase T_1. There is a Stokes' shift of 0.46 μm^{-1}. The high-temperature form is structureless and has a greater Stokes' shift (0.54 μm^{-1}). To interpret these results, it was suggested that the tryptophan, though largely in a hydrophobic environment, forms an exciplex above 220°K. The X-ray crystallographic analysis of the single crystals of the endonuclease has shown that the single tryptophan is indeed largely surrounded by hydrocarbon-like side-chains, but the enamine nitrogen proton is still available to the surface of the protein. Other proteins with similar behavior are elastase and the avidin–biotin complex (but not uncomplexed avidin). We suggest that the transition from a fine-structured low-temperature fluorescence to a red-shifted structureless fluorescence, with a greater Stokes' shift, at room temperature, with the preservation of an isoemissive point for part of this transition, indicates that the emission originates in partially exposed tryptophan residues. Exciplex formation, as indicated by the existence of an isoemissive point in temperature-dependent fluorescence spectra, is not restricted to partially exposed tryptophan residues. Isoemissive points can be found in the fluorescence spectra of denatured RNase T_1, denatured staphylococcal endonuclease, and both native and denatured human serum albumin.[381] Here the Stokes' shifts are 0.73 μm^{-1} for urea-denatured proteins and 0.62 μm^{-1} for native HSA. But the fluorescence never becomes distinctly fine structured at 77°K. We suggest that the fluorescence originates from tryptophan residues which are always involved in complexes, by analogy to tryptophan in polyalcohol glasses between 150 and 180°K. This is then another type of partially exposed residue.

Other proteins are known which behave indistinguishably from tryptophan monomers between 180 and 250°K. An example is native bovine serum albumin. The urea- or heat-denatured protein did show unique intersection points in temperature-dependent spectra. The simple polypeptide hormones corticotrophin and glucagon change their yield and Stokes' shift continuously with temperature, so it is not possible to be definitive on whether or not their emission is from exciplexes.

Another complication is found with lysozyme. This protein is known from its molecular model to have partially exposed tryptophan residues, though not all appear to be involved in the fluorescence. The temperature-dependent spectra can be interpreted by suggesting the presence of both exciplex formation below 220°K and solvation above 220–250K, i.e., perhaps a mixture of partially buried and exposed residues.

The temperature dependency of the Stokes' shift is found to be consistent with other excited-state studies such as absorption solvent perturbations and with the known molecular structure. A measurement of the temperature dependency of the spectra will allow a provisional assessment as to the gross nature of the environment of the emitting tryptophan residues.

5.20. Heterogeneity of Environment

Analysis of the molecular models of proteins shows that the tryptophan residues can exist in a variety of environments:

(a) Hydrophobic, buried in the interior of the protein structure and unavailable to the external solvent molecules.
(b) Hydrophobic, but the residue is complexed with either amide carbonyl or with a polar side-chain.
(c) Partially exposed—largely hydrophobic environment—there is the possibility of solvent complexes forming.
(d) Fully exposed, the residue lying on the exterior of the protein and completely available to the solvent.

Many group-specific chemical reactions have demonstrated that not all residues will react, or that not all residues react at the same rate, in native proteins. Excited-state studies, whereby the energy of the Frank–Condon excited level is changed by solvent dielectric, have shown that not all residues have their absorption influenced by changes in solvent composition.

There are no obvious indications of heterogeneity in either the fluorescence or phosphorescence spectra of proteins. The spectra of proteins with single tryptophan residues do not usually differ in either bandwidth or range of Stokes' shifts from those with several tryptophans. For example, the bandwidth of the first vibrational peak of either avidin or lysozyme is equal to that found for ribonuclease T_1 or subtilisin carlsberg, 10–12 mm^{-1}. The other extreme is shown by human serum albumin; the bandwidth is as broad as that shown by bovine or egg albumin, or denatured lysozyme (30 mm^{-1}).

5.20.1. Wavelength-Dependent Emission Spectra

The observation of a wavelength-dependent emission spectrum would indicate clearly that there is more than a single source of emission. The matrix analytic procedure has not shown any evidence for this with the albumins or pepsin. Phosphorescence, with its fine structure, is more suitable to study than the fluorescence at 298°K. One example is known which does exhibit the presence of two overlapping spectra, horse liver alcohol dehydrogenase. Purkey and Galley suggest that the phosphorescence comprises the summation of two separate spectra of the two trytophanyl residues, separated by 35 mm^{-1}.

Weinryb and Steiner[229] have found that the fluorescence of inactivated papain at 298°K originates entirely from the tryptophanyl residues. The Stokes' shift is dependent on the wavelength of excitation, and the wavelength maximum decreases from 340 to 330 nm, as the exciting wavelength changes from 295 to 270nm. Weinryb and Steiner propose that the cause of the change in Stokes' shift is a heterogeneity in the fluorescence spectra of individual tryphanyl residues. When light shorter than 295 nm was used to excite the fluorescence, an appreciable fraction of the absorbed energy was transferred to the tryptophan from tyrosine absorption, and they suggest that the transfer occurs preferentially to adjacent residues which are buried, or that as a result of the greater overlap integral, tyrosine also preferentially transfers to the buried residues. Either way leads to an enhancement of a contribution to the emission from residues with smaller Stokes' shift, hence a shift in the maximum wavelength. Steiner has found that oxidation of an exposed tryptophan in activated papain with N-bromosuccinimide lead to a large decrease in fluorescence yield. The residual fluorescence after oxidation had a small Stokes' shift. The fluorescence contribution removed by the oxidation reaction was that part of the papain fluorescence subject to quenching by the active-site histidine and cysteine residues. The same residue was also quenched (presumably through a heavy atom effect) by mercury bound to the active-site cysteine sulfhydryl group.

5.20.2. Differential Quenching

Heterogeneity has been demonstrated by studying the quenching constant at different wavelengths in emission. Binding of substrates, inhibitors, or analogues frequently leads to alterations in quantum yield, usually a reduction. This is also accompanied by an alteration in the spectrum. Two well-established examples are the binding of substrates to chymotrypsin and the binding of biotin by avidin. Leher and Fasman and Sharon and coworkers have investigated in detail the binding of substrates and analogues to hen's egg lysozyme. Lehrer and Fasman[383,384] found that, when N-acetylglucosamine oligomers are bound to lysozyme at neutral pH values, there is a small preferential quenching of residues with a large Stokes' shift and an enhancement of the yield of residues with small Stokes' shifts. The fluorescence difference spectrum obtained between free and complexed lysozyme was found to have a positive component, plus a smaller negative peak at 360 nm. This was attributable to tryptophan residues previously exposed becoming buried and having an increased yield with an intensity peak at 330 nm. The major component to the fluorescence peaked at 342 nm. The derivatives used by Lehrer and Fasman are known to complex with the lysozyme by binding at sites A, B, and C of the enzymatic site. These are adjacent to tryptophanyl residues 62 and 63.[385] Sharon[387] and coworkers found that the binding of larger oligomers than were used by Lehrer and Fasman caused much greater selective quenching of residues with a larger Stokes' shift, and there was pronounced enhancement of the smaller Stokes' shift components. The total fluorescence from the lysozyme complex had its intensity peak at 332 nm. The derivatives used are believed to involve complexing at the functional sites E and F of lysozyme together with sites A, B, and C. An additional tryptophan residue adjacent to the catalytic site is residue 108. One suggestion would be the burying of the tryptophan residue by substrate binding; another is the relief of a quenching reaction. Both account for the fluorescence effects. Recent chemical modification studies by Sharon[386] implicate Trp-108 as the predominant source of fluorescence, and the residue is less quenched in complexes than in the uncomplexed enzyme.

Pyruvate kinase of rabbit muscle complexes with either metal activator cations or with its substrate, and in each case fluorescence changes are found. The difference spectrum between free and complexed enzyme, unlike lysozyme, is a broad band, and is significantly wider than the emission from a single tryptophan. Suelter[388] had shown that the complex formation is accompanied by a burying of tryptophanyl residues. Therefore, there must be several interactions influencing different tryptophanyl residues in the complex formation, affecting residues with small and large Stokes' shifts. Another situation is exhibited when this enzyme complexes with its metal activator cation; there is an enhancement in the yield. The difference spectrum between complexed and uncomplexed enzyme shows that residues with larger Stokes' shifts are preferentially quenched, and replaced by residues with smaller Stokes' shifts. The absorption and emission data were consistent with the burying of tryptophan in the enzyme on complexing with the enzyme.

Transfer quenching has been used by Lehrer and Fasman[383] to demonstrate the heterogeneity of tryptophan fluorescence from lysozyme. Ionized tyrosine residues act as effective quenchers of tryptophan residues, and Lehrer and Fasman found that residues with smaller Stokes' shifts are quenched to a greater extent than exposed residues. Lehrer and Fasman felt that the reason for the larger quenching of the buried residues is the greater overlap between their emission and the tyrosinate absorption.

Other transfer quenching studies involve the binding of dyes to the protein. Elkana[389] used salicylamide as the quenching dye with both lysozyme and bovine serum albumin. She found that there is a different quenching constant at various emission wavelengths. The result was that the peak wavelength increased as there was more transfer quenching in both proteins.

When a specific quenching reaction involves ionizable groups, alteration of pH may disclose the presence of such reactions by causing preferential quenching of a tryptophan with distinctly different Stokes' shift from the remainder of the tryptophanyl residues of the protein. Lehrer and Fasman have reported a particularly clear example of this for lysozyme tri-N-acetyl glucosamine complexes. The complex, unlike the free enzyme, has a quantum yield which is pH dependent, which indicates formation of a carboxylate anion which relieves the quenching process. The difference spectra between different pH values show that the tryptophan involved has a small Stokes' shift. Sharon and cowokers have shown that the pH effect is absent in lysozyme complexes where Trp-108 has been oxidized by iodine. Both Lehrer and Fasman and Sharon and coworkers consider that Trp-108 is involved in a quenching reaction with Glu-35 and that fomation of the trimer complex weakens this interaction so that ionization of the carboxyl can take place. Independent studies, both X-ray crystallographic and NMR, show that the trimer does not directly complex with Trp-108, and so the effect of the binding of the trimer is a result of a conformation change induced into the enzyme.

Since the binding of dyes to many proteins is in itself a heterogeneous process, there are rarely unique binding sites for the dyes. Badley and Teale[229] adopted the use of chromophoric substrate analogues. Specific binding of chromophores had previously been shown to lead to transfer quenching. Galley and Stryer[390] had used chromophoric sulfonamide to bind to carbonic anhydrase, since sulfonamides were known to be effective competitive inhibitors of this enzyme. Chen and Kernohan[391] reported similar experiments with other sulfonamides, and later Galley and Stryer[342] studied the quenching of tryptophan by transfer to proflavin, which binds to the active site of chymotrypsin. Badley and Teale[128] synthesized several different chromophoric derivatives of phenylalanine which all bind to the active site of pepsin. They found that the singlet lifetime was reduced to a different extent than the yield by the different dyes, each with differing overlap integrals (i.e., different quenching effectiveness). They concluded from this that there is a heterogeneity of distances between the active site and individual tryptophan residues. A similar conclusion had been reached previously by Galley and Stryer with proflavin–chymotrypsin complexes.[390] They found that the triplet state of tryptophan sensitizes the singlet state of proflavin and studied the kinetics of this sensitization. Three groups of tryptophans were proposed. Residues close to the binding sites which transfer in 50 msec, residues at a greater distance which transfer with a rate of 450 msec, and finally a small number of tryptophans which do not transfer to the dye at the active site. Inspection of the atomic model of chymotrypsin disclosed that there are indeed two tryptophans within 0.5–0.7 nm of the active sites and two trytophans at a greater distance, 2.5 nm. The remaining are situated at intervening distances. Since the transfer rate is a sixth-power function of the distance, this is in good agreement with the quenching of the triplet states by the bound dye.

Other systems of potential value for investigating heterogeneity of distance and environment by transfer quenching to residues bound at unique positions are provided by metalloproteins. The binding of ferric and cupric cations to conalbumin and transferrin quenches the fluorescence from the tryptophans. Far smaller effects are found with copper proteins, azurin, plastocyanin, and hemocyanin.[392] In the case of conalbumin, tyrosine and tryptophan have been implicated in the binding site.[393] Tyrosinate is formed on a metal binding and thus acts as a quenching site.

Rather than use site specific quenchers, Vladimirov,[394] Barenboim,[395] and later Burstein,[396] used simple quenchers of tryptophan fluorescence such as iodide, O_2, and nitrate. All are effective quenchers of tryptophan fluorescence, and are effective in quenching the fluorescence of proteins too. Vladimirov and Zimina[219] discovered that the effectiveness of iodide was dependent on the type of protein and the conformation of the protein. Lehrer[397] and Burstein[398] and Teale and Badley have studied the iodide quenching of protein fluorescence; Kirby and Steiner[399] have used iodate in place of iodide. Not all tryptophans are equally quenched, and, in fact, not all can even be quenched by the presence of iodide anions. Lehrer found that the tryptophans with a greater Stokes' shift are quenched more readily than tryptophans with a lesser shift. The presence of substrate influenced the differential effectiveness of iodide as a quencher. More studies of this nature can be expected, since the iodide quenching provides data comparable to those previously obtained by investigating the effectiveness of natural organic solvents in altering the absorption spectrum.

5.20.3. Solvent Perturbation

Steiner and Edelhoch[269] found that neutral organic solvents will enhance the yield of tryptophan monomer systems and the yield of tryptophans in proteins. The effectiveness of neutral organic solvents in perturbing the yield can be used to investigate the heterogeneity of the protein tryptophans. A denatured conformer can be used as a control. Steiner and Edelhoch have found that the fluorescence yield enhancement, when studied as a function of molar fraction of organic cosolvent, provides similar data to those previously obtained by absorption solvent perturbation. Examples of proteins have been found which have completely exposed or completely buried tryptophans and situations where both types are present.

5.21. Heterogeneity of Phosphorescence

Purkey and Galley[295] have reported the only example of evidence for two components within the fine-structure phosphorescence from proteins. However, no intensive or systematic attempt has been made to investigate this aspect of the phosphorescence of globular proteins.

As triplet–triplet transfer involves an electron exchange, when a sensitizer binds to specific sites on the protein, it can be expected to selectively excite individual tryptophan triplets. A comparison with the triplets formed by direct absorption can be used to demonstrate heterogeneity in the phosphorescence spectrum. An example can be found in the work of Galley and Stryer, where an aminonaphthalene sulfonamide bound to the active site of carbonic anhydrase did not excite tyrosine but did excite tryptophan. The sensitizer used had a lower energy than the tyrosine triplet state. When Galley and Stryer[342] reacted a related dye to the active serine of chymotrypsin, no sensitized phosphorescence could be observed. We now know that there is not a tryptophan residue in the active site of this serine endoprotease.

Another example of heterogeneity has also been provided by the work of Galley and Stryer. Proflavin binds to the active site of chymotrypsin and quenches the triplet state by a triplet-to-singlet transfer. The delayed fluorescence produced has more than a single decay constant.

5.22. Transfer and Heterogeneity

The role that electronic energy transfer plays in influencing any search for heterogeneity will be complex, and this has not so far been adequately explored. Not all workers agree that energy transfer between tryptophans occurs in the majority of globular proteins; rather they feel that it is restricted to rare examples of residues in van der Waals contact.

Anderson and coworkers[295] report that aplysia apomyoglobin, a protein with only two tryptophan residues, has a heterogeneous fluorescence. One protein tryptophan lies in a polar environment and contributes predominantly to the blue region of the protein fluorescence; the second tryptophan lies in a polar region and leads to the red region of the fluorescence. Sperm whale apomyoglobin also has only two tryptophan residues, but both are located in nonpolar environments. Kirby and Steiner[399] found that the majority of the fluorescence from spermwhale apomyoglobin at pH 8 originates in a single tryptophan residue. Aplysia apomyoglobin differs from that of the sperm whale by the addition of a red component to its spectrum—here both residues contribute to the fluorescence, even though they possess different Stokes' shifts. Acid[339], urea, or guanidinium[400] denature sperm whale apomyoglobin, shift tryptophan fluorescence toward the red, and enhance tyrosine fluorescence. Independent solvent perturbation of the absorption discloses that under such solvent conditions both tryptophans are fully available to interact with the solvent. Anderson et al. determined the FAPS of aplysia apomyoglobin and of sperm whale apomyoglobin; they isolated the fluorescence with a color filter which passed predominantly the red portion of the spectrum. With sperm whale apomyoglobin, a typical FAP spectrum with indications of tyrosine transfer depolarization was obtained. With the aplysia protein there are no tyrosine residues, but the FAP spectrum was extensively depolarized and lacked fine structure details commonly observed. The proposal made to account for this most unusual FAPS was the transfer from nonpolar to polar tryptophan. The excitation spectra determined with 307 nm light (blue) and with 362 nm light (red) differ markedly. Since the nonpolar residue absorbs longer wavelengths (presumed to be H-bonded in a low-polarity environment) and the reverse applies in emission, transfer between the two tryptophans cannot be complete. It is only because a filter which isolates the red region is used, which biases the results toward indications of extensive intertryptophan transfers, that extensive depolarization can be found.

ACKNOWLEDGMENTS

I wish to thank the copyright holders for their kind permission to reprint several figures. Thanks especially to Nette Crowe for preparing the artwork and to Anne Skeel and Carole McCulley for their valuable contributions.

REFERENCES

1. C. Dhéré, in "La Fluorescence en Biochemie," Masson, Paris (1934).
2. J. B. Becconi, *Philosophical Transactions 44*, 81 (1746) (English translations).
3. O. Warburg and E. Negelein, *Biochem. Z. 202*, 202 (1928); 214, 64 (1929).
4. O. Warburg, in "Heavy Metal Prosthetic Groups and Enzyme Action," p. 144, Oxford University Press (1949).
5. O. Warburg, E. Negelein, and I. W. Christian, *Biochem. Z. 214*, 26 (1927).
6. T. Bucher and E. Negelein, *Biochem. Z. 311*, 162 (1942).
7. T. Bucher and L. Kaspers, *Naturwissenschaften 32*, 93 (1946).
8. T. Bucher, *Advan. Enzymol. 14*, 1 (1953).
9. R. Havemann and P. Wolff, *J. Physik. Chem. 241*, 246 (1952).
10. W. Broser and W. Lautsch, *Angew. Chemie 67*, 713 (1955).
11. W. Lautsch, W. Broser, W. Bidermann, and H. Gnichtel, *Angew. Chemie 66*, 123 (1954).
12. W. Lautsch and E. Schroeder, *Z. Naturforsch. 9b*, 277 (1954).
13. T. Bannister, *Arch. Biochem. Biophys. 49*, 222 (1954).
14. S. F. Velick, *J. Biol. Chem. 233*, 1455 (1958).
15. V. G. Shore and A. B. Pardee, *Arch. Biochem. Biophys. 62*, 355 (1956).
16. F. Kubowitz and E. Haas, *Biochem. Z. 257*, 337 (1933).
17. F. L. Gates, *J. Gen. Physiol. 18*, 265 (1934).
18. R. B. Setlow and B. Doyle, *Biochim. Biophys. Acta 24*, 27 (1957).
19. R. B. Setlow, *Biochim. Biophys. Acta 16*, 444 (1955).
20. P. Debye and J. O. Edwards, *Science 116*, 143 (1952).
21. R. S. Becker, "The Theory and Interpretation of Fluorescence and Phosphorescence," Wiley, New York (1969).
22. L. I. Grossweiner, *J. Chem. Phys. 24*, 1255 (1956).
23. L. I. Grossweiner and W. A. Mulac, *Radiation Res. 10*, 515 (1959).
24. S. Aksenkev, Y. A. Vladimirov, V. I. Ilenev, and Ye Ye Fesenko, *Biofizika 72*, 63 (1967).
25. R. Santus, A. Hélène, C. Hélène, and M. Ptak, *J. Phys. Chem. 74*, 550 (1970).
26. R. B. Setlow and B. Doyle, *Arch. Biochem. Biophys. 46*, 31 (1953).
27. G. Weber, *Advan. Protein Chem. 8*, 415 (1953).
28. V. G. Shore and A. B. Pardee, *Arch. Biochem. Biophys. 60*, 100 (1956).
29. R. L. Bowman, R. A. Caulfield, and S. Udenfriend, *Science 122*, 3157 (1955).
30. D. Duggan and S. Udenfriend, *J. Biol. Chem. 223*, 313 (1956).
31. F. W. J. Teale and G. Weber, *Biochem. J. 65*, 476 (1957).
32. G. Weber and F. W. J. Teale, *Trans. Farad. Soc. 53*, 646 (1957).
33. S. V. Konev, *Dokl. Akad. Nauk SSSR 116*, 594 (1957).

34. Y. A. Vladimirov, *Dokl. Akad. Nauk SSSR 116*, 780 (1957); Y. A. Vladimirov and E. A. Burstein, *Biofizika 5*, 385 (1960).
35. E. M. Brumburg, *Zh. Obshch. Biol. 17*, 401 (1956).
36. I. Y. Barskii and E. M. Brumburg, *Biokhimiya 23*, 791 (1958).
37. G. M. Barnboim, A. N. Domanskii, and K. K. Turoverov, in "Luminescence of Biopolymers and Cells," p. 168, Fig. 5.3, Plenum Press, New York (1969).
38. Y. A. Vladimirov, *Izvest. Akad. Nauk SSSR Ser. Fiz. 23*, 86 (1959).
39. S. V. Konev, *Izvest. Akad. Nauk SSSR Ser. Fiz. 23*, 93 (1959).
40. Y. A. Vladimirov and S. V. Konev, *Biofizika 4*, 533 (1959).
41. F. W. J. Teale and G. Weber, *Biochem. J. 72*, 15 (1959).
42. F. W. J. Teale, *Biochem. J. 76*, 381 (1960).
43. G. H. Beaven, *Advan. Spectroscopy 2*, 331 (1961).
44. D. B. Wetlaufer, *Advan. Protein Chem. 17*, 303 (1962).
45. G. D. Fasman, "Poly α-Amino Acids," Marcel Dekker, New York (1967).
46. M. Goodman, G. W. Davis, and E. Benedetti, *Accounts Chem. Res. 1*, 275 (1968).
47. N. Greenfield and G. D. Fasman, *Biochemistry 8*, 4108 (1969).
48. Y. P. Myer, *Biophys. J. 9*, A215 (1969); *Macromolecules 2*, 624 (1969).
49. J. F. Yan, G. Vanderkooi, and H. A. Scheraga, *J. Chem. Phys. 49*, 2713 (1968).
50. P. M. Bayley, E. B. Nielsen, and J. A. Schellman, *J. Phys. Chem. 73*, 228 (1969).
51. S. V. Madison and J. Schellman, *Biopolymers 9*, 511, 569 (1970).
52. M. L. Tiffany and S. Krim, *Biopolymers 8*, 367 (1969).
53. G. H. Fasman, E. Bodenheimer, and C. Lindblow, *Biochemistry 3*, 1665 (1964); M. Shiraki and K. Imahori, *Sci. Pap. Coll. Gen. Educ., Univ. Tokyo 16*, 25 (1966); F. Quadrifoglio and D. W. Urry, *Ann. Rev. Phys. Chem. 19*, 477 (1968); V. K. Damle, *Biopolymers 9*, 937 (1970); S. Friedman and *B.B.R.C. 42*, 510 (1971).
54. Y. H. Pao, R. C. Kornegay, and R. Longworth, *Biopolymers 3*, 516 (1965); A. K. Chen and R. W. Woody, *J. Am. Chem. Soc., 93*, 29 (1971).
55. I. Ooi, R. A. Scott, G. Vanderkooi, and H. A. Scheraga, *J. Chem. Phys. 46*, 4410 (1967).
56. G. W. Brady and R. Salovey, *Biopolymers 5*, 331 (1967).
57. T. C. Troxell and H. A. Scheraga, *Biochem. Biophys. Res. Commun. 35*, 319 (1969).
58. A. White, *Biochem. J. 71*, 217 (1959).
59. J. J. ten Bosch, J. W. Longworth, and R. O. Rahn, *Biochim. Biophys. Acta 175*, 10 (1969).
60. A. Cosani, E. Peggion, A. S. Verdini, and M. Terbojevich, *Biopolymers 6*, 963 (1968).
61. H. E. Auer and P. Doty, *Biochemistry 8*, 1708 (1966).
62. J. W. Longworth, *Biopolymers 4*, 1131 (1966).
63. J. W. Longworth, J. J. ten Bosch, J. A. Knopp, and R. O. Rahn, in "Molecular Luminescence" (E. Lim, ed.), p. 529, Benjamin, New York (1969).
64. J. W. Longworth and M. D. C. Battista, unpublished results.
65. J. A. Knopp, J. J. ten Bosch, and J. W. Longworth, *Biochim. Biophys. Acta 188*, 185 (1969).
66. J. W. Longworth and M. D. C. Battista, *Photochem. Photobiol. 10*, 825 (1970).
67. J. W. Longworth and R. O. Rahn, *Biochim. Biophys. Acta 147*, 526 (1967).
68. Y. A. Vladimirov and Li Chin-kuo, *Biofizika 7*, 270 (1962).
69. J. J. ten Bosch, J. W. Longworth, and R. O. Rahn, *Biochim. Biophys. Acta 175*, 10 (1969).

70. J. A. Knopp and J. W. Longworth, *Biochim. Biophys. Acta* **154**, 436 (1968).
71. A. Edelhoch, R. L. Periman, and M. Wilchick, *Biochemistry* **7**, 3893 (1968).
72. R. F. Steiner, *Biochem. Biophys. Res. Commun.* **30**, 502 (1968).
73. J. E. Maling, K. Rosenheck, and M. Weissbluth, *Photochem. Photobiol.* **4**, 241 (1965).
74. J. J. ten Bosch, R. O. Rahn, J. W. Longworth, and R. G. Shulman, *Proc. Nat. Acad. Sci.* **59**, 1003 (1968).
75. J. J. ten Bosch and J. A. Knopp, *Biochim. Biophys. Acta* **188**, 173 (1969).
76. J. W. Longworth and M. D. C. Battista, unpublished results.
77. A. Wada and Y. Ueno, *Biopolymers Symp.* **1**, 363 (1966).
78. C. Hélène, M. Ptak, and R. Santus, *J. Chim. Phys.* **65**, 160 (1968).
79. R. F. Steiner and R. Kolinsky, *Biochemistry* **7**, 1014 (1968).
80. D. Mantik, J. E. Maling, and M. Weissbluth, *Biophys. J.* **9**, A160 (1969).
81. G. D. Fasman, K. Norland, and A. Pesce, *Biopolymers Symp.* **7**, 325 (1964).
82. A. Pesce, E. Bodenheimer, K. Norland, and G. D. Fasman, *J. Am. Chem. Soc.* **86**, 5669 (1964).
83. K. Rosenheck and G. Weber, *Biochem. J.* **5**, 79, 29P (1961).
84. J. Feitelson, *J. Phys. Chem.* **68**, 391 (1964); *Photochem. Photobiol.* **9**, 401 (1969).
85. R. W. Cowgill, *Biochim. Biophys. Acta* **200**, 18 (1970).
86. G. Weber and K. Rosenheck, *Biopolymers Symp.* **1**, 333 (1964).
87. G. D. Fasman, E. Bodenheimer, and A. Pesce, *J. Biol. Chem.* **241**, 916 (1966).
88. A. D. McClaren and D. Shugar, "Photochemistry of Proteins and Nucleic Acids," Pergamon Press, New York (1964).
89. S. S. Lehrer and G. D. Fasman, *Biopolymers* **2**, 199 (1964).
90. S. S. Lehrer and G. D. Fasman, *Biochemistry* **6**, 757 (1967).
91. S. D. Andersen, *Acta Physiol. Scand.* **66** (Suppl.), 263 (1966).
92. G. Dobson and L. I. Grossweiner, *Trans. Farad. Soc.* **61**, 708 (1965).
93. H.-I. Joschek and S. I. Miller, *J. Am. Chem. Soc.* **88**, 3273 (1966).
94. E. E. Fessenko, E. A. Burstein, and Y. A. Vladimirov, *Biofizika* **12**, 616 (1967).
95. R. Santus, C. Hélène, and M. Ptak, *Photochem. Photobiol.* **7**, 341 (1968).
96. E. Schröder and K. Lübke, "The Peptides," Vol. II, Academic Press, New York (1966).
97. A. Bodanszky, M. A. Ondetti, V. Mutt, and M. Bodansky, *J. Am. Chem. Soc.* **91**, 966 (1969).
98. F. W. J. Teale, *Biochem. J.* **76**, 381 (1960).
99. J. W. Longworth, unpublished.
100. I. Weinryb and R. F. Steiner, *Biochemistry* **7**, 2488 (1968).
101. A. Stern, W. A. Gibbons, and L. C. Craig, *Proc. Nat. Acad. Sci.* **61**, 734 (1968); D. W. Urry and M. Ohnishi, "Spectroscopic Approaches to Biomolecular Conformation," A. M. A. (1970), p. 290; N. W. Correll and D. G. Guirey, *Biochem. Biophys. Commun.* **40**. 530 (1970).
102. M. W. Grimes, D. R. Graber, and A. Haug, *Biochem. Biophys. Res. Commun.* **37**, 853 1969.
103. S. M. Partridge, D. F. Elsden, and J. Thomas, *Nature* **197**, 1297 (1963).
104. H. Kallman, V. J. Krasnansky, and P. Person, *Ber. Bunsenges.* **72**, 340 (1968).
105. H. Kallman, V. J. Krasnansky, and P. Person, *Photochem. Photobiol.* **8**, 65 (1968).
106. E. Fujimori, *Biochemistry* **5**, 1034 (1966).
107. Y. A. Vladimirov, *Photochem. Photobiol.* **4**, 369 (1965), **11**, 227 (1970).
108. S. S. Lehrer and G. D. Fasman, *Biochemistry* **6**, 757 (1967).
109. R. W. Cowgill, *Arch. Biochem. Biophys.* **104**, 84 (1964).

110. J. W. Longworth, *Photochem. Photobiol.* 7, 587 (1968).
111. R. W. Cowgill, *Biochim. Biophys. Acta* 140, 37 (1967).
112. S. Beychock and E. Breslow, *J. Biol. Chem.* 243, 151 (1968).
113. L.-R. F. Johnson, I. L. Schwartz, and R. Walter, *Proc. Nat. Acad. Sci.* 64, 1269 (1969).
114. J. Wampler and J. E. Churchich, personal communication; J. Wampler, Ph.D. thesis, University of Tennessee (1969).
115. R. W. Cowgill, *Biochim. Biophys. Acta* 200, 18 (1970).
116. M. T. Franze de Fernandez, A. E. Delius, and A. C. Paladini, *Biochim. Biophys. Acta* 154, 223 (1968).
117. R. W. Cowgill, *Biochim. Biophys. Acta* 154, 231 (1968).
118. J. D. Young and F. H. Carpenter, *J. Biol. Chem.* 236, 743 (1961).
119. W. W. Bromer and R. E. Chance, *Biochim. Biophys. Acta* 133, 219 (1967).
120. D. A. Mercola, J. W. S. Morris, E. R. Arguilla, and W. W. Bromer, *Biochim. Biophys. Acta* 133, 224 (1967).
121. R. Sarges and B. Witkop, *J. Am. Chem. Soc.* 87, 202 (1965).
122. S. Laiken, M. Printz, and L. C. Craig, *J. Biol. Chem.* 264, 4454 (1969).
123. L. C. Craig, *Proc. Nat. Acad. Sci.* 61, 152 (1968).
124. A. Stern, W. A. Gibbons, and L. C. Craig, *J. Am. Chem. Soc.* 91, 2794 (1969).
125. R. S. Bernstein, M. Wilcheck, and H. Edelhoch, *J. Biol. Chem.* 244, 4398 (1968).
126. J. Eisinger, *Biochemistry* 8, 3902 (1969).
127. F. W. J. Teale and G. Weber, *Biochem. J.* 65, 476 (1957).
128. R. A. Badley and F. W. J. Teale, *J. Mol. Biol.* 44, 71 (1969).
129. R. F. Chen, *Anal. Letters* 1, 35 (1967).
130. H. C. Borresen, *Acta Chem. Scand.* 21, 920 (1967).
131. J. W. Bridges and R. T. Williams, *Biochem. J.* 107, 225 (1968).
132. A. N. Fletcher, *Photochem. Photobiol.* 9, 439 (1969).
133. J. Eisinger and G. Navon, *J. Chem. Phys.* 50, 2069 (1969).
134. J. J. Hermans and S. Levinson, *J. Opt. Soc. Am.* 41, 460 (1957).
135. A. Shepp, *J. Chem. Phys.* 25, 579 (1956).
136. M. Almgren, *Photochem. Photobiol.* 8, 231 (1968).
137. J. W. Longworth, *Biopolymers* 4, 1131 (1966).
138. F. Bishai, E. Kuntz, and L. Augenstein, *Biochim. Biophys. Acta* 140, 381 (1967).
139. Y. A. Vladimirov and Li Chin-Kuo, *Biofizika* 7, 270 (1962).
140. I. Weinryb and R. F. Steiner, *Biochemistry* 7, 2488 (1968).
141. J. W. Longworth, Ph.D. thesis, Sheffield, England (1962).
142. W. C. Galley and L. Stryer, *Biochemistry* 8, 1831 (1969).
143. M. Shinitzky and R. Goldman, *Europ. J. Biochem.* 3, 139 (1967).
144. H. Edelhoch and R. E. Lippoldt, *J. Biol. Chem.* 244, 3876 (1969).
145. H. Edelhoch, L. Brand, and M. Wilchek, *Biochemistry* 6, 547 (1967).
146. J. W. Longworth and M. D. C. Battista, unpublished results.
147. L. Stryer, *J. Am. Chem. Soc.* 88, 5708 (1966).
148. G. H. Beaven and E. R. Holiday, *Advan. Protein Chem.* 7, 319 (1952).
149. J. E. Bailey, G. H. Beaven, D. A. Chignell, and W. B. Gratzer, *Europ. J. Biochem.* 7, 5 (1968).
150. H. Edelhoch, *Biochemistry* 6, 1969 (1967).
151. T. T. Herskovitz and M. Sorensen, *Biochemistry* 7, 2523 (1968).
152. J. W. Donovan, *in* "Physical Principles and Techniques of Protein Chemistry" (S. J. Leach, ed.), p. 102, Academic Press, New York (1969).
153. F. W. J. Teale, *Biochem. J.* 76, 381 (1960).

154. F. W. J. Teale and G. Weber, *Biochem. J.* **65**, 476 (1957).
155. J. E. Wampler and J. E. Churchich, personal communication; J. E. Wampler, Ph.D. thesis, University of Tennessee (1970).
156. R. W. Cowgill, *Arch. Biochem. Biophys.* **106**, 84 (1964).
157. Y. A. Vladimirov, *Dokl. Akad. Nauk SSSR* **136**, 960 (1961).
158. J. W. Donovan, *Biochemistry* **6**, 3917 (1967).
159. Y. A. Vladimirov and E. A. Burstein, *Biofizika* **5**, 385 (1960).
160. C. J. R. Thorne and N. O. Kaplan, *J. Biol. Chem.* **238**, 1861 (1963).
161. S.-F. Wang, F. S. Kawahara, and P. Talalay, *J. Biol. Chem.* **238**, 576 (1963).
162. R. W. Cowgill, *Biochim. Biophys. Acta* **168**, 417 (1968).
163. A. Pesce, E. Bodenheimer, K. Norland, and G. D. Fasman, *J. Am. Chem. Soc.* **86**, 5669 (1964).
164. D. Shugar, *Biochem. J.* **52**, 142 (1952).
165. T. T. Herskovits and M. Laskowski, *J. Biol. Chem.* **243**, 2123 (1968).
166. J. Bello, *Biochemistry* **8**, 4542 (1969).
167. J. E. Bailey, G. H. Beaven, D. A. Chignell, and W. B. Gratzer, *Europ. J. Biochem.* **7**, 5 (1968).
168. H. Greenspan, J. Birnbaum, and J. Feitelson, *Biochim. Biophys. Acta* **126**, 13 (1966).
169. C. Y. Cha and H. A. Scheraga, *J. Biol. Chem.* **238**, 2958 (1963).
170. G. Weber and F. Wold, *Fed. Proc.* **22**, 1120 (1963).
171. J. Feitelson, *J. Phys. Chem.* **68**, 391 (1964).
172. G. Weber, and K. Rosenheck, *Biopolymers Symp.* **1**, 333 (1964).
173. R. W. Cowgill, *Biochim. Biophys. Acta* **154**, 231 (1968).
174. A. Y. Moon, D. C. Poland, and H. A. Scheraga, *J. Phys. Chem.* **69**, 2960 (1965).
175. R. W. Cowgill, *Biochim. Biophys. Acta* **109**, 536 (1965).
176. R. W. Cowgill, *Biochim. Biophys. Acta* **133**, 6 (1967); *Biochim. Biophys. Acta* **200**, 18 (1970).
177. R. W. Cowgill, *Biochim. Biophys. Acta* **140**, 37 (1967).
178. R. F. Steiner and E. P. Kirby, *J. Phys. Chem.* **73**, 4130 (1969).
179. G. Weber, *Biochem. J.* **75**, 365 (1960).
180. G. Weber, *Biochem. J.* **75**, 335 (1960).
181. R. W. Cowgill, *Biochim. Biophys. Acta* **96**, 81 (1965).
182. R. W. Cowgill, *Biochim. Biophys. Acta* **94**, 74 (1965).
183. C. L. Gemill, *Arch. Biochem. Biophys.* **63**, 2778 (1968).
184. R. L. Perlmann, A. van Zyl, and H. Edelhoch, *J. Am. Chem. Soc.* **90**, 2168 (1968).
185. R. F. Steiner, R. E. Lippoldt, H. Edelhoch, and V. Frattali, *Biopolymers Symp.* **1**, 355 (1964).
186. R. W. Cowgill, *Biochim. Biophys. Acta* **120**, 196 (1966).
187. C. C. Bigelow and T. Krinitsky, *Biochim. Biophys. Acta* **88**, 130 (1964).
188. R. W. Cowgill, in "Molecular Luminescence" (E. C. Lim, ed.), p. 589, Benjamin, New York (1969). *Biochim. Biophys. Acta* **214**, 228 (1970).
189. J. F. Riordan and B. L. Vallee, *Biochemistry* **2**, 1460 (1963).
190. E. A. Burstein, *Biofizika* **13**, 718 (1968).
191. H. Edelhoch, R. L. Perlman, and P. Wilcheck, *Biochemistry* **7**, 3893 (1968).
192. L. Weil, T. S. Seibles, and T. T. Herskovits, *Arch. Biochem. Biophys.* **111**, 308 (1965).
193. T. T. Herskovits, *J. Biol. Chem.* **240**, 628 (1965); C. L. Menendez, T. T. Herskovits, and M. Laskowski, *Biochemistry* **8**, 5042 (1969).
194. M. Ayoma, K. Kurihova, and K. Shibata, *Biochim. Biophys. Acta* **107**, 257 (1965).

195. C. L. Menedez and T. T. Herskovits, *Biochemistry 8*, 8080 (1969).
196. A. Massaglia, V. Rosa, G. Rialdi, and C. A. Ross, *Biochem. J. 115*, 11 (1969).
197. N. J. Adams, T. L. Blundell, E. J. Dodson, G. G. Dodson, M. Vijaijan, E. N. Barker, M. M. Harding, D. C. Hodgkins, B. Rommer, and S. Sheat, *Nature 244*, 491 (1969).
198. W. E. Blumberg, J. Eisinger, and G. Navon, *Biophys. J. 8*, A106 (1968).
199. J. W. Longworth and S. S. Stevens, unpublished results.
200. J. W. Longworth, *Photochem. Photobiol. 7*, 587 (1968).
201. J. A. Galley and G. M. Edelman, *Biochim. Biophys. Acta 60*, 499 (1962).
202. J. A. Galley and G. M. Edelman, *Biopolymers Symp. 1*, 367 (1964).
203. J. Herman and H. A. Scheraga, *J. Am. Chem. Soc. 83*, 3283 (1961).
204. J. W. Longworth and M. D. C. Battista, unpublished results.
205. C. E. Bodwell, N. M. Kominz, and P. J. Duntley, *Biochem. Biophys. Res. Commun. 21*, 210 (1965).
206. F. W. J. Teale, *Biochem. J. 76*, 381 (1960).
207. G. Weber, *Biochem. J. 79*, 29 (1961).
208. J. W. Longworth, unpublished results.
209. A. Finazzi Agro, G. Rotilio, L. Avigliano, P. Guerrieri, V. Boffi, and B. Mondovi, *Biochemistry 9*, 2009 (1970).
210. J. W. Longworth, *Photochem. Photobiol. 7*, 587 (1968); Y. Yamamoto and J. Tanaka, *Biochim. Biophys. Acta 207*, 522 (1970); contrast with C. O. Pongs, *Biochem. Biophys. Res. Commun. 38*, 431 (1970).
211. T. T. Herskovits, *Methods Enzymol. 11*, 748 (1967).
212. S. N. Timasheff and M. J. Gorbunoff, *Ann. Rev. Biochem. 36*, 13 (1967).
213. R. F. Chen, *Biochim. Biophys. Acta 120*, 169 (1966).
214. Y. A. Vladimirov and G. M. Zimina, *Biokhimiya 30*, 1105 (1966).
215. R. F. Chen, *Anal. Biochem. 14*, 497 (1966).
216. Y. A. Vladimirov and Li Chin-kuo, *Biofizika 7*, 270 (1962).
217. J. W. Longworth, Ph.D. thesis, University of Sheffield, U.K. (1962).
218. E. Kuntz, F. Bishai, and L. Augenstein, *Nature 212*, 980 (1966).
219. F. Bishai, E. Kuntz, and L. Augenstein, *Biochim. Biophys. Acta 100*, 381 (1967).
220. J. W. Longworth, *Biopolymers 4*, 1131 (1966).
221. I. Weinryb and R. F. Steiner, *Biochemistry 7*, 2488 (1968); R. W. Ricci, *Photochem. Photobiol. 12*, 67 (1970).
222. H. B. Steen, *Photochem. Photobiol. 8*, 47 (1968).
223. J. W. Longworth and S. S. Stevens, unpublished results.
224. Y. A. Chernitskii and I. D. Volotovski, *Biofizika 12*, 624 (1967).
225. R. F. Chen, G. G. Vurek, and N. Alexander, *Science 156*, 949 (1967).
226. W. E. Blumberg, J. Eisinger, and G. Navon, *Biophys. J. 8*, A106 (1968).
227. W. B. De Lauder and P. Wahl, *Biochemistry 9*, 2750 (1970).
228. L. F. Gladchenko, M. Y. Kostko, L. G. Pikulik, and A. N. Sevchenko, *Dokl. Akad. Nauk Belorussk. SSR 9*, 647 (1965).
229. I. Weinryb and R. F. Steiner, *Biochemistry 9*, 135 (1970); R. A. Badley and F. W. J. Teale, *J. Mol. Biol. 44*, 71 (1969); F. W. J. Teale and R. A. Badley, *Biochem J. 116*, 341 (1970).
230. R. H. Steele and A. Szent-Gyorgyi, *Proc. Nat. Acad. Sci. 43*, 477 (1957).
231. R. H. Steele and A. Szent-Gyorgyi, *Proc. Nat. Acad. Sci. 44*, 540 (1958).
232. S. Freed, J. A. Turnbull, and W. Salmre, *Nature 181*, 1731 (1958).
233. E. Fujimori, *Biochim. Biophys. Acta 40*, 251 (1960).
234. Y. A. Vladimirov and F. F. Litvin, *Biofizika 5*, 127 (1960).

235. Y. A. Vladimirov and E. A. Burstein, *Biofizika 5*, 385 (1960).
236. S. Freed and W. Salmre, *Science 128*, 1341 (1958).
237. G. Weber, *Nature 190*, 27 (1961); *Biochem. J. 79*, 29P (1961).
238. F. W. J. Teale, *Biochem. J. 80*, 14P (1961).
239. Y. A. Vladimirov, *Dokl. Akad. Nauk SSSR 136*, 960 (1961).
240. J. W. Longworth, *Biochem. J. 81*, 23P (1961).
241. T. Truong, R. Bersohn, P. Brumer, C. K. Luk, and T. Tao, *J. Biol. Chem. 242*, 2979 (1967).
242. T. T. Herskovits and M. Sorenson, *Biochemistry 7*, 2523 (1968).
243. R. F. Chen and P. F. Cohen, *Arch. Biochem. Biophys. 144*, 514 (1966).
244. J. E. Churchich, *Biochim. Biophys. Acta 102*, 280 (1965).
245. J. W. Longworth, *Photochem. Photobiol. 8*, 589 (1968).
246. R. F. Steiner and H. Edelhoch, *Nature 192*, 873 (1961).
247. R. F. Steiner and H. Edelhoch, *Biochim. Biophys. Acta 66*, 341 (1963).
248. R. F. Steiner and H. Edelhoch, *J. Biol. Chem. 238*, 925 (1963).
249. J. L. Cornog and W. R. Adams, *Biochim. Biophys. Acta 66*, 356 (1963).
250. J. W. Longworth and R. O. Rahn, *Biochim. Biophys. Acta 147*, 526 (1967).
251. P. Debye and J. O. Edwards, *J. Chem. Phys. 20*, 236 (1952).
252. J. W. Longworth and M. D. C. Battista, unpublished observations.
253. D. L. Dexter, *J. Chem. Phys. 21*, 836 (1953).
254. M. O. Dayhoff, "Atlas of Protein Sequence and Structure," Vol. 4, p. 188 (1969).
255. F. W. J. Teale and G. Weber, *Biochem. J. 65*, 476 (1957).
256. R. W. Cowgill, *Arch. Biochem. Biophys. 100*, 36 (1963).
257. A. White, *Biochem. J. 71*, 217 (1959).
258. J. W. Longworth and M. C. D. Battista, unpublished results.
259. G. D. Fasman, K. Norland, and P. Pesce, *Biopolymers Symp. 1*, 325 (1964).
260. G. Weber and K. Rosenheck, *Biopolymers Symp. 1*, 333 (1964).
261. K. Rosenheck and G. Weber, *Biochem. J. 79*, 29P (1961).
262. G. Weber and F. W. J. Teale, *in* "The Proteins" (H. Neurath, ed.), Vol. III, p. 473, Academic Press, New York (1965).
263. R. W. Cowgill, *Biochim. Biophys. Acta 133*, 61 (1967); *Biochim. Biophys. Acta 168*, 417 (1968); *Biochim. Biophys. Acta 200*, 18 (1970).
264. P. Cuatrecasas, H. Edelhoch, and C. B. Anfinsen, *Proc. Nat. Acad. Sci. 58*, 2083 (1967).
265. P. Cuatrecasas, S. Fuchs, and C. B. Anfinsen, *J. Biol. Chem. 242*, 4759 (1967).
266. P. Cuatrecasas, S. Fuchs, and C. B. Anfinsen, *J. Biol. Chem. 243*, 4787 (1968); *J. Biol. Chem. 244*, 406 (1969); P. Cuatrecasas, M. Wilcheck, and C. B. Anfinsen, *J. Biol. Chem. 244*, 4316 (1969).
267. A. Arnone, C. J. Bier, F. A. Cotton, E. E. Hazen, D. C. Richardson, and J. S. Richardson, *Proc. Nat. Acad. Sci. 64*, 420 (1969).
268. R. W. Cowgill, *Biochim. Biophys. Acta 200*, 18 (1970).
269. R. F. Steiner, R. E. Lippoldt, H. Edelhoch, and V. Frattali, *Bipolymers Symp. 1*, 355 (1964).
270. E. Yeargers, F. R. Bishai, and L. Augenstein, *Biochem. Biophys. Res. Commun. 23*, 570 (1966).
271. T. Cassen and D. R. Kearns, *Biochim. Biophys. Acta 194*, 203 (1969).
272. L. Augenstein and J. Nag-Chaudhuri, *Nature 203*, 1145 (1964); M. E. McCarville and S. P. McGlynn, *Photochem. Photobiol. 10*, 171 (1969).
273. G. Weber, *Biochem. J. 75*, 345 (1960); V. P. Bobrovich and S. V. Konev, *Dokl. Akad. Nauk Belorussk. SSR 9*, 399 (1965).

274. G. Weber and F. W. J. Teale, *Biochem. J.* 72, 15 (1959).
275. G. Weber, *Biochem. J.* 75, 345 (1960).
276. Y. A. Vladimirov. *Dokl. Akad. Nauk SSSR* 116, 780 (1957).
277. S. V. Konev, M. A. Katibnikov, and M. A. Petrova, *Biofizika* 6, 375 (1961); S. V. Konev and M. A. Katibnikov, *Biofizika* 8, 4 (1963); S. V. Konev, M. A. Katibnikov, and T. I. Lyskova, *Biofizika* 9, 121 (1964).
278. Y. A. Vladimirov, *Dokl. Akad. Nauk SSSR* 116, 780 (1957).
279. J. A. Knopp, J. ten Bosch, and J. W. Longworth, *Biochim. Biophys. Acta* 188, 185 (1969).
280. G. Weber and M. Shinitzky, *Proc. Nat. Acad. Sci.* 65, 823 (1970).
281. S. V. Konev and M. A. Katibnikov, *Dokl. Akad. Nauk SSSR* 136, 472 (1961).
282. E. A. Chernitskii and S. V. Konev, *Dokl. Akad. Nauk Belorussk. SSSR* 8, 258 (1964).
283. S. V. Konev and M. A. Katibnikov, *Biofizika* 8, 4 (1963).
284. S. V. Konev, V. P. Bobrovich, and E. A. Chernitskii, *Biofizika* 10, 42 (1965).
285. S. V. Konev, V. P. Bobrovich, and E. A. Chernitskii, *Dokl. Akad. Nauk SSSR* 165, 937 (1965).
286. R. F. Chen, *Anal. Biochem.* 19, 374 (1967).
287. V. P. Bobrovich and S. V. Konev, *Dokl. Akad. Nauk Belorussk. SSR* 9, 118 (1965).
288. E. A. Chernitskii, S. V. Konev, and V. P. Bobrovich, *Dokl. Akad. Nauk Belrussk. SSR* 7, 628 (1963).
289. S. V. Konev and V. P. Bobrovich, *Biofizika* 10, 813, 1965.
290. V. P. Bobrovich and S. V. Konev, *Dokl. Akad. Nauk SSSR* 155, 197 (1964).
291. E. A. Chernitskii and S. V. Konev, *Zh. Prikl. Spektrosk.* 2, 261 (1965).
292. J. Czekalla, W. Liptay, and E. Dollefeld, *Ber. Bunsenges.* 68, 80 (1964).
293. Y. A. Vladimirov and D. I. Roshchupkin, quoted in Y. A. Vladimirov, "Photochemistry and Luminescence of Proteins" (translated by the U.S. Department of Commerce), p. 23 (1969).
294. G. H. Beaven and E. R. Holiday, *Advan. Protein Chem.* 7, 319 (1952).
295. R. M. Purkey and W. C. Galley, *Biophys. J.* 10, 240a (1970); *Biochemistry* 9, 3569 (1970); S. R. Andersen, M. Brunori, and G. Weber, *Biochemistry*, 9, 4723 (1970).
296. Y. A. Vladimirov, "Photochemistry and Luminescence of Proteins" (translated by the U.S. Department of Commerce), p. 21 (1969).
297. J. Eisinger, *Biochemistry* 8, 3902 (1969).
298. S. V. Konev, "Fluorescence and Phosphorescence of Proteins and Nucleic Acids," p. 121, Plenum Press, New York (1967).
299. J. Eisinger and G. Navon, *J. Chem. Phys.* 50, 2069 (1969).
300. C. A. Parker and W. T. Rees, *Analyst* 87, 83 (1962).
301. G. Weber and F. W. J. Teale, *Trans. Farad. Soc.* 53, 646 (1957).
302. M. J. Kronman, *Fed. Proc.* 28, 2042 (1969).
303. J. A. Reynolds and C. Tanford, *Proc. Nat. Acad. Sci.* 66, 1002 (1970).
304. I. D. Volotovskii and S. V. Konev, *Biofizika* 12, 200 (1967).
305. A. White, Ph.D. thesis, University of Sheffield, U.K. (1959).
306. W. Sticks and I. M. Kolthoff, *Anal. Chem.* 25, 1050 (1953).
307. L.-O. Andersson, *Biochim. Biophys. Acta* 117, 115 (1966); 200, 363 (1970).
308. J. W. Donovan, *Biochem. Biophys. Res. Commun.* 29, 734 (1967).
309. D. Shugar, *Biochem. J.* 52, 142 (1952).
310. H. Edelhoch, L. Brand, and M. Wilcheck, *Biochemistry* 6, 547 (1967).
311. S. S. Lehrer and G. D. Fasman, *Biochem. Biophys. Res. Commun.* 23, 133 (1966).

312. J. W. Bridges and R. T. Williams, *Biochem. J. 107*, 225 (1968).
313. R. W. Cowgill, *Biochim. Biophys. Acta 200*, 18 (1970).
314. G. D. Fasman, E. Bodenheimer, and A. Pesce, *J. Biol. Chem. 201*, 916 (1966).
315. J. A. Galley and G. M. Edelman, *Biochim. Biophys. Acta 60*, 499 (1962).
316. M. Shinitzky and R. Goldman, *Europ. J. Biochem. 3*, 139 (1967).
317. M. Shinitzky and M. Fridkin, *Europ. J. Biochem. 9*, 176 (1969).
318. H. Edelhoch, P. G. Condliffe, R. E. Lippholdt, and H. G. Burger, *J. Biol. Chem. 241*, 5205 (1966).
319. P. B. Sigler, D. M. Blow, B. W. Matthews, and R. Henderson, *J. Mol. Biol. 35*, 143 (1968).
320. A. O. Barel and A. N. Glazer, *J. Biol. Chem. 244*, 268 (1969).
321. L. A. E. Sluyterman and M. J. M. De Graaf, *Biochim. Biophys. Acta 200*, 595 (1970); R. F. Steiner, *Biochemistry 10*, 771 (1971).
322. C. Hélène, F. Brun, and M. Yaniv, *Biochem. Biophys. Res. Commun. 37*, 393 (1969).
323. C. Hélène, F. Brun, and M. Yaniv, *J. Mol. Biol.* (in press).
324. J. G. Farrelly, M. P. Stulberg, and J. W. Longworth, *J. Biol. Chem.* (1970).
325. T. T. Herskovits and M. Laskowski, *J. Biol. Chem. 237*, 2481 (1962).
326. S. N. Timasheff, L. Mescanti, J. O. Basch, and R. Townsend, *J. Biol. Chem. 241*, 2496 (1966).
327. B. L. van Duuren, *J. Org. Chem. 26*, 2954 (1961).
328. J. W. Longworth and F. A. Bovey, *Biopolymers 4*, 1115 (1966).
329. R. W. Cowgill, *Biochim. Biophys. Acta 140*, 37 (1967).
330. J. E. Churchich, *Biochim. Biophys. Acta 92*, 194 (1964).
331. G. F. Perlman, *Biopolymers Symp. 1*, 383 (1964).
332. H. Edelhoch, V. Frattali, and R. F. Steiner, *J. Biol. Chem. 200*, 122 (1965).
333. R. W. Cowgill, *Biochim. Biophys. Acta 133*, 6 (1967).
334. L. Stryer, *J. Am. Chem. Soc. 88*, 5708 (1966); R. F. Steiner and E. P. Kirby, *J. Phys. Chem. 74*, 4480 (1970); S. S. Lehrer, *J. Am. Chem. Soc. 92*, 3459 (1970).
335. J. W. Longworth and M. D. C. Battista, *Photochem. Photobiol. 11*, 875 (1970).
336. R. F. Steiner and H. Edelhoch, *Nature 193*, 375 (1962).
337. J. A. Galley and G. M. Edelman, *Biopolymers Symp. 1*, 367 (1964).
338. J. G. Foss, *Biochim. Biophys. Acta 47*, 569 (1961).
339. J. Bello, *Biochemistry 8*, 4542 (1969).
340. G. M. Barenboim, A. N. Domanskii, and K. K. Turoverov, "Luminescence of Biopolymers and Cells," p. 51, Plenum Press, New York (1969); G. M. Barenboim, A. V. Sokolenko, and K. K. Turoverov, *Tsitologiya 10*, 636 (1968); K. K. Turoverov, *Opt. i Spektr. 26*, 310 (1969).
341. E. Leroy, H. Lami, and G. Laustriat, *Photochem. Photobiol. 13*, 411 (1970).
342. W. C. Galley and L. Stryer, *Biochemistry 8*, 1831 (1969).
343. T. R. Hopkins and R. Lumry, *Biophys. J. 8*, A216 (1968).
344. R. F. Steiner and E. P. Kirby, *J. Phys. Chem. 73*, 4130 (1969).
345. T. Shiga and L. H. Piette, *Photochem. Photobiol. 3*, 223 (1964).
346. S. A. Bernhard, B. Lee, and Z. H. Tashjian, *J. Mol. Biol. 18*, 405 (1966).
347. C. E. Singleterry and L. A. Weinberger, *J. Am. Chem. Soc. 73*, 4579 (1951).
348. F. Aurich and E. Lippert, *Spectrochim. Acta 22*, 1073 (1966).
349. M. Almgren, *Photochem. Photobiol. 8*, 231 (1968).
350. G. Weill and M. Calvin, *Biopolymers 1*, 401 (1963).
351. W. C. Galley, *Biopolymers 6*, 1279 (1968).
352. A. C. Albrecht, *J. Mol. Spectroscopy 6*, 84 (1961).

353. A. H. Kalantar, *J. Chem. Phys. 48*, 4992 (1968).
354. A. Shepp, *J. Chem. Phys. 25*, 579 (1956).
355. J. Koudelka and L. Augenstein, *Photochem. Photobiol. 7*, 613 (1968).
365. E. Kuntz, *Nature 217*, 845 (1968); E. Kuntz, R. Canada, R. Wagner, and L. Augenstein, *in* "Molecular Luminescence" (E. C. Lim, ed.) p. 551, Benjamin, New York (1969).
357. S. C. Tsai and G. W. Robinson, *J. Chem. Phys. 49*, 3184 (1968).
358. R. F. Chen, H. Edelhoch, and R. F. Steiner, *in* "Physical Principles and Techniques of Protein Chemistry" (S. U. Leach, ed.) Part 1A, p. 214, Academic Press, New York (1969).
359. L. G. Pikulik, M. Y. Kostko, S. V. Konev, and E. A. Chernitskii, *Dokl. Akad. Nauk Belorussk. SSSR 10*, 6 (1966).
360. S. V. Konev, M. Y. Kostko, L. G. Pikulik, and E. A. Chernitskii, *Biofizika 11*, 965 (1966).
361. S. V. Konev, "Fluorescence and Phosphorescence of Proteins and Nucleic Acids," p. 82, Plenum Press, New York (1967).
362. N. Mataga and Y. Murata, *J. Am. Chem. Soc. 91*, 3144 (1969); W. B. DeLauder and Ph. Wahl, *Biochem. Biophys. Res. Commun. 42*, 398 (1971); S. S. Lehrer, *Biophys. J. 11*, 72a (1971).
363. I. D. Volotovskii, S. V. Konev, and E. A. Chernitskii, *Biofizika 8*, 433 (1967).
364. P. Douzou and J. Francq, *J. Chem. Phys. 59*, 578 (1962).
365. B. I. Stepanov and V. P. Gribkovski, "Theory of Luminescence," p. 171, Iliffe Books, London (1966).
366. D. B. Siano and D. E. Metzler, *J. Chem. Phys. 51*, 1856 (1969).
367. J. W. Longworth, unpublished results.
368. J. A. Knopp and J. W. Longworth, unpublished results.
369. S. V. Konev, "Fluorescence and Phosphorescence of Proteins and Nucleic Acids," p. 78, Plenum Press, New York (1967); V. P. Bobrovich and S. V. Konev, *Dokl. Akad. Nauk SSSR 155*, 197 (1964).
370. P.-S. Song and W. E. Kurtin, *J. Am. Chem. Soc. 91*, 4892 (1969).
371. G. S. Kembrovskii, V. P. Bobrovich, and S. V. Konev, *Zh. Prikl. Spektrosk. 5*, 695 (1966); J. M. Hollas, *Spectrochim. Acta 19*, 753 (1963); E. H. Strickland, J. Horwitz, and C. Billups, *Biochemistry 9*, 4914 (1970); V. S. Ananthanarayan and C. C. Bigelow, *Biochemistry 8*, 3723 (1969).
372. E. Lippert, *Z. Elektrochem. 61*, 962 (1957).
373. N. Mataga, Y. Kaifu, and M. Koizumi, *Bull. Chem. Soc. Japan 29*, 465 (1956).
374. A. Kawski, *Acta Phys. Polon 29*, 507 (1966); K. Rotkiewicz and L. B. Grabowski, *Trans. Farad. Soc. 65*, 3263 (1969).
375. W. Liptay, *in* "Modern Quantum Chemistry" (O. Sinananoglu, ed.) Part 2, p. 342, Academic Press, New York (1965); W. Liptay, *Z. Naturforsch. 20a*, 1441 (1965).
376. G. A. Gerhold and E. Miller, *J. Phys. Chem. 72*, 2737 (1968).
377. M. S. Walker, T. W. Bednar, and R. Lumry, *J. Chem. Phys. 47*, 1020 (1967).
378. G. Weber, *quoted in* J. A. Lombardi, *J. Chem. Phys. 50*, 3780 (1969).
379. D. A. Chignell and W. A. Gratzer, *J. Phys. Chem. 72*, 2934 (1968).
380. A. Weller, *in* "Fast Reactions and Primary Processes in Chemical Kinetics" (S. Claesson, ed.) p. 413, Almqvist and Wiksell, Stockholm (1967).
381. J. W. Longworth and M. D. C. Battista, unpublished results.
382. T. D. S. Hamilton and K. Raz Naqvi, *Chem. Phys. Letters 2*, 374 (1968).
383. S. S. Lehrer and G. D. Fasman, *J. Biol. Chem. 242*, 4644 (1967).

384. B. Bablouzian, M. Grourke, and G. D. Fasman, *J. Biol. Chem.* **245**, 2081 (1970).
385. C. F. Blake, L. N. Johnson, G. A. Nair, A. C. T. North, D. C. Philips, and V. R. Sarma, *Proc. Roy. Soc. London Ser.* **B 167**, 378 (1967).
386. N. Sharon, *Proc. Roy. Soc. London Ser.* **B 167**, 402 (1967).
387. I. Teichberg and N. Sharon, *F.E.B.S. Letters* **7**, 171 (1970).
388. C. H. Suelter, *Biochemistry* **6**, 418 (1967).
389. Y. Elkana, *J. Phys. Chem.* **72**, 3654 (1968).
390. W. C. Galley and L. Stryer, *Proc. Nat. Acad. Sci.* **60**, 1081 (1968).
391. S. S. Lehrer, *J. Biol. Chem.* **244**, 3613 (1969).
392. N. Shaklai and E. Daniel, *Biochemistry* **9**, 564 (1970).
393. A. Y. Tan and R. C. Woodworth, *Biochemistry* **8**, 377 (1969).
394. Y. A. Vladimirov, "Photochemistry and Luminescence of Proteins" (translated by the U.S. Department of Commerce) p. 76 (1969).
395. G. M. Barenboim, *Biofizika* **8**, 154 (1963).
396. E. A. Burshtein, *Biofizika* **13**, 433 (1968).
397. S. S. Lehrer, *Biochem. Biophys. Res. Commun.* **29**, 767 (1967).
398. E. A. Burshtein, *Biofizika* **13**, 718 (1968).
399. E. P. Kirby and R. F. Steiner, *J. Biol. Chem.* **245**, 6300 (1970).
400. A. N. Schechter and C. J. Epstein, *J. Mol. Biol.* **35**, 567 (1968).

INDEX

Adiabatic approximation, 11, 20
Aggregates, 10, 92, 94
Anhydrides, 201-204
Antibiotics, 357-384
Aryl halides, 210
Arylsulfonyl chlorides, 205

Boltzmann distribution, 6
Born–Oppenheimer approximation, 20, 21

Carbodiimides, 212
Charge transfer, 132
Class A proteins, 378-396
Class B proteins, 396-474
Codon–anticodon binding, 191-194
Conjugation of proteins, 200-224

Dansyl chloride, 215
Diazoacetyl derivatives, 206, 207
Diazonium salts, 210, 211
Differential quenching, 469
Dinucleotides, 126-135
Dissipative processes, 2
DNA, 6, 136-139
DNA, singlet energy transfer in, 174-178
DNA, triplet energy transfer in, 178-185

Einstein coefficients, 3, 4, 14
Electronic energy transfer between amino acids, 420-422
Electronic relaxation, 19, 21, 22, 23
Emission spectra of protein conjugates, 226-235
Energy loss at 77° K, 450-453
Energy transfer between amino acids, 328
Energy transfer in polynucleotides, 165-184
Energy transfer in protein conjugates, 237-239
Epoxides, 205, 206
ESR, 114, 117
Europium, 140-143
Excimer, 108, 131
Exciplex, 108, 113, 130-135

Excitation spectra of protein conjugates, 226-235
Excited lifetime, 2, 4, 6, 19
Excited lifetimes of protein conjugates, 235-237
Excited state, 2, 4, 5, 6, 8
Excited states of nucleotides, 139-148
Excited states of proteins, 320-323

Fluorescein, 213, 214
Fluorescence lifetimes of amino acids, 402-403
Fluorescence lifetimes of proteins, 451-455
Fluorescence of amino acids, 325-328
Fluorescence of polynucleotides, 134-137
Fluorescence of proteins, 323-325, 329-334
Fluorescence polarization spectra of amino acids, 422-427
Fluorescence spectra of protein tryptophans, 456-459
Fluorescent antibodies, 253-258
Fluorescent probes, 229-233
Fluorescent protein conjugates, 199-312
Forster transfer in polynucleotides, 167, 169-171
Franck-Condon factor, 23, 24, 25

Gamma globulin, 247, 248

Haloacetyl derivatives, 207, 208
Hamiltonian, 11
Herzberg–Teller coefficients, 11
Heterocyclic molecules, 6
Heterocyclics, 18, 26
Heterogeneity of fluorescence, 467-474
Heterogeneity of phosphorescence, 472, 473
Hormones, 357-364
2-Hydroxy-5-nitrobenzyl bromide, 209
Hypochromicity, 125

Imidoesters, 211, 212
Intersystem crossing, 19, 24
Ionized polytyrosine, 345-350
Isocyanates, 204
Isothiocyanates, 204, 205
Isotope effect, 443-444

Labelling procedures, 216-224
Ligand binding, 248-250

Mercurials, 210
Microcrystals, 108, 112
Molecular orbital calculations, 123, 124, 125
Molecular orbitals, 7
Multiple-state decay, 6
Multiplicity, 4, 15
Mustards, 209

N-bromosuccinimide, 211
Noncovalently bound labels, 258-262
Nonradiative deactivation of nucleotides, 140-148
Nuclear charges, 15
Nuclear displacements, 12, 13, 15
Nuclear motions, 8
Nucleic acids, 107-198
N-substituted maleimides, 208

Oligonucleotides, 125-135

Perturbations, 8, 11
Phenylalanyl transfer RNA, 189-194
Phosphorescence, 5
Phosphorescence lifetimes of amino acids, 400-402
Phosphorescence lifetimes of proteins, 455-456
Phosphorescence of dinucleotides, 133-135
Phosphorescence of polynucleotides, 137-139
Phosphorescence polarization of amino acids, 427-428
Phosphorescence spectra of tryptophan, 460-462

Photochemistry of polytyrosine, 355, 357
Photochemistry of proteins, 320-323
Photodimers of pyrimidines, 157-165
Photolesions (in DNA), 120
Photoproducts of nucleotides, 154-165
Polarization, 14, 19
Polarization of fluorescence, 239-253
Polyadenylic acid, singlet energy transfer in, 171-174
Polynucleotides, 134-139
Polypeptides, 334-357
Polyphenylalanine, 336-345
Polytryptophan, 336-344
Polytyrosine, 336-345
Purines, 107-198
Pyrimidines, 107-198

Quantum yields of protein conjugates, 235-237
Quantum yields of proteins, 330-334
Quantum yields of tryptophan residues, 436-443
Quasi-continuum, 5
Quenching of tryptophan residues, 436-443
Quinones, 208, 209

Relaxation processes, 20
Rhodamine, 214, 215

Selection rules, 15
Sensitized phosphorescence, 120
Singlet state, 5, 94
Solvent perturbation, 443
Solvent perturbation of fluorescence, 472
Solvent reorientation, of purines, 140
Spin functions, 16
Spin–orbit interactions, 8, 17, 19, 25
Spin polarization, 27
Stokes' shift of indole, 462-467
Strong coupling, in polynucleotides, 107-108
Subtilisin, 398

Temperature dependence of quantum yields of amino acids, 445-449

Thermal quenching, of nucleotides,
    148-154
TNS, 215, 229, 230
Transfer RNA, luminescence of, 185-194
Transition dipole moment, 8, 9, 10, 12
Triazine derivatives, 212
Triplet energy transfer, 114
Triplet – singlet transitions, 15, 16
Triplet state, 5, 15, 16, 18, 24, 25, 87, 92, 94, 99, 100, 101, 102
Triplet – triplet transfer, 139
Tryptophan-containing polypeptides, 364-379
Tryptophan excitation spectra, 428-436
Tyrosine excitation spectra, 414-417
Tyrosine fluorescence, 403-417
Tyrosine – glutamate copolymers, 354-356
Tyrosine – lysine copolymers, 354-356
Tyrosine phosphorescence, 403-411, 418-420
Tyrosine quantum yield, 411-414
Tyrosine – tryptophan copolymers, 350-354

Very weak coupling, in polynucleotides, 168-169
Vibrational – electronic interactions, 11
Vibrational relaxation, 19, 20
Vibronic effects, 14
Visible fluorescent tracing, 253-257

Wave-functions (purines and pyrimidines), 121, 122, 123
Weak coupling, in polynucleotides, 168

Y base, 189-194